Semiconductor Device Modeling for VLSI

Kwyro Lee
Michael Shur
Tor A. Fjeldly
Trond Ytterdal

Prentice Hall Series in Electronics and VLSI
Charles S. Sodini, Series Editor

Prentice Hall, Englewood Cliffs, New Jersey 07632

Library of Congress Cataloging-in-Publication Data

Semiconductor device modeling for VLSI : with the AIM-spice circuit
  simulator / Kwyro Lee ... [et al.].
      p.   cm.
  Includes bibliographical references and index.
  ISBN 0-13-805656-0
  1. Semiconductors--Design and construction--Data processing.
2. SPICE (Computer file) 3. Integrated circuits--Very large scale
integration--Design and construction--Data processing.   I. Lee,
Kwyro.
TK7871.85.S445   1993
621.3815'2--dc20   92-37399
                                                                CIP

Publisher: Alan Apt
Production Editor: Mona Pompili
Cover Designer: Design Source
Prepress Buyer: Linda Behrens
Manufacturing Buyer: Dave Dickey
Editorial Assistant: Shirley McGuire

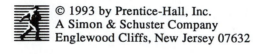

© 1993 by Prentice-Hall, Inc.
A Simon & Schuster Company
Englewood Cliffs, New Jersey 07632

The author and publisher of this book have used their best efforts in preparing this book. These efforts include the development, research, and testing of the theories and programs to determine their effectiveness. The author and publisher shall not be liable in any event for incidental or consequential damages in connection with, or arising out of, the furnishing, performance, or use of these programs.

Printed in the United States of America

10 9 8 7 6 5 4 3 2 1

ISBN 0-13-805656-0

Prentice-Hall International (UK) Limited, *London*
Prentice-Hall of Australia Pty. Limited, *Sydney*
Prentice-Hall Canada, Inc., *Toronto*
Prentice-Hall Hispanoamericana, S.A., *Mexico*
Prentice-Hall of India Private Limited, *New Delhi*
Prentice-Hall of Japan, Inc., *Tokyo*
Simon & Schuster Asia Pte. Ltd, *Singapore*
Editora Prentice-Hall do Brasil, Ltda., *Rio de Janeiro*

To our families and to our friends, colleagues and students
on many continents.

Contents

PREFACE xiii

CHAPTER 1. BASIC SEMICONDUCTOR PHYSICS - A RESUME

1.1. Introduction 1

1.2. Material Properties and Band Structure 2

1.3. Electrons and Holes. Semiconductor Statistics 13

1.4. Electron and Hole Mobilities and Drift Velocities 25

1.5. Basic Semiconductor Equations 30

1.6. Hall Effect and Magnetoresistance 38

1.7. Quasi Fermi levels. Generation and Recombination of Carriers 45

1.8. *p-n* Junctions 55

    1.8.1. *p-n* Junctions Under Zero Bias 55

    1.8.2. Current-Voltage Characteristic of a *p-n* Diode 61

    1.8.3. Capacitance and Equivalent Circuit of a *p-n* Junction 71

    1.8.4. *p-n* Junction Breakdown 76

1.9. Metal Semiconductor Contacts 79

    1.9.1. Schottky Barriers 79

    1.9.2. Ohmic Contacts 91

1.10. Heterojunctions 93

    1.10.1. Basic Concepts 93

    1.10.2. Depletion Capacitance 97

    1.10.3. Current-Voltage Characteristic 98

    1.10.4. Diffusion Capacitance 103

References   104
Problems   108

CHAPTER 2. BIPOLAR JUNCTION TRANSISTOR

2.1.   Introduction   113
2.2.   Principle of Operation   114
2.3.   Overview of Bipolar Junction Transistor Technology   118
    2.3.1. Silicon Homojunction Technology   119
    2.3.2. Heterojunction Bipolar Technology   122
2.4.   BJT Modeling   124
    2.4.1. Ebers-Moll Model   124
    2.4.2. Gummel-Poon Model   128
2.5.   Breakdown in Bipolar Junction Transistors   134
2.6.   Small Signal Equivalent Circuit and Cutoff Frequencies   140
2.7.   Heterojunction Bipolar Transistor (AIM-Spice Model HBTA1)   151
    2.7.1. Principle of Operation   151
    2.7.2. Technological Aspects   156
    2.7.3. HBT Theory   161
    2.7.4. AIM Spice HBT Model (HBTA1)   173
References   177
Problems   182

CHAPTER 3. METAL OXIDE SEMICONDUCTOR FIELD EFFECT TRANSISTORS

3.1.   Introduction   188
3.2.   Overview of CMOS Technology   191
3.3.   Surface Charge and the Metal Insulator Semiconductor Capacitor   196
    3.3.1. Surface Charge   196
    3.3.2. MIS Capacitance   205
    3.3.3. Unified Charge Control Model (UCCM) for MIS Capacitors 211
    3.3.4. Analytical Unified MIS Capacitance Model   217
    3.3.5. Quantum Theory of the Two Dimensional Electron Gas 219
3.4.   Basic MOSFET Theory   229

3.4.1. Gradual Channel Approximation   229
3.4.2. A Simple Charge Control Model   238
3.4.3. Unified Charge Control Model for MOSFETs   241
3.4.4. Effect of Source and Drain Series Resistances   243
3.5. Saturation Regime   245
3.5.1. Introduction   245
3.5.2. A Simple Velocity Saturation Model   248
3.5.3. The Region of the Channel with Velocity Saturation   252
3.6. Subthreshold Regime   257
3.6.1. Subthreshold Current in Long Channel MOSFETs   258
3.6.2. Short Channel Effects and the Charge Sharing Model   263
3.6.3. Drain Induced Barrier Lowering (DIBL)   266
3.6.4. Subthreshold Current in Short Channel MOSFETs   274
3.6.5. Generalized UCCM   276
3.7. Threshold Voltage Engineering   277
3.8. Carrier Mobility and Velocity in MOSFET Channels   290
3.8.1. Velocity-Field Relationships   290
3.8.2. Mobility in the NMOS Channel   292
3.8.3. Mobility in the PMOS Channel   298
3.9. Universal Modeling of MOSFETs (AIM-Spice Model MOSA1)   301
3.9.1. Unified Channel Capacitance   301
3.9.2. Current-Voltage Characteristics   303
3.9.3. Parameter Extraction   310
3.10. Current-Voltage Characteristics of PMOS (AIM-Spice Models)   316
3.10.1. Basic Assumptions and Equations   316
3.10.2. Intrinsic Unified Long Channel PMOS Model (PMOSA1)   318
3.10.3. Intrinsic Strong Inversion PMOS model   322
3.10.4. Extrinsic Strong Inversion PMOS Model (PMOSA2)   325
3.10.5. Parameter Extraction Using Strong Inversion PMOS Model   327
3.10.6. Intrinsic Unified Short Channel PMOS Model (PMOSA3)   331
3.10.7. Parameter Extraction for Unified Short Channel PMOS Model 335
3.11. Current-Voltage Characteristics of NMOS (AIM-Spice Models)   340
3.11.1. Basic Assumptions and Equations   340
3.11.2. Intrinsic Unified Long Channel NMOS Model (NMOSA1)   342
3.11.3. Intrinsic Strong Inversion NMOS Model   345
3.11.4. Extrinsic Strong Inversion NMOS Model (NMOSA2)   347

3.11.5. Parameter Extraction Using Strong Inversion NMOS Model   349

3.11.6. Intrinsic Unified Short Channel NMOS Model (NMOSA3)   354

3.11.7. Parameter Extraction for Unified Short Channel NMOS Model 356

3.12. Unified Quasi-static Capacitance Modeling   359

3.12.1. The Inversion Sheet Charge Density   360

3.12.2. The Bulk Sheet Charge Density   361

3.12.3. The Three Terminal MOS Capacitances   365

3.12.4. The Four Terminal MOS Capacitances   365

3.13. Hot Electron Effects   370

References   375

Problems   381

CHAPTER 4. COMPOUND SEMICONDUCTOR FIELD EFFECT TRANSISTORS

4.1.   Introduction   393

4.2.   Device Technology   397

4.3.   Basic MESFET Models   404

4.3.1. The Shockley Model   406

4.3.2. Velocity Saturation Model   408

4.4.   Universal MESFET Model (AIM-Spice Model)   413

4.4.1. Unified Channel Capacitance   414

4.4.2. Unified Current-Voltage Characteristics   418

4.4.3. Model for Uniformly Doped Channel (MESA1)   423

4.4.4. Model for Channel with Delta Doping (MESA2)   425

4.4.5. Parameter Extraction   431

4.4.6. Temperature and Frequency Dependence of MESFET Parameters 437

4.4.7. Backgating Effects in MESFET   439

4.5.   Basic HFET Model   441

4.6.   Universal HFET Model (AIM-Spice model HFETA)   450

4.6.1. Capacitance Model   451

4.6.2. Current-Voltage Characteristics   454

4.7.   Current-Voltage Characteristics of HFETs   458

4.7.1. Basic Assumptions   458

4.7.2. Intrinsic Current-Voltage Characteristics   460

4.7.3. Extrinsic Current-Voltage Characteristics   465
4.7.4. Parameter Extraction   468
4.8. Gate Leakage Current   471
References   479
Problems   487

CHAPTER 5. THIN FILM TRANSISTORS

5.1. Introduction   494
5.2. Amorphous Silicon Thin Film Transistors   495
      5.2.1. Material Properties of Amorphous Silicon   495
      5.2.2. Fabrication of a-Si TFTs   497
      5.2.3. Regimes of Operation of a-Si TFTs   498
      5.2.4. Densities of Free and Trapped Charges. Field Effect Mobility 500
      5.2.5. Current-Voltage Characteristics (AIM-Spice Model ASIA1)   505
      5.2.6. Capacitance-Voltage Characteristics   511
5.3. Polysilicon Thin Film Transistors   512
      5.3.1. Fabrication of Polysilicon TFTs   512
      5.3.2. Model for Polysilicon TFTs   514
      5.3.3. Characterization of Polysilicon TFTs   523
References   528
Problems   530

CHAPTER 6. AIM-Spice USERS MANUAL
(WINDOWS™3.1 VERSION)

6.1. Introduction   534
      6.1.1. Conventions Used in This Chapter   535
6.2. Introduction to Microsoft WINDOWS™-3.1   536
      6.2.1. Basics of the WINDOWS™-3.1 Environment   536
      6.2.2. Working with Windows   539
      6.2.3. Working with Menus   543
      6.2.4. Dialog Boxes   546
6.3. AIM-Spice   549
      6.3.1. Getting Started   551

6.3.2. Editing the Circuit Description (Netlist)   551
6.3.3. Working with AIM-Spice Circuit Files   555
6.3.4. Circuit Description in AIM-Spice   559
6.3.5. Circuit Analysis with AIM-Spice   566
6.4. AIM-Postprocessor   585
6.4.1. AIM-Postprocessor Tutorial   585
6.4.2. Command Reference   602
6.5. Simulation Examples Using AIM-Spice   613
6.5.1. Heterostructure Diode   614
6.5.2. Heterostructure Bipolar Transistor (HBTA)   616
6.5.3. MOSFET (MOSA1)   617
6.5.4. MESFET (MESA1)   619
6.5.5. HFET (HFETA)   624
6.5.6. Thin Film Transistors (ASIA1)   626
References   633
Problems   633

APPENDICES

A1. Physical Constants   634
A2. Properties of Silicon (Si)   635
A3. Properties of Gallium Arsenide (GaAs)   636
A4. Properties of $Al_xGa_{1-x}As$ 637
A5. Properties of Amorphous Silicon (a-Si)   638
A6. Properties of SiO_2 and Si_3N_4 639
A7. Approximate Solution of the Non-Ideal Diode Equation   640
A8. Summaries of AIM-Spice Models   646
A8.1. Heterostructure Diode Model (HDIA)   646
A8.2. HBT Model (HBTA)   647
A8.3. Universal MOSFET Model (MOSA1)   650
A8.4. PMOS Models (PMOSA1 and PMOSA2)   651
A8.5. NMOS Models (NMOSA1 and NMOSA2)   654
A8.6. MESFET Models (MESA1 and MESA2)   657
A8.7. Universal HFET Model (HFETA)   661
A8.8. a-Si TFT Model (ASIA1)   663
A8.9. Poly-Si TFT Model (PSIA)   665

A9. SPICE Reference   667
Background on SPICE   667
Notation Used in This Appendix   667
AC Analysis   668
DC Operating Point   668
DC Transfer Curve Analysis   668
Noise Analysis   669
Pole/Zero Analysis   670
Transfer Function Analysis   670
Transient Analysis   671
Options   672
Title Line   673
Comment Line   673
A. Heterostructure Field Effect Transistors (HFETs)   674
B. Non-Linear Dependent Sources   676
C. Capacitors   678
D. Diodes   679
E. Linear Voltage-Controlled Voltage Sources   681
F. Linear Current-Controlled Current Sources   682
G. Linear Voltage-Controlled Current Sources   682
H. Linear Current-Controlled Voltage Sources   683
I. Independent Current Sources   683
J. Junction Field Effect Transistors (JFETs)   688
K. Coupled Inductors (Transformers)   689
L. Inductors   690
M. MOSFETs   690
N. Heterojunction Bipolar Transistors (HBTs)   702
O. Lossy Transmission Lines (LTRA)   705
Q. Bipolar Junction Transistors (BJTs)   708
R. Resistors   711
S. Voltage-Controlled Switch   713
T. Transmission Lines (Lossless)   714
U. Uniform Distributed RC Lines (URC)   715
V. Independent Voltage Sources   716
W. Current Controlled Switch   721
Z. MESFETs   722

.SUBCKT Statement   726
.ENDS Statement   727
Call to Subcircuits   727
.NODESET Statement   727
.IC Statement   728
Bugs Reported by Berkeley   728
A10. Adding New Device Models to AIM- Spice   730
Software Requirements   730
Adding New Device Models   730
Data Structures Specific for a Device   730
Parameter Descriptors   732
The Main Device Structure   732
Device Specific Functions   733
Step One: Adding New Routines   733
Step Two: Modification to Existing Interface Routines   733
Step Three: Integration into the Main Loops of the Simulator   735
Step Four: Compiling and Relinking   735

INDEX   737

Preface

This book is intended to serve as a reference book for device and electronics engineers and scientists working in research, development and manufacturing of VLSI (Very Large Scale Integrated Circuits), as a text book for senior and graduate courses on semiconductor device modeling, and as a manual to our circuit simulator, Automatic Integrated Circuit Modeling Spice (AIM-Spice).

The book presents to the readers a combination of background device physics and technology, a review of existing device models, and, more importantly, a set of new and improved models compatible with the most advanced technology. We use these models to establish new device characterization techniques which are accurate, unambiguous, and fast. These techniques and the implementation of our models in AIM-Spice allow the readers to use these models for device and circuit design, circuit simulation, parameter extraction, statistical yield analysis, and other tasks which have been difficult before. We also included into the book a clear description of how one can add new device models to AIM-Spice which may be especially helpful to researchers working on advanced semiconductor devices. Still, AIM-Spice is a version of SPICE with standard SPICE parameters. AIM-Spice runs under the WINDOWS^{TM}-3.1 environment, taking full advantage of its graphics user interface. AIM-Spice will run on IBM PCs and compatible computers. With the introduction of the NT (New Technology) operating system from Microsoft, the Windows version of AIM-Spice will also run on work stations. Hence, the use of AIM-Spice presents no barrier for VLSI engineers or designers. The inclusion of AIM-Spice makes our book unique and quite different from all other books in this field.

The device feature size in VLSI has been scaled down into the deep submicron range. We can expect that quarter micron feature size will become typical for VLSI only a few years from now and that a 0.1 μm feature size will be used in VLSI by the start of the new century. This is a great challenge for device and circuit designers. New device models have become vital for accurate device modeling and characterization and for optimized device design. The importance of modeling software is commensurate with the increase in the integration scale and the device sophistication, and device models implemented into circuit simulation programs have already become an integral part of semiconductor technology, as essential as fabrication equipment.

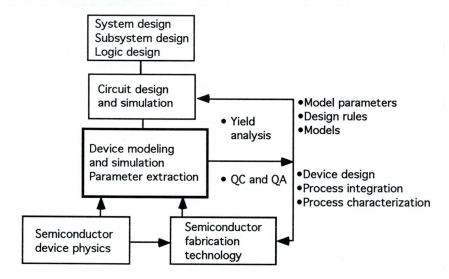

The figure shows a schematic diagram of the VLSI technological hierarchy. This technology is often divided into "top" and "bottom" parts. The "top" part includes system, subsystem, and logic design. The "bottom" part includes device and circuit design and fabrication technology. Semiconductor device modeling is an important component of this "bottom" part. In addition to device and circuit design, modeling helps to establish fabrication technology and process integration, and provides process characterization data required to maintain proper yield, Quality Control (QC), and Quality Assurance (QA). The models establish parameter lists for circuit simulation, design rules and, sometimes, in-house circuit simulation models. The information extracted from yield, QC and QA is then reflected in the next generation of device and circuit design.

These tasks require physics-based, simple, and accurate device models. Device parameters should be extractable in an easy and unambiguous manner. This last requirement is especially important for statistical yield analysis, QC and QA. At the present time, the parameter extraction for most of the models relies on numerical optimization. However, such an approach makes it difficult to trace the observed electrical characteristics back to the fabrication process. Moreover, using numerical optimization, it is almost impossible to obtain statistical device data.

In this book, we describe semiconductor device models with emphasis on applications for VLSI modeling. We try to meet challenging demands for fast computation time and guaranteed conversion combined with ease and reproducibility of parameter extraction. The emphasis of this book is on Field Effect Transistor (FET) models. For most FETs, AIM-Spice and this book contain three sets of models: basic, intermediate, and advanced. The basic models illustrate basic device physics and can be used for fast and relatively crude simulation. These models are based on well-known analytical models. The intermediate set are unified models which describe the device operation in the entire range of bias voltages. These models are suitable for device characterization and parameter extraction. The advanced models are the most detailed and accurate but also the most complex. They should be used when a high degree of accuracy is especially important.

Our intermediate and advanced FET models in general, and Complementary Metal Oxide Semiconductor (CMOS) models in particular, are based on the new concept of a unified charge control model which allows us to describe different regimes of transistor operation, including the subthreshold regime, using one basic equation. In turn, this minimizes the number of device parameters required for accurate modeling and makes the parameter extraction easier and much more straightforward. We have also developed new analytical approaches which allow us to incorporate parasitic resistances into our device models. Direct incorporation of the parasitic resistances may be important for high frequency analog and digital circuits, especially for small device sizes and for statistical and reliability modeling. The extrinsic models lead to significant savings of computer time. This is useful for simulating large circuits.

A large part of this book is devoted to silicon n-channel and p-channel FETs for the purpose of CMOS modeling. However, we also describe new accurate models for compound semiconductor transistors, such as GaAs MEtal

Semiconductor Field Effect Transistors (MESFETs) and AlGaAs/GaAs and AlGaAs/InGaAs/GaAs Heterostructure Field Effect Transistors (HFETs), and for amorphous silicon and polysilicon Thin Film Transistors (TFTs). We also present new analytical circuit models for heterostructure diodes and Heterostructure Bipolar junction Transistors (HBTs) – the transistors which have the highest speed of operation.

We have incorporated all our device models in a new version of the popular circuit simulator SPICE (Version 3e.1) developed at Berkeley. The student version of AIM-Spice is available gratis to adopters of the book from your local Prentice Hall representative and runs under WINDOWS™-3 on IBM and compatible PCs. We have used this program for the device and circuit simulation examples included in the book. The professional version is available from the authors at a nominal cost.

The book is organized as follows: In Chapter 1 we give a brief summary of basic semiconductor physics. This summary includes basic material properties of important semiconductors, basic semiconductor equations, and discussion of p-n junctions, Schottky barriers, ohmic contacts, and heterojunctions. The purpose of Sections 1.1 through 1.7 is to give the readers a concise resume of the semiconductor physics which is required for the understanding of the subsequent chapters. These sections will be particularly useful for practicing device engineers and scientists who may like to use the opportunity to review the physics background of device modeling. However, they can be omitted from courses on semiconductor device modeling and used as material for additional reading. Sections 1.8 through 1.10 on p-n junctions, Schottky barriers, and heterojunctions constitute a necessary introduction to the subsequent discussion of transistor models. In Chapter 2, we consider Ebers-Moll and Gummel-Poon models for bipolar junction transistors, and present a new model for the Heterojunction Bipolar Transistor (HBT). Chapters 3 and 4 deal with Metal Oxide Semiconductor Field Effect Transistors (MOSFETs) and compound semiconductor FETs, respectively, and introduce new intrinsic and extrinsic models for the most important field effect transistors such as n-MOSFET, p-MOSFET, MESFET, and HFET. These chapters also include new, physically based calculations of the saturation and the subthreshold regimes of operation, with emphasis on short-channel phenomena. Chapter 5 deals with Thin Film Transistors (TFTs) which are becoming increasingly important for display technology and consumer electronics. We include new models for both amorphous silicon and polysilicon TFTs. In Chapters 3, 4, and 5, we also discuss characterization techniques for field-effect transistors based on the new models. All these characterization and parameter extraction techniques are illustrated using experimental data obtained in our laboratories and from the literature. Chapter 6 contains basic information about the WINDOWS™-3 environment, a detailed AIM-Spice manual, and examples of using AIM-Spice for modeling different semiconductor devices. The appendices include lists of relevant material properties, a summary of the device models incorporated into AIM-Spice, and a reference section for SPICE.

In order to make the book useful as a reference, we made the sections devoted to the different models to be fairly self-contained with sufficient explanations of modeling assumptions and notation. This should also help researchers and students who may use these sections as "prototypes" for developing their own AIM-Spice models for novel semiconductor devices.

The combination of a user friendly and powerful software package with detailed description of underlying device and device models makes this book very useful not only as a reference and manual for scientists and engineers, but also as a text for seniors and first year graduate electrical engineering students. The AIM Spice package allows students to solve engineering problems bridging the gap between theory and practice. The book can be used as a text for either one semester senior level or two semester graduate level courses. We recommend that the one semester course covers the material included in Sections 1.8 through 1.10, Chapter 2, and Sections 3.1 through 3.5. The two semester course should also cover Sections 3.6 through 3.13, Chapter 4, and Chapter 5. The remaining sections of the book are primarily for reference purposes and additional reading. These courses are primarily engineering design courses. Students using this text will have at their disposal a practical engineering design tool – AIM-Spice – and should be strongly encouraged to solve real engineering design problems. The models used in AIM-Spice and used in this book are linked to the state-of-the-art technology and use typical values of device parameters as the default values. Many problems included in the book are design problems which do not have unique solutions, and which involve tradeable parameter values. The design problems at the end of each chapter are marked with asterisks. For courses based on this book, we recommend that two thirds of the credit hours given to students be allocated to engineering design and one third to engineering science.

The problems in the book range from very simple to fairly advanced. Their number is sufficient for regular homework assignments and exams, including take home exams suitable for a design oriented engineering course. Some of the problems are fairly traditional and self-contained. Others will require a student to look up certain material and device parameters and/or make a reasonable guess. Many of the problems can be solved (or their analytical solutions can be checked) using AIM-Spice. We believe that students should be encouraged to do just that since AIM-Spice is a state of the art VLSI design tool and since real life problems often do not have analytical solutions.

A detailed solution manual and sets of transparencies for lectures based on this book will be made available for instructors.

The authors of this book are a Korean, an American who emigrated from Russia, and two Norwegians. We have been working with graduate students and research associates from Korea, USA, Norway, Germany, China, India, and Russia. We have greatly enjoyed working with each other. We belong to the same international scientific community and feel that cultural differences present no barrier for mutual understanding and cooperation.

For more than three years, we concentrated most of our efforts on this book and related work. During all this time, our families provided us with support and encouragement.

Our present and former graduate and undergraduate students helped us in doing research for this book. We are especially grateful to Drs. B. J. Moon, Y. Byun, C. K. Park, G. U. Jensen, Mr. K. M. Rho, Mr. T. Steen, Mr. S. H. Kim, Mr. K. S. Min, and Mr. E. M. Fjeldly for their invaluable research contributions, dedication, and hard work. We would like to thank Dr. H. G. Lee of GoldStar Electron, Ltd.; Dr. W. S. Min of Hyundai Electronics Industries; Ltd., Dr. Boris Gelmont of the University of Virginia; Dr. Michael Hack of Xerox PARC; Dr. Tayo Akinwande of Honeywell, Inc.; and Dr. Krishna Pande of COMSAT, Inc. for their help and encouragement.

Kwyro Lee, Taejon, Korea
Michael Shur, Charlottesville, Virginia, USA
Tor A. Fjeldly, Trondheim, Norway
Trond Ytterdal, Trondheim, Norway

1

Basic Semiconductor Physics - A Resume

1-1. INTRODUCTION

The purpose of this Chapter is to give a brief review of basic equations and the most important results of the semiconductor physics relevant to modeling and simulation of modern semiconductor devices and integrated circuits. Today, most semiconductor devices are made of silicon. However, submicron devices made of compound semiconductors, such as gallium arsenide, successfully compete for applications in microwave and ultra-fast digital circuits. Therefore, this Chapter describes material properties of silicon and compound semiconductors. Advances in fabrication technology have led to a shrinking of the minimum device feature size from about 20 micrometers in the early sixties to submicron dimensions in the late eighties. In shorter devices, electrons take less time to travel across the device, leading to higher speeds and operating frequencies. In addition, a smaller active device volume translates into lower operating power. In novel device structures, the dimensions are so small that quantum effects become important or even dominant, and we include some discussion of quantum mechanical phenomena. However, the emphasis of this Chapter is on the semiconductor equations used in modeling of semiconductor devices. More detailed discussion of this material may be found, for example, in monographs of Seeger (1985), Hess (1988), or Shur (1990).

1-2. MATERIAL PROPERTIES AND BAND STRUCTURE

The most important semiconductor material is silicon. In a silicon crystal, an atom forms four bonds with four other atoms (four nearest neighbors) and shares two valence electrons with each of them. Hence, it shares eight valence electrons with all four nearest neighbors. In other words, in a silicon or germanium crystal, each atom is tetrahedrally coordinated (see Fig. 1.2.1) in order to share eight electrons (two electrons per bond), corresponding to complete *p* and *s* valence subshells.

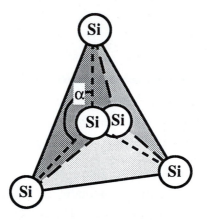

Fig. 1.2.1. Tetrahedral atomic configuration.

The valence configurations of Si and Ge are very similar (in both cases we have two *s* electrons and two *p* electrons), and we can expect that their physical and chemical properties should be similar too. Indeed, both elements are semiconductors. But semiconductor compounds can also be formed by combining different elements such as those shown in the section of the Periodic Table in Fig. 1.2.2. Thus, if we combine Ga and As into a GaAs compound, then each atom, on average, will have two *s* electrons and two *p* electrons, like in Si and Ge. In a similar way, we can form many semiconductor compounds, combining other elements of column III of the Periodic Table (having three valence electrons, two *s* electrons and one *p* electron) with elements from column V (having five valence electrons, two *s* electrons and three *p* electrons). Some examples of such so-called III-V compounds are GaAs, InAs, InP, AlAs, GaP, AlP, InSb, BN, AlN, GaSb, and GaN.

IIb	IIIa	IVa	Va	VIa
	B	C	N	O
	Al	Si	P	S
Zn	Ga	Ge	As	Se
Cd	In	Sn	Sb	Te
Hg		Pb		

Fig. 1.2.2. Section of the Periodic Table with elements from atomic groups IIb to VIa.

Elements from the second and sixth columns of the periodic table (having two and six valence electrons respectively) can also be combined to form II-VI compound semiconductors. CdS, ZnS, CdTe, and CdSe are examples of II-VI compound semiconductors. Moreover, many III-V and/or II-VI compound semiconductors may form solid state solutions, such as $Al_xGa_{1-x}As$ where x is a molar fraction of Al. Such materials may be composed of three atomic species and are called ternary compounds. By varying x from 0 to 1, one can change the properties of $Al_xGa_{1-x}As$ continuously from those of GaAs to those of AlAs. This particular compound is especially important because the lattice constants of GaAs and AlAs are very close (5.653 Å and 5.661 Å at 300 K for GaAs and AlAs, respectively). That is why $Al_xGa_{1-x}As$ can be easily grown on GaAs forming a heterostructure, i.e., a structure including two different semiconductor materials in intimate contact. Heterostructures have found numerous and increasingly important applications in novel semiconductor devices. Other important examples of ternary compounds include $In_xGa_{1-x}As$, $In_xGa_{1-x}P$, $Al_xIn_{1-x}As$, etc. Moreover, so-called quaternary compounds, such as $In_xGa_{1-x}As_yP_{1-y}$ can also be formed. This "material engineering" approach allows us to design semiconductor materials with desired properties, for instance with a given band gap. Fig. 1.2.3 shows the band gap, E_g, and the lattice constant, a, for some important III-V compound systems (see also Table 1.2.1). Many of these new materials, however, demand further study and refinement before they find applications in the new and exciting high-performance devices of the future.

Material	Lattice constant a (Å) at 25°C	Nearest neighbor distance, $a\sqrt{3}/4$,(Å)	Sum of covalent radii (Å)
Si	5.434	2.353	2.34
Ge	5.657	2.450	2.44
A_3B_5			
AlAs	5.661	2.451	2.440
AlP	5.451	2.360	2.360
AlSb	6.136	2.657	2.620
BAs	4.776	2.068	2.060
BN	3.615	1.565	1.580
BP	4.538	1.965	1.980
BSb	5.170	2.239	2.240
GaAs	5.653	2.448	2.440
GaP	5.451	2.360	2.360
GaSb	6.095	2.639	2.620
InAs	6.058	2.623	2.620
InP	5.867	2.540	2.540
InSb	6.479	2.805	2.800
A_2B_6			
CdTe	6.482	2.807	2.950
HgS	5.841		
HgSe	6.084		
HgTe	6.462	2.798	2.950
ZnS	5.415		
ZnSe	5.653		
ZnTe	6.101	2.642	2.780

Table 1.2.1. Lattice constants (from Shur (1990)).

Each atom in a silicon crystal lattice forms four bonds with its nearest neighbors. This corresponds to the so-called tetrahedral configuration in which each atom is located at the center of the tetrahedron formed by four silicon atoms (see Fig. 1.2.1). In a silicon crystal, this tetrahedral bond configuration is repeated, forming the same crystal structure as in the diamond crystal shown in Fig. 1.2.4. This structure (called the diamond structure) is formed by two interpenetrating face-centered cubic sublattices of atoms, shifted with respect to each other by one fourth of the body diagonal. (Another important semi-conductor – germanium (Ge) – has exactly the same crystal configuration as well.)

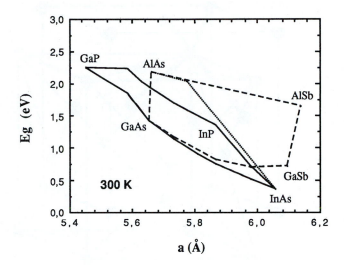

Fig. 1.2.3. Energy band gaps and lattice constants for some III-V compound systems.

In compound semiconductors, the chemical bond between the nearest neighbors is partially heteropolar. Nevertheless, in many such compounds this bond is still more or less covalent, leading to a tetrahedral bond configuration similar to that in a silicon crystal (see Fig. 1.2.1). As a consequence, most III-V compounds crystallize in the so-called zinc blende crystal structure (see Fig. 1.2.5), which is very similar to the diamond structure. The primitive cell of the zinc blende structure contains two atoms, A and B, which are repeated in space, with each species forming a face-centered cubic lattice. It can be described as consisting of two mutually interpenetrating face-centered cubic (fcc) lattices of element A and element B (gallium and arsenic in the case of GaAs), shifted relative to each other by a quarter of the body diagonal of the unit cell cube.

If atoms A and B are identical, this crystal structure reduces to the diamond structure. The diamond lattice has an inversion symmetry. This means that the crystal will remain unchanged if all the atoms of the crystal are moved in such a way that their space coordinates x, y, z (counted with respect to the point called the center of the inversion) are changed to -x, -y, -z. The center of inversion for the diamond structure is at the midpoint between two nearest neighbors. Clearly, the zinc blende structure lacks such inversion symmetry.

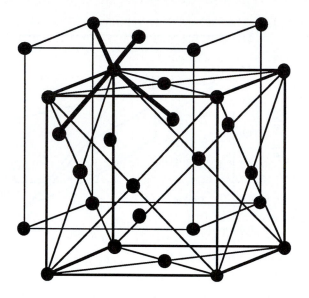

Fig. 1.2.4. Diamond structure.

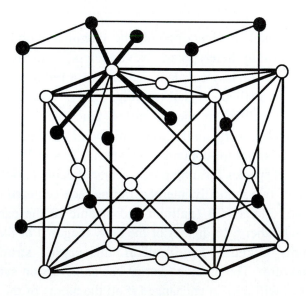

Fig. 1.2.5. Zinc blende structure.

The lattice constants of several important semiconductors with diamond or zinc blende structure are given in Table 1.2.1. (For a cubic lattice the lattice constant is equal to the side of the unit cell cube.) We notice that the lattice constants of GaAs and AlAs are very close. Hence, AlAs can be grown on top of a GaAs wafer with very few imperfections, such as dislocations or surface states. Another way to "match" lattice constants is to use a heterostructure consisting of a ternary compound and a binary compound. The lattice constant, a_{ter}, of a ternary compound $A_xC_{1-x}B$ varies roughly linearly with composition; i.e.,

$$a_{ter} \approx a_{bin1}x + a_{bin2}(1 - x)$$

(1-2-1)

where a_{bin1} is the lattice constant of the binary compound AB and a_{bin2} is the lattice constant of the binary compound CB. Hence, for $In_{0.47}Ga_{0.53}As$, for example, we find from eq. (1-2-1) and Table 1.2.1 $a_{ter} \approx 5.84$ Å, matching quite well the lattice constant of InP ($a = 5.86$ Å). A consequence of this good match is that $In_{0.47}Ga_{0.53}As$ can be grown with few imperfections on InP substrates.

A very important characteristic of a semiconductor material is its band structure – the dependence of allowed electronic energies on the electron wave vector, **k**. Many features of the band structure are determined by the crystal symmetry. The crystal has a translational symmetry which means that the whole crystal can be reproduced by repeating one primitive cell (with an atomic basis) in space. This crystal symmetry determines unique allowed values of electronic wave vectors. These values are defined in the so-called reciprocal or wave vector space where all distances are measured in inverse units of length, such as l/m. In the reciprocal space, this translational symmetry is accounted for by introducing a reciprocal lattice – an array of points in the reciprocal space with a translational symmetry determined by the real space symmetry of the crystal. The reciprocal lattice can be reproduced by repeating a primitive cell in the reciprocal space. This primitive cell contains complete information about the reciprocal lattice. All points in the reciprocal lattice may be obtained from just one primitive cell by adding different reciprocal lattice vectors. Usually, such a primitive cell is chosen as a so-called Wigner-Seitz cell of the reciprocal lattice and is called the first Brillouin zone (BZ). This cell is formed by bisecting all reciprocal lattice vectors emanating from a lattice site by planes and choosing the smallest volume formed by these intersecting planes. Hence, we can obtain the first Brillouin zone by constructing planes which bisect perpendicularly the reciprocal lattice vectors emanating from the origin in **k**-space. All values of **k** in the first Brillouin zone are such that subtracting or adding any reciprocal lattice vector from a vector **k** inside the zone does not give another vector also inside this zone. As shown below, physically significant and unique solutions of the Schrödinger equation for an electron in a crystal may be assigned to the wave vectors **k** in the first Brillouin zone.

The first Brillouin zone for the face-centered cubic, diamond and the zinc blende structures is shown in Fig. 1.2.6. It is the same for all three structures because it can be formed from the same face-centered cubic lattice, placing a different atomic basis into the lattice nodes. The reciprocal lattice for these crystals is body centered cubic. The electron energy is a function of k_x, k_y, and k_z and varies within the first Brillouin zone. Most important semiconductor properties are determined by regions in the first Brillouin zone corresponding to lines and points of symmetry. These symmetry points and lines are marked in Fig. 1.2.6. Points Γ (in the center of the first Brillouin zone), X, and L, and directions Δ and Λ are particularly important. The dependencies of the electronic energy on the wave vector (energy bands) are usually calculated for the lines of highest symmetry in the first Brillouin zone.

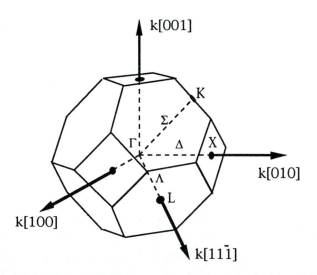

Fig. 1.2.6. The first Brillouin zone for the face-centered cubic, diamond and the zinc blende structures. Also shown are symmetry points and directions of the first Brillouin zone.

The most important features of the band structures are represented by the minimum of the lowest conduction band, E_c, and by the maximum of the highest valence band, E_V, since the states with energies much higher than the minimum of the conduction band are empty and the states with energies well below the maximum of the valence band are completely filled. A model which summarizes the most important features of the band structure of the cubic semiconductors is shown in Fig. 1.2.7. These bands are typical for Si, Ge, III-V, and most II-VI semiconductors.

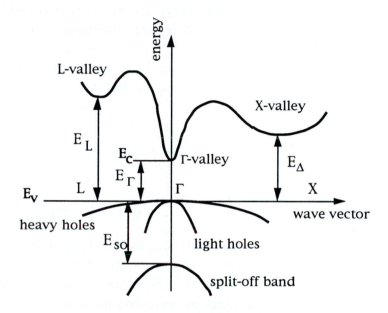

Fig. 1.2.7. Schematic band diagram for cubic semiconductors.

A typical feature is that the top valence band and the lowest conduction band are separated by an energy gap. In all important cubic semiconductors, the top of the highest filled (valence) band is located at the Γ point of the first Brillouin zone (see Fig. 1.2.6), i.e., at the point $\mathbf{k} = (0,0,0)$. Two of the valence bands, the light and the heavy hole bands, have the same energy at this point (i.e., are degenerate), while a third band is separated from the other two and is called the split-off band. The lowest minimum of the conduction band in different semiconductors are located at different points in the first Brillouin zone. For example, in germanium, the L minimum (corresponding to the L-point in the first Brillouin zone) is the lowest. The lowest minimum of the conduction band in silicon is found along the Δ axis (corresponding to the direction $(k,0,0)$), close to the X-point of the first Brillouin zone. In GaAs, the lowest minimum of the conduction band is at the Γ point; i.e. at the same value of the wave vector **k** as the top of the valence band. GaAs is therefore called a direct gap semiconductor. Silicon and germanium, however, are indirect gap semiconductors.

Gray tin and HgTe are examples of so-called gapless semiconductors. The absence of the energy gap leads to very interesting and unusual properties of these materials. Utilizing, for example, CdHgTe alloys one may create materials with narrow energy gaps. Such alloys are of interest as sensitive infrared detectors since they can be used to register interband transitions caused by infrared photons.

Near the minimum of the conduction band, the dependence of kinetic energy versus wave vector can be approximated by one of the following functions:

$$E_n(k) = E_c - \hbar^2 k^2 / (2m_n) \quad \text{(spherical)}$$

(1-2-2)

$$E_n(k) = E_c - \frac{\hbar^2}{2}\left(k_x^2 / m_x + k_y^2 / m_y + k_z^2 / m_z\right) \quad \text{(ellipsoidal)}$$

(1-2-3)

The terms "spherical" and "ellipsoidal" describe the shape of the surfaces of equal energy in the k-space. Wave vector **k** and kinetic energy E_n in eqs. (1-2-2) and (1-2-3) are measured from the minimum E_c of the conduction band. Eq. (1-2-2) represents a parabolic dependence of energy on the wave vector, similar to the dependence for an electron in free space, $E_n = \hbar^2 k^2/(2m_e)$. However, the value of m_n (which is called the effective mass of the electron) can be very much different from the free electron mass, m_e. For example, in gallium arsenide $m_n \approx 0.067\ m_e$. This difference is caused by the periodic crystal potential. Eq. (1-2-2) corresponds to a band with a single scalar effective mass and spherical surfaces of equal energy. It is an appropriate model for the Γ minimum of the conduction band. A similar equation

$$E_p(k) = E_v - \hbar^2 k^2/(2 m_p) \quad \text{(spherical)} \qquad (1-2-4)$$

can be used for the valence band. (Here E_v is the energy corresponding to the top of the valence band.) Eq. (1-2-2) is also the simplest possible model of the band structure and is frequently used for crude estimates of transport properties. The similarity with the free electron motion is very useful for providing an insight into the physics of electronic motion. Due to the rotational symmetry of the ellipsoids of equal energy for the L and X minima of a conduction band, eq. (1-2-3) can be rewritten in a simpler form:

$$E_n(k) = E_c - \frac{1}{2} \hbar^2 \left(\frac{k_l^2}{m_l} + \frac{k_t^2}{m_t} \right) \quad \text{(ellipsoidal)} \qquad (1\text{-}2\text{-}5)$$

where $1/m_l$ and $1/m_t$ are the longitudinal and transverse components of the inverse effective mass tensor and k_l and k_t are the longitudinal and transverse components of the wave vector. Energy band gaps and effective masses for important cubic semiconductors are given in Table 1.2.2.

More realistic models should take into account the deviation from parabolicity of the E vs. $\mathbf{k}$ relationship. In the conduction band this can be done, in the simplest way, by substituting E_n in eq. (1-2-2) with

$$\gamma(E) = E(1 + \alpha E) \qquad (1\text{-}2\text{-}6)$$

which gives the following equation

$$E_n(1 + \alpha E_n) = \hbar^2 k^2 / (2 m_n) \qquad (1\text{-}2\text{-}7)$$

Eq. (1-2-7) takes into account an increase of the electron effective mass with energy which can be noticed as a decrease in the curvature of the E_n versus k relationship with increasing E_n (see Fig. 1.2.7).

All parameters characterizing the band structure are temperature-dependent. According to Blakemore (1982), the temperature dependencies of the energy gaps can be described by the following phenomenological equation

$$E_g(T) = E_{go} - \alpha_{temp} T^2 / (T + \beta_{temp}) \qquad (1\text{-}2\text{-}8)$$

where the values of the energy gap E_{go} at $T = 0$ and the coefficients α_{temp} and β_{temp} are given for Si, Ge and GaAs in Table 1.2.3.

	E_Γ (eV)	E_L (eV)	E_Δ (eV)	E_{so} (eV)	Electron effective masses			Hole effective masses	
					m_l	m_n	m_t	m_{ph}	m_{pl}
Si	4.08	1.87	1.13	0.04	0.98	--	0.19	0.53	0.16
Ge	0.89	0.76	0.96	0.29	1.64	--	0.082	0.35	0.043
AlP	3.3	3.0	2.1	0.05	--	--	--	0.63	0.2
AlAs	2.95	2.67	2.16	0.28	2.0	--	--	0.76	0.15
AlSb	2.5	2.39	1.6	0.75	1.64	--	0.23	0.94	0.14
GaP	2.24	2.75	2.38	0.08	1.12	--	0.22	0.79	0.14
GaAs	1.42	1.71	1.90	0.34	--	0.067	--	0.62	0.074
GaSb	0.715	1.07	1.30	0.77	--	0.045	--	0.49	0.046
InP	1.35	2.0	2.3	0.13	--	0.080	--	0.85	0.089
InAs	0.35	1.45	2.14	0.38	--	0.023	--	0.6	0.027
InSb	0.17	1.5	2.0	0.81	--	0.014	--	0.47	0.015
ZnS	3.8	5.3	5.2	0.07	--	0.28	--	--	--
ZnSe	2.9	4.5	4.5	0.43	--	0.14	--	--	--
ZnTe	2.56	3.64	4.26	0.92	--	0.18	--	--	--
CdTe	1.80	3.40	4.32	0.91	--	0.096	--	--	--

Table 1.2.2. Energy band gaps and effective masses (in units of m_e) of some cubic semiconductors (after Jacoboni and Reggiani (1979)).

Semiconductor	E_{go} (eV)	α_{temp} (meV/K^2)	β_{temp} (K)
Si	1.170	0.473	63
Ge	0.7437	0.477	235
GaAs	1.519	0.5405	204

Table 1.2.3. Temperature dependence of the energy gaps (from Blakemore (1982)).

1-3. ELECTRONS AND HOLES. SEMICONDUCTOR STATISTICS

Electron wave functions in a crystal are described by so-called Bloch waves

$$\psi_{\mathbf{k}}(\mathbf{r}) = e^{\,i\mathbf{k}\mathbf{r}} u_{\mathbf{k}}(\mathbf{r}) \qquad (1\text{-}3\text{-}1)$$

where $u_{\mathbf{k}}(\mathbf{r})$ is periodic in real space with the period of the crystal lattice. Each allowed value of $\mathbf{k}$ with coordinates k_x, k_y, and k_z occupies a volume $(2\pi)^3/V$ in the reciprocal space. V is the crystal volume. According to the Pauli exclusion principle applied to a crystal, only two electrons (with opposite spins) in the same energy band may have the same wave vector $\mathbf{k}$. For a finite temperature T, the probability of having an energy state occupied is given by the Fermi-Dirac occupation function, f_n, given by

$$f_n = \frac{1}{1 + \exp\left[\left(E - E_F\right)/k_B T\right]} \qquad (1\text{-}3\text{-}2)$$

Here $E(\mathbf{k})$ is the energy of the electrons and E_F is the Fermi level. The function f_n is plotted in Fig. 1.3.1 for different values of temperature. As can be seen from this figure, all states below approximately $E_F - 3k_BT$ are essentially filled, all states above approximately $E_F + 3k_BT$ are practically empty, and $f_n(E_F) = 1/2$. When $E - E_F \geq 3k_BT$, the Fermi function may be approximated by

$$f_n \approx \exp\left(\frac{E_F - E}{k_B T}\right) \qquad (1\text{-}3\text{-}3)$$

If the Fermi level is in the conduction band and the temperature is low, then $f_n \approx 1$ for filled states corresponding to wave vectors smaller than

$$k_F = \sqrt{2 m_n\left(E_F - E_c\right)}/\hbar \qquad (1\text{-}3\text{-}4)$$

(using eq. (1-2-2)), and $f_n \approx 0$ for empty states with wave vectors larger than k_F. The wave vector k_F is called the Fermi vector of the conduction band electrons and is measured from the bottom, E_c, of the conduction band. An n-type semiconductor with a Fermi level E_F such that $E_F - E_c > 3k_BT$ is called a degenerate semiconductor. Similarly, a p-type semiconductor is said to be degenerate if $E_v - E_F > 3k_BT$, where E_v is the top of the valence band.

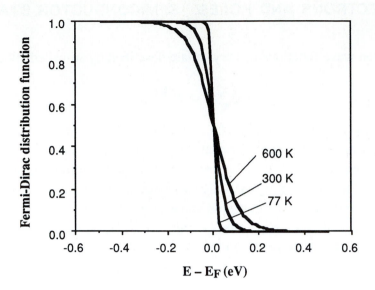

Fig. 1.3.1. Fermi-Dirac distribution function.

For parabolic bands the density of allowed energy states (including spin degeneracy) per unit volume is

$$g_n(E) = 4\pi \left(\frac{2 m_n}{h^2} \right)^{3/2} (E - E_c)^{1/2} \qquad (1\text{-}3\text{-}5)$$

(A derivation of this equation can be found, for example, in Shur (1990).)

The concentration of electrons in the conduction band can be expressed in terms of the position of the Fermi level as follows

$$n = \int_{E_c}^{\infty} g_n(E) f_n(E) \, dE = N_c \, F_{1/2}(\eta_n) \qquad (1\text{-}3\text{-}6)$$

where

$$N_c = 2 \left(\frac{m_n k_B T}{2\pi \hbar^2} \right)^{3/2} \qquad (1\text{-}3\text{-}7)$$

is the so-called effective density of states for the conduction band,

$$\eta_n = (E_F - E_c)/k_B T \tag{1-3-8}$$

and

$$F_{1/2}(\eta_n) = \frac{2}{\sqrt{\pi}} \int_0^\infty \frac{x^{1/2}dx}{[1 + \exp(x - \eta_n)]} \tag{1-3-9}$$

is the Fermi integral. When $\eta_n \leq -3$,

$$F_{1/2} \approx \exp(\eta_n) \tag{1-3-10}$$

When $\eta_n \geq 3$,

$$F_{1/2} \approx \frac{4\eta_n^{3/2}}{3\sqrt{\pi}} \tag{1-3-11}$$

in which case the Fermi vector (eq. (1-3-4)) can be expressed as

$$k_F = \left(3\pi^2 n\right)^{1/3} \tag{1-3-12}$$

For $-10 < \eta_n < 10$, the Fermi integral, $F_{1/2}(\eta_n)$, can be interpolated by the following expression (see Shur (1990))

$$F_{1/2}(\eta_n) = \exp(-0.32881 + 0.74041\eta_n - 0.045417\eta_n^2 \\ - 8.797\text{x}10^{-4}\eta_n^3 + 1.5117\text{x}10^{-4}\eta_n^4) \tag{1-3-13}$$

A **hole** can be represented as a positive "particle" (i.e., the absence of a negatively charged electron), and the Fermi-Dirac occupation function for a hole can be calculated as

$$f_p(E) = 1 - f_n(E) = \frac{1}{1 + \exp[(E_F - E)/k_B T]} \tag{1-3-14}$$

In full analogy with the above discussion for electrons, we find for holes the following density of allowed energy states per unit volume in the valence band (parabolic approximation):

$$g_p(E) = 4\pi \left(\frac{2\,m_p}{h^2} \right)^{3/2} (E_v - E)^{1/2} \qquad (1\text{-}3\text{-}15)$$

The total hole density is

$$p = \int_{-\infty}^{E_v} g_p(E)\, f_p(E)\, dE = N_v\, F_{1/2}(\eta_p) \qquad (1\text{-}3\text{-}16)$$

where

$$N_v = 2 \left(\frac{m_p\, k_B\, T}{2\pi h^2} \right)^{3/2} \qquad (1\text{-}3\text{-}17)$$

is the effective density of states for the valence band and

$$\eta_p = (E_v - E_F)/k_B T \qquad (1\text{-}3\text{-}18)$$

In semiconductors, such as Si, Ge, GaAs, etc., the energy bands $E_n(\mathbf{k})$ and $E_p(\mathbf{k})$ are quite complicated. In particular, in Si and Ge the electronic effective mass is anisotropic so that an effective mass tensor should be introduced. Also, there are several equivalent minima in the conduction band (see Section 1.2). In this case, the effective electron mass, m_n, should be replaced by the so-called density-of-states effective mass

$$m_{dn} = Z^{2/3} (m_x\, m_y\, m_z)^{1/3} \qquad (1\text{-}3\text{-}19)$$

where Z is the number of the equivalent minima.

In nearly all cubic semiconductors there are "light" and "heavy" holes (see Section 1.2) and, as a consequence, the hole effective mass, m_p, in the equation for the density of states should be substituted by the density of states effective mass

$$m_{dp} = \left(m_{ph}^{3/2} + m_{pl}^{3/2} \right)^{1/3} \qquad (1\text{-}3\text{-}20)$$

When the Fermi level is in the energy gap and is separated by more than several thermal energies from the edges of the conduction and valence bands ($\eta_n < -3$, $\eta_p <- 3$), the semiconductor is said to be non-degenerate. In this important case, the approximation in eq. (1-3-10) for the Fermi integral applies so that

$$n = N_c \exp\left(\frac{E_F - E_c}{k_B T}\right) \qquad (1\text{-}3\text{-}21)$$

$$p = N_v \exp\left(\frac{E_v - E_F}{k_B T}\right) \qquad (1\text{-}3\text{-}22)$$

In the opposite limiting case, when the Fermi level enters the conduction or the valence band, the semiconductor is degenerate and the approximation in eq. (1-3-11) applies for the Fermi integral. In this case, when the Fermi level enters the conduction band,

$$n \approx \frac{1}{3\pi^2}\left(\frac{2 m_{dn}(E_F - E_c)}{\hbar^2}\right)^{3/2} \qquad (1\text{-}3\text{-}23)$$

and, similarly, with the Fermi level in the valence band

$$p \approx \frac{1}{3\pi^2}\left(\frac{2 m_{dp}(E_v - E_F)}{\hbar^2}\right)^{3/2} \qquad (1\text{-}3\text{-}24)$$

For a non-degenerate semiconductor, we find from eqs. (1-3-21) and (1-3-22)

$$np = N_c N_v \exp\left(- E_g / k_B T\right) \qquad (1\text{-}3\text{-}25)$$

where $E_g = E_c - E_v$ is the energy gap. This equation can be re-written as

$$np = n_i^2 \qquad (1\text{-}3\text{-}26)$$

where

$$n_i = (N_c N_v)^{1/2} \exp\left(- \frac{E_g}{2 k_B T}\right) \qquad (1\text{-}3\text{-}27)$$

is called the intrinsic carrier concentration.

As can be seen from eqs. (1-3-26) and (1-3-27) for a non-degenerate semi-conductor, the np product is independent of the position of the Fermi level and is determined by the densities of states in the valence and conduction bands, the energy gap and the temperature. Eq. (1-3-26) is called the *mass-action law* and is valid at thermal equilibrium.

Eqs. (1-3-6) and (1-3-16) for the electron and hole concentrations can be generalized for arbitrary band shapes

$$n = \int_{E_c}^{E_{tn}} (dn/dE)\, dE = \int_{E_c}^{E_{tn}} f_n(E)\, g_n(E)\, dE \tag{1-3-28}$$

$$p = \int_{E_{bp}}^{E_v} (dp/dE)\, dE = \int_{E_{bp}}^{E_v} f_p(E)\, g_p(E)\, dE \tag{1-3-29}$$

where E_{tn} and E_{bp} are energies corresponding to the highest energy in the conduction band and to the lowest energy in the valence band, respectively; $g_n(E)$ and $g_p(E)$ are now densities of states in the conduction and valence band per unit energy, respectively, including effects of nonparabolicities.

Fig. 1.3.2 shows the dependencies of the density of states, $g_n(E)$, the distribution function, $f_n(E)$, and the electron density, dn/dE, per unit energy for a non-degenerate and for a degenerate semiconductor

As will be discussed in Chapter 4, so-called quantum well structures have found important applications in novel semiconductor devices. In such structures, a thin region of a narrow gap semiconductor is sandwiched between layers of a wide band gap semiconductor. If the narrow gap semiconductor layer is thin enough, the motion of carriers in the direction perpendicular to the heterointerfaces is quantized. For electrons, the lowest energy levels can be estimated as follows, using the quantum formalism for a one-dimensional square well potential (see for example Shur (1990)):

$$E_{ni} - E_c = \frac{\pi^2 \hbar^2}{2 m_n d^2} i^2 \tag{1-3-30}$$

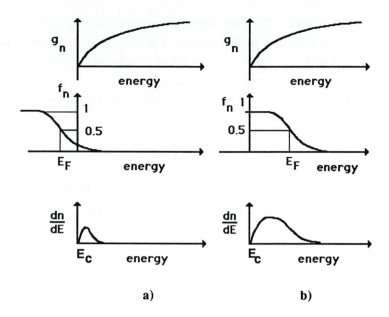

a)   b)

Fig. 1.3.2. The density of states, distribution function, and electron density per unit energy for a non-degenerate (a) and a degenerate (b) semiconductor.

(as long as E_{ni} is well below the bottom of the conduction band in the wide band material). Here i is the quantum number labeling the levels, and d is the thickness of the quantum well. For the quantization, the difference between the levels should be much larger then the thermal energy k_BT, i.e.,

$$\frac{\pi^2 \hbar^2}{2m_n d^2} \gg k_B T \tag{1-3-31}$$

Using this condition, we find, for example, that the levels are quantized at room temperature when $d \ll 147$ Å in GaAs where $m_n/m_e \approx 0.067$.

In the direction parallel to the heterointerfaces, the wave function can be described as a two-dimensional Bloch function with the dispersion relation for the conduction band, i.e.,

$$E_n - E_i = \frac{\hbar^2 \left(k_x^2 + k_y^2 \right)}{2 m_n} \tag{1-3-32}$$

(notice the absence of a k_z-component – the motion in the z-direction is quantized so that the wave function in this direction is completely different from the plane wave proportional to $\exp(ik_zz)$). In other words, each quantum level given by eq. (1-3-30) corresponds to an energy subband.

The density of states, D, for one subband is given by

$$D = \frac{m_n}{\pi \, \hbar^2} \qquad (1\text{-}3\text{-}33)$$

The states of the first (bottom) subband overlap with the states of the second subband for energies larger than the second (from the bottom) energy level, etc. As a consequence, the overall density of states has a "stair-case" shape as shown in Fig. 1.3.3. By increasing the well thickness, d, the steps in Fig. 1.3.3 gradually decrease and merge into an envelope parabolic function which is equal to the three dimensional density of states function multiplied by d (as shown by the thin line in Fig. 1.3.3).

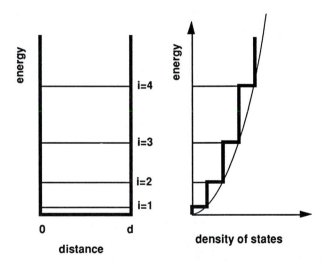

Fig. 1.3.3. Energy levels (bottoms of subbands) and density of states for quantum well structure (from Shur (1990)).

The equations derived above relate carrier concentrations to the position of the Fermi level, E_F, which may be found from the requirement of electric neutrality for the semiconductor. For a pure semiconductor, this means that the concentration of negatively charged electrons must be equal to the concentration of positively charged holes. Hence, we find from eq. (1-3-27)

$$n = p = n_i = \left(N_c N_v \right)^{1/2} \exp\left(- \frac{E_g}{2 k_B T} \right) \qquad (1\text{-}3\text{-}34)$$

(this equation applies only to a non-degenerate semiconductor). In this case, the semiconductor is said to be intrinsic, and the corresponding Fermi level is called the intrinsic Fermi level, $E_F = E_i$, which is given by

$$E_i = \frac{E_c + E_v}{2} + \frac{3}{4} k_B T \ln\left(\frac{m_{dp}}{m_{dn}} \right) \qquad (1\text{-}3\text{-}35)$$

The concentrations of electrons and holes, the condition of neutrality and the position of the Fermi level may all be changed by doping, i.e., by introducing impurities into a semiconductor. Impurity atoms belonging to the fifth column of the Periodic Table (see Fig. 1.2.2) have five valence electrons and will act as n-type dopants (donors) in silicon which belongs to the fourth column of the periodic table and has four valence electrons. Antimonide (Sb), phosphorus (P), and arsenic (As) are examples of n-type dopants in silicon. Donors are neutral when occupied by electrons and become positively charged when they "donate" their excess valence electron to the conduction band.

Atoms belonging to the third column of the periodic table act as p-type dopants (acceptors) in silicon. Examples are boron (B), aluminum (Al), gallium (Ga), and indium (In). Acceptors are negatively charged when filled and neutral when empty. If both donors and acceptors are present, the semiconductor is said to be compensated.

In an energy band scheme, dopants are characterized by the discrete energy levels which they introduce in the forbidden gap. (It is interesting to notice that the appearance of new energy levels can be linked to the breaking of translational symmetry in the host crystal.) The energy levels of so-called "shallow" donors are located below and close to the bottom of a conduction band. The energy levels of shallow acceptors are above and close to the top of a valence band. This means that a very small energy (in many cases less than a thermal energy $k_B T$) is needed to ionize a shallow impurity and supply an electron from a donor into the conduction band or a hole from an acceptor into the valence band.

For a population N_d of donor atoms with a donor energy level E_d and a degeneracy factor g_d, the number of ionized donors are given by

$$N_d^+ = \frac{N_d}{1 + g_d \exp\left[(E_F - E_d)/(k_B T)\right]}$$ (1-3-36)

and the ratio of the concentrations of the filled (N_d^o) and empty (N_d^+) donors is given by

$$\frac{N_d^o}{N_d^+} = g_d \exp\left(\frac{E_F - E_d}{k_B T}\right)$$ (1-3-37)

In the simplest case, $g_d = 2$ (because of two possible values for the electron spin), and E_d can be calculated using the equation for the ground state of a hydrogen atom-like impurity:

$$E_c - E_d \approx \left(\frac{\varepsilon_o}{\varepsilon_s}\right)^2 \left(\frac{m_n}{m_e}\right) \frac{q^4 m_e}{32\pi^2 \varepsilon_o^2 \hbar^2}$$ (1-3-38)

or

$$E_c - E_d \text{ (eV)} \approx 13.6 \left(\frac{\varepsilon_o}{\varepsilon_s}\right)^2 \left(\frac{m_n}{m_e}\right)$$ (1-3-39)

This means that the difference between the bottom of the conduction band and the donor energy level is estimated as the Bohr energy where the free electron mass m_e is substituted by the electron effective mass, m_n, and the free space permittivity, ε_o, is substituted by the static dielectric permittivity, ε_s, of the semiconductor crystal. (In reality, the experimental values of E_d and g_d may be quite different from these simple estimates.)

The Bohr radius of an electron on a shallow donor is given by

$$a_{Bd} = \frac{4\pi\varepsilon_o \hbar^2}{m_e q^2} \left(\frac{\varepsilon_s}{\varepsilon_o}\right)\left(\frac{m_e}{m_n}\right)$$ (1-3-40)

Similarly, we have for acceptor levels

$$\frac{N_a^-}{N_a^o} = \frac{1}{g_a} \exp\left(\frac{E_F - E_a}{k_B T}\right) \tag{1-3-41}$$

where E_a is the acceptor level energy. In the simplest case of a shallow hydrogen-like acceptor level, we have

$$E_a - E_v \approx \left(\frac{\varepsilon_o}{\varepsilon_s}\right)^2 \left(\frac{m_p}{m_e}\right) \frac{q^4 m_e}{32\pi^2 \varepsilon_o^2 \hbar^2} \tag{1-3-42}$$

and $g_a = 4$ (a factor of 2 in g_a is due to the electron spin and another factor of 2 is due to the double degeneracy of the valence band at the Γ point in cubic semiconductors). Again, more reliable values of E_a and g_a should be found from the experimental data

The position of the Fermi level, E_F , can be found from the condition of neutrality, i.e.,

$$p + \sum_j Z_j N_j - n = 0 \tag{1-3-43}$$

where Z_j is the charge (in units of the electronic charge) of an impurity of type j and N_j is the concentration of such impurities. When a semiconductor is doped by shallow donors to a concentration N_d, we find from eq. (1-3-43)

$$n = p + N_d^+ \tag{1-3-44}$$

This equation may be solved together with eq. (1-3-26) for the *mass-action law* to yield the concentration of electrons, n_n, in the n-type semiconductor

$$n_n = \frac{1}{2}\left(\sqrt{N_d^{+2} + 4 n_i^2} + N_d^+\right) \tag{1-3-45}$$

For shallow donors $N_d^+ \approx N_d$ and eq. (1-3-45) may be simplified as follows:

$$n_n \approx \frac{1}{2}\left(\sqrt{N_d^2 + 4 n_i^2} + N_d\right) \tag{1-3-46}$$

The concentration of holes in the n-type material is found from eq.(1-3-26):

$$p_n = n_i^2/n_n \qquad (1\text{-}3\text{-}47)$$

In many practical cases, the temperature is high enough so that $N_d^+ \approx N_d \gg n_i$ in an n-type semiconductor and the equations given above are reduced to:

$$n_n \approx N_d \qquad (1\text{-}3\text{-}48)$$

$$p_n \approx n_i^2/N_d \qquad (1\text{-}3\text{-}49)$$

$$E_F \approx E_c - k_B T \ln\left(\frac{N_c}{N_d}\right) \qquad (1\text{-}3\text{-}50)$$

In the general case, for a p-type semiconductor, the charge neutrality condition is given by

$$n + N_a^- = p \qquad (1\text{-}3\text{-}51)$$

Performing calculations similar to those for the n-type material, we find

$$p_p = \frac{1}{2}\left(\sqrt{N_a^{-2} + 4n_i^2} + N_a^-\right) \qquad (1\text{-}3\text{-}52)$$

For shallow acceptors, $N_a^- \approx N_a$, but in the general case

$$N_a^- = \frac{N_a}{1 + g_a \exp\left[(E_a - E_F)/(k_B T)\right]} \qquad (1\text{-}3\text{-}53)$$

At relatively high temperatures, $N_a^- \approx N_a \gg n_i$ and

$$p_p \approx N_a \qquad (1\text{-}3\text{-}54)$$

$$n_p \approx n_i^2/N_a \qquad (1\text{-}3\text{-}55)$$

$$E_F \approx E_v + k_B T \ln\left(\frac{N_v}{N_a}\right) \qquad (1\text{-}3\text{-}56)$$

(The above analysis is valid only for non-degenerate semiconductors where $F_{1/2}(\eta) \approx \exp(\eta)$.) Practically all semiconductors, even those of extreme purity, contain many different impurities. The presence of both donors and acceptors results in compensation and the conductivity type (n or p) is determined by the larger concentration of ionized impurities. If $N_d > N_a$, the effective donor density becomes

$$N_{deff} = N_d - N_a \qquad (1\text{-}3\text{-}57)$$

In the opposite case, when $N_a > N_d$, we have

$$N_{aeff} = N_a - N_d \qquad (1\text{-}3\text{-}58)$$

We can now use the equations describing the extrinsic carrier concentration given above if we substitute N_d by N_{deff} or N_a by N_{aeff}.

1-4. ELECTRON AND HOLE MOBILITIES AND DRIFT VELOCITIES

In low electric fields the carrier drift velocity is $\mathbf{v} = \mu\mathbf{F}$ where μ the low field mobility. This equation can be obtained from Newton's second law of motion for an electron moving in an electric field $\mathbf{F}$:

$$m_n\frac{d\mathbf{v}}{dt} = q\mathbf{F} - m_n\frac{\mathbf{v}}{\tau_p} \qquad (1\text{-}4\text{-}1)$$

Here m_n is the electron effective mass. The first term in the right-hand side of eq. (1-4-1) represents the acceleration of the electron by the electric field, the second term describes collisions caused by lattice vibrations, impurities, and crystal imperfections. This term limits the electron drift velocity and electron drift momentum. Therefore, the time constant, τ_p, is called the momentum relaxation time. Usually, τ_p is of the order of 10^{-12} to 10^{-14} s. At low frequencies, the left-hand side of eq.(1-4-1) is small compared to either term on the right-hand side of this equation so that

$$q\mathbf{F} \approx m_n\frac{\mathbf{v}}{\tau_p} \qquad (1\text{-}4\text{-}2)$$

and, hence,

$$\mathbf{v} \approx q\tau_p \mathbf{F}/m_n \tag{1-4-3}$$

Electrons in a semiconductor move with random velocities due to a chaotic thermal motion. The drift velocity caused by an applied electric field is superimposed on this thermal motion. At room temperature, the electron velocity due to the thermal motion is usually greater or at least comparable to the drift velocity. Therefore, an exact description of the electronic motion in a semiconductor has to account for the randomness of the electron velocity. However, it can be shown that with a judicial choice of the constants m_n and τ_p in eq. (1-4-1), eqs. (1-4-1) to (1-4-3) can describe an average drift velocity quite accurately.

In a low electric field, τ_p and m_n are independent of the electric field, so that $\mathbf{v} = \mu\mathbf{F}$ where $\mu = q\tau_p/m_n$. The linear dependence of the electron drift velocity does not hold in high electric fields when electrons may gain a considerable energy from the electric field. This process can be described by eq. (1-4-1) in combination with the following equation:

$$\frac{dE}{dt} = q\mathbf{F} \cdot \mathbf{v} - \frac{E - E_o}{\tau_E} \tag{1-4-4}$$

Here, E is the electron energy, $E_o = 3k_BT/2$ is the electron energy under thermal equilibrium conditions, and τ_E is the effective energy relaxation time. Usually, the values of τ_E are of the order of 10^{-11} to 10^{-13} s. Eqs. (1-4-1) and (1-4-4) are, in effect, conservation laws for the average momentum and energy, respectively, and can readily be derived from the Boltzmann transport equation (see, for example, Constant (1985)). In a steady state, when $dE/dt = 0$, we have

$$E = E_o + q\tau_E \mathbf{F} \cdot \mathbf{v} \tag{1-4-5}$$

When the electric field, F, is small, $E \approx E_o$. However, in high electric fields, the electron energy can greatly exceed E_o. In this case, the electrons are referred to as hot electrons. For hot electrons, τ_p, τ_E, and m_n depend on the electric field, and the electron drift velocity is no longer proportional to F. Instead, the drift velocity becomes nearly independent of the electric field in most semiconductors (see Fig. 1.4.1). In many semiconductors, such as GaAs, InP, and InGaAs, the electron velocity in a certain range of electric fields may actually decrease with the increase of the electric field due to intervalley transfer of hot electron from the central (Γ) valley of the conduction band into the satellite (X and L) valleys. As can be also seen from Fig. 1.4.1, compound semiconductors have a potential for a higher speed operation than silicon because electrons in these materials may move faster.

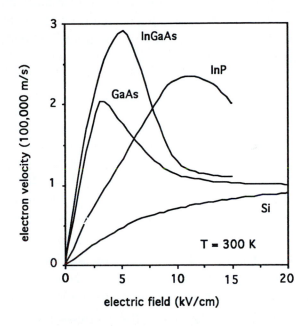

Fig. 1.4.1. Velocity versus electric field for several semiconductors.

The above discussion assumes that the electron drift velocity depends on the electric field alone, so that this velocity should be the same in a very long and in a very short sample, provided that the electric field is the same. However, this is only true for relatively long samples when the transit times $t = L/v$ and $t_{th} = L/v_{th}$ are much greater then the momentum and energy relaxation times. Here L is the sample length and

$$v_{th} = \left(3 k_B T / m_n \right)^{1/2} \tag{1-4-6}$$

is the average electron thermal velocity. These conditions are frequently violated in modern-day short semiconductor devices where critical dimensions become comparable to the mean free path of the electrons. Then, transient phenomena associated with acceleration of carriers become important, giving rise to phenomena such as velocity overshoot (Ruch 1972). In the limiting case of very short devices, the electron transit time may become so small that electrons will not have time to experience any collisions during the transit. Such a mode of the electron transport is called ballistic transport (see Shur and Eastman (1979)).

The low field mobility, μ, is determined by electron collisions with phonons and impurities. The momentum relaxation time, τ_p, can be approximately expressed as

$$\frac{1}{\tau_p} = \frac{1}{\tau_{ii}} + \frac{1}{\tau_{ni}} + \frac{1}{\tau_{ac}} + \frac{1}{\tau_{npo}} + \frac{1}{\tau_{po}} + \frac{1}{\tau_{pe}} + \dots \qquad (1\text{-}4\text{-}7)$$

where the terms on the right-hand side represent relaxation times due to different scattering processes such as ionized impurity scattering (τ_{ii}), neutral impurity scattering (τ_{ni}), acoustic deformation potential scattering (τ_{ac}), non-polar optical scattering (τ_{npo}), polar optical scattering (τ_{po}) and piezoelectric scattering (τ_{pe}). The last two scattering mechanism are present only in partially heteropolar crystals, such as GaAs. Typically, only two or so scattering mechanisms are dominant for a given value of temperature and impurity concentration. In non-polar semiconductors, such as silicon and germanium, scattering by acoustic phonons and ionized impurities determine the electron and hole mobilities in low electric fields. Scattering by non-polar optical phonons becomes dominant in these materials in high electric field leading to a saturation of the electron and hole velocities. Scattering by polar optical phonons and ionized impurities dominate for electrons and holes in GaAs in low electric fields. Scattering by acoustic phonons (deformation potential scattering) and scattering caused by piezoelectric properties (piezoelectric scattering) become important in pure GaAs samples at low temperatures.

Relaxation times determined by different scattering mechanisms are considered, for example, in Seeger (1985). Scattering mechanisms for holes in semiconductors have recently been reviewed by Brudevoll et al. (1990).

The electron and hole drift velocities in high electric fields depend on temperature and on the total concentration of charged impurities, $N_T = N_a + N_d$. According to Caugley and Thomas (1967), Thornber (1982), Sze (1981), Arora et al. (1982), and Yu and Dutton (1985), the electron velocity in bulk silicon,

$v_n(F)$, can be approximated by

$$v_n = \frac{\mu_n F}{\sqrt{1 + (\mu_n F / v_s)^2}} \tag{1-4-8}$$

where

$$\mu_n (N_T, T) = \mu_{mn} + \frac{\mu_{on}}{1 + (N_T / N_{cn})^{\nu}} \tag{1-4-9}$$

$$\mu_{mn} = 88 (T/300)^{-0.57} \quad (cm^2/Vs) \tag{1-4-10}$$

$$\mu_{on} = 1.25 \times 10^3 (T/300)^{-2.33} \quad (cm^2/Vs) \tag{1-4-11}$$

$$\nu = 0.88 (T/300)^{-0.146} \tag{1-4-12}$$

$$N_{cn} = 1.26 \times 10^{17} (T/300)^{2.4} \quad (cm^{-3}) \tag{1-4-13}$$

The saturation velocity, v_s, for electrons and holes is given by (see Sze (1981), Yu and Dutton (1985))

$$v_s = \frac{2.4 \times 10^7}{1 + 0.8 \exp(T/600)} \quad (cm/s) \tag{1-4-14}$$

According to Caugley and Thomas (1967), Thornber (1982), Sze (1981), Arora et al. (1982), and Yu and Dutton (1985), the hole velocity in silicon, $v_p(F)$, is given by

$$v_p = \frac{\mu_p F}{1 + (\mu_p F / v_s)} \tag{1-4-15}$$

where

$$\mu_p(N_T,T) = \mu_{mp} + \frac{\mu_{op}}{1 + (N_T/N_{cp})^\nu} \qquad (1\text{-}4\text{-}16)$$

$$\mu_{mp} = 54\,(T/300)^{-0.57} \quad (cm^2/Vs) \qquad (1\text{-}4\text{-}17)$$

$$\mu_{op} = 230\,(T/300)^{-2.33} \quad (cm^2/Vs) \qquad (1\text{-}4\text{-}18)$$

$$N_{cp} = 2.35 \times 10^{17}\,(T/300)^{2.4} \quad (cm^{-3}) \qquad (1\text{-}4\text{-}19)$$

$$\nu = 0.88\,(T/300)^{-0.146} \qquad (1\text{-}4\text{-}20)$$

i.e., the exponent ν is the same for holes as for electrons. The saturation velocity for holes is also approximately the same as for electrons.

The expressions for the hole and electron velocity given above are obtained for majority carriers in bulk silicon. The mobility and velocity of minority carriers may be quite different because of electron-hole scattering. Nevertheless, the same equations are frequently used for the minority carriers as well because of the lack of the sufficient experimental data (see Yu and Dutton (1985)).

1-5. BASIC SEMICONDUCTOR EQUATIONS

In a low electric field, the current densities for electrons and holes are given by

$$\mathbf{j}_n = q(n\mu_n\,\mathbf{F} + D_n\nabla n) \qquad (1\text{-}5\text{-}1)$$

$$\mathbf{j}_p = q\left(p\mu_p\,\mathbf{F} - D_p\nabla p\right) \qquad (1\text{-}5\text{-}2)$$

The first and second terms in the parenthesis on the right-hand-sides of eqs. (1-5-1) and (1-5-2) represent drift and diffusion current densities, respectively. In low electric fields, the diffusion coefficients are related to the mobilities via the Einstein relationships

$$D_n = \mu_n \, k_B T / q \tag{1-5-3}$$

$$D_p = \mu_p \, k_B T / q \tag{1-5-4}$$

Eqs. (1-5-3) and (1-5-4) are valid for non-degenerate semiconductors. At high carrier concentrations, these relationships have to be modified (see Smith (1978)).

In a high electric field, carriers are heated by the field and the carrier energy becomes larger than the average thermal energy, $3k_BT/2$, where T is the sample temperature. This changes the conditions for scattering. Scattering by ionized impurities, for example, becomes less important since the carriers will be traveling at a higher average speed, thus spending less time in the vicinity of scattering centers. On the other hand, the scattering involving emission of phonons becomes more important since the probability of a carrier having enough energy to emit a phonon increases. In addition, high carrier energies cause a redistribution of electrons between the conduction band valleys and a redistribution of holes between the light and heavy hole valence bands, thereby changing the transport conditions. As a consequence, the electron and hole velocities are no longer proportional to the electric field when the electric field is high (see Fig. 1.4.1). The diffusion coefficients also become dependent on the electric field (see Ruch and Kino (1968)).

The following phenomenological equations are frequently used in device modeling in order to describe the electron and hole transport in both low and high electric field

$$\mathbf{j}_n = q[- n\mathbf{v}_n(\mathbf{F}) + D_n(\mathbf{F})\nabla n] \tag{1-5-5}$$

$$\mathbf{j}_p = q[p\mathbf{v}_p(\mathbf{F}) - D_p(\mathbf{F})\nabla p] \tag{1-5-6}$$

Here, $\mathbf{v}_n(\mathbf{F})$, $\mathbf{v}_p(\mathbf{F})$, $D_n(\mathbf{F})$, and $D_p(\mathbf{F})$ are assumed to be the same functions of electric field as computed or measured for the uniform sample under steady state conditions. In a low electric field, these equations reduce to eqs. (1-5-1) and (1-5-2). However, the equations may lead to considerable errors in describing hot electron behavior or even near equilibrium transport in simple systems with large built-in electric fields such as p-n junctions (see Bløtekjær (1970)).

At high frequencies, comparable to the inverse energy relaxation time (which is of the order of 2 ps for electrons in the central minimum of the conduction band in GaAs), the velocity and diffusion do not follow instantaneously the variations of the electric field. Therefore, the effective differential mobility, for example, becomes frequency-dependent.

The advantage of using eqs. (1-5-5) and (1-5-6), or even similar equations with field-independent diffusion coefficients, is the relative simplicity of the analysis – an analysis which allows one to achieve some insight into the device physics.

Eqs. (1-5-5) and (1-5-6) (or eqs. (1-5-1) and (1-5-2) in a low electric field) should be solved together with Poisson's equation

$$\nabla \cdot \mathbf{F} = \frac{\rho}{\varepsilon} \tag{1-5-7}$$

and the continuity equations

$$\frac{\partial n_t}{\partial t} + \frac{\partial n}{\partial t} = \frac{1}{q} \nabla \cdot \mathbf{j}_n + G - R \tag{1-5-8}$$

$$\frac{\partial p_t}{\partial t} + \frac{\partial p}{\partial t} = -\frac{1}{q} \nabla \cdot \mathbf{j}_p + G - R \tag{1-5-9}$$

Here, ε is the dielectric permittivity,

$$\rho = q(N_d - N_a - n + p) + q(p_t - n_t) \tag{1-5-10}$$

is the space charge density, p_t and n_t are the densities of trapped holes and electrons, respectively, N_d and N_a are the concentrations of ionized donors and acceptors, respectively, G is the generation rate of electron-hole pairs (due to the light excitation or impact ionization), and R is the recombination rate.

Let us now consider the application of these basic semiconductor equations to a nearly neutral n-type region which may contain both electrons and holes at the concentrations n_n and p_n that are larger than the equilibrium concentrations n_{no} and p_{no}. We assume that there are no traps ($p_t = 0$, $n_t = 0$). An example would be a piece of an n-type semiconductor where extra carriers are generated by light. For simplicity, we consider a one-dimensional steady state situation where $\partial \mathbf{j}_n/\partial t = 0$ and $\partial \mathbf{j}_p/\partial t = 0$, in which case the continuity equations (1-5-8) and (1-5-9) can be rewritten as

$$D_n \frac{\partial^2 n_n}{\partial x^2} + \mu_n F \frac{\partial n_n}{\partial x} + \mu_n n_n \frac{\partial F}{\partial x} + G - R = 0 \tag{1-5-11}$$

$$D_p \frac{\partial^2 p_n}{\partial x^2} - \mu_p F \frac{\partial p_n}{\partial x} - \mu_p p_n \frac{\partial F}{\partial x} + G - R = 0 \tag{1-5-12}$$

With the assumption that the semiconductor is almost neutral (the space charge density is nearly zero), we have

$$n_n - n_{no} \approx p_n - p_{no} \tag{1-5-13}$$

By eliminating the $\partial F/\partial x$ term from eqs. (1-5-11) and (1-5-12), we obtain the following equation:

$$\mu_p \, p_n D_n \frac{\partial^2 n_n}{\partial x^2} + \mu_n \, n_n D_p \frac{\partial^2 p_n}{\partial x^2} + \mu_n \mu_p \, p_n F \frac{\partial n_n}{\partial x} +$$
$$\mu_n \mu_p \, n_n F \frac{\partial p_n}{\partial x} + (G - R)(\mu_n \, n_n + \mu_p \, p_n) = 0 \tag{1-5-14}$$

The quasi neutrality condition, eq. (1-5-13), yields

$$\frac{\partial n_n}{\partial x} \approx \frac{\partial p_n}{\partial x} \tag{1-5-15}$$

and

$$\frac{\partial^2 n_n}{\partial x^2} \approx \frac{\partial^2 p_n}{\partial x^2} \tag{1-5-16}$$

Using eqs. (1-5-15) and (1-5-16), we find from eq. (1-5-14)

$$D_a \frac{\partial^2 p_n}{\partial x^2} - \mu_a F \frac{\partial p_n}{\partial x} + G - R = 0 \tag{1-5-17}$$

where

$$D_a = \frac{\mu_p \, p_n D_n + \mu_n \, n_n D_p}{\mu_n \, n_n + \mu_p \, p_n} \tag{1-5-18}$$

is called the ambipolar diffusion coefficient and

$$\mu_a = \frac{\mu_n \mu_p (n_n - p_n)}{\mu_n n_n + \mu_p p_n} \tag{1-5-19}$$

is called the ambipolar mobility. When $n_n \gg p_n$, we obtain $D_a \approx D_p$ and $\mu_a \approx \mu_p$ and eq. (1-5-17) reduces to

$$D_p \frac{\partial^2 p_n}{\partial x^2} - \mu_p F \frac{\partial p_n}{\partial x} + G - R = 0 \tag{1-5-20}$$

Eq. (1-5-20) is the continuity equation for minority carriers (holes) in an n-type sample under steady state conditions. This equation is extremely useful for the analysis of different semiconductor devices, such as p-n junctions, bipolar junction transistors, and solar cells.

Another useful semiconductor equation may be derived from the expressions derived above when the trapped charge, $\rho_T = (p_t - n_t)$ in eq. (1-5-10), can be neglected. Indeed, subtracting eq. (1-5-8) from eq. (1-5-9), we obtain

$$q \frac{\partial}{\partial t}(p - n) + \nabla \cdot \left(\mathbf{j}_n + \mathbf{j}_p \right) = 0 \tag{1-5-21}$$

From eq. (1-5-10), we find

$$q \frac{\partial}{\partial t}(p + p_t - n - n_t) = \frac{\partial \rho}{\partial t} \tag{1-5-22}$$

Differentiating Poisson's equation (eq. (1-5-7)) with respect to time and substituting $\partial \rho / \partial t$ derived from eqs. (1-5-21) and (1-5-22), we obtain

$$\varepsilon \frac{\partial}{\partial t} \nabla \cdot \mathbf{F} + \nabla \cdot \left(\mathbf{j}_n + \mathbf{j}_p \right) = 0 \tag{1-5-23}$$

Integration of eq. (1-5-23) over the space coordinates leads to

$$\mathbf{j}(t) = \mathbf{j}_n + \mathbf{j}_p + \varepsilon \frac{\partial}{\partial t} \mathbf{F} \tag{1-5-24}$$

where $\mathbf{j}(t)$ is the total current density, including the displacement current $\varepsilon \partial F / \partial t$. Furthermore, by integrating the current density over the sample cross section, we obtain the total current

$$I = \int_S \mathbf{j} \, ds \tag{1-5-25}$$

Thus, for a sample with a constant cross section S and a uniform current density, we have

$$j = I / S \tag{1-5-26}$$

In low electric fields, when the Einstein relationship is valid, the electron current density may be expressed through the electron quasi-Fermi level, E_{Fn}, (see Section 1.7) as follows:

$$
\begin{aligned}
j_n &= q\mu_n \, nF + qD_n \frac{\partial n}{\partial x} \\
&= q\mu_n \, n\frac{1}{q}\frac{\partial E_c}{\partial x} + qD_n \frac{\partial}{\partial x}\left[N_c \exp\left(\frac{E_{Fn} - E_c}{k_B T} \right)\right] \\
&= \mu_n \, n\frac{\partial E_{Fn}}{\partial x} \tag{1-5-27}
\end{aligned}
$$

In this case, the semiconductor is assumed to be non-degenerate and the approximation in eq. (1-3-21) has been used for n in the diffusion term. In a similar fashion, we find

$$j_p = \mu_p \, p\frac{\partial E_{Fp}}{\partial x} \tag{1-5-28}$$

Modern semiconductor devices may have very short feature sizes (as small as 0.1 μm or less). Typical operating voltages are usually of the order of several volts, giving rise to extremely large internal electric fields (10^5 V/cm and higher). These devices may operate at very high frequencies (up to 100 GHz). As was already mentioned above, eqs. (1-5-5) and (1-5-6) are not valid under such conditions. Within certain limitations, they still may provide some useful insight into the device operation. However, a more accurate analysis requires either a numerical solution of the Boltzmann transport equation together with Poisson's equation or, for very small devices, even a direct solution of the Schrödinger equation. The difficulties in such a "brute force" approach are related not only to the large computational resources required for the calculations but also to the lack of detailed information about numerous material parameters needed for a realistic description of the high-field transport. Therefore, several simple (but less accurate) approaches for dealing with high-field transport have been proposed.

As we pointed out earlier, conventional semiconductor equations based on field dependent velocity and diffusion are inadequate for modeling small semiconductor devices. Insight into the physics of such devices is provided by the Monte Carlo method (see, for example, Jacoboni and Lugli (1989), Reggiani (1985) and Jensen et al. (1991)). However, approximate analytical equations can be extremely useful for the analysis of the device behavior. Traditionally, the electron temperature approximation has been used for deriving approximate equations describing "hot" electrons. In this approach, it is assumed that the symmetrical part of the electron distribution function may be approximated as

$$f = C \exp\left(-E_n/k_B T_n\right) \tag{1-5-29}$$

where T_n is an effective electron temperature.

Eq. (1-15-29) is valid when the electrons exchange energy through electron-electron collisions at a faster rate than they lose it through electron-phonon scattering. At very high electron concentrations, the electron-electron collisions redistribute both electron energy and momentum, leading to the so-called displaced Maxwellian distribution function

$$f(\mathbf{p}) = C \exp\left(-\frac{|\mathbf{p} - \mathbf{p}_o|^2}{2 m_n k_B T_n}\right) \tag{1-5-30}$$

where $\mathbf{p}$ is the momentum and $\mathbf{p}_o$ is the drift momentum.

For this approximation to be valid, the average time between the electron-electron collisions should be much smaller than the momentum relaxation time. If the number of electrons in the conduction band is determined by the concentration of the ionized donors, this can never be true and, therefore, the displaced Maxwell distribution function can be considered only as a crude approximation.

If, however, the displaced Maxwellian distribution function is assumed, phenomenological transport equations can be derived by substituting eq. (1-5-30) into the Boltzmann equation (see Bløtekjær (1970)). The resulting equations are fairly complicated and may be inaccurate if the electron distribution function is very non-Maxwellian. In this section, we describe two simple models based on the results of steady-state Monte Carlo calculations and the application of these results to short semiconductor devices under non-equilibrium conditions. These models are empirical and can only be justified by comparing the results of the calculations (based on these models) with more rigorous calculations and experimental data. This is the price to pay for their relative simplicity and ease of use.

   The first model (proposed by Shur (1976) and further developed by Carnez et al. (1980) and Cappy et al. (1980)) can be called an energy-dependent relaxation time model. It describes the electron transport in short samples and/or in rapidly varying electric fields based on the results of the Monte Carlo simulations for long samples under steady state conditions. When the diffusion is neglected, the equations of this model are given by

$$\frac{dm(E)\mathbf{v}}{dt} = -q\mathbf{F} - \frac{m\,\mathbf{v}}{\tau_p(E)} \tag{1-5-31}$$

$$\frac{dE}{dt} = \frac{\mathbf{j}\cdot\mathbf{F}}{n} - \frac{E-E_o}{\tau_E(E)} \tag{1-5-32}$$

$$\mathbf{j} = -qn\mathbf{v} \tag{1-5-33}$$

Here, E is the average electronic energy, τ_p is the effective momentum relaxation time, τ_E is the effective energy relaxation time, $E_o = 3k_B T_o/2$, where T_o is the lattice temperature, and $m(E)$ is the energy dependent effective mass of the electrons. $m(E)$, $\tau_p(E)$, and $\tau_E(E)$ are determined from the steady state Monte Carlo calculations from the requirement that, in steady state, eqs. (1-5-31) and (1-5-32) give the same results as the Monte Carlo calculations. This requirement leads to the following expressions:

$$\tau_p(E) = \left\{ \frac{m[F(E)]\,v[F(E)]}{qF(E)} \right\}_{\text{steady state}} \tag{1-5-34}$$

$$\tau_E(E) = \frac{E-E_o}{q\{F(E)v[F(E)]\}_{\text{steady state}}} \tag{1-5-35}$$

where the steady state expressions in the curly brackets are taken from the Monte

Carlo calculations. This empirical approach has been shown to be in a very good agreement with direct Monte Carlo simulations for non-equilibrium conditions (see Shur (1976), Carnez et al. (1980), Cappy et al. (1980)). The agreement is good not only for GaAs but also for InP and $Ga_{.47}In_{.53}As$ samples (see Cappy et al. (1980)).

In short semiconductor devices the diffusion effects may be very important. These effects have been incorporated into this model by using the following equation for the electron current density

$$\mathbf{j}_n = - qn\mathbf{v} + qD_n(E)\nabla n \qquad (1\text{-}5\text{-}36)$$

where the diffusion coefficient $D(E)$ is determined from steady-state Monte Carlo simulations.

In an extension of the above model, Fjeldly and Johnsen (1988) proposed to base the discussion of hot electron transport on transient responses in the average velocity, the electron temperature, etc., produced by step changes in the electric field in uniform semiconductors. These transients are unique for a given semiconductor with a given doping and temperature, and can be obtained once and for all by performing relatively simple Monte Carlo calculations. Based on such transients, a self-consistent formalism for the steady state hot electron transport in semiconductor devices can be developed in terms of the first few conservation equations derived from the Boltzmann transport equation (see Section 1.4), in combination with Poisson's equation. Connecting ohmic contacts are well described in terms of low-field drift and diffusion theory.

1-6. HALL EFFECT AND MAGNETORESISTANCE

Standard experimental techniques of measuring mobilities include the studies of transport properties in a magnetic field. Let us consider a p-type semiconductor sample placed into a magnetic field as shown in Fig. 1.6.1. When the voltage V is applied to the sample, holes move from contact 2 to contact 1 with the velocity

$$v_p = \mu_p F = \mu_p V / L \qquad (1\text{-}6\text{-}1)$$

where F is the electric field, L is the sample length, and μ_p is the hole mobility. The magnetic field, B, exerts a force $\mathbf{f}_L = q\mathbf{v}_p\mathbf{x}\mathbf{B}$ (called the Lorentz force), acting on the holes and deflecting them towards side 3 as shown in Fig. 1.6.1.

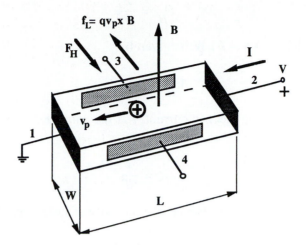

Fig. 1.6.1. Hall effect in *p*-type semiconductor sample placed into a magnetic field (from Shur (1990)).

As a consequence, holes accumulate on this side and create a net positive charge there. Side 4 becomes depleted with holes and is negatively charged. The magnitude of these charges is such that the electric field F_H, created by the charges, exactly counterbalances the Lorentz force under stationary conditions, i.e.,

$$F_H = v_p\, B \tag{1-6-2}$$

(Here we assume that the magnetic field is small so that $\mu_p B \ll 1$.) Hence, the voltage difference

$$V_H = v_p\, B W \tag{1-6-3}$$

develops between contacts 3 and 4.

This effect is called the Hall effect, contacts 3 and 4 are called the Hall contacts, and the electric field F_H and the voltage V_H are called the Hall electric field and the Hall voltage, respectively.

The electric current density is

$$j = qp\mu_p\, F = qpv_p \tag{1-6-4}$$

and, hence,

$$F_H = \frac{Bj}{qp} \tag{1-6-5}$$

The Hall voltage, $V_H = F_H W$, is given by

$$V_H = RBI/t \tag{1-6-6}$$

where $I = jtW$ is the electric current, t is the sample thickness, W is the sample width, and

$$R = \frac{1}{qp} \tag{1-6-7}$$

is called the Hall constant. Hence, the mobility of holes can be written as

$$\mu_p = \sigma_p R \tag{1-6-8}$$

where

$$\sigma_p = j/F = qp\mu_p \tag{1-6-9}$$

is the conductivity of the sample. Once R is calculated from V_H and I (see eq. (1-6-6)), the concentration of holes can be found from eq. (1-6-7) and μ_p from eq. (1-6-8). Note that σ_p can be determined from the measured values of I and V (see eq. (1-6-9)).

This approach allows us to understand the physics of the Hall effect. However, it is not totally accurate. Indeed, we have assumed that all holes in the sample move with the same velocity v_p. In fact, this velocity is the average drift velocity superimposed on the random thermal motion which means that some holes move slower and some move faster. The Hall electric field is the same for all the holes, counterbalancing the Lorentz force which acts on individual holes with different velocities only on the average. A more accurate analysis of this problem shows that eqs. (1-6-7) and (1-6-8) have to be modified as follows

$$R = \frac{r_H}{qp} \tag{1-6-7a}$$

$$\mu_{Hp} = \sigma_p R \tag{1-6-8a}$$

where r_H is called the Hall factor and

$$\mu_{Hp} = r_H \mu_p \tag{1-6-10}$$

is called the Hall mobility of holes. The value of r_H depends on temperature, doping, magnetic field, and other factors. When the relaxation time approximation is valid, we can write

$$r_H = \frac{<\tau^2>}{<\tau>^2} \tag{1-6-11}$$

(see, for example, Seeger (1985)) where for a non-degenerate carrier distribution,

$$<\tau> = \frac{\frac{4}{3\sqrt{\pi}} \int_0^\infty \tau(E) E^{3/2} \exp(-E/k_B T)\, dE}{\int_0^\infty E^{3/2} \exp(-E/k_B T)\, dE} \tag{1-6-12}$$

$$<\tau^2> = \frac{\frac{4}{3\sqrt{\pi}} \int_0^\infty \tau^2(E) E^{3/2} \exp(-E/k_B T)\, dE}{\int_0^\infty E^{3/2} \exp(-E/k_B T)\, dE} \tag{1-6-13}$$

$\tau(E)$ is here the momentum relaxation time which is a function of the electron kinetic energy, E. In the particular case when τ may be represented as

$$\tau = \tau_o \left(\frac{E}{k_B T} \right)^{-s} \tag{1-6-14}$$

where τ_o and s are constants, the integrals in eqs. (1-6-12) and (1-6-13) can be expressed through the gamma function (Γ):

$$<\tau> = \frac{4}{3\sqrt{\pi}} \tau_o \Gamma\left(\frac{5}{2} - s\right) \tag{1-6-15}$$

$$<\tau^2> = \frac{4}{3\sqrt{\pi}} \tau_o^2 \Gamma\left(\frac{5}{2} - 2s\right) \tag{1-6-16}$$

Similar expressions can be derived for an n-type sample:

$$R = - \frac{r_H}{qn} \qquad (1\text{-}6\text{-}17)$$

$$\mu_{Hn} = \sigma_n R = r_H \mu_n \qquad (1\text{-}6\text{-}18)$$

where μ_{Hn} is the Hall mobility of electrons. When impurity scattering is dominant, $r_H \approx 1.93$. When phonon scattering is dominant, r_H is closer to unity (typically, 1.2 to 1.4).

The Hall electric field develops as a result of the carrier flux in the direction perpendicular to the current flow. This perpendicular motion is caused by the magnetic field when it is turned on. At first, the carriers move under the angle

$$\theta = \tan^{-1}\left(\frac{v_y}{v_x}\right) = \tan^{-1}\left(\frac{\mu v_x B}{v_x}\right) = \tan^{-1}(\mu B) \qquad (1\text{-}6\text{-}19)$$

called the Hall angle. Here v_y and v_x are components of the carrier velocity in x and y directions, respectively (see Fig. 1.6.2). Then carriers accumulate at one of the sample side surfaces, depleting the other side and building up the Hall electric field. This field compensates the Lorentz force and stops further motion of carriers in the direction perpendicular to the current flow.

This implies, however, that the sample dimensions should satisfy the requirement $L \gg W$ for the Hall electric field to fully develop. In the opposite limiting case, $W \gg L$, most of the carriers moving under the Hall angle arrive to the contact, not to the sample side, as illustrated by Fig. 1.6.2b and, hence, the Hall electric field is almost totally shorted by the contact. As a result, carriers travel the distance

$$L_H = L\left(1 + \tan^2\theta\right)^{1/2} \qquad (1\text{-}6\text{-}20)$$

which exceeds the sample length L. Also, the component of their velocity in the direction of the applied electric field decreases due to the action of the Lorentz force. Assuming a zero Hall field, we find

$$\mathbf{v} = \mu(\mathbf{F} + \mathbf{v} \times \mathbf{B}) \qquad (1\text{-}6\text{-}21)$$

where μ is the low field mobility. Hence,

$$v_x = \mu\left(F_x + v_y\, B \right) \qquad (1\text{-}6\text{-}22)$$

$$v_y = -\, \mu v_x\, B \qquad (1\text{-}6\text{-}23)$$

(because $F_y = 0$). Substituting eq. (1-6-23) into eq. (1-6-22), we find

$$v_x = \frac{\mu F_x}{1 + \mu^2 B^2} \qquad (1\text{-}6\text{-}24)$$

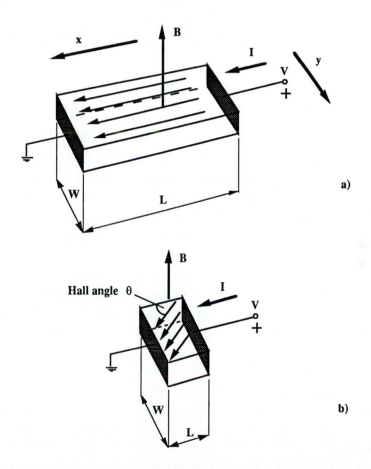

Fig. 1.6.2. Streamlines of electric current in semiconductor samples with
$W \ll L$ (a) and $W \gg L$ (b) (from Shur (1990)).

Eq. (1-6-24) means that the effective low field mobility, μ_B, for a sample with

$L/W \ll 1$ in the magnetic field B, is given by

$$\mu_B = \frac{\mu}{1 + \mu^2 B^2}$$ (1-6-25)

The related increase in the low-field resistivity, $\Delta\rho$, in the magnetic field (the so-called geometric magnetoresistance) becomes

$$\frac{\Delta\rho}{\rho_o} = (\mu B)^2$$ (1-6-26)

where

$$\rho_o = \frac{1}{qp\mu}$$ (1-6-27)

is the low-field resistivity and p is the carrier concentration.

Measuring the geometric magnetoresistance allows one to measure the low-field mobility, μ, (see, for example, Jarvis and Johnson (1970)). Eq. (1-6-26) for the geometric magnetoresistance is valid in the case when the Hall field is completely shorted by the contacts. This may only be achieved when the ratio L/d tends to zero. The expressions that allow one to calculate the geometric magnetoresistance for arbitrary values of L/d are given, for example, by Madelung (1964).

In addition to the geometric magnetoresistance, a so-called "physical" magnetoresistance is also observed in semiconductors, independent of geometry. One of the reasons for this magnetoresistance is the difference in carrier velocities related to random thermal motion. The Hall electric field compensates the Lorentz force acting on an "average" carrier. Faster carriers are slowed down as a result of the magnetic field, slower carriers are accelerated. However, the faster carriers are slowed down relatively more than the slower carriers are accelerated. As a consequence, there is some increase in the sample resistance. The magnitude of this resistance increase depends on the dominant scattering mechanisms.

In the particular case when the momentum relaxation time depends on energy according to $\tau \sim E^{-s}$ (see eq. (1-6-14)) and the scattering mechanisms are assumed to be the same for electrons and hole, the "physical" magnetoresistance is given by the following expression:

$$\frac{\Delta\rho}{\rho_o} = B^2 \left\{ \left(\frac{\Gamma^2\left(\frac{5}{2}\right)\Gamma\left(\frac{5}{2} - 3s\right)}{\Gamma^3\left(\frac{5}{2} - s\right)} \right) \left(\frac{\mu_n^3\, n + \mu_p^3\, p}{\mu_n\, n + \mu_p\, p} \right) \right.$$

$$\left. - \left(\frac{\Gamma^2\left(\frac{5}{2}\right)\Gamma\left(\frac{5}{2} - 2s\right)}{\Gamma^2\left(\frac{5}{2} - s\right)} \right)^2 \left(\frac{\mu_n^2\, n - \mu_p^2\, p}{\mu_n\, n + \mu_p\, p} \right)^2 \right\} \qquad (1\text{-}6\text{-}28)$$

(see Seeger (1985)). (This expression is valid when the scattering mechanisms for electrons and holes are the same.) For $n \gg p$ we find $\Delta\rho/\rho_o \sim (\mu_n B)^2$, and for $p \gg n$, we have $\Delta\rho/\rho_o \sim (\mu_p B)^2$. However, when $L/d \ll 1$, the geometric magnetoresistance is much greater than the "physical" magneto-resistance.

1-7. QUASI-FERMI LEVELS. GENERATION AND RECOMBINATION OF CARRIERS

In Section 1.3, we considered concentrations of electrons and holes under the conditions of thermal equilibrium when the probability of occupancy of an electronic state is given by the Fermi-Dirac distribution function. However, the distribution function changes quite dramatically when a high electric field is applied to a semiconductor sample. Under such non-equilibrium conditions, the electron and hole concentrations are no longer related by

$$pn = n_i^2 \qquad (1\text{-}7\text{-}1)$$

as for a non-degenerate semiconductor in thermal equilibrium, and the very concept of a Fermi level is no longer applicable. The same is true when non-equilibrium conditions are created by generating extra electron-hole pairs in a semiconductor by absorption of light. Photons with energies greater than the energy gap may promote electrons from the valence into the conduction band, creating electron-hole pairs.

Under such non-equilibrium conditions, it may still be useful to represent the distribution function for electrons and holes, f_n and f_p, as

$$f_n = \frac{1}{1 + \exp\left[(E_n - E_{Fn})/k_B T \right]} \tag{1-7-2}$$

$$f_p = \frac{1}{1 + \exp\left[(E_{Fp} - E_p)/k_B T \right]} \tag{1-7-3}$$

Eqs. (1-7-2) and (1-7-3) may be considered as definitions of E_{Fp} and E_{Fn} which are called the electron and hole quasi-Fermi levels, respectively (sometimes they are also called Imrefs, i.e., Fermi spelled backwards). Under equilibrium conditions, $E_{Fp} = E_{Fn} = E_F$. However, under non-equilibrium conditions, E_{Fp} is not equal to E_{Fn} and they both may be functions of position. Actually, the difference $E_{Fn} - E_{Fp}$ can be used as the measure of the deviation from equilibrium. In the non-degenerate case, we obtain from eqs. (1-7-2) and (1-7-3)

$$f_n \approx \exp\left[(E_{Fn} - E_n)/k_B T \right] \tag{1-7-4}$$

$$f_p \approx \exp\left[(E_p - E_{Fp})/k_B T \right] \tag{1-7-5}$$

Using these expressions, we obtain for parabolic bands (see Section 1.3)

$$n = N_c \exp\left[(E_{Fn} - E_c)/k_B T \right] \tag{1-7-6}$$

$$p = N_v \exp\left[(E_v - E_{Fn})/k_B T \right] \tag{1-7-7}$$

In other words, in this case, the quasi-Fermi levels are proportional to the logarithms of electron and hole concentrations.

We can try to extend this concept to situations where the applied electric field causes a substantial increase in average energies of the random motion of electrons or holes, by introducing effective electron and hole temperatures, T_n and T_p. The effective electron temperature can then be expressed as $T_n = 2E_n/3k_B$ where E_n is the average electron kinetic energy. In this case, eqs. (1-7-2) and (1-7-3) become

$$f_n = \frac{1}{1 + \exp\left[(E_n - E_{Fn})/k_B T_n \right]} \tag{1-7-8}$$

$$f_p = \frac{1}{1 + \exp\left[\left(E_{Fp} - E_p \right)/k_B T_p \right]} \qquad (1\text{-}7\text{-}9)$$

However, the concept of effective electron and hole temperatures may not be very accurate, as discussed in Section 1.5.

The introduction of quasi-Fermi levels is very useful because carrier concentrations in a practical semiconductor device may vary as functions of position and/or bias by many orders of magnitude, whereas the quasi-Fermi levels change only within the energy gap or just inside the bands near the band edges. This variation is much easier to visualize.

Let us now consider an example. Assume that we shine light on an n-type GaAs sample with doping density N_d, and that the light is uniformly absorbed such that electron-hole pairs with density P are produced. Then, the electron concentration in the sample equals

$$n \approx P + N_d \qquad (1\text{-}7\text{-}10)$$

and the hole concentration is

$$p \approx P + n_i^2/N_d \qquad (1\text{-}7\text{-}11)$$

The electron and hole quasi-Fermi levels, calculated using eqs. (1-7-6), (1-7-7), (1-7-10), and (1-7-11) are shown in Fig. 1.7.1 versus the electron-hole pair density. Fig. 1.7.1 shows that when P varies from 10^{11} to 10^{17} cm^{-3}, E_{Fn} varies from 1.31 to 1.35 eV and E_{Fp} varies from 0.52 eV to approximately 0.15 eV.

A semiconductor with excess charge carriers will tend to re-establish thermal equilibrium through electron-hole recombination processes. These processes become more intensive as the electron-hole concentration increases. Steady state values of carrier concentration are reached when the generation rate G is balanced by the recombination rate R, i.e.,

$$G = R \qquad (1\text{-}7\text{-}12)$$

Recombination processes include direct (band-to-band) radiative recombination, radiative band-to-impurity recombination (see Fig. 1.7.2), non-radiative recombination via impurity (trap) levels, and surface recombination. The radiative band-to-band recombination rate is proportional to the np product.

Hence, for a non-degenerate semiconductor, we have

$$R = G_{th}\, np/n_i^2 \qquad (1\text{-}7\text{-}13)$$

where G_{th} is the thermal generation rate (so that, at equilibrium, $np = n_i^2$ as expected).

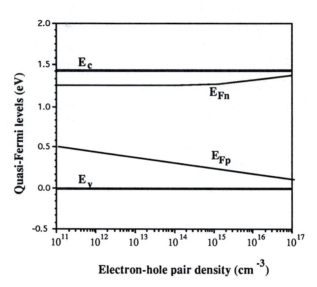

Fig. 1.7.1. Electron and hole quasi-Fermi levels versus concentration of light-generated electron-hole pairs in n-type GaAs. Parameters used: $T = 300$ K, $N_d = 10^{15}$ cm^{-3}, $N_c = 4.7 \times 10^{17}$ cm^{-3}, $N_v = 7 \times 10^{18}$ cm^{-3}, $n_i = 1.79 \times 10^6$ cm^{-3}.

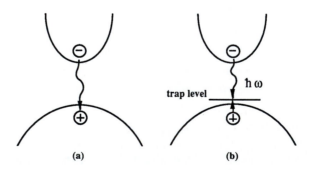

Fig. 1.7.2. Mechanisms of radiative recombination: a) band-to-band recombination, b) band-to-impurity recombination

The expression for G_{th} was first derived by van Roosbroeck and Shockley (1954) who related G_{th} to the index of refraction, n_r, and to the absorption coefficient, α:

$$G_{th} = 32\pi^2 \left(\frac{k_B T}{h} \right)^4 \int \frac{\xi(x)\, n_r^3 x^3 dx}{\exp(x) - 1} \qquad (1\text{-}7\text{-}14)$$

where $x = h\nu/(k_B T)$, $\xi(x) = c\alpha h/(4\pi k_B Txn_r)$, ν is the frequency, and c is the speed of light in vacuum. The constant in front of the integral in eq. (1-7-14) is equal to 1.785×10^{22} $[T(\text{K})/300]^4$ (cm^{-3}/s).

Eq. (1-7-13) may be rewritten as

$$R = C_r np \qquad (1\text{-}7\text{-}15)$$

In steady state, we find

$$G = C_r np = C_r(n_o + \Delta n)(p_o + \Delta p) \qquad (1\text{-}7\text{-}16)$$

where Δn and Δp are excess concentrations of electrons and holes, and n_o and p_o are the corresponding equilibrium concentrations ($n_o p_o = n_i^2$). When the generation of electron-hole pairs is caused by light, $\Delta n = \Delta p$ and the generation rate G is proportional to the light intensity I. Let us consider, for example, an n-type semiconductor with the concentration N_d of shallow donors where $n_o = N_d$ and $p_o = n_i^2/N_d$ under equilibrium conditions. At low light intensities, when $\Delta n \ll N_d$ but $\Delta n \gg n_i^2/N_d$, we find from eq. (1-7-16)

$$\Delta n = G \tau_r \qquad (1\text{-}7\text{-}17)$$

where $\tau_r = 1/(C_r N_d)$ is called the radiative band-to-band recombination lifetime. When the light intensity is small, Δn is proportional to G and, hence, to I. At high intensities, when $\Delta n \gg n_o$ and $\Delta p \gg p_o$, $G \approx C_r \Delta n \Delta p = C_r(\Delta n)^2$, and Δn becomes proportional to $\sqrt{G}$, and, hence, to $\sqrt{I}$.

In practical light-emitting semiconductor devices, radiative band-to-impurity recombination (see Fig. 1.7.2b) is often more important than radiative band-to-band recombination. The radiative band-to-impurity recombination lifetime is given by

$$\tau_r = 1/(B_r N_A) \qquad (1\text{-}7\text{-}18)$$

where B_r is the radiative recombination coefficient and N_A is the concentration of impurities involved in this recombination process (see Goodfellow et al. (1985)).

In many cases, the dominant recombination mechanism is recombination via traps, especially in indirect semiconductors such as silicon. The theory of this process was developed by Shockley and Read (1952) and later by Sah et al. (1957). When only one trap level is involved, four distinct electron and hole transitions are possible in this recombination mechanism as shown in Fig. 1.7.3.

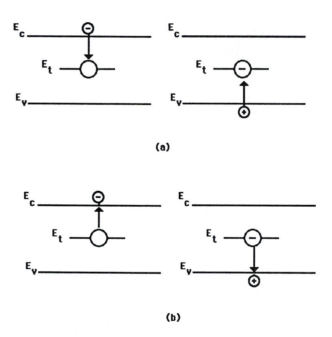

Fig. 1.7.3. Electron and hole transitions involved in recombination via traps (from Shur (1990)); a) an electron captured by an empty trap and a hole captured by a filled trap result in recombination of an electron-hole pair, b) inverse processes – an electron generated from a filled trap into the conduction band and a hole generated from an empty trap into the valence band.

For example, when an electron is captured by an empty trap and then a hole is captured by the trap filled by the electron, the electron-hole pair recombines. Inverse processes are the emission of an electron from a filled trap into the conduction band and the emission of a hole from an empty trap into the valence band. The rate of electron capture by the traps, R_{nc}, is proportional to the number of electrons and to the number of empty traps, $(1 - f_t)N_t$:

$$R_{nc} = C_n n (1 - f_t) N_t \qquad (1\text{-}7\text{-}19)$$

Here, f_t is the occupancy function of the trap level. The coefficient C_n may be presented as

$$C_n = \sigma_n v_{thn} \qquad (1\text{-}7\text{-}20)$$

where σ_n is called the capture cross-section for electrons and

$$v_{thn} = \left(3 k_B T / m_n\right)^{1/2} \qquad (1\text{-}7\text{-}21)$$

is the electron thermal velocity, and m_n is the electron effective mass.

The rate of the electron emission from the traps, R_{ne}, is proportional to the number of filled traps, $f_t N_t$:

$$R_{ne} = e_n f_t N_t \qquad (1\text{-}7\text{-}22)$$

Under equilibrium conditions, the principle of detailed balance can be invoked such that

$$R_{nc} = R_{ne} \qquad (1\text{-}7\text{-}23)$$

and, hence,

$$C_n n_o = \frac{e_n f_{to}}{1 - f_{to}} \qquad (1\text{-}7\text{-}24)$$

where

$$n_o = N_c \exp \left[(E_F - E_c) / k_B T \right] \qquad (1\text{-}7\text{-}25)$$

is the equilibrium electron concentration. The ratio $f_{to}/(1 - f_{to})$ can be found using the Fermi-Dirac occupation function

$$\frac{f_{to}}{1 - f_{to}} = \exp \left(\frac{E_F - E_t}{k_B T} \right) \qquad (1\text{-}7\text{-}26)$$

Here, E_t is the energy of the trap level. Hence, from eq. (1-7-24), we find

$$e_n = n_1 C_n \tag{1-7-27}$$

where

$$n_1 = N_c \exp\left(\frac{E_t - E_c}{k_B T}\right) = n_i \exp\left(\frac{E_t - E_i}{k_B T}\right) \tag{1-7-28}$$

The difference between the electron capture and electron emission rates is given by

$$R_n = R_{nc} - R_{ne} = C_n N_t [(1 - f_t)n - f_t n_1] \tag{1-7-29}$$

A similar derivation yields the following expression for the difference between the hole capture and the hole emission rates

$$R_p = R_{pc} - R_{pe} = C_p N_t [f_t p - (1 - f_t)p_1] \tag{1-7-30}$$

where

$$p_1 = N_v \exp\left(\frac{E_v - E_t}{k_B T}\right) = n_i \exp\left(\frac{E_i - E_t}{k_B T}\right) \tag{1-7-31}$$

and

$$C_p = \sigma_p v_{thp} \tag{1-7-32}$$

Here, σ_p is the capture cross-section,

$$v_{thp} = \left(3 k_B T / m_p\right)^{1/2} \tag{1-7-33}$$

is the thermal velocity, and m_p is the effective mass for holes. Under steady-state conditions, there is no net accumulation of charge and, hence, electrons and holes must recombine in pairs. Thus,

$$R_p = R_n = R \tag{1-7-34}$$

where R is the net recombination rate. The occupation function, f_t, can be found from this condition, i.e.,

$$f_t = \frac{n\,C_n + p_1 C_p}{C_n(n + n_1) + C_p(p + p_1)} \tag{1-7-35}$$

Substituting this expression into eq. (1-7-29) using eq. (1-7-34) and noting that $n_1 p_1 = n_i^2$, we find

$$R = \frac{pn - n_i^2}{\tau_{p1}(n + n_1) + \tau_{n1}(p + p_1)} \tag{1-7-36}$$

Here, n_i is the intrinsic concentration, and τ_{n1} and τ_{p1} are electron and hole lifetimes given by

$$\tau_{n1} = \frac{1}{v_{thn}\,\sigma_n\,N_t} \tag{1-7-37}$$

$$\tau_{p1} = \frac{1}{v_{thp}\,\sigma_p\,N_t} \tag{1-7-38}$$

In particular, when electrons are minority carriers ($n \ll p \approx N_a$ where N_a is the concentration of shallow ionized acceptors, $p \gg p_1$, $p \gg n_1$), eq. (1-7-36) reduces to

$$R = \frac{n - n_o}{\tau_{n1}} \tag{1-7-39}$$

where $n_o = n_i^2/N_a$. When holes are minority carriers ($p \ll n \approx N_d$ where N_d is the concentration of shallow ionized donors, $n \gg n_1$, $n \gg p_1$), eq. (1-7-36) reduces to

$$R = \frac{p - p_o}{\tau_{p1}} \tag{1-7-40}$$

where $p_o = n_i^2/N_d$. The electron lifetime, τ_{n1}, in p-type silicon and the hole lifetime, τ_{p1}, in n-type silicon decrease with increasing doping. At low doping levels, this decrease may be explained by higher trap concentrations in the doped semiconductor. If the trap concentration is proportional to the concentration of dopants, we should expect $\tau_{n1} \sim 1/N_a$ and $\tau_{p1} \sim 1/N_d$. At relatively high doping levels, the lifetimes decrease faster than the inverse doping concentration. The reason is that a different recombination mechanism, called Auger recombination, becomes important at very high doping levels. In an Auger process, an electron and a hole recombine without involving trap levels and the released energy (of the order of the energy gap) is transferred to another carrier (a hole in p-type material and an electron in n-type material). Auger recombination is the inverse of impact ionization where energetic carriers cause the generation of electron-hole pairs. Because two electrons (in n-type material) or two holes (in p-type material) are involved in the Auger recombination process, the recombination lifetime associated with this process is inversely proportional to the square of the majority carrier concentration, i.e.,

$$\tau_{n2} = \frac{1}{G_p N_a^2} \qquad (1\text{-}7\text{-}41)$$

for p-type material and, equivalently, for n-type material

$$\tau_{p2} = \frac{1}{G_n N_d^2} \qquad (1\text{-}7\text{-}42)$$

where, $G_p = 9.9\text{x}10^{-32} \text{ cm}^6/\text{s}$ and $G_n = 2.28\text{x}10^{-32} \text{ cm}^6/\text{s}$ for silicon (see Dziewior and Schmid (1982)).

In many semiconductor devices, the recombination rate is very high near the surface where extra defects and traps are present. As a consequence, the diffusion flux of minority carriers at the surface is determined by the surface recombination processes. For example, when minority carriers are holes (n-type material), we can write

$$D_p \frac{\partial p_n}{\partial x}\bigg|_{x=0} = - S_p [p_n (x = 0) - p_{no}] \qquad (1\text{-}7\text{-}43)$$

where D_p is the hole diffusion coefficient, p_n is the hole concentration, $p_{no} = n_i^2/N_d$ is the equilibrium hole concentration, and $x = 0$ corresponds to the surface of the sample. S_p is called the surface recombination rate and is given by

$$S_p = \sigma_p v_{thp} N_{st} \qquad (1\text{-}7\text{-}44)$$

where N_{st} is the density of the surface traps.

1-8.   p-n JUNCTIONS

1.8.1. *p-n* Junctions Under Zero Bias

An *n*-type and a *p*-type semiconductor at a large distance from each other, such that no exchange of carriers can take place, will, in general, have different Fermi energies. However, when brought close together, forming what is called a *p-n* junction, electrons will be exchanged in an attempt to establish thermal equilibrium. Equilibrium is reached when the Fermi level becomes constant throughout the entire system. In the process, a dipole layer is established due to the transfer of a net amount of electronic charge across the boundary.

The energy band diagram of a *p-n* junction is shown in Fig. 1.8.1.

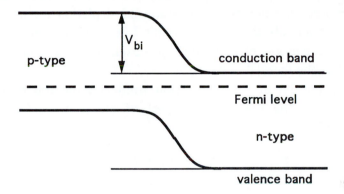

Fig. 1.8.1. Energy band diagram of a *p-n* junction.

Well within the *n*-region, $n \approx N_d$, $p_n = n_i^2/N_d$, and in the *p*-type region, $p \approx N_a$, $n_p = n_i^2/N_A$, and the energy bands are flat. Here, N_d and N_a are the concentrations of shallow (completely ionized) donors and acceptors. The equilibrium electron and hole concentrations change by a factor of e when the bands shift by a thermal energy $qV_{th} = k_BT$ (or 25.8 meV at $T = 300$ K). The total band bending is typically several tenths of an electron volt or more (depending on doping levels, the energy gap, and the effective densities of states in the conduction and valence bands), i.e., much greater than qV_{th}. As a consequence, we have $n << N_d$ and $p << N_a$ in almost the entire transition region. Hence, the charge densities are

$$\rho_n \approx qN_d \tag{1-8-1}$$

in the n-type section of the transition region and

$$\rho_p \approx - qN_a \qquad (1\text{-}8\text{-}2)$$

in the p-type section of the transition region. Eqs. (1-8-1) and (1-8-2) correspond to the so-called depletion approximation. The transition region is frequently called the depletion region or space charge region.

The depletion approximation does not imply that the depletion region is devoid of carriers. There are electrons and holes in the region but their concentrations are such that $N_d \gg n \gg n_{po}$ and $N_a \gg p \gg p_{no}$, where n_{po} and p_{no} are the equilibrium minority carrier concentrations in the p- and n- type regions, respectively. The depletion approximation is not valid in the boundary layers between the neutral sections and the depletion regions. However, these layers are relatively thin as long as the total band bending, i.e., the difference between the bottoms of the conduction band in the p- and n- regions, is much greater than the thermal voltage V_{th}.

The bottom of the conduction band and the top of the valence band correspond to the potential energies of electrons and holes respectively. Consequently,

$$E_c = - q\phi + const. \qquad (1\text{-}8\text{-}3)$$

$$E_v = - q\phi - E_g + const. \qquad (1\text{-}8\text{-}4)$$

where $E_g = E_c - E_v$ is the energy gap, and ϕ is the electric potential. If we arbitrarily choose $E_c = 0$ and $\phi = 0$ in the n-type region far from the junction, we find

$$n = N_d \exp\left(\frac{\phi(x)}{V_{th}}\right) = N_d \exp\left(- \frac{E_c(x)}{qV_{th}}\right) \qquad (1\text{-}8\text{-}5)$$

$$p = N_a \exp\left[- \frac{V_{bi} + \phi(x)}{V_{th}}\right] = N_a \exp\left[\frac{- qV_{bi} + E_c(x)}{qV_{th}}\right] \qquad (1\text{-}8\text{-}6)$$

where x is the spatial coordinate and V_{bi} is the built-in voltage (see Fig. 1.8.1)). V_{bi} is found from eqs. (1-8-5) and (1-8-6) by, for example, considering a position deep inside the *n*-type region where $n = N_d$ and $p = N_a \exp(-V_{bi}/V_{th})$. Using $pn = n_i^2$ we find that

$$V_{bi} = V_{th} \ln\left(\frac{N_d N_a}{n_i^2} \right)$$

(1-8-7)

Note that for the present choice of potential reference, we have $V_{bi} = -\phi(-\infty)$ where $\phi(-\infty)$ is the potential of the *p*-region far from the junction.

The distribution of the electric potential in the junction is determined from Poisson's equation

$$\frac{d^2\phi}{dx^2} = -\frac{\rho}{\varepsilon_s}$$

(1-8-8)

where ε_s is the dielectric permittivity of the semiconductor and

$$\rho = q\left(p - n + N_d - N_a \right)$$

(1-8-9)

is the space charge density.

Using the depletion approximation, we can rewrite Poisson's equation as follows:

$$\frac{d^2\phi}{dx^2} = \begin{cases} \dfrac{qN_a}{\varepsilon_s}, & \text{for} \quad -x_p < x < 0 \\[2ex] -\dfrac{qN_d}{\varepsilon_s}, & \text{for} \quad 0 < x < x_n \end{cases}$$

(1-8-10)

Here, $x = 0$ corresponds to the metallurgical junction between the *p*- and *n*- type regions, and x_p and x_n are the depletion widths on the two sides of the junction. A typical charge distribution according to the depletion approximation is shown in Fig. 1-8-2.

The electric field

$$F = -\frac{d\phi}{dx} = \frac{1}{q}\frac{dE_c}{dx}$$

(1-8-11)

is given by the first integral of eq. (1-8-10).

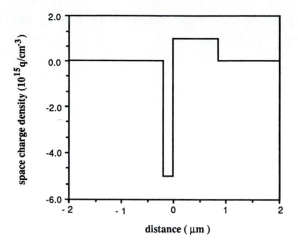

Fig. 1.8.2. Charge density profile (in units of electronic charge per cm³) for a silicon *p-n* junction (from Shur (1990)).

Using the conditions $F(x = x_n) = 0$ and $F(x = -x_p) = 0$, we obtain

$$F = \begin{cases} - F_m \left(1 + \dfrac{x}{x_p}\right), & \text{for} \quad -x_p < x < 0 \\[2mm] - F_m \left(1 - \dfrac{x}{x_n}\right), & \text{for} \quad 0 < x < x_n \end{cases} \tag{1-8-12}$$

(see Fig. 1.8.3). The maximum magnitude of the electric field, F_m, in the junction (reached at the *p-n* interface, i.e., at $x = 0$) is given by

$$F_m = \frac{q N_d \, x_n}{\varepsilon_s} = \frac{q N_a x_p}{\varepsilon_s} \tag{1-8-13}$$

The potential distribution is found by integrating eq. (1-8-12), i.e.,

$$\phi = \begin{cases} - V_{bi} + \dfrac{q N_a (x + x_p)^2}{2\varepsilon_s}, & \text{for} \quad -x_p < x < 0 \\[3mm] - \dfrac{q N_d (x - x_n)^2}{2\varepsilon_s}, & \text{for} \quad 0 < x < x_n \end{cases} \tag{1-8-14}$$

where we used the conditions $\phi(-x_p) = -V_{bi}$ and $\phi(x_n) = 0$.

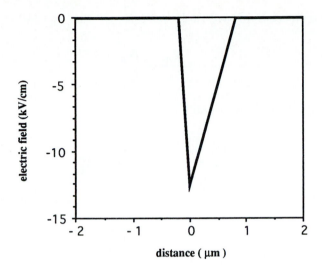

Fig. 1.8.3. Field distribution in a silicon *p-n* junction at zero bias at room temperature (from Shur (1990)).

The potential distribution for a silicon *p-n* junction at zero external bias is shown in Fig. 1.8.4.

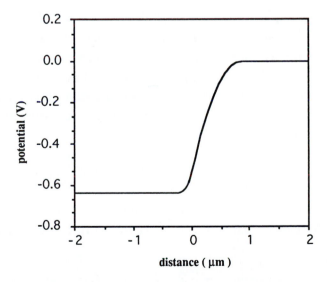

Fig. 1.8.4. Potential distribution in a silicon *p-n* junction at zero bias. The potential reference $(\phi = 0)$ was chosen in the *n*-type region far from the junction (from Shur (1990)).

From eq. (1-8-13) we find

$$\frac{x_n}{x_p} = \frac{N_a}{N_d} \tag{1-8-15}$$

and from eq. (1-8-14) we obtain

$$\frac{qN_d\, x_n^2}{2\varepsilon_s} + \frac{qN_a\, x_p^2}{2\varepsilon_s} = V_{bi} \tag{1-8-16}$$

Hence,

$$x_n = \sqrt{\frac{2\varepsilon_s\, V_{bi}}{qN_d\left(1 + N_d/N_a\right)}} \tag{1-8-17}$$

$$x_p = \sqrt{\frac{2\varepsilon_s\, V_{bi}}{qN_a\left(1 + N_d/N_a\right)}} \tag{1-8-18}$$

These equations for x_n and x_p show that for a one sided n^+-p junction, where $N_d \gg N_a$ and $x_p \gg x_n$, the electric field distribution looks like a right triangle.

As mentioned above, the depletion approximation breaks down in the thin boundary layers between the depletion layer and neutral layers where the carrier concentrations are comparable with the dopant concentrations. In the boundary layer in the n-type region we can rewrite Poisson's equation as follows

$$\frac{d^2\phi}{dx^2} = -\frac{qN_d}{\varepsilon_s} + \frac{qN_d}{\varepsilon_s}\exp\!\left(\frac{\phi}{V_{th}}\right) \tag{1-8-19}$$

When $n \approx N_d$ ($\phi \ll V_{th}$), the second term on the right hand side of eq. (1-8-19) can be expanded in a Taylor series, $(\exp(\phi/V_{th}) \approx 1 + \phi/V_{th})$, yielding

$$\frac{d^2\phi}{dx^2} \approx \frac{qN_d\phi}{\varepsilon_s V_{th}} = \frac{\phi}{L_{Dn}^2} \tag{1-8-20}$$

where the characteristic scale of the potential variation

$$L_{Dn} = \sqrt{\frac{\varepsilon_s V_{th}}{qN_d}} \tag{1-8-21}$$

is called the Debye length. The width of the boundary layer is of the order of a few Debye lengths. This is confirmed by numerical calculations of the potential distribution. Similarly, for the *p*-type region, the width of the boundary region is determined by the Debye length

$$L_{Dp} = \sqrt{\frac{\varepsilon_s V_{th}}{qN_a}} \tag{1-8-22}$$

1.8.2. Current-Voltage Characteristic of a *p-n* Diode

The most resistive portion of a *p-n* junction is the space charge (depletion) region. When a negative voltage is applied to the *p*-region with respect to the *n*-region, the holes and electrons are drawn away from the junction increasing the width of the depletion region and, hence, the resistance of the device. This situation corresponds to a reverse bias. A voltage of the opposite polarity (forward bias – positive potential applied to the *p*-region) pushes the electrons and holes towards the junction. This decreases the width of the depletion region and the device resistance.

Under reverse bias and small forward voltages, the resistance of the space charge region is much greater than the resistance of the neutral sections of the device and, as a consequence, almost all applied voltage drops across the depletion region where the electric field is very high. At zero bias, there is an exact balance between the drift component of the electric current (proportional to the electric field) and the compensating diffusion current. The net current is zero, and the Fermi level is constant. When an external voltage is applied to a *p-n* junction, this balance between the drift and diffusion components of the electric current is lifted. However, for small voltages, the net current is still much smaller than the drift and diffusion components of the current in the depletion region evaluated separately. This means that the carrier distribution is still close to the equilibrium state and, consequently, the electron and hole quasi–Fermi levels remain nearly constant throughout the depletion region (see Fig. 1.8.5).

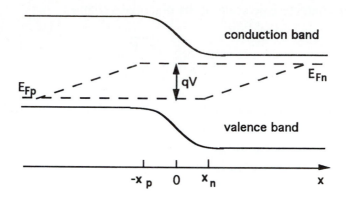

Fig. 1.8.5. Quasi-Fermi levels in a *p-n* junction under forward bias.

Indeed, numerical simulations of *p-n* junctions at reverse bias and low forward biases show that the concentrations of electrons and holes in the depletion region are approximately given by

$$n = N_c \exp\left(\frac{E_{Fn} - E_c}{k_B T}\right) \tag{1-8-23}$$

$$p = N_v \exp\left(\frac{E_v - E_{Fp}}{k_B T}\right) \tag{1-8-24}$$

where the electron and hole quasi-Fermi levels, E_{Fn} and E_{Fp}, remain practically the same as the equilibrium quasi-Fermi levels in the *n*-type and *p*-type region, respectively.

The applied forward voltage reduces the potential barrier between the *p*-type and *n*-type regions from the built-in voltage, V_{bi}, to $V_{bi} - V$ (assuming that all applied voltage drops across the space charge region with negligible voltage drops across neutral regions). Hence, the difference between E_{Fn} and E_{Fp} in the depletion region is equal to qV (see Fig. 1.8.5), i.e.,

$$qV = E_{Fn} - E_{Fp} \tag{1-8-25}$$

From the above equations, we obtain

$$pn = n_i^2 \exp\left(\frac{E_{Fn} - E_{Fp}}{k_B T}\right) = n_i^2 \exp\left(\frac{V}{V_{th}}\right) \qquad (1\text{-}8\text{-}26)$$

Here, we let V be the voltage applied to the *p*-region with respect to the *n*-region, such that a forward bias corresponds to a positive applied voltage. Eq. (1-8-26) is called *the law of the junction*. One can show that this relationship is valid even at high injection conditions (i.e., when $p(x_n) > N_d$ and/or $n(-x_p) > N_a$) for a short diode where the lengths of the *n*- and *p*- type regions are comparable to the respective diffusion lengths for the minority carriers.

Let us now consider the boundary between the depletion region and the neutral region at the *n*-side of the junction. At this boundary $n \approx N_d$. Hence, from eq. (1-8-26) we obtain

$$p \approx \frac{n_i^2}{N_d} \exp\left(\frac{V}{V_{th}}\right) = p_{no} \exp\left(\frac{V}{V_{th}}\right) \qquad (1\text{-}8\text{-}27)$$

Similarly, at the boundary between the depletion and neutral region at the *p*-side of the junction, we find

$$n \approx \frac{n_i^2}{N_a} \exp\left(\frac{V}{V_{th}}\right) = n_{po} \exp\left(\frac{V}{V_{th}}\right) \qquad (1\text{-}8\text{-}28)$$

Here, n_{po} and p_{no} are the concentrations of the minority carriers in the neutral regions at zero applied bias. Eqs. (1-8-27) and (1-8-28) show that the concentrations of the minority carriers at the boundaries of the depletion region increase exponentially with increasing forward bias (and decrease exponentially with increasing reverse bias).

As was shown in Section 1.4, the concentration of minority carriers in the *n*-type neutral region (holes) can be found by solving the continuity equation

$$D_p \frac{\partial^2 p_n}{\partial x^2} - \mu_p F \frac{\partial p_n}{\partial x} + G - R = 0 \qquad (1\text{-}8\text{-}29)$$

In the case of a *p-n* junction with a generation rate $G = G_{th}$ (thermal generation), the net recombination rate may be expressed as

$$U_R = R - G = \frac{p_n - p_{no}}{\tau_p} \qquad (1\text{-}8\text{-}30)$$

where τ_p is the ambipolar lifetime (coinciding with the hole lifetime for low injection level when $p_n \ll n_n$). The second term in eq. (1-8-29) is negligible since the electric field is very small in the neutral region. Hence,

$$D_p \frac{\partial^2 p_n}{\partial x^2} - \frac{p_n - p_{no}}{\tau_p} = 0 \qquad (1\text{-}8\text{-}31)$$

For a long neutral region, eq. (1-8-31) has to be solved with the two boundary conditions:

$$p_n(x_n) = p_{no} \exp\left(\frac{V}{V_{th}}\right) \qquad (1\text{-}8\text{-}32)$$

$$p_n(x \rightarrow \infty) = p_{no} \qquad (1\text{-}8\text{-}33)$$

A general solution of eq. (1-8-31) is given by

$$p_n(x) - p_{no} = A \, \exp\left(\frac{x - x_n}{L_p}\right) + B \, \exp\left(-\frac{x - x_n}{L_p}\right) \qquad (1\text{-}8\text{-}34)$$

where the constants A and B are to be determined from the boundary conditions, and

$$L_p = \sqrt{D_p \tau_p} \qquad (1\text{-}8\text{-}35)$$

is called the hole diffusion length. Using the boundary conditions, eqs. (1-8-32) and (1-8-33), in combination with eq. (1-8-34), we find

$$p_n(x) - p_{no} = p_{no} \left[\exp\left(\frac{V}{V_{th}}\right) - 1 \right] \exp\left(-\frac{x - x_n}{L_p}\right) \qquad (1\text{-}8\text{-}36)$$

The hole current density in the neutral n–type region, J_p, is primarily caused by diffusion (J_{pD}), i.e.,

$$J_p \approx J_{pD} = -qD_p \frac{\partial p_n}{\partial x} = \frac{qD_p \, p_{no}}{L_p}\left[\exp\left(\frac{V}{V_{th}}\right)-1\right]\exp\left(-\frac{x-x_n}{L_p}\right) \quad (1\text{-}8\text{-}37)$$

and will be a function of distance. The electron diffusion current in the same region, J_{nD}, follows from the quasi neutrality condition (see eqs. (1-5-13) and (1-5-15))

$$\begin{aligned} J_{nD} &= qD_n \frac{\partial n_p}{\partial x} \approx qD_n \frac{\partial p_n}{\partial x} \\ &= -\frac{qD_n \, p_{no}}{L_p}\left[\exp\left(\frac{V}{V_{th}}\right)-1\right]\exp\left(-\frac{x-x_n}{L_p}\right) \end{aligned} \quad (1\text{-}8\text{-}38)$$

From eqs. (1-8-37) and (1-8-38) we find

$$J_{nD} = -\frac{D_n}{D_p}J_{pD} \quad (1\text{-}8\text{-}39)$$

so that the total diffusion current J_D in the *n*-type neutral region becomes

$$J_D = J_{pD} + J_{nD} \approx \frac{q p_{no}}{L_p}(D_p - D_n)\left[\exp\left(\frac{V}{V_{th}}\right)-1\right]\exp\left(-\frac{x-x_n}{L_p}\right)$$

$$(1\text{-}8\text{-}40)$$

(This, of course, comes in addition to the electron drift current in the neutral *n*-region.)

The hole and electron diffusion current densities (and hence total electron and hole current densities) vary with distance due to recombination. The characteristic scales of this variation are the hole diffusion length L_p (see eq. (1-8-35)) for the *n*-type region and the electron diffusion length

$$L_n = \sqrt{D_n \tau_n} \quad (1\text{-}8\text{-}41)$$

for the *p*-type neutral region. Typically,

$$L_n \gg x_n + x_p \quad (1\text{-}8\text{-}42)$$

$$L_p \gg x_n + x_p \quad (1\text{-}8\text{-}43)$$

In this case, relatively little recombination occurs in the depletion region, and the hole and electron current densities remain nearly constant throughout this region. Therefore, the total current density through the *p-n* junction can be written as:

$$J = J_p\Big|_{x=x_n} + J_n\Big|_{x=-x_p}$$
(1-8-44)

where J_n and J_n are the electron and hole current densities in the neutral *p*- and *n*-type regions, respectively. Hence, from eq. (1-8-37) and from a similar equation for the electrons in the *p*-type neutral region, we find

$$I = I_s\left[\exp\left(\frac{V}{V_{th}}\right) - 1\right]$$
(1-8-45)

where $I = SJ$ is the diode current, S is the cross-section, and I_s is the so-called diode saturation current given by

$$I_s = S\left(\frac{qD_p\, p_{no}}{L_p} + \frac{qD_n\, n_{po}}{L_n}\right)$$
(1-8-46)

Eq. (1-8-45) is called the Shockley equation or the Diode equation. This equation is in good agreement with the experimental data for silicon and germanium diodes but does not describe very well the current–voltage characteristics of gallium arsenide *p-n* junctions at low forward biases.

Very often, the diode is constructed as a one-sided abrupt p^+-*n* or n^+-*p* junction. For example, in a p^+-*n* diode, $p_{no} \gg n_{po}$ and the second term in the parenthesis of eq. (1-8-46) can be neglected.

In a real semiconductor diode, the parasitic series resistance, R_s, of the device contacts and the semiconductor neutral regions may play an important role, and eq. (1-8-45) has to be modified to:

$$I = I_s\left[\exp\left(\frac{V - IR_s}{V_{th}}\right) - 1\right]$$
(1-8-47)

An approximate analytical solution of eq. (1-8-47) is given in Appendix A7 (see also Fjeldly et al. (1991)).

In most practical semiconductor diodes, the total lengths, X_n and X_p, of the *n*-type and *p*-type regions are comparable to or smaller then the diffusion length of minority carriers, and the continuity equation has to be solved with new boundary conditions. If it can be assumed that all excess carriers are extracted at $x = X_n$ in the n-region (and at $x = -X_p$ in the p-region), we have to replace eq. (1-8-33) by

$$p_n(x = X_n) = p_{no} \tag{1-8-48}$$

Using eqs. (1-8-32) and (1-8-48), we find from eq. (1-8-34)

$$p_n - p_{no} = p_{no}\left[\exp\left(\frac{V}{V_{th}}\right) - 1\right]\left[a_1\exp\left(\frac{x - x_n}{L_p}\right) + b_1\exp\left(-\frac{x - x_n}{L_p}\right)\right] \tag{1-8-49}$$

where

$$a_1 = -\frac{\exp\left[-(X_n - x_n)/L_p\right]}{2\sinh\left[(X_n - x_n)/L_p\right]} \tag{1-8-50}$$

$$b_1 = \frac{\exp\left[(X_n - x_n)/L_p\right]}{2\sinh\left[(X_n - x_n)/L_p\right]} \tag{1-8-51}$$

The hole current density J_p at $x = x_n$ is then given by

$$J_p = -qD_p\frac{\partial p_n}{\partial x}\bigg|_{x = x_n} = \frac{qD_p\,p_{no}}{L_p}\left[\exp\left(\frac{V}{V_{th}}\right) - 1\right]\coth\left(\frac{X_n - x_n}{L_p}\right) \tag{1-8-52}$$

A similar expression can be derived for the electron current density in the p-region at the boundary between the neutral and depletion region.

It is also interesting to consider the case of a short p^+-n diode such that the length of the n-section, X_n, is much smaller than the hole diffusion length L_p. The solution for this case can be found from eq. (1-8-49) using a Taylor series expansion

$$p_n(x) \approx p_{no} + p_{no}\left[\exp\left(\frac{V}{V_{th}}\right) - 1\right]\left(\frac{X_n - x}{X_n - x_n}\right) \tag{1-8-53}$$

which corresponds to a linear distribution of the minority carriers in the n-region. Hence, the total current density is given by

$$J \approx - qD_p \frac{\partial p_n}{\partial x}\bigg|_{x=x_n} = \frac{qD_p \, p_{no}}{X_n - x_n}\left[\exp\left(\frac{V}{V_{th}}\right) - 1\right] \tag{1-8-54}$$

which is similar to the corresponding equation for a long p^+-n junction except that L_p is substituted by $X_n - x_n$. This discussion of a short p^+-n diode is very useful for the consideration of the carrier distributions in a bipolar junction transistor (see Chapter 2).

In addition to the diffusion current components considered above, so-called generation and recombination current components may play an important or even dominant role. These currents depend on the concentrations, distribution and energy levels of traps in the depletion region. Traps associated with different impurities are always present in any semiconductor material. The simplest model describing the generation and recombination currents is based on the assumption that there is just one type of traps, uniformly distributed across the device, that dominates recombination and generation processes. In reality, traps may be non-uniformly distributed and more than one type of traps may be involved.

The generation current becomes important at reverse bias voltages and can be described by the following equation

$$J_{gen} = q \int_{-x_p}^{x_n} \left|U_R\right| dx = \frac{qn_i \, x_d}{\tau_{gen}} \tag{1-8-55}$$

where U_R is the net recombination rate given by eq. (1-7-36), now negative since $n_i^2 \gg pn$ in the depletion region. The width of the depletion region is

$$x_d = x_n + x_p \tag{1-8-56}$$

and τ_{gen} is called the generation life time. With very few carriers in the depletion region, we can assume that $n \ll n_t$ and $p \ll p_t$. Hence, τ_{gen} can be expressed as

$$\tau_{gen} = \tau_{p1} \exp\left(\frac{E_t - E_i}{k_B T}\right) + \tau_{n1} \exp\left(\frac{E_i - E_t}{k_B T}\right) \tag{1-8-57}$$

using eq. (1-7-36) in combination with eqs. (1-7-28) and (1-7-31). Here, τ_{n1} and τ_{p1} are the electron and hole lifetimes, respectively, given by eqs. (1-7-37) and

(1-7-38). Note that there is practically no recombination in the depletion region since few carriers are available in reverse bias. The generation process tries to restore equilibrium ($pn \rightarrow n_i^2$) but the generated carriers are swept away leading to the generation current density J_{gen} given by eq. (1-8-55). For an abrupt junction, x_d and, hence, J_{gen} are proportional to $(V_{bi} - V)^{1/2}$.

The total reverse current density, J_R, is given by:

$$J_R = J_s + J_{gen} \tag{1-8-58}$$

where the diffusion component J_s can be written as (see eq. (1-8-46)):

$$J_s = \left(\frac{qD_p}{N_d L_p} + \frac{qD_n}{N_a L_n} \right) n_i^2 \tag{1-8-59}$$

Since J_s is proportional to n_i^2 and J_{gen} is proportional to n_i, the generation current will be dominant when n_i is sufficiently small. In practice, this is always the case for GaAs and for Si at room temperature and below at reverse bias.

Under the forward bias conditions ($V > 0$) there is an excess of carriers in the depletion region. At its two boundaries with the neutral regions, we have, according to eqs. (1-8-27) and (1-8-28),

$$pn = n_i^2 \exp\left(\frac{V}{V_{th}} \right) \tag{1-8-60}$$

Assuming that eq. (1-8-60) is valid for the entire depletion region, the net recombination rate becomes positive for forward bias and can be written as

$$U_R = \frac{n_i^2 \left[\exp\left(\dfrac{V}{V_{th}} \right) - 1 \right]}{\tau_{p1}(n + n_t) + \tau_{n1}(p + p_t)} \tag{1-8-61}$$

The recombination current density is given by

$$J_{rec} = q \int_{-x_p}^{x_n} U_R \, dx \tag{1-8-62}$$

This integral can be evaluated by expanding U_R near its maximum value, leading to the following expression (see, for example, van der Ziel (1976)):

$$J_{rec} = J_{recs} \exp\left(\frac{V}{2V_{th}}\right)$$

(1-8-63)

where

$$J_{recs} = \sqrt{\frac{\pi}{2}} \frac{q n_i V_{th}}{\tau_{rec} F_{max}}$$

(1-8-64)

Here, F_{max} is the maximum electric field in the depletion layer. We have also assumed that $\tau_{n1} = \tau_{p1} = \tau_{rec}$. However, the following empirical expression for the forward bias recombination current may be more accurate for practical devices

$$J_{rec} = J_{recs} \exp\left(\frac{V}{m_r V_{th}}\right)$$

(1-8-65)

where m_r may differ from the value 2 (see Lee and Nussbaum (1980)). The total forward current density now becomes

$$J_F = J_s \exp\left(\frac{V}{V_{th}}\right) + J_{recs} \exp\left(\frac{V}{m_r V_{th}}\right)$$

(1-8-66)

However, it is frequently more convenient to use the empirical formula

$$J_F = J_{seff} \exp\left(\frac{V}{\eta V_{th}}\right)$$

(1-8-67)

where η is called the ideality factor and J_{seff} is the effective saturation current. The deviation of η from unity may be considered a measure of the importance of the recombination current.

Since J_s is proportional to n_i^2 and J_{recs} is proportional to n_i, the recombination current is dominant when n_i is small enough. In practice, this is always the case for relatively small applied voltages for large gap materials such as GaAs at or below room temperature. At large voltages, the diffusion component is dominant because it is proportional to $\exp(V/V_{th})$, whereas the recombination current density varies with the forward bias only as $\exp[V/(m_r V_{th})]$ where $m_r \approx 2$.

1.8.3. Capacitance and Equivalent Circuit of a *p-n* Junction

At reverse bias, zero bias or small forward bias, the capacitance of a *p-n* diode is primarily determined by the depletion layer capacitance, i.e.,

$$C_d = \varepsilon_s S / x_d \qquad (1\text{-}8\text{-}68)$$

where $x_d = x_n + x_p$ is the total width of the depletion layer. When an external bias, V, is applied, x_n and x_p are given by

$$x_n = \sqrt{\frac{2\varepsilon_s (V_{bi} - V)}{qN_d(1 + N_d/N_a)}} \qquad (1\text{-}8\text{-}69)$$

$$x_p = \sqrt{\frac{2\varepsilon_s (V_{bi} - V)}{qN_a(1 + N_a/N_d)}} \qquad (1\text{-}8\text{-}70)$$

Using the above equations, we obtain

$$C_d = S \sqrt{\frac{q\varepsilon_s N_{eff}}{2(V_{bi} - V)}} \qquad (1\text{-}8\text{-}71)$$

where

$$N_{eff} = \left\{ \frac{1}{\sqrt{N_d\left(1 + \dfrac{N_d}{N_a}\right)}} + \frac{1}{\sqrt{N_a\left(1 + \dfrac{N_a}{N_d}\right)}} \right\}^{-2} = \frac{N_a N_d}{N_a + N_d} \qquad (1\text{-}8\text{-}72)$$

is the effective doping density.

As can be seen from eq. (1-8-71), C_d is inversely proportional to $(V_{bi} - V)^{1/2}$ for an abrupt junction considered above. For an arbitrary doping profile, the depletion capacitance may be determined from

$$C_d = S \left| \frac{dQ_d}{dV} \right| \qquad (1\text{-}8\text{-}73)$$

where Q_d is the charge per unit area in the depletion layer of the *n*-type section of the device. In particular, for a linearly graded profile $(N_d - N_a = ax)$, we find

$$C_d = S\left[\frac{qa\varepsilon_s^2}{12(V_{bi} - V)}\right]^{1/3}$$

(1-8-74)

The built-in voltage for the linearly graded junction is given by:

$$V_{bi} = 2V_{th}\ln\left(\frac{ax_d}{2n_i}\right)$$

(1-8-75)

One can show (see, for example, Shur (1990)) that for an arbitrary doping profile

$$C_d = S\left|\frac{dQ_d}{dV}\right| = S\left|\frac{dQ_d}{dx_d}\frac{dx_d}{dV}\right| = S\frac{\varepsilon_s}{x_d}$$

(1-8-76)

This is the same expression as the one we used for the abrupt junction. Differentiating eq. (1-8-76) with respect to V, we obtain:

$$N_d(x_d) = \frac{C_d^3}{q\varepsilon_s S^2\left|dC_d/dV\right|} = \frac{2}{q\varepsilon_s S^2}\left|\frac{1}{d(1/C_d^2)/dV}\right|$$

(1-8-77)

Eq.(1-8-77) may be used to deduce the doping profile from the measured capacitance-voltage characteristics at reverse and small forward bias.

The p-n junction also has a capacitance related to the excess charge of the minority carriers, the so–called diffusion capacitance. This capacitance becomes important or even dominant under forward bias conditions. In order to find the diffusion capacitance, we will calculate the device impedance using the continuity equations for the minority carriers. These equations have to be modified to include the time derivative terms $\partial p/\partial t$ and $\partial n/\partial t$. Thus, for the n-region, we have

$$D_p\frac{\partial^2 p_n}{\partial x^2} - \frac{p_n - p_{no}}{\tau_p} = \frac{\partial p_n}{\partial t}$$

(1-8-78)

We assume that the applied voltage, $V(t)$, and the current density, $J(t)$, are given by a large signal stationary term and a small signal time dependent term:

$$V(t) = V_o + v^* \exp(i\omega t) \tag{1-8-79}$$

$$J(t) = J_o + j^* \exp(i\omega t) \tag{1-8-80}$$

where $v^* \ll V_o$ and $j^* \ll j_o$. We seek a solution of eq. (1-8-78) of the following form

$$p_n(x,t) = p_{ns}(x) + p^* \exp(i\omega t) \tag{1-8-81}$$

where $p_{ns}(x)$ is the solution for the steady state continuity equation and the second term on the right-hand side represents the small signal time dependent component. Substituting eq. (1-8-81) into eq. (1-8-78), we obtain

$$D_p \frac{\partial^2 p^*}{\partial x^2} - \frac{p^*}{\tau_p} = i\omega\, p^* \tag{1-8-82}$$

The boundary conditions for eq. (1-8-81) are:

$$p^*(x \to \infty) = 0 \tag{1-8-83}$$

$$p^*(x = x_n) = \frac{p_{no}v^*}{V_{th}} \exp\left(\frac{V_o}{V_{th}}\right) \tag{1-8-84}$$

where $x = x_n$ corresponds to the boundary of the depletion region on the n-type side of the junction. Eq. (1-8-84) is obtained by expanding the exponent on the right-hand side of the following equation for the hole density at $x = x_n$ (see eq. (1-8-32))

$$p_n(x = x_n) = p_{no} \exp\left[\frac{V_o + v^* \exp(i\omega t)}{V_{th}}\right] \tag{1-8-85}$$

into a Taylor series.

Eq. (1-8-82) coincides with the steady state continuity equation for $p_n - p_{no}$ (see eq. (1-8-36)) if the lifetime τ_p is replaced by $\tau_p/(1 + i\omega\tau_p)$. Hence, the previously derived relationship between p_n and the hole component of the diode current, J_p, can be adopted for the amplitude of the small-signal hole current density, $j_p{}^*$, (at $x = x_n$)

$$j_p^* \approx \frac{v^*}{V_{th}} \frac{qD_p\, p_{no}}{L_p} \sqrt{1 + i\omega\tau_p}\; \exp\!\left(\frac{V_o}{V_{th}}\right) \tag{1-8-86}$$

Here, we have assumed forward bias, i.e., $V_o \gg V_{th}$. A similar expression may be obtained for the electron component in the p-region, $j_n{}^*$, so that the total small signal current amplitude can be expressed as

$$j^* = j_p^* + j_n^* = \frac{v^*}{V_{th}} \left[\frac{qD_p\, p_{no}}{L_p} \sqrt{1 + i\omega\tau_p} + \frac{qD_n\, n_{po}}{L_n} \sqrt{1 + i\omega\tau_n} \right] \exp\!\left(\frac{V_o}{V_{th}}\right) \tag{1-8-87}$$

The small-signal admittance, Y, is defined as

$$Y = S j^*/v^* \tag{1-8-88}$$

At relatively low frequencies, when $\omega\tau_p \ll 1$ and $\omega\tau_n \ll 1$, we obtain

$$Y = G_d + i\omega C_{dif} \tag{1-8-89}$$

where

$$G_d \equiv 1/R_d = S J_o/V_{th} \tag{1-8-90}$$

$$J_o = \left(\frac{qD_p\, p_{no}}{L_p} + \frac{qD_n\, n_{po}}{L_n} \right) \exp\!\left(\frac{V_o}{V_{th}}\right) \tag{1-8-91}$$

and

$$C_{dif} = \frac{Sq}{2V_{th}} \left(L_p\, p_{no} + L_n\, n_{po} \right) \exp\!\left(\frac{V_o}{V_{th}}\right) \tag{1-8-92}$$

For a p^+-n diode, $p_{no} \gg n_{po}$ and eq. (1-8-92) may be simplified to

$$C_{dif} = \frac{S J_o \tau_p}{2V_{th}} = \frac{1}{2} G_d \tau_p \tag{1-8-93}$$

This equation shows that the characteristic time constant, $C_{dif}R_d$, of a forward biased *p-n* junction is of the order of the recombination time, as could be expected.

The characteristic time constant for a short p^+-n diode (where the length of the *n*-section, X_n, is much smaller than the hole diffusion length) is determined by the hole transit time, $(X_n - x_n)^2/(2D_p)$. Hence, the time response of such a diode is considerably faster than for a long diode.

The equivalent circuit of a *p-n* junction is shown in Fig.1.8.6.

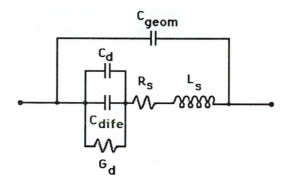

Fig. 1.8.6. The small signal equivalent circuit of a *p-n* junction.

In this equivalent circuit, we have taken into account that, at large forward bias, eq. (1-8-92) for the diffusion capacitance is no longer valid because the densities of minority carriers are no longer proportional to $\exp(V_o/V_{th})$. This can be taken into account by introducing an effective diffusion capacitance, C_{dife}:

$$C_{dife} = \frac{1}{\dfrac{1}{C_{dif}} + \dfrac{1}{C_{difmax}}}$$

(1-8-94)

where the maximum diffusion capacitance, C_{difmax} is estimated as

$$C_{difmax} \approx \frac{\varepsilon_s S}{L_{Dn} + L_{Dp}}$$

(1-8-95)

and L_{Dn} and L_{Dp} are Debye lengths in the *n*-type and *p*-type regions, respectively. In addition to the depletion and diffusion capacitances, C_d and C_{dife}, and the differential resistance of the *p-n* junction, R_d, the equivalent circuit includes a series resistance, R_s, (comprised of the contact resistances and of the resistance in the neutral regions of the semiconductor) as well as a parasitic inductance, L_s, and the geometric capacitance of the sample

$$C_{geom} = \varepsilon_s \, S / L \qquad (1\text{-}8\text{-}96)$$

where L is the sample length.

Practical measurements of the small signal junction capacitance are usually done at a finite frequency (typically 2 MHz or so). At such frequencies, the parasitic inductance, L_s, may become very important. At large forward bias, the diffusion capacitance, C_{dife}, is much greater than the depletion capacitance, C_d, and the geometric capacitance, C_{geom}. Hence, to first order, the measured impedance is given by

$$Z_{meas} \approx R_s + R_d / (1 + i\omega C_{dife} \, R_d) + i\omega \, L_s \qquad (1\text{-}8\text{-}97)$$

When the differential resistance, R_d, becomes small enough at large forward bias, the second term on the right-hand side of eq. (1-8-97) becomes small, and the impedance changes from reactive to inductive. This introduces a practical complication, especially when one needs to estimate the equivalent capacitance, C_{eq}, of a *p-n* junction under forward bias for large signal applications. Even when the parasitic inductance is small, the diffusion capacitance under large forward bias may decrease because the voltage drop will primarily occur across the diode series resistance, and an incremental increase in the voltage drop will not lead to an additional injection of carriers.

1.8.4. *p-n* Junction Breakdown

When a reverse bias applied to a *p-n* junction exceeds some critical value, the reverse current rises rapidly with further increase in the reverse bias voltage (see Fig. 1.8.7). Two basic mechanisms, avalanche or tunneling breakdown, may be responsible for this rapid increase in current. Avalanche breakdown is caused by the impact ionization process. During this process, an electron (or a hole) gains so much energy from the electric field that it initiates a transition of an electron from the valence band to the conduction band, thus creating an additional electron-hole pair. Newly created carriers are, in turn, accelerated by the electric field and create new electron-hole pairs via the impact ionization process. If the applied voltage is high enough, this will lead to an uncontrolled rise of current (until the current is either limited by an external load or the sample is destroyed).

A crude estimate of the critical voltage, V_{abr}, causing the avalanche breakdown may be obtained by assuming that the breakdown occurs when the electric field in the reverse biased *p-n* junction exceeds a certain critical field, F_{br}, for the impact ionization process.

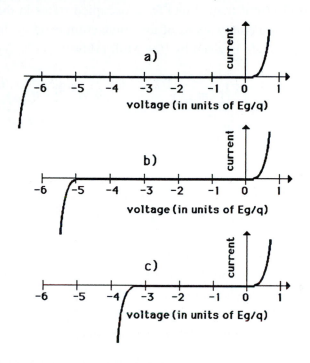

Fig. 1.8.7. Current-voltage characteristics of *p⁺-n* diodes for a relatively weakly doped *n* -region (a), an intermediate level of doping in the *n*-region (b), and a relatively strongly doped, degenerate *n*-region (c) (from Shur (1990)).

For a *p⁺-n* junction, this model leads to the following estimate (see, for example, Sze (1985))

$$V_{abr} = \frac{\varepsilon_s F_{br}^2}{2qN_d} \qquad (1\text{-}8\text{-}98)$$

The breakdown field at 300 K is of the order of 100 kV/cm for germanium, 300 kV/cm for silicon, 400 kV/cm for GaAs, and 2300 kV/cm for silicon carbide. Hence, V_{abr} can be quite large for diodes with low-doped n-regions. In some devices, V_{abr} can exceed two thousand volts.

Tunneling breakdown usually occurs in fairly highly doped semiconductors when the maximum electric field in the depletion layer approaches values of the order of 10^6 V/cm. Under such conditions, the width of the depletion layer is so narrow that electrons may tunnel from occupied states in the valence band of the p-type region into empty states of the conduction band in the n-type region. The tunneling current, I_t, is given by (see Moll (1964))

$$I_t = \frac{SVF_{av} \, q^{5/2} \left(2m_{pn}\right)^{1/2}}{h^2 E_g^{1/2}} \exp\left[- \frac{8\pi q^{1/2}\left(2m_{pn}\right)^{1/2} E_g^{3/2}}{3h \, F_{av}} \right] \qquad (1\text{-}8\text{-}99)$$

where

$$F_{av} = \frac{F_m}{2} = \sqrt{\frac{q(V_{bi} - V) N_a N_d}{2\varepsilon_s (N_a + N_d)}} \qquad (1\text{-}8\text{-}100)$$

is the average electric field in the junction, F_m is the maximum electric field occurring at the metallurgical interface, E_g is the energy gap (in eV), V_{bi} is the built-in voltage (which may be assumed approximately equal to E_g/q for highly doped p-n junctions), V is the applied voltage, N_a and N_d are acceptor and donor concentrations, m_{pn} is the reduced effective mass

$$m_{pn} = 2 \left(1/m_n + 1/m_{lh}\right)^{-1} \qquad (1\text{-}8\text{-}101)$$

m_n is the conduction band effective mass and m_{lh} is the effective mass for light holes. In terms of the free electron mass, m_e, the effective light hole masses are $m_{lh} = 0.044 \, m_e$, $m_{lh} = 0.16 \, m_e$ and $m_{lh} = 0.082 \, m_e$ for Ge, Si and GaAs, respectively. The effective electron mass is $m_n = 0.067 \, m_e$ for GaAs. For Si and Ge m_n may be estimated as

$$m_n = \left(\frac{m_{nl}}{3} + \frac{2 m_{nt}}{3} \right) \qquad (1\text{-}8\text{-}102)$$

where $m_{nl} = 0.98 \, m_e$, $m_{nt} = 0.19 \, m_e$ for Si and $m_{nl} = 1.64 \, m_e$, $m_{nt} = 0.082 \, m_e$ for Ge. (Strictly speaking, the appropriate effective masses may depend on the direction of tunneling.)

   The exponential rise of the tunneling current in highly doped p-n junctions with reverse voltage leads to tunneling breakdown. The critical voltage for tunneling breakdown, V_{tbr}, may be estimated by defining the onset of breakdown as the reverse voltage which gives a ten-fold increase in the reverse current over the saturation current, i.e.,

$$I\left(V_{brt}\right) \approx 10\, S J_s \qquad\qquad (1\text{-}8\text{-}103)$$

According to Sze (1981), the critical voltage for tunneling breakdown is usually less than approximately $4E_g/q$. When the breakdown voltage is higher than $6E_g/q$, the breakdown is typically caused by the avalanche effect.   Both mechanisms may play a role for breakdown voltages between $4E_g/q$ and $6E_g/q$ (see Fig. 1.8.7).

   In p-n diodes with a large power dissipation, a so-called thermal breakdown may lead to a run-away increase in the reverse current and to an S-type negative differential resistance. This mechanism is especially important in devices made from relatively narrow gap materials (such as Ge) where the device temperature may increase appreciably even at relatively low current densities.

   A more detailed discussion of the avalanche breakdown may be found, for example, in the text by Shur (1990).

1-9. METAL-SEMICONDUCTOR CONTACTS

1.9.1. Schottky Barriers
   A metal and a semiconductor at a large distance from each other such that no exchange of carriers can take place will, in general, have different Fermi energies. However, when brought close together, forming what is called a Schottky contact, electrons will be exchanged in an attempt to establish thermal equilibrium. Equilibrium is reached when the Fermi level becomes constant throughout the entire metal-semiconductor system. (This is similar to the situation considered for a p-n junction, see Section 1.8.) In the process, the interface becomes polarized due to the transfer of a net amount of electronic charge across the boundary. The net charge on the metal side resides in a thin surface layer while the opposite charge on the semiconductor side is distributed

between surface or interface states (absent in the case of an ideal Schottky contact) and a space charge region much like that found in the depletion layer of a *p-n* junction.

A very much simplified energy diagram of a metal-semiconductor (*n*-type) contact in thermal equilibrium is shown in Fig. 1.9.1. Here Φ_m and Φ_s are called the metal and the semiconductor work functions. The work function is defined as the energy separation between the vacuum level and the Fermi level. (Note that the vacuum level is subject to the same effect of the interface polarization as the energy bands of the materials constituting the junction.)   The electron affinity of the semiconductor, X_S, corresponds to the energy separation between the vacuum level and the conduction band edge. This quantity is nearly independent of the doping (i.e., position of the Fermi level) in the semiconductor, at least at low doping densities.

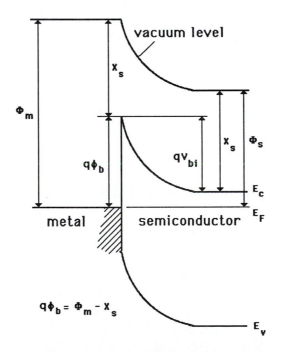

Fig. 1.9.1. Simplified energy diagram of metal-semiconductor barrier; $q\phi_b$ is the barrier height, X_s is the electron affinity in the semiconductor, Φ_s and Φ_m are the semiconductor and the metal work functions, and V_{bi} is the built-in voltage (from Shur (1990)).

In the idealized picture of the Schottky junction shown in Fig. 1.9.1, an energy barrier

$$q\phi_b = \Phi_m - X_s \qquad (1\text{-}9\text{-}1)$$

appears between the metal and the semiconductor. This barrier height is independent of the doping in the semiconductor. In most cases, $\Phi_m > \Phi_s$ and the metal will be charged negatively. The positive net space charge in the semiconductor gives rise to a total band bending

$$qV_{bi} = \Phi_m - \Phi_s \qquad (1\text{-}9\text{-}2)$$

where V_{bi} is called the built-in voltage, in analogy with the corresponding quantity in a p-n junction. Note that qV_{bi} is also identical to the difference between the Fermi levels in the metal and the semiconductor when separated by a large distance (no exchange of charge).

In practice, the Schottky barrier height is quite different from that predicted by eq. (1-9-1). Instead, the barrier height is only weakly dependent on Φ_m (it increases by 0.1 to 0.3 eV when Φ_m increases by 1 to 2 eV). Bardeen (1947) developed a model explaining this difference by including the effects of the surface states at the boundary between the semiconductor and a thin oxide layer which is almost always present at the surface. (The oxide is so thin that electrons can easily tunnel through.) These surface states are continuously distributed in energy within the energy gap of the semiconductor. They are characterized by a "neutral" level, $q\phi_0$, such that the states below $q\phi_0$ are neutral when filled by electrons and states above $q\phi_0$ are neutral when empty, as indicated in Fig. 1.9.2. It can be shown that in this case the barrier height is given by (see Rhoderick (1977))

$$q\phi_b = \gamma_s(\Phi_m - X_s) + (1 - \gamma_s)(E_g - \phi_0) - \gamma_s(\varepsilon_s/\varepsilon_i) F_m \delta_{ox} \qquad (1\text{-}9\text{-}3)$$

where E_g is the energy gap, ε_s and ε_i are the dielectric permittivities of the semiconductor and the interfacial layer, respectively, δ_{ox} is the thickness of the interfacial layer, N_s is the density of the surface states, and

$$\gamma_s = \frac{\varepsilon_i}{\varepsilon_i + qN_s\delta_{ox}} \qquad (1\text{-}9\text{-}4)$$

$$F_m = \sqrt{2qN_d V_{bi}/\varepsilon_s}$$ (1-9-5)

F_m is the maximum electric field inside the space charge region of the semiconductor, N_d is the donor concentration (*n*-type semiconductor) and the built-in voltage can be written as (see Fig. 1.9.2)

$$V_{bi} = \phi_b - (E_c - E_F)/q$$ (1-9-6)

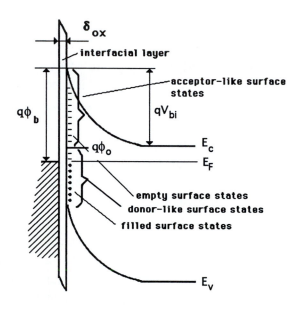

Fig. 1.9.2. Surface states at metal-semiconductor boundary (from Shur (1990)).

The last term on the right-hand side of eq. (1-9-3) is small in most cases. Hence, eq. (1-9-3) reduces to eq. (1-9-1) when the density of the surface states is zero. In the opposite limiting case when $N_s \rightarrow \infty$, we have $\gamma_s \rightarrow 0$ such that

$$q\phi_b = E_g - q\phi_o$$ (1-9-7)

Indeed, in this case, the Fermi level in the semiconductor is "pinned" to the neutral level because any deviation of E_F from this position will result in an infinitely large charge at the interface.

Bardeen's model is in better agreement with experimental data than the simplistic model that assumes that the barrier height is given by eq. (1-9-7). Still, it cannot explain many properties of the Schottky barrier diodes, nor can it explain the mechanism of the surface states formation. Spicer et al. (1979) related the formation of the surface states to defects formed during metal deposition (the so-called Unified Defect Model). Tersoff (1984) proposed that the Schottky barrier heights (as well as band discontinuities at heterointerfaces, see Section 1.10) are controlled by electrons tunneling from one material into the other forming an interfacial dipole. Even though a detailed and accurate understanding of Schottky barrier formation still remains a challenge, many properties of Schottky barriers may be understood independently of the exact mechanism determining the barrier height. In other words, we can simply determine the effective barrier height from experimental data.

Once the Schottky barrier height is known, the variation of the space charge density, ρ, the electric field, F, and the potential, ϕ, in the semiconductor space charge region can be found using the depletion approximation. Consider an n-type semiconductor with a donor concentration N_d and a depletion layer width, x_n. Defining $x = 0$ as the metal-semiconductor interface, we have inside the depletion region $(0 < x < x_n)$:

$$\rho = qN_d \tag{1-9-8}$$

$$F = -\frac{qN_d(x_n - x)}{\varepsilon_s} \tag{1-9-9}$$

$$\phi = -\frac{qN_d(x_n - x)^2}{2\varepsilon_s} = -V_{bi}\left(1 - \frac{x}{x_n}\right)^2 \tag{1-9-10}$$

The depletion layer width with no voltage applied is given by

$$x_n = \sqrt{\frac{2\varepsilon_s V_{bi}}{qN_d}} \tag{1-9-11}$$

The shape of the depletion region under reverse bias and small forward bias may be obtained by substituting V_{bi} by $V_{bi} - V$, where V is the applied voltage. Forward bias corresponds to a positive voltage applied to the metal with respect to the semiconductor.

The thermionic model of the electron transport in Schottky barriers is valid when the interfacial barrier presents an important impediment to the current flow. The situation is then very similar to that in a *p-n* junction (see Section 1.8). Under reverse bias or relatively small forward bias, the electron quasi-Fermi level in the depletion region near the interface remains practically constant with distance. Just like in a *p-n* junction under similar conditions, drift and diffusion current components in the depletion region are both much greater than their sum which is the net current.

In a relatively lightly doped semiconductor with low mobility, the current through the Schottky barrier may be limited more by diffusion and drift processes in the space charge region than by the barrier at the metal-semiconductor interface. Under such conditions, the electron quasi-Fermi level will vary in the depletion region, and the thermionic model will not apply. The quantitative criterion for the validity of the thermionic model is obtained by comparing the mean free path of electrons, λ, with the distance

$$d_T = V_{th}/F_m \qquad (1\text{-}9\text{-}12)$$

over which the potential in the depletion region near the interface decreases by $V_{th} = k_BT/q$ (see Rhoderick (1977)). Here, F_m is the maximum electric field at the metal semiconductor interface given by eq. (1-9-5) and the mean free path is

$$\lambda = \frac{\mu_n}{q}\sqrt{3\,k_B\,T\,m_n} \qquad (1\text{-}9\text{-}13)$$

where μ_n is the low-field electron mobility and m_n is the effective mass. (Eq. (1-9-13) is obtained by substituting the thermal velocity $v_{th} = (3k_BT/m_n)^{1/2}$ and the momentum relaxation time $\tau = \mu_n m_n/q$ into the equation for the mean free path, $\lambda = v_{th}\tau$.) We assume that the thermionic model is valid if

$$\lambda > d_T \qquad (1\text{-}9\text{-}14)$$

In this case, electrons near the metal-semiconductor interface that have sufficient energy to cross over the barrier, have a fair chance to do so before experiencing scattering. Of course, the momentum relaxation time of such high energy electrons may be quite different from the low electric field momentum relaxation time, τ, that enters into the low-field mobility. Therefore eq. (1-9-14) is a fairly crude criterion. Nevertheless, it gives us some idea about conditions for the validity of the thermionic model.

Substituting eqs. (1-9-5), (1-9-12) and (1-9-13) into eq. (1-9-14), we find

$$N_d \mu_n^2 > \frac{\varepsilon_s V_{th}}{6 m_n V_{bi}} \tag{1-9-15}$$

or

$$N_d \mu_n^2 \left(\frac{1}{V^2 \text{cm s}^2} \right) > 4.19 \times 10^{18} \left(\frac{T}{300} \right) \left(\frac{\varepsilon_s}{\varepsilon_o} \right) \left(\frac{m_e}{m_n} \right) \frac{1}{V_{bi}} \tag{1-9-16}$$

For Si and GaAs at room temperature, the inequality eq. (1-9-16) is fulfilled for N_d greater than approximately $10^{14}\,\text{cm}^{-3}$.

In the case when $\lambda \ll d_T$, the position of the quasi-Fermi level in the depletion region becomes dependent on distance, and the thermionic model is no longer applicable. This situation corresponds to the so-called diffusion model (see Rhoderick (1977)). It occurs, for example, in Schottky barriers based on amorphous silicon where the electron mobility is very small (mobility $\mu \le 1\,\text{cm}^2/\text{Vs}$).

When the thermionic model is valid, the current-voltage characteristics of the Schottky diode can be found by calculating the electronic fluxes in and out of the semiconductor at the metal-semiconductor interface. The electronic current flux, $J_{s \to m}$, out of the semiconductor may be taken to be proportional to the density, n_s, of electrons in the semiconductor bulk with sufficient energy to cross the barrier. With an applied voltage, V, this flux can be written as

$$J_{s \to m} \propto n_s = N_d \exp\left(-\frac{V_{bi} - V}{V_{th}} \right) = N_c \exp\left(-\frac{\phi_b - V}{V_{th}} \right) \tag{1-9-17}$$

Here, N_c is the effective density of states in the conduction band defined in Section 1.3 (see eq. (1-3-7)). The flux in the opposite direction, $J_{m \to s}$, – from the metal into the semiconductor – is independent of the applied voltage (assuming that the barrier height is independent of the applied voltage). The total flux, J, must be equal to zero when $V = 0$. Hence,

$$J_{m \to s} = J_{s \to m} \qquad (V = 0) \tag{1-9-18}$$

and we obtain the following expression for the electric current density

$$J = J_{ss}\left[\exp\left(\frac{V}{V_{th}}\right) - 1\right]$$

(1-9-19)

where

$$J_{ss} = A^*T^2\exp\left(-\frac{\phi_b}{V_{th}}\right)$$

(1-9-20)

Here, the Richardson constant

$$A^* = \alpha\frac{m_n q k_B^2}{2\pi^2 \hbar^3} \approx 120\alpha\frac{m_n}{m_e}\left(\frac{A}{cm^2 K^2}\right)$$

(1-9-21)

includes an empirical factor α that accounts for deviations from the simple theory. According to Crowell and Sze (1966), $\alpha \approx 0.5$. This equation is valid for parabolic bands and spherical constant energy surfaces (which is approximately true for the conduction band in GaAs where $m_n \approx 0.067\, m_e$). For ellipsoidal surfaces of equal energy (as for the conduction band of silicon), A^* depends on the direction of the current flow. According to Crowell (1965), m_n in eq. (1-9-21) should then be replaced by

$$m^* = l_\theta^2 m_y m_z + m_\theta^2 m_z m_x + n_\theta^2 m_x m_y$$

(1-9-22)

where m_x, m_y, and m_z are components of the effective mass tensor for the directions coinciding with the principal axes of the ellipsoidal equal energy surfaces and l_θ, m_θ, and n_θ are cosines of angles formed by the normal to the metal-semiconductor interface and the three principal axes of the ellipsoid. This should be done for each valley (i.e., each equivalent minimum of the conduction band) and the resulting Richardson constants should be added up. In silicon, there are six equivalent valleys and such a calculation yields (see Rhoderick (1977))

$$m^* = 2m_t + 4\sqrt{m_l m_t} \approx 2.05\, m_e \qquad \text{for <100> directions}$$

(1-9-23)

$$m^* = 6\sqrt{\frac{1}{3}m_t^2 + \frac{2}{3}m_l m_t} \approx 2.15\, m_e \qquad \text{for <111> directions}$$

(1-9-24)

For germanium, we find

$$m^* = 4\sqrt{\frac{1}{3}m_t^2 + \frac{2}{3}m_l m_t} \approx 1.19 m_e \qquad \text{for <100> directions} \qquad (1\text{-}9\text{-}25)$$

$$m^* = m_t + \sqrt{m_t^2 + \frac{8}{3}m_l m_t} \approx 1.07\, m_e \quad \text{for <111> directions} \qquad (1\text{-}9\text{-}26)$$

Here, m_t and m_l are the transverse and longitudinal effective masses respectively and m_e is the free electron mass. For {111} surfaces of Si and GaAs, A^* is equal to 96 A/(cm^{2}K^2) and 4.4 A/(cm^{2}K^2), respectively.

In practical devices, the current-voltage characteristic is more accurately described by the following equation

$$J = J_{ss}\left[\exp\left(\frac{V}{\eta V_{th}}\right) - 1\right] \tag{1-9-27}$$

The parameter η in eq. (1-9-27) is called the ideality factor. A review of different mechanisms, including the dependence of the barrier height on bias voltage leading to this dependence, was given by Rhoderick (1977).

It is instructive to compare the value of the saturation current density for the Schottky barrier, J_{ss}, with the saturation current density, J_s, for a p^+-n junction:

$$\frac{J_{ss}}{J_s} \approx \frac{\alpha}{\sqrt{2\pi}} \frac{N_d}{N_c} \frac{m_n}{m_p} \sqrt{\frac{\tau_p}{\tau_h}} \exp\left(\frac{E_g - q\phi_b}{k_B T}\right) \tag{1-9-28}$$

where τ_p is the hole life time in the n-type region, τ_h is the hole momentum relaxation time, E_g is the energy gap in the p-n junction, $q\phi_b$ is the Schottky barrier height, N_d is the donor density, and N_c is the effective density of states in the conduction band (see Section 1.3). Typically, $q\phi_b$ is of the order of $0.6E_g$. The ratio τ_p/τ_h can be of the order of 10^4 in GaAs and even higher in Si. Hence, the saturation current in a Schottky diode is larger than that in a p^+-n junction by many orders of magnitude, and the turn-on voltage is smaller by several tenths of a Volt.

In Schottky barriers based on highly doped semiconductors, the depletion region becomes so narrow that electrons can tunnel through the barrier near the top where the barrier is thin (see Fig. 1.9.3).

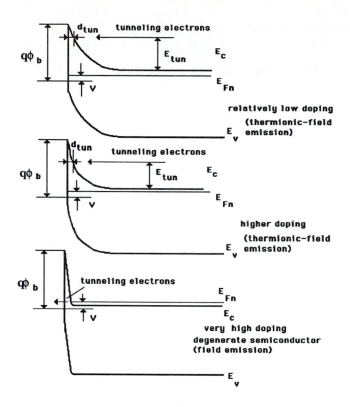

Fig. 1.9.3. Thermionic-field and field emission under forward bias; d_{tun} is the characteristic tunneling length. At low doping levels, electrons tunnel across the barrier closer to the top of the barrier. With increasing doping, the characteristic tunneling energy, E_{tun}, decreases. In highly doped degenerate semiconductors, electrons near the Fermi level tunnel across a very thin depletion region.

This process is called thermionic-field emission. The number of electrons with a given energy E decreases exponentially with energy as $\exp[-E/(k_BT)]$. On the other hand, the barrier transparency increases exponentially with the decrease in the barrier width. Hence, the dominant electron tunneling path occurs at lower energies as the doping increases and the barrier becomes thinner. In degenerate semiconductors, especially in semiconductors with a small electron effective mass such as GaAs, electrons can tunnel through the barrier near the Fermi level and the tunneling current is dominant. This mechanism is called field emission. The current-voltage characteristic of a Schottky diode in the case of thermionic-field emission or field emission can be calculated by evaluating the product of the tunneling transmission coefficient and the number of electrons as a function of energy and integrating over the states in the conduction band. Such a calculation yields (see Padovani and Stratton (1966)):

$$J = J_{stf} \exp\left(\frac{qV}{E_o}\right) \tag{1-9-29}$$

where

$$E_o = E_{oo} \coth\left(\frac{E_{oo}}{k_B T}\right) \tag{1-9-30}$$

and

$$E_{oo} = \frac{qh}{4\pi}\sqrt{\frac{N_d}{m_n \varepsilon_s}} = 1.85 \times 10^{-11} \left[\frac{N_d(\text{cm}^{-3})}{(m_n/m_o)(\varepsilon_s/\varepsilon_o)}\right]^{1/2} \quad (\text{eV}) \tag{1-9-31}$$

The pre-exponential term, J_{tsf}, was calculated by Crowell and Rideout (1969):

$$J_{stf} = \frac{A^* T \sqrt{\pi E_{oo} q(\phi_b - V - \Delta V_{Fn})}}{k_B \cosh\left(E_{oo}/k_B T\right)} \exp\left[-\frac{\Delta V_{Fn}}{V_{th}} - \frac{q(\phi_b - \Delta V_{Fn})}{E_o}\right] \tag{1-9-32}$$

Here, $\Delta V_{Fn} = (E_c - E_{Fn})/q$. Note that ΔV_{Fn} is negative for a degenerate semiconductor. In GaAs, the thermionic-field emission occurs roughly for $N_d > 10^{17}$ cm^{-3} at 300 K and for $N_d > 10^{16}$ cm^{-3} at 77 K. In silicon, the corresponding values of N_d are several times larger.

The field emission takes place at very high doping levels where the width of the depletion region becomes so small that direct tunneling from the semiconductor to the metal may take place as shown in Fig. 1.9.3. This happens when E_{oo} becomes much greater than $k_B T$. The current-voltage characteristics in this regime are given by

$$J = J_{sf} \exp\left(\frac{qV}{E_{oo}}\right) \tag{1-9-33}$$

where

$$J_{sf} = J_{sto} \exp\left(-\frac{q\phi_b}{E_{oo}}\right) \qquad (1\text{-}9\text{-}34)$$

and J_{sto} depends on temperature and applied voltage (see Padovani and Stratton (1966) for details).

The effective resistance of the Schottky barrier in the field emission regime is quite low. Therefore metal - n^+ contacts are used as ohmic contacts.

The small signal equivalent circuit of a Schottky barrier includes the parallel combination of the differential resistance of the Schottky barrier

$$R_d = \frac{dV}{dI} \qquad (1\text{-}9\text{-}35)$$

and the differential capacitance of the space charge region (which may be estimated using the depletion approximation, just like we did for a p^+-n junction, see Section 1.8):

$$C_d = S \sqrt{\frac{qN_d \varepsilon_s}{2(V_{bi} - V)}} \qquad (1\text{-}9\text{-}36)$$

The parallel circuit comprised of these elements comes in addition to the series resistance, R_s, which includes the contact resistance and the resistance of the neutral semiconductor region between the ohmic contact and the depletion region (and, at very high frequencies, also an equivalent inductance). The device geometric capacitance, C_{geom}, is given by

$$C_{geom} = \varepsilon_s \, S \,/L \qquad (1\text{-}9\text{-}37)$$

where L is the device length.

The major difference between the equivalent circuit of a Schottky diode and that of a p-n junction is the absence of the diffusion capacitance in the case of the Schottky diode. This leads to a much faster response under forward bias conditions and allows us to use Schottky diodes as microwave mixers, detectors, etc. (see, for example, Maas (1986)).

In most cases, $R_s \ll R_d$ and $C_d \gg C_{geom}$ so that the characteristic time constant limiting the frequency response of a Schottky diode is given by

$$\tau_{Schottky} = R_s C_d \tag{1-9-38}$$

A good review of early work on Schottky contacts was given by Rhoderick (1977). A more recent bibliography was given by Sharma and Gupta (1980). Gelmont et al. (1991) calculated capacitances of small Schottky diodes of different geometries.

1.9.2. Ohmic Contacts

Current–voltage characteristics of a Schottky barrier diode and of an ohmic contact are compared in Fig. 1.9.4.

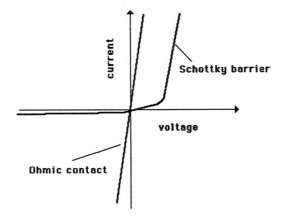

Fig. 1.9.4. Current–voltage characteristics of metal-semiconductor contacts (from Shur (1990)).

An ohmic contact has a linear current-voltage characteristic and a very small resistance that is negligible compared to the resistance of the active region of the semiconductor device. Ideally, the ohmic contact to an *n*-type semiconductor should be made using a metal with a lower work function than that of the semiconductor. Unfortunately, very few practical material systems satisfy this condition and, usually, metals form Schottky barriers at semiconductor interfaces. Therefore, a practical way to obtain a low resistance ohmic contact is to increase the doping near the metal-semiconductor interface to a very high value so that the depletion layer caused by the Schottky barrier becomes very thin and the current transport through the barrier is enhanced by tunneling (field emission regime).

Band diagrams of a metal-n^+-n ohmic contact and a Schottky contact are compared in Fig. 1.9.5. This comparison clearly illustrates the role played by the n^+ layer.

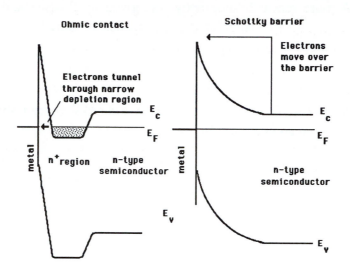

Fig. 1.9.5. Band diagrams of metal-semiconductor contacts (from Shur (1990)).

As was discussed above, in the case when the tunneling mechanism is dominant, the current density, J, is given by eqs. (1-9-33) and (1-9-34). From these equations, we find that the specific contact resistance, r_c, defined as

$$ r_c = \frac{dV}{dJ}\bigg|_{V \to 0} \tag{1-9-39} $$

varies according to

$$ r_c \propto \exp\left(\frac{4\pi\phi_b}{h} \sqrt{\frac{\varepsilon_s\, m_n}{N_d}} \right) \tag{1-9-40} $$

At relatively low doping levels, when thermionic-field emission and thermionic emission play a dominant role, the contact resistance should be much less dependent on the doping concentration. This conclusion is in good agreement with experimental data.

A conventional approach to reducing the specific contact resistance is to form a very highly doped region near the surface by using alloyed ohmic contacts. Ion implantation or diffusion have also been used to create a highly doped region near the surface to facilitate the formation of ohmic contacts. The doping concentration is limited by the impurity solubility and may exceed 10^{20} cm^{-3} for n-type GaAs and 10^{21} cm^{-3} for p-type GaAs (see figure 22 in Sze (1981), p. 33). Using ion implantation, doping concentrations in access of the solubility limit may be reached near the surface (Robinson (1983)). Good alloyed ohmic contacts to n-type GaAs have a low contact resistance (less than 10^{-6} Ωcm^2) and high reliability. A thorough review of metal-semiconductor contacts to III-V compounds was published by Robinson (1983) (see also Ghandhi (1983)).

1-10. HETEROJUNCTIONS

1.10.1. Basic Concepts

Conventional p-n diodes utilize doping profiles in order to control electron and hole currents. The charges of ionized donors and acceptors in the depletion region near the boundary between the n-type and the p-type semiconductors create an electric field which results in a potential barrier for electrons and holes. The potential energy for electrons is given by the bottom of the conduction band and the potential energy for holes is given by the top of the valence band. In the presence of the electric field, the bands are tilted but remain parallel to each other as illustrated to the left in Fig. 1.10.1.

Changing the material composition as function of distance provides new opportunities for governing the behavior of electrons and holes as was clearly explained by Kroemer (1982). Following Kroemer's argument, we compare in Fig. 1.10.1 qualitative band diagrams of semiconductors with uniform and graded composition and, hence, with uniform and graded energy gaps, respectively. In a graded semiconductor, the built-in field pushes electrons and holes in the same direction.

If the change in material composition is not gradual but abrupt, the change in energy gap is also abrupt as illustrated in Fig. 1.10.2. The band diagram shown in this figure corresponds to a layer of a narrow gap semiconductor sandwiched between two layers of a wide band gap semiconductor.

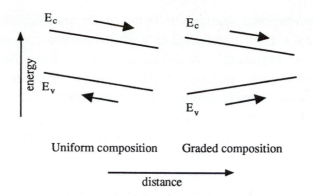

Uniform composition   Graded composition

distance

Fig. 1.10.1. Qualitative energy band diagrams for a uniform semi-conductor and for a semiconductor with graded composition.

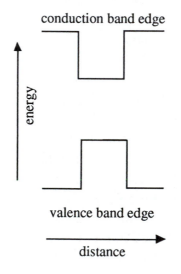

distance

Fig. 1.10.2. Band diagram for a layer of narrow gap semiconductor sandwiched between two layers of a wide band gap semiconductor.

If the narrow energy gap layer is fairly thin, it forms a quantum potential well for charge carriers. The abrupt changes in the conduction and valence bands are created by chemical forces produced by the contact between the two dissimilar materials forming the heterostructure. Ionized donors or acceptors can never produce such abrupt variations. (The band diagram shown in Fig. 1.10.2 may be the band diagram of a double heterostructure quantum well laser at the flat band condition.   The high quantum efficiency in these devices is related to the localization of electrons and holes in the quantum well active region.)

The two examples depicted in Fig. 1.10.1 and 1.10.2 clearly show that heterostructure systems open up new and intriguing possibilities for novel semiconductor devices. Such devices include diodes, transistors, solar cells, light emitting diodes, lasers, and other optoelectronic devices. In this book, we will only consider the heterojunction diode, the heterojunction bipolar transistor, and the heterostructure field effect transistor. (The models for these devices have been implemented in AIM-Spice.)

When two different semiconductors are joined together, the atoms at the heterointerface have to form chemical bonds. If the lattice constants of the two materials are different, atoms at the heterointerface have to adjust by developing strain. If this strain exceeds some critical value, it results in dislocations which are crystal imperfections propagating across many crystalline layers. These dislocations act as scattering and recombination centers for electrons and holes, limiting carrier mobilities and lifetimes. The result may be very poor device properties. One possible way to avoid this problem is to use materials with nearly equal lattice constants such as GaAs and AlAs (or the ternary compound $Al_xGa_{1-x}As$). Another approach (used, for example, in growing Si-Ge hetero-structures) is to use thin alternating layers of the two semiconductor materials producing a structure that is called a superlattice. This distributes the strain more evenly between the two materials, inhibiting the formation of dislocations. Still another technique is to choose the substrate surface to be slightly offset from a major crystal plane so that the distance between the atoms on the surface approximates the distance between the atoms in the deposited film. This may also lead to a deflection of dislocations, confining them primarily to a region near the heterointerface. Such an approach is used for growing GaAs films on silicon substrates (see, for example, Fisher et al. (1986)).

Another consequence of the lattice mismatch in heterostructures is the appearance of surface states similar to those at the interface in metal-semiconductor contacts (see Section 1.9). These surface states also act as scattering and recombination centers for charge carriers, limiting mobilities and lifetimes. In short, heterojunction interfaces may be far from ideal, especially when the lattice mismatch is noticeable. Nevertheless, a model for the ideal heterojunction, first developed by Anderson (1962), provides some useful insight into the device behavior.

According to this model, the energy bands in each of the materials comprising a heterostructure are not affected. Hence, the problem reduces to a proper alignment of the band edges at the heterointerface. Anderson assumed that the vacuum energy level is continuous and that the conduction band discontinuity is determined by the difference of the electron affinities, X_1 and X_2, in the two regions:

$$\Delta E_c = X_1 - X_2 \tag{1-10-1}$$

The valence band discontinuity is then given by

$$\Delta E_v = \Delta E_g - \Delta E_c = \Delta E_g - X_1 + X_2 \tag{1-10-2}$$

where ΔE_g is the energy gap discontinuity. However, this model is in disagreement with experimental data (see, for example, Bauer and Margaritonto (1987)) and different alternative models have been proposed (see, for example, the discussion by Kroemer (1983) and Terzoff (1984)).

Terzoff (1984) postulated that the band discontinuities in the heterojunctions are controlled by the same mechanism that controls the barrier height in real Schottky diodes. In addition to the exchange of electrons that takes place between the two materials in the ideal case, this mechanism also accounts for the carrier exchange with interface states (see Section 1.9). According to this model, the conduction band discontinuity can be related to the difference in the Schottky barrier heights in vacuum, $q\phi_{b1}$ and $q\phi_{b2}$, for the two materials considered separately, i.e.,

$$\Delta E_c = q\left(\phi_{b1} - \phi_{b2}\right) \tag{1-10-3}$$

Eizenberg et al. (1986) tested this model experimentally by measuring the Schottky barrier heights and band discontinuities for $Al_xGa_{1-x}As$ (for different values of x) and GaAs using the internal photoemission technique and found a good correlation.

In most models, however, the only change from the ideal Anderson model is a different choice of $\Delta E_c / \Delta E_g$. In principle, this parameter may be determined from experiments. However, different experiments yield somewhat different values even for the most widely studied heterojunction system, $GaAs/Al_xGa_{1-x}As$.

When two semiconductors form a heterojunction, the Fermi level must be constant throughout the system in equilibrium. Just as for a conventional *p-n* junction, this requirement leads to a band-bending as shown in Fig. 1.10.3.

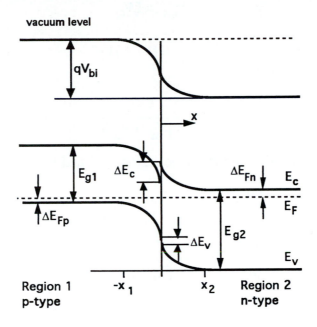

Fig. 1.10.3. Energy band diagram of a *p-n* heterojunction (lower part) shown together with the variation in the vacuum level (upper part).

If the conduction band discontinuity, ΔE_c, is known, the built-in voltage, V_{bi}, may be found from this figure:

$$qV_{bi} = E_{g1} - \Delta E_{Fn} - \Delta E_{Fp} + \Delta E_c \qquad (1\text{-}10\text{-}4)$$

Here, E_{g1} is the energy gap of the narrower gap material (region 1), which we assume to be doped *p*-type, ΔE_{Fp} is the difference between the Fermi level and the top of the valence band in this material, and ΔE_{Fn} is the difference between the bottom of the conduction band in the wide gap *n*-type material (region 2) and the Fermi level far from the heterointerface.

1.10.2. Depletion Capacitance

Using the depletion approximation the same way as for the conventional *p-n* junction (see Section 1.8), we find that the built-in potential is divided between the p-region and the n-region as follows:

$$V_{bi1} = \xi V_{bi} \tag{1-10-5}$$

$$V_{bi2} = (1 - \xi)V_{bi} \tag{1-10-6}$$

where

$$\xi = \frac{\varepsilon_1 N_d}{\varepsilon_1 N_d + \varepsilon_2 N_a} \tag{1-10-7}$$

The depletion widths in the p-type region, x_{d1}, and in the n-type region, x_{d2}, are given by

$$x_{d1} = \sqrt{\frac{2\varepsilon_1 \xi V_{bi}}{qN_a}} \tag{1-10-8}$$

$$x_{d2} = \sqrt{\frac{2\varepsilon_2(1 - \xi)V_{bi}}{qN_d}} \tag{1-10-9}$$

Hence, the depletion capacitance per unit area becomes

$$c_d = \sqrt{\frac{q\varepsilon_1\varepsilon_2 N_a N_d}{2V_{bi}(\varepsilon_1 N_a + \varepsilon_2 N_d)}} \tag{1-10-10}$$

When a reverse or a small forward bias, V, is applied to the heterojunction, we have to substitute V_{bi} by $V_{bi} - V$ in the above equations. (A forward bias corresponds to a positive voltage applied to the p-type region relative to the n-type region.)

1.10.3. Current-Voltage Characteristic

For an abrupt heterojunction with a spike ΔE_c in the conduction band, as indicated in Fig. 1.10.3, the transport of electrons through the depletion region will be impeded by thermionic emission across the spike. Typically, this effect will be important only for $\Delta E_c \geq k_B T$. For $\Delta E_c \leq k_B T$, as in a conventional p-n junction, the transport in the depletion region itself is fully accounted for in terms of the built-in voltage (neglecting generation and recombination processes). Hence, for a heterojunction as shown in Fig. 1.10.3, the electron current density across the heterointerface can be described in terms of the electron fluxes in the two opposite directions (see Grinberg et al. (1984)), i.e.,

$$J_n = J_{1 \to 2} - J_{2 \to 1} \qquad (1\text{-}10\text{-}11)$$

where

$$J_{2 \to 1} = q v_n \, n(x = 0^+) \qquad (1\text{-}10\text{-}12)$$

is the electron current density corresponding to the electron flux from the n-type material (region 2) to the p-type material (region 1) across the interface (which is assumed to be located at $x = 0$). Here, $n(x = 0^+)$ is the electron concentration at the interface in region 2. The corresponding electron concentration in region 1 is denoted $n(x = 0^-)$, and the electron current density corresponding to the electron flux from region 1 to region 2 can written as

$$J_{1 \to 2} = - q v_n \, n(x = 0^-) \exp\left(- \frac{\Delta E_c}{q V_{th}} \right) \qquad (1\text{-}10\text{-}13)$$

where $V_{th} = k_B T / q$ is the thermal voltage, $v_n = [q V_{th}/(2 \pi m_n)]^{1/2}$ is the mean electron thermal velocity in the negative x-direction (from region 2 towards region 1) , and m_n is the lighter of the electron effective masses of the two regions (see Grinberg (1986)). The interface electron concentrations are related to those at the boundaries of the depletion regions of the p-type material $(x = -x_1)$ and the n-type material $(x = x_2)$ as follows:

$$n(x = 0^-) = n(- x_1) \exp\left(\frac{V_{d1}}{V_{th}} \right) \qquad (1\text{-}10\text{-}14)$$

$$n(x = 0^+) = n(x_2) \exp\left(- \frac{V_{d2}}{V_{th}} \right) = N_d \exp\left(- \frac{V_{d2}}{V_{th}} \right) \qquad (1\text{-}10\text{-}15)$$

where, according to the results derived above, the potential drops across the depletion regions on the p-type and the n-type side are

$$V_{d1} = \xi(V_{bi} - V) \qquad (1\text{-}10\text{-}16)$$

$$V_{d2} = (1 - \xi)(V_{bi} - V) \qquad (1\text{-}10\text{-}17)$$

respectively, when a voltage V is the applied across the junction.

The electron current density given by eq. (1-10-11) is equal to the electron current density due to diffusion, $J_n(-x_1)$, at the boundary of the space charge region in the p-type material (neglecting interface recombination). Following the derivation for a conventional p-n junction (see Section 1.8), we find for an infinitely long p-type region,

$$J_n(-x_1) = \frac{qD_n}{L_n}\left[n(-x_1) - n_{po}\right] \qquad (1\text{-}10\text{-}18)$$

Here, n_{po}, D_n, and L_n are the equilibrium concentration, the diffusion coefficient and the diffusion length, respectively, of the electrons in region 1. The excess electron concentration at the boundary of the p-type depletion region, $n(-x_1) - n_{po}$, can now be determined by equating eqs. (1-10-11) and (1-10-18), i.e.,

$$n(-x_1) - n_{po} = \frac{n_{po}}{\zeta_n}\left[\exp\left(\frac{V}{V_{th}}\right) - 1\right] \qquad (1\text{-}10\text{-}19)$$

where

$$\zeta_n = 1 + \frac{D_n}{v_n L_n}\exp\left(\frac{\Delta E_n}{qV_{th}}\right) \qquad (1\text{-}10\text{-}20)$$

$$\Delta E_n = \Delta E_c - qV_{d1} \qquad (1\text{-}10\text{-}21)$$

In a typical diode, the p-type region is very highly doped. Hence, V_{d1} is small so that $\Delta E_n > 0$, in which case ΔE_n represents an additional barrier for the electrons entering region 1 from region 2. When $\Delta E_n < 0$, the conduction band discontinuity does not create an additional barrier because the conduction band spike at the heterointerface is below the bottom of the conduction band in the neutral section of the p-type region.

Hence, the electron current density across the interface can be written as

$$J_n(-x_1) = \frac{qD_n\,n_{po}}{L_n\zeta_n}\left[\exp\left(\frac{V}{V_{th}}\right) - 1\right]$$ (1-10-22)

In this expression, the effect of the conduction band spike is contained in the factor ζ_n. Usually, as for the p-GaAs/n-Al$_x$Ga$_{1-x}$As diode, the second term in ζ_n (see eq. (1-10-20)) will be dominant. A similar derivation can be made for the hole current for material combinations which give a spike in the valence band. However, for the p-GaAs/n-Al$_x$Ga$_{1-x}$As system, there is no such spike and the hole current can be calculated as for a conventional p-n junction (see Section 1.8). In this case, the effect of the band gap discontinuity will enter only through the equilibrium minority carrier concentration p_{no} in the expression for the hole component of the current density:

$$J_p(x_2) = \frac{qD_p\,p_{no}}{L_p\zeta_p}\left[\exp\left(\frac{V}{V_{th}}\right) - 1\right]$$ (1-10-23)

Here, D_p and L_p are the diffusion coefficient and diffusion length, respectively, of the electrons in region 2 and, for the p-GaAs/n-Al$_x$Ga$_{1-x}$As diode,

$$\zeta_p = 1$$ (1-10-24)

(Of course, if a certain heterojunction has a spike in the valence band at the heterointerface, the expression for ζ_p will be given by an equation similar to eq. (1-10-20).) From eqs. (1-10-22) and (1-10-23), we obtain the following total current density in a long heterojunction diode (neglecting generation and recombination in the depletion region):

$$J = \left(\frac{qD_n\,n_{po}}{L_n\zeta_n} + \frac{qD_p\,p_{no}}{L_p\zeta_p}\right)\left[\exp\left(\frac{V}{V_{th}}\right) - 1\right]$$ (1-10-25)

In practical diodes, the length X_p of the p-type region and the length X_n of the n-type region are comparable to or smaller than the diffusion lengths of the minority carriers. In this case, eq. (1-10-25) is still valid if the factors ζ_n and ζ_p are changed to:

$$\zeta_n = \tanh\left(\frac{X_p - x_1}{L_n}\right) + \frac{D_n}{v_n L_n}\exp\left(\frac{\Delta E_n}{qV_{th}}\right)$$ (1-10-26)

$$\zeta_p = \tanh\left(\frac{X_n - x_2}{L_p}\right) \qquad (1\text{-}10\text{-}27)$$

For very short diodes, the hyperbolic tangents can be expanded to first order in $(X_p - x_1)/L_n$ and $(X_n - x_2)/L_p$. Furthermore, a practical hetero-structure diode construction would be of the type p^+-n or n^+-p. In the case of a p^+-n diode, we will normally have $V_{d1} \ll \Delta E_c$, $\xi \ll 1$ and $\Delta E_n \gg V_{th}$. Hence, the current density of such a diode can be written as

$$J \approx \left[qn_{po}\, v_n \exp\left(-\frac{\Delta E_n}{qV_{th}}\right) + \frac{qD_p\, p_{no}}{X_n - x_2} \right]\left[\exp\left(\frac{V}{V_{th}}\right) - 1 \right] \qquad (1\text{-}10\text{-}28)$$

Fig 1.10.4 shows the AIM-Spice characteristic of a typical heterostructure diode according to the present theory.

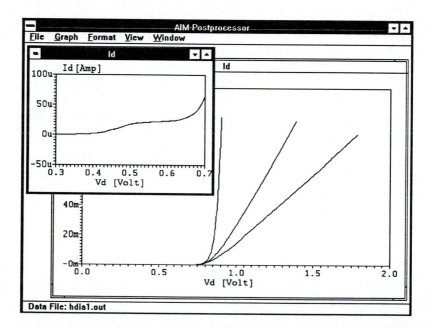

Fig. 1.10.4. AIM-Spice current-voltage relationship of an ideal hetero-structure p^+-n junction. The plateau between 0.5 and 0.6 V indicated in the small window is caused by the presence of the spike in the conduction band edge (see Fig. 1.10.3).

In a real heterostructure diode, the current level is usually much higher than that predicted above. The reason may be that the effective additional barrier, ΔE_n, for the electron transport is less than that expected for an ideal junction. Such a barrier lowering can result from an intentional or unintentional grading of the material composition near the interface and/or interdiffusion of dopants. The effect may be further compounded by some degree of inhomogeneity over the cross-section, which will cause the current to flow where the barrier is lowest. In addition, with strongly doped diodes, tunneling through the barrier will also be a significant factor. (Such tunneling may be strongly enhanced by the presence of impurities in the barrier.) Consequently, ΔE_n can be considered as an effective additional barrier which should be adjusted to give the correct current level. Finally, the experimental *I-V* characteristics are often derived under bias conditions which may cause a significant temperature increase in the samples. Clearly, raising the temperature to, for example, 100 °C, strongly enhances the diode current at a given bias condition. (Such an increase is quite typical for normal operation of Heterojunction Bipolar Transistors, see Section 2.7.)

1.10.4. Diffusion Capacitance

As for the conventional *p-n* junction, the diffusion capacitance in a heterostructure diode may become important or even dominant under forward bias conditions. We determine this capacitance from the small signal device impedance using the continuity equations for the minority carriers in the two regions. Following the derivation in Section 1.8, we apply a small signal time dependent voltage, $v^* \exp(i\omega t)$, in addition to a large stationary voltage, V_o, and obtain a small signal current density with the amplitude:

$$j^* = \frac{qD_n n^*}{\zeta_{no} L_n}\sqrt{1 + i\omega\tau_n} \; + \; \frac{qD_p p^*}{\zeta_{po} L_p}\sqrt{1 + i\omega\tau_p} \qquad (1\text{-}10\text{-}29)$$

where the additional "*o*" in parameter subscripts denotes stationary values (corresponding to $V = V_o$), τ_p and τ_n are the minority carrier lifetimes, and the amplitudes of the variation in the minority carrier densities, n^* at $x = x_2$ and p^* at $x = -x_1$, are given by

$$n^* \approx \frac{n_{po}\, v^*\, v_n\, L_n}{V_{th}\, D_n} \exp\left(-\frac{\Delta E_{no}}{qV_{th}}\right)\exp\left(\frac{V_o}{V_{th}}\right) \qquad (1\text{-}10\text{-}30)$$

$$p^* \approx \frac{p_{no} v^* L_p}{V_{th}(X_n - x_2)} \exp\left(\frac{V_o}{V_{th}}\right) \qquad (1\text{-}10\text{-}31)$$

Here, we have again considered a short heterojunction p^+-n diode and limited the discussion to forward bias.

The small signal admittance, y, per unit area is defined as

$$y = j^*/v^* \qquad (1\text{-}10\text{-}32)$$

At relatively low frequencies, when $\omega \tau_p \ll 1$ and $\omega \tau_n \ll 1$, we obtain the following expression for the diffusion capacitance per unit area:

$$
\begin{aligned}
c_{dif} &\approx \frac{q D_p \tau_p\, p^*}{2 \zeta_{po} L_p v^*} + \frac{q D_n \tau_n\, n^*}{2 \zeta_{no} L_n v^*} \\
&\approx \left[\frac{q p_{no} L_p^3}{2 V_{th}\left(X_n - x_1\right)^2} + \frac{q n_{po} v_n^2 \tau_n^2}{2 V_{th} L_n} \exp\left(-\frac{2 \Delta E_n}{q V_{th}}\right) \right] \exp\left(\frac{V_o}{V_{th}}\right) \qquad (1\text{-}10\text{-}33)
\end{aligned}
$$

Here, we have used $D_p = L_p^2/\tau_p$ and $D_n = L_n^2/\tau_n$.

We should emphasize that the diffusion capacitance deviates from the dependence given by eq. (1-10-33) at high bias voltages when the effective diffusion capacitance should be used as was discussed in detail in Section 1.8.

REFERENCES

R. A. ANDERSON, *Solid State Electronics*, 5, pp. 341-351 (1962)

N. D. ARORA, J. R. HAUSER and D. J. RULSON, *IEEE Trans. Electron Devices*, ED-29, p. 292 (1982)

J. BARDEEN, *Phys. Rev.*, 71, p. 717 (1947)

R. S. BAUER and G. MARGARITONTO, *Physics Today*, p. 3, Jan. (1987)

S. BLAKEMORE, "Semiconductor and Other Major Properties of GaAs", *J. Appl. Phys.*, 53, No. 10, pp. R123-R181, Oct. (1982)

K. BLØTEKJÆR, "Transport Equations for Two-Valley Semiconductors", *IEEE Trans. Electron Devices*, ED-17, No. 1, pp. 38-47, Jan. (1970)

T. BRUDEVOLL, T. A. FJELDLY, J. BAEK, M. SHUR, "Scattering Rates for Holes near the Valence-Band Edge in Semiconductors", *J. Appl. Phys.*, 67, No. 12, pp. 7373-7382 (1990)

A. CAPPY, B. CARNEZ, R. FAUQUEMBERGUE, G. SALMER and E. CONSTANT, "Comparative Potential Performance of Si, GaAs, GaInAs, InAs Submicrometer-Gate FET's", *IEEE Trans. on Electron Devices*, ED-27, No. 11, p. 2158, Nov. (1980)

B. CARNEZ, A. CAPPY, A. KASZINSKI, E. CONSTANT and G. SALMER, "Modeling of Submicron Gate Field-Effect Transistor Including Effects of Non-Stationary Electron Dynamics", *J. Appl. Phys.*, 51, No. 1 (1980)

D. M. CAUGLEY and R. E. THOMAS, *Proc. IEEE*, 55, pp. 2192-2193 (1967)

E. CONSTANT, "Non-Steady-State Carrier Transport in Semiconductors in Perspective with Submicrometer Devices", in *Hot-Electron Transport in Semiconductors*, ed. L. Reggiani, Springer-Verlag, Berlin Heidelberg (1985), p. 227

C. R. CROWELL, *Solid State Electronics*, 8, p. 395 (1965)

C. R. CROWELL and V. L. RIDEOUT, *Solid State Electronics*, 12, p. 89 (1969)

C. R. CROWELL and S. M. SZE, *Solid State Electronics*, 9, p. 695 (1966)

J. DZIEWIOR and W. SCHMID, *Appl. Phys. Lett.*, 31, p. 346 (1982)

M. EIZENBERG, M. HEIBLUM, M. I. NATHAN, N. BRASLAU, and P. M. MOONEY, "Barrier- Heights and Electrical Properties of Intimate Metal-AlGaAs Junctions", *J. Appl. Phys.*, 61, pp. 1516-1522 (1987)

T. A. FJELDLY and L. JOHNSEN, "Self-Consistent Theory Including Non-Stationary Phenomena for Carrier Transport in the AlGaAs/GaAs High Electron Mobility Transistor", *J. Appl. Phys.*, 31, p. 1768 (1988)

T. A. FJELDLY, B. MOON and M. SHUR, "Analytical Solution of Generalized Diode Equation", *IEEE Trans. Electron Devices*, ED-38, No. 8, pp. 1976-1977, August (1991)

R. J. FISHER, W. P. KOPP, J. S. GEDYMIN and H. MORKOÇ, "Properties of MODFET's Grown on Si Substrates at DC and Microwave Frequencies", *IEEE Trans. Electron Devices*, ED-33, No. 10, pp. 1407-1412, October (1986)

B. GELMONT, M. SHUR and R. J. MATTAUCH, "Capacitance-Voltage Characteristics of Microwave Schottky Diodes", *IEEE Trans. Microwave Theory and Technique*, 39, p. 857, May (1991)

S. K. GHANDHI, *VLSI Fabrication Principles, Silicon and Gallium Arsenide*, John Wiley and Sons, New York (1983)

R. C. GOODFELLOW, B. T. DEBNEY, G. J. REES and J. BUUS, *IEEE Trans. Electron Devices*, ED-32, No. 12, p. 2562, Dec. (1985)

A. A. GRINBERG, M. S. SHUR, R. FISHER and H. MORKOÇ, "An Investigation of the Effect of Graded Layers and Tunneling on the Performance of AlGaAs/GaAs Heterojunction Bipolar Transistors", IEEE *Trans. Electron Devices*, ED-31, No. 12, pp. 1758-1765, Dec. (1984)

A. A. GRINBERG, "Thermionic Emission in Heterosystems with Different Effective Electronic Masses", *Phys. Rev.*, B33, No. 10, pp. 7256-7258 (1986)

K. HESS, *Advanced Theory of Semiconductor Devices*, Prentice Hall, New York (1988)

C. JACOBONI and L. REGGIANI, "Bulk Hot-Electron Properties of Cubic Semiconductors", *Advances in Physics*, 28, No. 4, pp. 493-553 (1979)

C. JACOBONI and P. LUGLI, *The Monte Carlo Method for Semiconductor Simulation*, Springer Series on Computational Microelectronics, ed. S. Selberherr, Springer-Verlag, Wien, New York (1989)

T. R. JARVIS and E. F. JOHNSON, "Geometric Magnetoresistance and Hall Mobility in Gunn Effect Devices", *Solid State Electronics*, 13, pp. 181-189 (1970)

G. U. JENSEN, B. LUND, T. A. FJELDLY and M. SHUR, "Monte Carlo Simulation of Semiconductor Devices", *Computer Physics Communications*, 67, pp. 1-61, August (1991)

H. KROEMER, "Heterostructure Bipolar Transistors and Integrated Circuits", *Proc. IEEE*, 70, pp. 13-25 (1982)

H. KROEMER, "Heterostructure Bipolar Transistors: What Should We Build?", *J. Vac. Sci. Technol.*, B1 (2) , pp. 112-130 (1983)

K. LEE and A. NUSSBAUM, "The Influence of Trap Levels on Generation and Recombination in Silicon Diodes", *Solid State Electronics*, 23, p. 655 (1980)

S. A. MAAS, *Microwave Mixers*, Artech House, Dedham, MA (1986)

O. MADELUNG, *Physics of III-V Compounds*, John Wiley and Sons, New York (1964)

J. L. MOLL, *Physics of Semiconductors*, McGraw Hill, New York (1964)

A. PADOVANI and R. STRATTON, *Solid State Electronics*, 9, p. 695 (1966)

L. REGGIANI, in *Hot Electron Transport in Semiconductors*, *Topics in Physics*, Vol. 58, ed. L. Reggiani, Springer-Verlag, Berlin (1985), p. 7

G. Y. ROBINSON, "Schottky Diodes and Ohmic Contacts for the III-V Semiconductors", in *Physics and Chemistry of III-V Semiconductor Interfaces*, ed. C. W. Wilmsen, Plenum Press (1983)

E. H. RHODERICK, *Metal-Semiconductor Contacts*, Clarendon Press, Oxford (1977)

W. VAN ROOSBROECK and W. SHOCKLEY, *Phys. Rev.*, 94, p. 1558 (1954)

J. G. RUCH and W. FAWCETT, "Temperature Dependence of the Transport Properties of Gallium Arsenide Determined by a Monte Carlo Method", *J. Appl. Phys.*, 41, No. 9, pp. 3843-3849, Aug. (1970)

G. RUCH and G. S. KINO, *Phys. Rev.*, 174, p. 921 (1968)

C. T. SAH, R. N. NOYCE and W. SHOCKLEY, "Carrier Generation and Recombination in *p-n* Junction and *n-p* Junction Characteristics," *Proc. IRE*, 45, p. 1228 (1957)

K. SEEGER, *Semiconductor Physics, An Introduction,* Springer Verlag, Series on Solid-State Sciences, Vol. 40, Springer Verlag, Berlin, Heidelberg, New York (1985), 3rd Edition

B. L. SHARMA and S. C. GUPTA, "Metal-Semiconductor Barrier Junctions", *Solid State Technology*, 23, pp. 90-95 (1980)

W. SHOCKLEY and W. T. READ, *Phys. Rev.*, 87, p. 835 (1952)

M. SHUR, "Influence of Non-Uniform Field Distribution on Frequency Limits of GaAs Field-Effect Transistors", *Electron Letters*, 12, No. 23, pp. 615-616, Nov. (1976)

M. SHUR and L. F. EASTMAN, "Ballistic Transport in Semiconductors at Low-Temperatures for Low Power High Speed Logic," *IEEE Trans. Electron Devices*, ED-26, No. 11, pp. 1677-1683, Nov. (1979)

M. SHUR, *Physics of Semiconductor Devices*, Prentice Hall, New Jersey (1990)

R. A. SMITH, *Semiconductors*, 2nd Edition, Cambridge University Press, Cambridge, London, New York, Melbourne, Sydney (1978)

W. E. SPICER, P. W. CHYE, P. R. SKEATH, C. Y. SU and I. LINDAU, *J. Vac. Sci. Tech.*, 16, p. 1422 (1979)

S. M. SZE, *Physics of Semiconductor Devices*, John Wiley & Sons, New York (1981)

S. M. SZE, *Semiconductor Devices. Physics and Technology*, John Wiley & Sons, New York (1985)

J. TERSOFF, *Phys. Rev.*, B-30, p. 4879 (1984)

K. K. THORNBER, *IEEE Electron Device Lett.*, EDL-3, No. 3, pp. 69-71 (1982)

Z. YU and R. W. DUTTON, *Sedan III - A General Electronic Material Device Analysis Program*, Program manual, Stanford University, July (1985)

A. VAN DER ZIEL, *Solid State Physical Electronics*, Prentice Hall, New Jersey (1976)

PROBLEMS

1-2-1. In narrow gap semiconductors, such as InSb (energy gap $E_g = 0.22$ eV compared to $E_g = 1.1$ eV for Si), the non-parabolicity in the conduction band is very important. According to the Kane model, the dispersion relation (for small values of the wave vector, k) is given by

$$E(k) = \frac{1}{2}\left(\sqrt{E_g^2 + \frac{2\hbar^2 k^2 E_g}{m_n}} - E_g \right)$$

where the energy is counted from the bottom of the conduction band at $k = 0$. Find the relationship between the Fermi level and carrier concentration, n, for this dispersion relationship when the semiconductor is degenerate.

1-2-2. Consider a two-dimensional band structure with the dispersion relation for the conduction band given by

$$E(k) = E_c + \frac{\hbar^2 \left(k_x^2 + k_y^2 \right)}{2m_n}$$

(no k_z-component – this is a fairly realistic model for electron subbands in heterostructure quantum wells.)

Derive the expression linking the density of electrons in the conduction band to the position of the Fermi level, E_F, and to the temperature, T

(a) for an arbitrary position of the Fermi level with respect to the bottom of the conduction band, E_c

(b) for a non-degenerate distribution of electrons in the conduction band

(c) for a degenerate distribution of electrons in the conduction band, i.e., assuming that $E_F - E_c \gg k_B T$ where E_c is the bottom of the conduction band.

1-2-3. Consider a one-dimensional quantum wire where electron motion in two directions is quantized and in one direction electrons are free to move. Assuming that the effective electron mass is equal to m_n, that the dispersion relation for the electron energy in a subband is parabolic $(E(k) = \hbar^2 k^2/(2m_n))$, find the density of states, $\Omega(E)$, in the subband as a function of energy E.

Hint: $dN = \Omega(E)dE$ where dN is the number of states in the energy interval between E and $E + dE$.

1-2-4. For a simple parabolic band, the relationship between the electron energy, E, and the wave vector, k, is given by $E(k) = \hbar^2 k^2/(2m_n)$ where m_n is the effective electron mass. The non-parabolicity can be accounted for by changing this relationship to $(E + \alpha E) = \hbar^2 k^2/(2m_n)$ where α is a constant. However, this equation does not satisfy the requirement $dE/dk|_{k=\pm K_B/2} = 0$ where $K_B = 2\pi/a$ is the vector of the reciprocal lattice and a is the lattice constant (i.e., at the boundaries of the first Brillouin zone. Propose an analytical function which coincides with the above non-parabolic expression for $k \ll K_B$ and satisfies this requirement.

Hint: The problem may have many different solutions. E must be an even function of k.

1-2-5. Consider a free electron in the quantum well heterostructure shown in Fig. P1.2.5. Calculate the energies of the two lowest subbands and write down approximate expressions for the electron wave functions in these subbands.

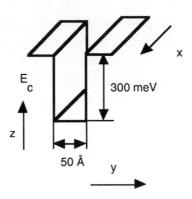

Fig. P1.2.5

Hint: Calculate the position of the subband energies using the approximation of an infinitely deep potential well and argue why this is a good approximation for deep subbands.

1-2-6. Using eqs. (1-2-2) and (1-2-7) plot the energy E (in eV) versus the wave vector k for direction [001] of the reciprocal lattice between $k = 0$ and the boundary of the Brillouin zone for GaAs. Compare the resulting plot with the parabolic dependence for the same effective mass. The lattice constant of GaAs is 5.65 Å. The electron effective mass in the central valley of the conduction band is $m = 0.067 m_e$ where $m_e = 9.11 \times 10^{-31}$ kg is the free electron mass. The energy gap is $E_g = 1.42$ eV. Superimpose on this plot the dependence of E versus k for the L-valleys assuming that the Γ-L valley separation is $E_{\Gamma L} = 0.32$ eV and the effective mass in the L valley is $m_L = 0.85 m_e$.

1-3-1. The effective density of states in the conduction band is N_c. The donor ionization energy is E_d, the donor concentration is N_d, the acceptor concentration is $N_a < N_d$, and the donor degeneracy factor is g_d. Derive the formula describing the temperature dependence of the carrier concentration, n, on temperature, T.

Assume that the semiconductor is not degenerate and the hole concentration is very small compared to n.

Hint: Express n through N_d, N_a, and $n^* = \dfrac{N_c}{g_d} \exp\left(-\dfrac{E_d}{k_B T}\right)$

1-3-2. Consider a silicon sample doped by donors with the ionization energy of 0.01 eV. The degeneracy factor for the donors, $g_d = 2$. What is the smallest donor concentration at which the Fermi level at some temperature will coincide with the bottom of the conduction band?

1-4-1. For an arbitrary carrier concentration, n, the Einstein relation for electrons can be expressed as

$$D_n = \frac{\mu_n n}{\dfrac{qdn}{d(E_F - E_c)}}$$

where E_F is the Fermi energy and E_c is the bottom of the conduction band. Derive this equation and show that it reduces to the conventional Einstein relation $(D_n = \mu_n k_B T/q)$ for a non-degenerate semiconductor.

1-6-1. Consider a Hall sample. The current is $I = 10$ mA. The sample length, width, and thickness are 100 μm, 5 μm, and 0.5 μm, respectively. The applied voltage is 1 V, the magnetic field is $B = 0.2$ T, the Hall factor is $r_H = 1.2$, and the Hall voltage is $V_H = 2.5$ mV. Find the Hall constant and the Hall mobility.

1-6-2. Derive the following equation describing the electron current density, $\mathbf{j_n}$, in a semiconductor in a weak electric field, $\mathbf{F}$, and magnetic field, $\mathbf{B}$:

$$\mathbf{j_n} = \frac{qn\mu_n}{1+(\mu_n B)^2}\left(\mathbf{F} - \mu_n[\mathbf{F}\times\mathbf{B}]\right)$$

Hint: Assume that the electron drift velocity, $\mathbf{v}$, is given by:

$$\mathbf{v} = -\mu_n\left(\mathbf{F} + [\mathbf{v}\times\mathbf{B}]\right)$$

1-7-1. Consider an n-type semiconductor sample with a concentration N_d of shallow donors. Light shining on the sample creates a uniform distribution of electron-hole pairs. Assuming that the electric field is zero, the generation rate is G, and the recombination rate is $b(np - n_i^2)$ where n and p are the electron and hole concentrations, respectively, and n_i is the intrinsic carrier concentration.
a) Calculate n and p in the stationary state.

b) At time $t = 0$, the light is turned off. Derive the equations describing how n and p vary with time for $t > 0$.

(See R. A. Anderson, *Solid State Electronics*, 5, pp. 341-351 (1962).)

1-8-1. Consider the total excess minority carrier charge (diffusion charge) in the diffusion regions on the p-side (Q_n) and on the n-side (Q_p) of a p-n junction. Often, the diffusion capacitance is calculated as follows:

$$C_{dif} = \frac{dQ_p}{dV} + \frac{d|Q_n|}{dV}$$

Show that this leads to a result which is twice the value found in eq. (1-8-93). Give a physical explanation for this difference.

Hint: Show that $Q_n = qSL_n(p_n - p_{no})$ and $Q_p = qSL_p(n_p - n_{po})$ where S is the junction cross section, L_n and L_p are the diffusion lengths for minority electrons and holes, respectively, and $(p_n - p_{no})$ and $(n_p - n_{po})$ are the excess minority carrier concentrations on the p- and n-side, respectively.

1-10-1. Derive the expression in eq. (1-10-33) for the diffusion capacitance of a short heterostructure p^+-n diode under forward bias, using the formalism in Sections 1.8 and 1.10. Discuss the special features and limitations of this result.

2

Bipolar Junction Transistors

2-1. INTRODUCTION

Since John Bardeen and Walter Brattain (1948) published the first paper on the Bipolar Junction Transistor (BJT), this device has been developed into one of the most essential components of modern electronics. Even though Field Effect Transistors, in particular, Complementary Field Effect Transistors (described in Chapter 3), have become dominant in most applications, the bipolar technology is very important in high speed applications, including high-speed computers and communication systems. New Heterojunction Bipolar Transistors utilizing the AlGaAs/GaAs and AlInAs/InP material systems open new horizons for bipolar technology, competing with the speediest field effect transistors.

The purpose of this chapter is to give basic information about BJTs and describe the BJT models implemented in AIM-Spice. In Section 2.2, we consider the principle of operation of the bipolar junction transistor. Section 2.3 gives a brief overview of bipolar transistor technology. Section 2.4 deals with the basic BJT models – the Ebers-Moll and Gummel-Poon models. Breakdown in bipolar junction transistors is considered in Section 2.5. Section 2.6 contains basic information about BJT small-signal equivalent circuits and cutoff frequencies. Finally, Section 2.7 describes the principle of operation and the AIM Spice model for Heterojunction Bipolar Transistors.

2-2. PRINCIPLE OF OPERATION

A bipolar junction transistor (BJT) is comprised of two back-to-back *p-n* junctions as shown schematically in Fig. 2.2.1a. An emitter region and a base region form the first junction. The second *p-n* junction is formed between the base and a collector region. The base is made to be much shorter than the diffusion length of the minority carriers. Therefore, the emitter-base and collector-base junctions affect each other, an interaction which is crucial for BJT operation. The equivalent circuit of an ideal BJT is shown in Fig. 2.2.1b, where controlled current sources account for the interaction between the emitter-base and the collector-base junctions.

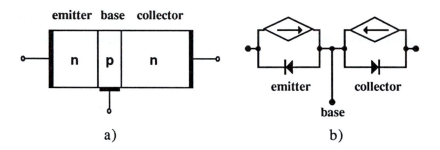

Fig. 2.2.1. Schematic of an *n-p-n* BJT and its equivalent circuit.

Under typical operating conditions, the emitter-base junction is forward-biased while the collector-base junction is reverse-biased. (This regime is called the active mode of operation.) Hence, the minority carrier (electron) concentration in the base at the emitter-base junction is much larger than the equilibrium concentration which is $n_{bo} = n_{ib}^2/N_{ab}$. Here n_{ib} is the intrinsic carrier concentration and N_{ab} is the acceptor doping density in the base region. As shown in Section 1.8, the concentration of minority carriers at the edge of the depletion region in the base, n_{be}, is related to the voltage drop across the junction and to the equilibrium concentration of the minority carriers (electrons) in the base:

$$n_{be} = n_{bo} \exp\left(V_{be}/V_{th}\right) \gg n_{bo} \qquad (2\text{-}2\text{-}1)$$

Here, $V_{th} = k_B T/q$ is the thermal voltage and V_{be} is the emitter-base voltage ($V_{be} \gg V_{th}$). Correspondingly, the electron concentration in the base at the depletion zone edge of the collector-base junction, n_{bc}, is much smaller than the equilibrium concentration:

$$n_{bc} = n_{bo} \exp\left(V_{bc}/V_{th}\right) \ll n_{bo} \qquad (2\text{-}2\text{-}2)$$

V_{bc} is here the collector-base voltage $(V_{bc} < 0)$.

Just as in a regular n^+-p junction (see Section 1.8), the diffusion component of the electron current is dominant in the p-region (the base) of an n-p-n BJT. If the width, W, of the base is small compared to the diffusion length, L_{nb}, of electrons in the base, the carrier recombination in the base is also small. To first order, the electron concentration in the base then varies linearly with distance. Assuming that the widths of the depletion regions at the emitter-base and collector-base interfaces are sufficiently small and can be neglected compared to the base width, we have

$$n_b \approx n_{be}(1 - x/W) \qquad (2\text{-}2\text{-}3)$$

where $x = 0$ corresponds to the position of the emitter-base junction. This result is based on the solution of the continuity equation for electrons in the base and is obtained in the same way as for a short p-n diode (see Section 1.8), i.e., by expanding $\exp(x/L_{nb})$ and $\exp(-x/L_{nb})$ into Taylor series.

The electron diffusion current in the base

$$I_n = SqD_n\frac{\partial n_b}{\partial x} \qquad (2\text{-}2\text{-}4)$$

remains practically constant throughout the base and equal to both the emitter and the collector current. (Here, D_n is the electron diffusion coefficient, and S is the device cross-section.) The reason is that practically all electrons injected into the base region from the emitter diffuse to the collector. In the collector, they are carried away towards the collector contact by the strong electric field of the reverse-biased collector-base p-n junction. Of course, some electrons do recombine with the holes in the base region. However, the number of such recombinations is small since the base region is much shorter than the electron diffusion length.

Substituting eq. (2-2-1) into eq. (2-2-3) and the resulting equation into eq. (2-2-4), we obtain for the emitter current, I_e, and the collector current, I_c,

$$I_e \approx I_c \approx |I_n| \approx \frac{S q D_n \, n_{bo}}{W} \exp\left(\frac{V_{be}}{V_{th}}\right) \tag{2-2-5}$$

Hence, a change in the emitter-base voltage by V_{th} (approximately 26 mV at room temperature) leads to an increase of both the emitter and the collector current by a factor $e = \exp(1)$.

Let us now consider how we can operate a bipolar junction transistor as an amplifier. The input signal may be represented by an ac voltage source connected across the emitter-base junction as shown in Fig. 2.2.2a. An external load resistance, R_c, can be connected between the collector terminal and the power supply terminal (this corresponds to the common-base configuration).

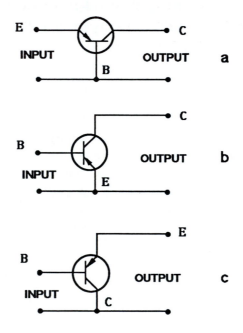

Fig. 2.2.2. Common-base (a), common-emitter (b) and common-collector (c) circuit configurations for the bipolar junction transistor.

A small variation of the emitter-base voltage, ΔV_{be}, caused by the signal, leads to a nearly equal variation of the emitter and the collector currents $(\Delta I_c \approx \Delta I_e)$. However, the collector current flows in the loop containing the power supply (bias voltage $V_{cc} \gg V_{be}$) and the variation of the voltage drop, $\Delta V_c = \Delta I_c R_c$, across R_c can be much greater than ΔV_{be}. (With an appropriate value of R_c, the maximum value of ΔV_c is limited by the collector voltage supply, V_{cc}.) Hence, we can have a voltage gain of $\Delta V_c/\Delta V_{be}$. The ac power supplied by the input signal is of the order of $\Delta I_e \Delta V_{be}$ and the ac power generated in the output loop is of the order of $\Delta I_c \Delta V_c \gg \Delta I_e \Delta V_{be}$. Hence, this circuit also provides power gain. This gain occurs because electrons are injected from the emitter region through the base into the collector region where the voltage drop is large compared to the small forward emitter-base voltage.

Two other possible circuit configurations of the bipolar junction transistor – the common-emitter and the common-collector configurations – are illustrated by Figs. 2.2.2b and 2.2.2c, respectively. The principle of the transistor operation in the common-emitter configuration may be explained by considering what happens when we increase the base current by increasing the emitter-base voltage. The base current, I_b, is primarily the hole current that flows from the base into the emitter region, I_{pe}. The hole current in the n-type emitter is a diffusion current, just like the electron current in the p-type base. Hence, we have, in full analogy with eq. (2-2-5),

$$I_b \approx I_{pe} \approx SqD_p \left.\frac{\partial p_e}{\partial x}\right|_{x=0} \approx \frac{SqD_p \, p_{eo}}{X_e} \exp\left(\frac{V_{be}}{V_{th}}\right) \qquad (2\text{-}2\text{-}6)$$

Here, X_e is the width of the emitter region and $p_{eo} = n_{ie}^2/N_{de}$ is the equilibrium minority carrier concentration, where N_{de} is the donor density and n_{ie} is the intrinsic carrier concentration in the emitter. In eq. (2-2-6), it is assumed that $X_e \ll L_{pe}$, where L_{pe} is the diffusion length of holes in the emitter region. This is always true for practical BJTs. Let us now compare eqs. (2-2-5) and (2-2-6). Usually, X_e is of the same order of magnitude as W. However, the emitter region is much more heavily doped than the base region and, hence, $p_{eo} \ll n_{bo}$. This means that the base current is much smaller than the emitter current, and the transistor in the common emitter configuration has a current gain. From eqs. (2-2-5) and (2-2-6), we can estimate the common-emitter current gain, $\beta = I_c/I_b \approx I_e/I_b$, to be

$$\beta \approx \frac{D_n \, N_{de} \, X_e}{D_p \, N_{ab} \, W} \qquad (2\text{-}2\text{-}7)$$

Here, β is much greater than one since $N_{de} \gg N_{ab}$. Since $D_n/D_p = \mu_n/\mu_p$

(from the Einstein relation) is approximately 2.5 in silicon, it is also clear that Si $n\text{-}p\text{-}n$ BJTs should, in general, have higher β than Si $p\text{-}n\text{-}p$ BJTs.

In modern bipolar junction transistors, I_{pe} is the dominant component of the base current. The base current also includes other components, such as the recombination current in the depletion region of the emitter-base junction, the generation current in the depletion region of the collector-base junction, and the recombination current in the neutral base region. However, for a crude estimate of the transistor current gain, all components of the base current, except I_{pe}, may be neglected.

As shown above, the operation of a BJT is based on the exponential variation of the injected carrier density with the height of the potential barrier between the emitter and base, controlled by the emitter-base voltage. This leads to the exponential dependence of the emitter and collector currents on the emitter-base voltage (see eq. (2-2-5)) and to a very high transconductance

$$g_m = \frac{\partial I_c}{\partial V_{be}} \qquad (2\text{-}2\text{-}8)$$

compared with the corresponding parameter for a field effect transistor, which utilizes a capacitive modulation of charge in the conducting channel (see Chapter 3). High transconductance and current swing make the bipolar junction transistor a device of choice for many high speed and high power applications both in discrete and in integrated circuits.

2-3. OVERVIEW OF BIPOLAR TRANSISTOR TECHNOLOGY

As was explained in Section 2.2, the BJT is a vertical device which includes a collector, a base, and an emitter layer. In most cases, the collector is at the bottom, and the collector contact occupies a fairly large area of the wafer surface (see Fig. 2.3.1). Thus, the ratio of the active region to the total device area, including isolation shared with neighboring transistors, is usually much smaller than for Field Effect Transistors (FETs) considered in Chapters 3 and 4. Hence, the parasitic components are even more important for BJTs than for FETs. Reducing the value of the parasitic components without sacrificing too much of yield, is essential for high performance and high packing density bipolar circuits.

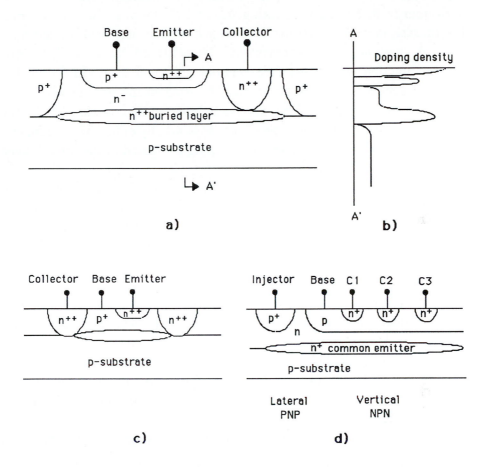

Fig. 2.3.1. Different BJT technologies: Cross section (a) and vertical doping profile (b) of the Standard Buried Collector (SBC) technology; cross section of the Collector Diffusion Isolation (CDI) technology (c), and the Integrated Injection Logic (I^2L) technology (d).

The key parasitic components to minimize are the base spreading resistance and the extrinsic collector capacitance. Also, the collector series resistance must be minimized as much as possible. This can be achieved by using a collector buried layer (see Fig. 2.3.1a). The use of self-alignment techniques and the merging of transistor structures in the integrated circuits (as will be discussed below) can be used to reduce both the base spreading resistance and the parasitic collector capacitance. A proper scaling of the transistor dimensions leads to both a better performance and a higher packing density. However, for a given minimum feature size, there is a trade-off between performance and packing density.

A qualitative improvement can be achieved by using heterostructure bipolar junction technology (see Sections 1.10 and 2.7). In this technology, different semiconductor materials may be used in the emitter and in the base and/or in the collector. This approach leads to a reduction of the parasitic components, such as the base spreading resistance, by orders of magnitude since it allows us to greatly increase the base doping compared to that of a conventional BJT.

An accurate modeling of both the intrinsic device and the parasitic components is very important. In this book we focus mainly on intrinsic bipolar device modeling, because parasitic components are very dependent on the specific technology. Furthermore, we limit ourselves to modeling vertical transistors. Typically, *n-p-n* transistors are vertical devices as indicated in Fig. 2.3.1.

In complementary circuits and in so-called Integrated Injection Logic (I^2L), there is also a need for *p-n-p* transistors. These are mostly lateral devices, since vertical *p-n-p* BJTs are difficult to fabricate together with vertical *n-p-n* transistors. However, the lateral *p-n-p* BJTs show an order of magnitude lower performance compared to the vertical *n-p-n* transistors and are seldom being used in high performance circuits. Because of the two dimensional nature of the lateral device operation, *p-n-p* device modeling is very difficult and very technology dependent.

In this Section, we briefly review vertical homojunction and heterojunction bipolar technologies.

2.3.1. Silicon Homojunction Technology

The description of different silicon bipolar technologies can be found in books by Sze (1981), Ghandhi (1983), Warner and Grung,(1983), Sze (1988) (see Chapter 11 by Hillenius in Sze (1988)), and in the Proceedings of the annual IEEE Bipolar Circuits and Technology Meetings (see, for example, the review paper by Nakamae (1987)).

Here, we will describe three important silicon bipolar technologies for the purpose of illustrating the issues important in VLSI modeling. The most popular is the so-called Standard Buried Collector (SBC) technology (see Fig. 2.3.1 a and b). The doping densities are approximately $10^{20}\,\mathrm{cm^{-3}}$ in the emitter, $10^{18}\,\mathrm{cm^{-3}}$

in the base, and 10^{16} cm^{-3} in the collector. The emitter and the base doping are done by double diffusion (or implantation) from the surface. A grading in the base profile creates a built-in electric field which accelerates the minority carriers (see, for example, Shur (1990), Chapter 3). Thereby, the transit time is reduced and a higher cut-off frequency is obtained. This technology gives a high performance. However, it requires large silicon areas for contacts to the buried collector layer and for the device isolation, especially when junction isolation is used. Oxide isolation can drastically reduce the device size. Even further reduction can be obtained using trench isolation (see, for example, Yamazaki et al. (1990)).

The device structure shown in Fig. 2.3.1c is called CDI (Collector Diffusion Isolation). The CDI technology has an inherent advantage in terms of packing density because of the built-in isolation. However, both performance and uniformity are inferior to SBC. The higher doping in the collector than that in the base causes serious base modulation and leads to a low breakdown voltage. The uniformity of the base width, determined by epitaxy, is also quite poor. The thickness and doping of the base region can be much more uniformly controlled when the base is doped by ion implantation and successive drive-in compared to either diffusion or epitaxial growth.

The structure shown in Fig. 2.2.1d represents the so-called Integrated Injection Logic (I^2L). This technology has the highest packing density among the bipolar technologies available at present. In I^2L, the vertical n-p-n BJTs have a buried common emitter, multiple collector outputs are available without any separate isolation scheme, and the lateral p-n-p BJTs are merged with the vertical n-p-n transistors. Hence, only the base and collector terminals need to be connected by metal lines. All these features make this technology uniquely suitable for bipolar VLSI. In addition, power consumption can be controlled by changing the injection current. A higher speed can be obtained by increasing the injection current, but at the expense of a larger power consumption. Unfortunately, however, the performance of this circuit is inferior to other technologies because the emitter down n-p-n transistor has a low current gain and a high collector doping (the same problem as in CDI). Modifications of the I^2L technology utilizing Schottky barriers such as STL (Schottky Transistor Logic) and ISL (Integrated Schottky Logic) have a higher speed (see, for example, Etiemble and Mohssine (1987)).

As stated earlier, it is very important to have a compact device structure.

The most critical issue is to make the base terminal as close to the emitter as possible. There are many examples of this, such as the "washed" emitter, the "walled" emitter, the "poly" emitter, and "double poly" silicon transistors. As will be discussed in Chapter 3, polysilicon has been extensively used in self-aligned MOS technologies for many decades. By using this mature MOS technology for the emitter and/or base formation, very high performance BJTs have been obtained. The most sophisticated device at present has been fabricated using self-aligned double polysilicon, one for emitter and the other for base. This device holds the highest speed record with a cut-off frequency of more than 40 GHz (see Yamazaki et al. (1990)). This is the result of a very short base-emitter spacing of around 0.3 μm, yielding a very small base resistance. The importance of the base resistance will be discussed later in Sections 2.4 and 2.6.

2.3.2. Heterojunction Bipolar Technology

The Heterojunction Bipolar Transistors (HBTs) have a much better performance than conventional homojunction transistors. W. Shockley (1951) proposed the concept of a wide band gap emitter HBT in order to achieve a very high current gain transistor. However, the gain can be traded off in favor of a higher circuit speed by emphasizing the reduction of parasitic components, such as the base resistance and the collector capacitance. A wider band gap in the emitter than in the base, allows us to dope the base more strongly than the emitter while retaining a reasonable gain. Thus, the base resistance and the base width modulation can be drastically reduced. This becomes more important as the base width is scaled down.

There are two types of HBTs – Single and Double Heterojunction Bipolar Transistors (SHBT and DHBT). Typical SHBTs utilize material combinations such as AlGaAs/GaAs/GaAs, InAlAs/InGaAs/InGaAs, and InP/InGaAs/InGaAs (the material order is emitter/base/collector). An SHBT does not have a barrier for the transport of electrons from the base to the collector. Hence, most of the electrons injected into the base from the emitter reach the collector. However, the lack of symmetry between the base-emitter and the base-collector junctions may result in a large collector-emitter offset voltage in the current voltage characteristics. This is detrimental to the power dissipation in SHBT circuits (see Section 2.7 for further discussion of this issue).

Typical DHBTs utilize material combinations such as Si/SiGe/Si, AlGaAs/GaAs/AlGaAs, InAlAs/InGaAs/InP, and InP/InGaAs/InP. Sometimes,

DHBT technologies suffer from a low current gain because of the low base transport factor caused by the heterojunction barrier (the "spike") at the base-collector junction. However, this spike can be reduced by a proper compositional grading. As a result of the symmetry between the two junctions, the offset voltage can be quite low.

InP based HBTs (see, for example, Nottenburg et al. (1986) and Section 2.7)) are particularly promising for the following reasons: The material is compatible with long wavelength laser technology. The wider band gap of the InP substrate than the active base region allows the use of back-side illumination. The wider band gap of the InP substrate also makes it possible to use an emitter down structure. (The emitter down structure has a potential speed advantage for VLSI owing to the lower collector capacitance and emitter reactance, and the higher packing density when using I^2L, as further discussed in Section 2.7). Both InP and InGaAs have large Γ-X or Γ-L valley separations which enhance velocity overshoot and ballistic transport. An order of magnitude smaller contact resistance can be obtained in InP compared to GaAs for n-type ohmic contacts.

All in all, InP based HBTs have the highest cutoff frequency, f_T, of all HBTs. See, for example, Chen et al. (1989) who reported $f_T = 165$ GHz. This figure is expected to rise to more than 500 GHz. However, Si/SiGe/Si DHBTs have drawn much attention recently because they can be fabricated relatively easily by some modifications of the mature silicon technology. Cut-off frequencies of more than 75 GHz have been obtained and are expected to reach more than 100 GHz in the near future.

As can be seen from this brief overview of bipolar technology, the transistor structures are quite complex and are very different from simplistic one-dimensional models. An accurate BJT and HBT device modeling has to be linked to the process modeling and involve two dimensional or even three dimensional calculations. Needless to say, the circuit modeling of bipolar transistors is not even close to this degree of accuracy or complexity. The present day models discussed in this chapter rely on semi-empirical equations and parameters which have to be determined from experimental data.

2-4. BJT MODELING

The simplest large signal equivalent circuit of an ideal ("intrinsic") bipolar junction transistor consists of two diodes and two current-controlled current sources describing the interaction between the emitter-base and collector-base junctions, as shown in Fig. 2.2.1. Current sources representing the recombination and generation currents have to be added to this circuit to represent a real BJT, and the resulting equivalent circuit for an *n-p-n* transistor is as shown in Fig. 2.4.1.

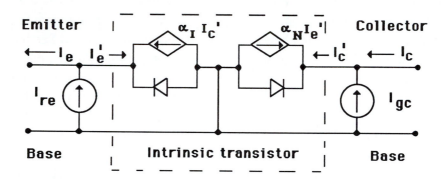

Fig. 2.4.1. Equivalent circuit for an *n-p-n* bipolar transistor (from Shur (1990)).

2.4.1. Ebers-Moll Model

The simple model for the "intrinsic" transistor, called the Ebers-Moll model after Ebers and Moll (1954), is described in terms of the theory for *p-n* junctions developed in Section 1.8. For the narrow base region, the continuity equation now has to be solved using the minority carrier concentrations at the edges of the neutral base as boundary conditions (see eqs. (2-2-1) and (2-2-2) for the *n-p-n* transistor). By straightforward calculations, the following "intrinsic" emitter current, I_e', and collector current, I_c', are obtained:

$$I_e' = a_{11}\left[\exp\left(\frac{V_{be}}{V_{th}}\right) - 1\right] + a_{12}\left[\exp\left(\frac{V_{bc}}{V_{th}}\right) - 1\right] \qquad (2\text{-}4\text{-}1)$$

$$I_c' = a_{21}\left[\exp\left(\frac{V_{be}}{V_{th}}\right) - 1\right] + a_{22}\left[\exp\left(\frac{V_{bc}}{V_{th}}\right) - 1\right] \tag{2-4-2}$$

where, again, $V_{th} = k_B T/q$ is the thermal voltage. The coefficients a_{11}, a_{12}, a_{21}, and a_{22} may be related to the material parameters, the transistor dimensions and the doping levels. For n-p-n transistors and for the sign convention for the currents I_e' and I_c' used in Fig. 2.4.1, we find

$$a_{11} = - qS\left[\frac{D_n n_{bo}}{L_{nb}} \coth\left(\frac{W_{eff}}{L_{nb}}\right) + \frac{D_p P_{eo}}{x_e}\right]$$

$$\approx - qS n_i^2\left[\frac{D_n}{N_{ab} W_{eff}} + \frac{D_p}{N_{de} x_e}\right] \tag{2-4-3}$$

$$a_{12} = a_{21} = \frac{qS D_n n_{bo}}{L_{nb} \sinh\left(W_{eff}/L_{nb}\right)} \approx \frac{qS D_n n_i^2}{N_{ab} W_{eff}} \tag{2-4-4}$$

$$a_{22} = - qS\left[\frac{D_n n_{bo}}{L_{nb}} \coth\left(\frac{W_{eff}}{L_{nb}}\right) + \frac{D_p P_{co}}{x_c}\right]$$

$$\approx - qS n_i^2\left[\frac{D_n}{N_{ab} W_{eff}} + \frac{D_p}{N_{dc} x_c}\right] \tag{2-4-5}$$

Here, W_{eff} is the effective base width, reduced in comparison to the metallurgical width, W, owing to the finite extents of the junction depletion regions into the base; x_e and x_c are the widths of the emitter and collector regions (assumed to be much less than the corresponding minority carrier diffusion lengths); L_{nb} is the minority carrier diffusion length in the base; N_{de}, N_{ab} and N_{dc} are the doping densities of the emitter, the base and the collector, respectively; and D_n and D_p are the diffusion constants for minority electrons and minority holes. The other symbols are defined in Section 2.2. The approximate results shown in eqs. (2-4-3) to (2-4-5) are obtained by assuming that $W_{eff} \ll L_{nb}$.

The emitter-base voltage may be expressed in terms of the emitter current of the intrinsic transistor, I_e', using eq. (2-4-1)

$$\exp\left(\frac{V_{be}}{V_{th}}\right) - 1 = \frac{I_e'}{a_{11}} - \frac{a_{12}}{a_{11}}\left[\exp\left(\frac{V_{bc}}{V_{th}}\right) - 1\right] \tag{2-4-6}$$

Substituting eq.(2-4-6) into eq. (2-4-2) we find

$$I_c' = \frac{a_{12}}{a_{11}}I_e' + \left(a_{22} - \frac{a_{12}a_{21}}{a_{11}}\right)\left[\exp\left(\frac{V_{bc}}{V_{th}}\right) - 1\right] \tag{2-4-7}$$

This equation may be rewritten as

$$I_c' = -\alpha_N I_e' - I_{co}\left[\exp\left(\frac{V_{bc}}{V_{th}}\right) - 1\right] \tag{2-4-8}$$

Here,

$$\alpha_N = -a_{12}/a_{11} \tag{2-4-9}$$

is the normal (or forward) common-base current gain and

$$I_{co} = a_{12}a_{21}/a_{11} - a_{22} \tag{2-4-10}$$

is the common-base collector reverse saturation current. Note that $\alpha_N I_e'$ is identical to the current source on the collector side of the "intrinsic" Ebers-Moll model of Fig. 2.4.1. For the active forward mode, V_{bc} is negative, $|V_{bc}|$ is much larger than V_{th}, and eq. (2-4-8) reduces to

$$I_c = \alpha I_e + I_{cbo} \tag{2-4-11}$$

where I_{cbo} is the collector current with open emitter. Here, we neglect the recombination current, I_{re}, and the generation current, I_{gc}, such that $I_e \approx -I_e'$, $\alpha_N \approx \alpha$ and $I_{co} \approx I_{cbo}$. However, eq. (2-4-11) may account for the recombination current, I_{re}, (via the dependence of α on the collector current) and, hence, the current gains α_N and α and the saturation currents I_{co} and I_{cbo} do not have to be exactly equal.

    In a similar manner, we can derive from eqs. (2-4-1) and (2-4-2) an equation relating the emitter current of the "intrinsic" transistor to the collector current:

$$I'_e = \frac{a_{21}}{a_{22}} I'_c + \left(a_{11} - \frac{a_{12}a_{21}}{a_{22}} \right) \left[\exp\left(\frac{V_{be}}{V_{th}} \right) - 1 \right] \qquad (2\text{-}4\text{-}12)$$

This equation may be rewritten as

$$I'_e = - \alpha_I I'_c - I_{eo} \left[\exp\left(\frac{V_{be}}{V_{th}} \right) - 1 \right] \qquad (2\text{-}4\text{-}13)$$

Here, α_I is the inverse current gain (sometimes also called inverse alpha) and I_{eo} is the emitter reverse saturation current. Eqs. (2-4-12) and (2-4-13) describe the inverse mode of operation when the collector-base junction is forward biased and the emitter-base junction is reverse biased. Note that $\alpha_I I_c'$ represents the current source on the emitter side of the "intrinsic" Ebers-Moll model of Fig. 2.4.1.

The Ebers-Moll model for the intrinsic transistor has four parameters: a_{11}, a_{12}, a_{21}, and a_{22}, or α_N, α_I, I_{eo} and I_{co}. The reciprocity relationship for an ideal two-port device requires

$$a_{12} = a_{21} \qquad (2\text{-}4\text{-}14)$$

and, hence, only three parameters (for example, α_N, I_{eo}, and I_{co}) are required for this basic transistor model. In practice, however, the reciprocity is not precisely satisfied for real transistors.

This simple version of the Ebers-Moll model does not take into account the dependence of the current gain on the injection level. Hence, it may only be applied for moderate emitter and collector currents. Also, the dependence of the effective base width, W_{eff}, on the reverse bias, V_{bc}, (forward active mode) has not been explicitly explored. This gives rise to the so-called base-width modulation or the Early effect (Early (1952)) which manifests itself as a finite slope in the active region of the common-emitter characteristics of BJTs.

The basic Ebers-Moll model may be somewhat improved by adding current sources accounting for the emitter-base and collector-base recombination currents (see Fig. 2.4.1). A feed-back loop with an additional current source is sometimes added between the collector and the emitter to account for the base-width modulation effect (see Sze (1981)). However, even this improved model does not account for the high injection effects and for effects related to Auger recombination and emitter current crowding. All these mechanisms are very important for practical regimes of transistor operation, and the simplified Ebers-Moll model predicts values of the current gain that are too high. A more accurate and realistic model has been proposed by Gummel and Poon (1970).

2.4.2. Gummel-Poon Model

Gummel and Poon (1970) developed an improved model that is applicable in the high injection regime. They started from rewriting the Ebers-Moll equations in an equivalent but more symmetrical form. Following their approach and using equations of the Ebers-Moll model given above, we can express the coefficients a_{11}, $a_{12} = a_{21}$, and a_{22} through the emitter reverse saturation current, I_{eo}, the normal common-base current gain, α_N, and the inverse common-base current gain, α_I:

$$a_{11} = - a_{12}/\alpha_N \qquad (2\text{-}4\text{-}15)$$

$$a_{22} = - a_{12}/\alpha_I \qquad (2\text{-}4\text{-}16)$$

$$- a_{11} + a_{12}^2/a_{22} = I_{eo} \qquad (2\text{-}4\text{-}17)$$

Solving these three equations with respect to a_{11}, a_{12}, and a_{22}, we obtain

$$a_{11} = - \frac{I_{eo}}{1 - \alpha_N \alpha_I} \qquad (2\text{-}4\text{-}18)$$

$$a_{12} = \frac{\alpha_N I_{eo}}{1 - \alpha_N \alpha_I} \qquad (2\text{-}4\text{-}19)$$

$$a_{22} = - \frac{\alpha_N I_{eo}}{\alpha_I (1 - \alpha_N \alpha_I)} \qquad (2\text{-}4\text{-}20)$$

We now introduce a parameter called the intercept current, I_i:

$$I_i = - a_{12} = - \frac{\alpha_N I_{eo}}{1 - \alpha_N \alpha_I} \qquad (2\text{-}4\text{-}21)$$

From eqs. (2-4-1) and (2-4-21), it can be seen that $\ln(I_i)$ is determined by extrapolating $\ln(I_e)$ versus V_{bc} from $V_{bc} \gg V_{th}$ back to $V_{bc} = 0$, with the base-emitter shorted. Using this notation and the relationships between the common-emitter current gain and the common-base current gain

$$\beta_N = \frac{\alpha_N}{1 - \alpha_N} \tag{2-4-22}$$

$$\beta_I = \frac{\alpha_I}{1 - \alpha_I} \tag{2-4-23}$$

we can rewrite the Ebers-Moll equations (2-4-1) and (2-4-2) as follows:

$$I'_e = \left(1 + \frac{1}{\beta_N}\right) I_i \left[\exp\left(\frac{V_{be}}{V_{th}}\right) - 1\right] - I_i \left[\exp\left(\frac{V_{bc}}{V_{th}}\right) - 1\right] \tag{2-4-24}$$

$$I'_c = - I_i \left[\exp\left(\frac{V_{be}}{V_{th}}\right) - 1\right] + \left(1 + \frac{1}{\beta_I}\right) I_i \left[\exp\left(\frac{V_{bc}}{V_{th}}\right) - 1\right] \tag{2-4-25}$$

Following Gummel and Poon, we may now represent the emitter and collector currents for the "intrinsic" transistor as

$$I'_e = I_{cc} + I_{be} \tag{2-4-26}$$

$$I'_c = - I_{cc} + I_{bc} \tag{2-4-27}$$

where

$$I_{cc} = I_i \left[\exp\left(\frac{V_{be}}{V_{th}}\right) - \exp\left(\frac{V_{bc}}{V_{th}}\right)\right] \tag{2-4-28}$$

is the principal component of the emitter and collector currents, and

$$I_{be} = \frac{I_i}{\beta_N} \left[\exp\left(\frac{V_{be}}{V_{th}}\right) - 1\right] \tag{2-4-29}$$

$$I_{bc} = \frac{I_i}{\beta_I}\left[\exp\left(\frac{V_{bc}}{V_{th}}\right) - 1\right] \tag{2-4-30}$$

The current component I_{be} is small compared to I_e' in the important active mode of operation because β_N is much larger than unity. The idea of the Gummel-Poon model is to simplify the mathematics of the problem by accounting for high injection effects only for the most important current component, I_{cc}. In this model, eq. (2-4-28) is replaced by

$$I_{cc} = \frac{qn_G I_i}{Q_b}\left[\exp\left(\frac{V_{be}}{V_{th}}\right) - \exp\left(\frac{V_{bc}}{V_{th}}\right)\right] \tag{2-4-31}$$

where Q_b is the total charge of the majority carriers in the base per unit area and n_G is the number of impurities per unit area in the base, the so-called Gummel number (see, for example, Sze (1981)). This equation can be derived assuming a constant electron quasi-Fermi level in the base and neglecting the recombination in the base. When the injection level is low and the Early effect may be neglected, we have

$$Q_b = qn_G \tag{2-4-32}$$

In the general case, however, the base hole charge, Q_b, is affected by the emitter-base and collector-base space charge regions and by the charge of the holes injected into the base region:

$$Q_b = qn_G + Q_{be} + Q_{bc} + Q_{dife} + Q_{difc} \tag{2-4-33}$$

where Q_{dife} and Q_{difc} are the charges of holes per unit area injected into the base, associated with the emitter-base and collector-base diffusion capacitances, respectively. Furthermore,

$$Q_{be} = - qx_b N_{ab} = -\sqrt{\frac{2\,q\varepsilon_s\, N_{ab}\, N_{de}\,(V_{bibe} + V_{eb})}{(N_{ab} + N_{de})}} \tag{2-4-34}$$

is the charge in the depletion region on the base side of the emitter-base junction, and

$$Q_{bc} = -qx_b, N_{ab} = -\sqrt{\frac{2q\varepsilon_s N_{ab} N_{dc} (V_{bibc} + V_{cb})}{(N_{ab} + N_{dc})}} \qquad (2\text{-}4\text{-}35)$$

is the charge in the depletion region on the base side of the collector-base junction. Here, the extensions of the emitter-base and collector-base depletion regions into the base region, x_b and $x_{b'}$, are found using the depletion approximation (see Section 1.8). V_{bieb} and V_{bicb} are the built-in voltages of the emitter-base and collector-base junctions, respectively. It has been assumed that $N_{de} \gg N_{ab}$ and $N_{dc} \ll N_{ab}$. Typically, for the forward active mode, $|Q_{bc}| \gg |Q_{be}|$ because $V_{bc} (= -V_{cb})$ is negative and $|V_{bc}| \gg |V_{bel}|$. Eqs. (2-4-34) and (2-4-35) are valid for uniform doping profiles.

For practical transistors with non-uniform doping in the base, an empirical linearized expression for Q_{bc} and Q_{be} is used:

$$Q_{be} + Q_{bc} = -qn_G V_{ce}/|V_A| \qquad (2\text{-}4\text{-}36)$$

where V_A is the Early voltage and V_{ce} is the collector-emitter voltage (positive for n-p-n transistors in the active forward mode). The Early voltage is approximately the collector-emitter voltage found by extrapolating the active mode common-emitter characteristics of the BJT to zero collector current. $|V_A|$ is roughly proportional to the Gummel number and to the base width:

$$|V_A| = k_A qn_G W/\varepsilon_s \qquad (2\text{-}4\text{-}37)$$

where k_A is a numerical constant of the order of unity. Using eqs. (2-4-33) and (2-4-36), we find from eq. (2-4-31)

$$I_{cc} = I_i \frac{qn_G [\exp(V_{be}/V_{th}) - \exp(V_{bc}/V_{th})]}{qn_G (1 - |V_{ce}/V_A|) + Q_{dife} + Q_{difc}} \qquad (2\text{-}4\text{-}38)$$

The high injection effects caused by the majority carriers (holes) injected into the base are accounted for by the charges Q_{dife} and Q_{difc} in eq. (2-4-38). The changes in these charges describe the dependence of the transistor parameters on the injection level. These charges may be expressed as (see, for example, Sze (1981))

$$Q_{dife} = B\tau_F I_F/S \qquad (2\text{-}4\text{-}39)$$

$$Q_{difc} = \tau_R I_R / S \tag{2-4-40}$$

where B is an empirical factor which is equal to unity at relatively low injection levels but may be greater than unity at high injection levels (B describes the changes in Q_{dife} caused by the Kirk effect, see Sze (1981)). τ_F and τ_R are effective minority carrier lifetimes associated with the forward and reverse currents, I_F and I_R, respectively, where

$$I_F = I_i \frac{qn_G}{Q_b} \left[\exp\left(\frac{V_{be}}{V_{th}}\right) - 1 \right] \tag{2-4-41}$$

$$I_R = I_i \frac{qn_G}{Q_b} \left[\exp\left(\frac{V_{bc}}{V_{th}}\right) - 1 \right] \tag{2-4-42}$$

Using eqs. (2-4-39) to (2-4-42) in combination with eq. (2-4-33), we obtain a quadratic equation that yields the following expression for Q_b

$$Q_b = \frac{(qn_G + Q_{be} + Q_{bc})}{2} + \left[\frac{(qn_G + Q_{be} + Q_{bc})^2}{4} + \frac{qn_G I_i}{S} \left\{ B \tau_F \left[\exp\left(\frac{V_{be}}{V_{th}}\right) - 1 \right] + \tau_R \left[\exp\left(\frac{V_{bc}}{V_{th}}\right) - 1 \right] \right\} \right]^{1/2} \tag{2-4-43}$$

The dynamic base current may now be written as

$$I_b = S \frac{dQ_b}{dt} + I_r + \frac{I_F}{\beta_N} + \frac{I_R}{\beta_I} \tag{2-4-44}$$

where the total recombination current

$$I_r = I_{rbe} + I_{rbc} + I_{re} + I_{rc} \tag{2-4-45}$$

includes the terms

$$I_{rbe} = I_{1e} \left[\exp\left(\frac{V_{be}}{V_{th}}\right) - 1 \right] \tag{2-4-46}$$

$$I_{rbc} = I_{1c} \left[\exp\left(\frac{V_{bc}}{V_{th}} \right) - 1 \right] \tag{2-4-47}$$

which are the contributions to the recombination current in the base region (outside of space charge regions),

$$I_{re} = I_2 \left[\exp\left(\frac{V_{be}}{m_{re} V_{th}} \right) - 1 \right] \tag{2-4-48}$$

which is the recombination current in the emitter-base space charge region (see Section 1.8), and

$$I_{rc} = I_3 \left[\exp\left(\frac{V_{bc}}{m_{rc} V_{th}} \right) - 1 \right] \tag{2-4-49}$$

which is the recombination current in the collector-base space charge region. The ideality factors m_{re} and m_{rc} typically range from 1 to 2.

The dynamic emitter and collector currents are given by

$$I_e = -I_{cc} - I_{be} + I_{re} + I_{rbe} + \tau_F \frac{dI_F}{dt} + C_{de} \frac{dV_{be}}{dt} \tag{2-4-50}$$

$$I_c = -I_{cc} + I_{be} - I_{rc} - I_{rbc} - \tau_R \frac{dI_R}{dt} + C_{dc} \frac{dV_{bc}}{dt} \tag{2-4-51}$$

where

$$C_{de} = \varepsilon_s S / x_{deb} \tag{2-4-52}$$

$$C_{dc} = \varepsilon_s S / x_{dcb} \tag{2-4-53}$$

are depletion capacitances for the emitter-base and collector-base junctions, respectively (see Section 1.8).

The above equations form a set of equations for the Gummel-Poon model. Some important effects, such as the Early effect or the high injection effects, are directly included into this model. Other effects, such as Auger recombination and band gap narrowing, may be accounted for indirectly by an appropriate choice of the model parameters, i.e., I_i, m_{re}, m_{rc}, I_{1e}, I_{1c}, I_2, and I_3. In addition, parasitic emitter and collector series resistances and the base spreading resistance have to be included for a realistic modeling of a bipolar junction transistor. In order to take into account emitter and collector series resistances, r_{es} and r_{cs}, we have to substitute the "extrinsic" emitter-base and collector-base voltages, V_{eb} and V_{cb}, for the "intrinsic" emitter-base and collector-base voltages, V_{eb}' and V_{cb}', in the equations of the Gummel-Poon model, i.e.,

$$V_{be}' = V_{be} - |I_e| r_{es} \qquad\qquad (2\text{-}4\text{-}54)$$

$$V_{cb}' = V_{cb} - |I_c| r_{cs} \qquad\qquad (2\text{-}4\text{-}55)$$

The base spreading resistance is more difficult to account for as it leads to a non-uniform distribution of the emitter and collector current densities and to the effect of emitter current crowding (see, for example, Sze (1985)). However, in the frame of the Gummel-Poon model, the degradation of the common-emitter current gain, β, caused by the emitter current crowding at high emitter currents may be reproduced to some extent by choosing an appropriate effective value of τ_F.

Fig. 2.4.2 shows the dependence of the common-emitter current gain, β, on the collector current calculated using the Ebers-Moll and Gummel-Poon models implemented in AIM-Spice. These models give similar results at low collector currents, except for an offset which results from including the effects of depletion charges ($Q_{be} + Q_{bc}$) in the Gummel-Poon model. At large collector currents, the Gummel-Poon model predicts a decrease of β caused by high injection effects as discussed above.

2-5. BREAKDOWN IN BIPOLAR JUNCTION TRANSISTORS

There are two important mechanisms of breakdown in bipolar junction transistors - the avalanche (or Zener) breakdown of the collector-base junction and the "punch-through" breakdown of the base. Punch-through occurs when the reverse collector-base voltage becomes so large that the collector-base depletion region merges with the emitter-base depletion region. The mechanism of avalanche breakdown in bipolar junction transistors is similar to that in *p-n* diodes.

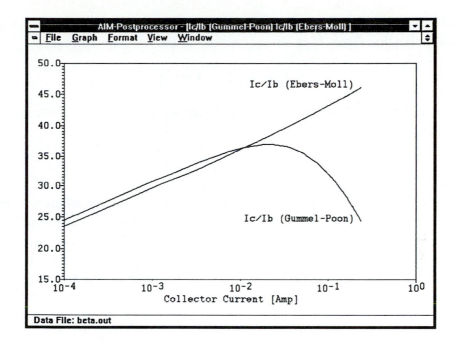

Fig. 2.4.2. AIM-Spice dependencies of the common-emitter current gain on the collector current calculated using the Ebers-Moll and Gummel-Poon models.

However, the critical voltages of the avalanche breakdown in bipolar junction transistors depend on the transistor circuit configuration (i.e., whether the device operates in a common-emitter or a common-base configuration) and on the external circuit (for example, on the external resistances connected to the transistor terminals).

When the collector-base voltage exceeds some critical value, BV_{cb}, avalanche breakdown occurs in the collector-base junction leading to an increase in the collector current. A crude estimate for BV_{cb} may be obtained assuming that the avalanche breakdown takes place when the maximum electric field at the collector-base interface exceeds the breakdown field, F_{br}, which is approximately 3×10^7 V/m for Si and 4×10^7 V/m for GaAs (see Section 1.8):

$$BV_{cb} \approx \frac{\varepsilon_s F_{br}^2}{2q}\left(\frac{1}{N_{ab}} + \frac{1}{N_{dc}}\right) \approx \frac{\varepsilon_s F_{br}^2}{2q\,N_{dc}} \qquad (2\text{-}5\text{-}1)$$

The increase in current for voltages higher than BV_{cb} can be accounted for by a multiplication factor M in the expression for the collector current

$$I_c = M\left(I_{cbo} + \alpha I_e\right) \tag{2-5-2}$$

where I_{cbo} is the common-base collector current with an open emitter (see Section 2.4),

$$\alpha = \gamma \alpha_T \tag{2-5-3}$$

is the common-base current gain, γ is the emitter injection efficiency, and α_T is the base transport factor. The factor M is equal to unity under normal operating conditions and exceeds unity when avalanche breakdown occurs. When the emitter is open, the multiplication factor due to avalanche breakdown in the collector-base junction, M_{cb}, may be approximated by the following empirical expression

$$M_{cb} = \frac{1}{1 - \left(\dfrac{V_{cb}}{BV_{cb}}\right)^{m_b}} \tag{2-5-4}$$

where V_{cb} is the collector-base voltage. The constant m_b depends on the doping profile in the collector region and on temperature. Typically, m_b is between 2 and 5 for silicon transistors.

When the base is open, the collector current under avalanche multiplication conditions is obtained by substituting $I_e = I_c$ into eq. (2-5-2) using $M \approx M_{cb}$, i.e.,

$$I_c = I_{cbo} \frac{M_{cb}}{1 - \alpha M_{cb}} \tag{2-5-5}$$

From this expression, we obtain the following condition for breakdown in the common emitter configuration:

$$\alpha M_{cb} = 1 \tag{2-5-6}$$

Usually, when breakdown is reached, the reverse collector-base voltage is much greater than the forward voltage bias across the emitter-base junction. Hence, at breakdown $V_{ce} \approx V_{cb}$. Substituting V_{cb} in eq. (2-5-4) by BV_{ce} and combining the resulting equation with eq. (2-5-6) , we find

$$BV_{ce} = BV_{cb}(1 - \alpha)^{1/m_b}$$

(2-5-7)

Since α is very close to unity and $(1 - \alpha) << 1$, BV_{ce} is much smaller than BV_{cb}. The smaller value of BV_{ce} is related to the amplification of the avalanche base current in the common-emitter configuration (see Fig. 2.5.1).

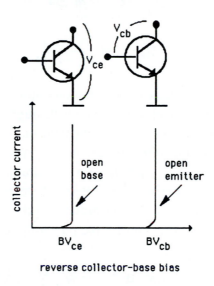

Fig. 2.5.1. Comparison of multiplication factors for open base and open emitter configurations.

Under normal operating conditions, when $V_{ce} < BV_{ce}$, eq. (2-5-2) becomes

$$I_c = I_{cbo} + \alpha I_e$$

(2-5-8)

When the base current is equal to zero (open base), we have $I_e = I_c$ and I_e can be written as

$$I_e = I_{ceo}$$

(2-5-9)

where I_{ceo} is the common-emitter current with an open base. Substituting this into eq. (2-5-8), we obtain

$$I_{ceo} = I_{cbo}/(1 - \alpha) \tag{2-5-10}$$

Hence, in the common-emitter configuration with an open base, the current is much larger than in the common-base configuration with an open emitter. Again, this is related to a large current gain in the common-emitter configuration. By substituting I_{cbo} from eq. (2-5-10) into eq. (2-5-5), we find

$$I_c = M_{ce} I_{ceo} \tag{2-5-11}$$

where

$$M_{ce} = M_{cb} \frac{1 - \alpha}{1 - \alpha M_{cb}} \tag{2-5-12}$$

is the multiplication factor for the common-emitter configuration.

Let us now consider the punch-through breakdown. This type of breakdown occurs when the reverse collector-base voltage becomes so large that the collector-base and the emitter-base depletion regions merge, and the effective base width, W_{eff}, becomes equal to zero. Assuming that most of the voltage drop at punch-through occurs across the collector-base junction and using the depletion approximation for uniform doping profiles in the collector and base region, we obtain the following expression for the punch-through voltage, V_{pth}, (see Section 1.8)

$$V_{pth} \approx \frac{q N_{ab} W^2}{2\varepsilon_S}\left(1 + \frac{N_{ab}}{N_{dc}}\right) \approx \frac{q N_{ab}^2 W^2}{2\varepsilon_S N_{dc}} \tag{2-5-13}$$

Here, the built in voltage of the base-collector junction, V_{bibc}, is assumed to be much smaller than V_{pth}, and the depletion region on the base side of the emitter-base junction has been neglected. The ratio of the avalanche breakdown voltage for the common base configuration, BV_{cb} (see eq. (2-5-1)), and the punch-through voltage, V_{pth}, is given by

$$\frac{BV_{cb}}{V_{pth}} \approx \left(\frac{\varepsilon_S F_{br}}{q n_G}\right)^2 \tag{2-5-14}$$

where $n_G = N_{ab}W$ is the Gummel number (see Section 2.4). For silicon, $F_{br} \approx 3 \times 10^7$ V/m, $\varepsilon_s \approx 1.05 \times 10^{-10}$ F/m, and eq. (2-5-14) yields

$$\frac{BV_{cb}}{V_{pth}} \approx \left(\frac{2 \times 10^{12}}{n_G \, (\text{cm}^{-2})} \right)^2 \tag{2-5-15}$$

If we assume that the base doping level is $N_{ab} = 10^{17}$ cm^{-3}, then V_{pth} is larger than BV_{cb} for base widths larger than 0.2 μm and, hence, avalanche breakdown is dominant. For thinner bases, punch-through breakdown occurs at voltages smaller than BV_{cb}.

According to the results obtained above, the breakdown voltages of both the avalanche and the punch-through mechanisms increase when the collector region doping is lowered. In practical power transistors, there is, however, another important effect called the "second breakdown". This effect, illustrated by Fig. 2.5.2, was first reported by Thorton and Simmons (1958). It occurs at relatively large collector currents. The origin of this effect seems to be a thermal instability.

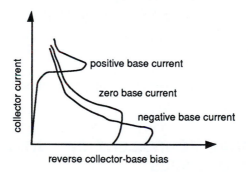

Fig. 2.5.2. Qualitative shape of current-voltage characteristics explaining the effect of second breakdown (after Shur (1990)).

Indeed, the intrinsic carrier concentration in silicon increases exponentially with temperature. If the collector current is kept constant, the temperature increase leads to a decrease of the emitter-base voltage with a negative temperature coefficient of approximately 2 mV/°C for a typical silicon transistor. If the emitter-base voltage is kept constant and the collector current is allowed to change, it increases with temperature since it is proportional to n_i. A local increase in the collector current density leads to an increase of the temperature caused by Joule heating. This, in turn, leads to a higher collector current density until a run-off condition causes destruction of the device.

Typically, short circuits between the emitter and collector are produced by local melting of the Al-Si system in the vicinity of the contacts at approximately 577 °C (see Villa (1986)). In order to guarantee a safe transistor operation, the collector current and the collector-emitter voltage should remain within the so-called Safe Operating Area (SOA) in the I_c-V_{ce} plane. This area is limited by the lines corresponding to different failure modes (see Fig. 2.5.3). SOA is an important characteristic of a power bipolar junction transistor.

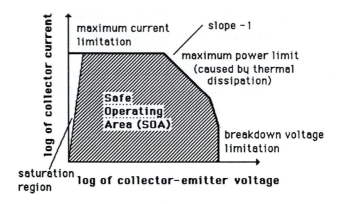

Fig. 2.5.3. Safe operating area (after Shur (1990)).

2-6. SMALL SIGNAL EQUIVALENT CIRCUIT AND CUTOFF FREQUENCIES

One of the important modes of transistor operation is the small signal (or linear) regime when the ac input signal is relatively small so that the input current, the output current, and the output voltage vary only in the vicinity of the operating point. In Fig. 2.6.1, we show the common-emitter output characteristics of a bipolar transistor with superimposed small signal variations.

With an applied dc collector voltage, V_{cc}, and a load resistance, R_c, the load line can be expressed as

$$I_c = (V_{cc} - V_{ce})/R_c \qquad (2\text{-}6\text{-}1)$$

where I_c is the dc collector current and V_{ce} is the dc collector-emitter voltage. The operating point is defined as the intersection between the load line and the characteristic defined by the applied dc base current, I_b.

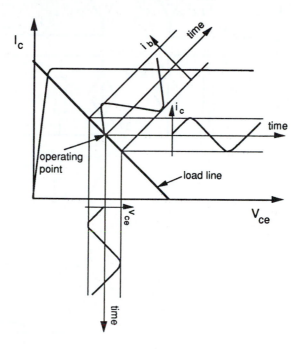

Fig. 2.6.1. Small signal (linear) regime of transistor operation.

The ac transistor response may be described in terms of a linear two-port network, allowing the transistor to be represented by several equivalent circuits depending on the choice of dependent and independent current and voltage variables. In the following, lower case symbols are used to denote the small signal current and voltage variables (see Fig. 2.4.1). Hence, we denote the input ac current and voltage as i_1 and v_1, respectively, and the output ac current and voltage as i_2 and v_2, respectively. Correspondingly, we denote the total input current and voltage as I_1 and V_1, and the total output current and voltage as I_2 and V_2. Four frequently used choices of the independent and dependent variables are illustrated in Fig. 2.6.2. In a sense, all these choices are equivalent and the elements of the two by two matrices of the z, y, h, and g parameters introduced in this figure can be expressed through each other. The so-called h-parameters are most frequently used for transistor characterization.

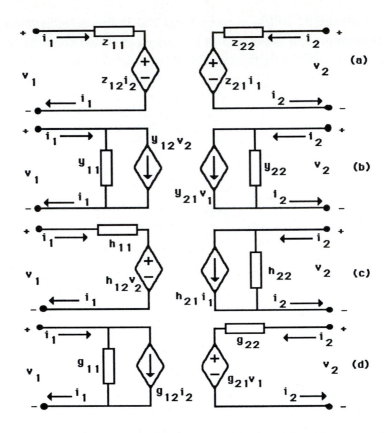

Fig. 2.6.2. Two-port linear networks for BJTs; a) z-parameters, b) y-parameters, c) h-parameters, d) g-parameters.

Using these parameters, the following equations describe the relationship between the dependent and independent variables:

$$v_1 = h_{11}i_1 + h_{12}v_2 \qquad (2\text{-}6\text{-}2)$$

$$i_2 = h_{21}i_1 + h_{22}v_2 \qquad (2\text{-}6\text{-}3)$$

Similar equations can be written for the z, y, and g parameters by substituting v_1 and i_2 in eqs. (2-6-2) and (2-6-3) by the corresponding dependent variables from Fig. 2.6.2 and by substituting i_1 and v_2 in eqs. (2-6-2) and (2-6-3) by corresponding independent variables from Fig. 2.6.2.

All these small signal parameters may be determined from different short circuit or open-circuit measurements at the input and output ports. For example,

$$h_{11} = \frac{v_1}{i_1}\bigg|_{v_2 = 0} \tag{2-6-4}$$

$$h_{12} = \frac{v_1}{v_2}\bigg|_{i_1 = 0} \tag{2-6-5}$$

$$h_{21} = \frac{i_2}{i_1}\bigg|_{v_2 = 0} \tag{2-6-6}$$

$$h_{22} = \frac{i_2}{v_2}\bigg|_{i_1 = 0} \tag{2-6-7}$$

The h-parameter equivalent circuit is frequently used for bipolar junction transistors at relatively low frequencies (below 100 MHz or so). The parameter h_{11} is called the short-circuit input impedance (h_i), h_{12} is called the open circuit reverse voltage ratio (h_r), h_{21} is called the short-circuit forward current ratio (h_f), and h_{22} is called the open-circuit output admittance (h_o). In the alternate notation shown in parentheses, a second subscript is often used to denote the transistor configuration. For example, the h-parameters for the common-emitter transistor circuit configuration are denoted as h_{ie}, h_{re}, h_{fe}, and h_{oe}. Out of a total of twelve h-parameters for three transistor configurations, the four common-emitter parameters are usually provided by transistor manufacturers on transistor data sheets. Approximate relationships allowing us to express other h-parameters in terms of these are given in Table. 2.6.1.

The parameters of the small signal equivalent circuits vary with temperature, measuring frequency, and operating point. Some of the h-parameters can be directly related to the parameters of the Ebers-Moll or the Gummel-Poon model. For example, at low frequencies, h_{fe} is equal to β and h_{fb} is equal to $-\alpha$. For other h-parameters, such a relationship is less straightforward. For the common-emitter configuration, the so-called hybrid-π equivalent circuit is frequently used (see Fig. 2.6.3). The parameters of this circuit are easier to relate to the device parameters.

Common-emitter h-parameters	Common-base h-parameters	Common-collector h-parameters
h_{ie}	$h_{ib} = h_{ie}/(h_{fe} + 1)$	$h_{ic} = h_{ie}$
h_{re}	$h_{rb} = h_{ie}h_{oe}/(h_{fe} + 1) - h_{re}$	$h_{rc} = 1$
h_{fe}	$h_{fb} = -h_{fe}/(h_{fe} + 1)$	$h_{fc} = -h_{fe} - 1$
h_{oe}	$h_{ob} = h_{oe}/(h_{fe} + 1)$	$h_{oc} = h_{oe}$

Table 2.6.1. Approximate relationships expressing different h-parameters in terms of h_{ie}, h_{re}, h_{oe}, and h_{fe} (from Shur (1990)).

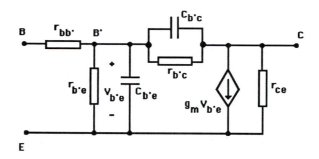

Fig. 2.6.3. Hybrid-π equivalent circuit.

The transconductance, g_m, in Fig. 2.6.3 may be related to the dynamic (differential) resistance, r_e, of the forward-biased emitter-base junction in the following way:

$$r_e = \frac{\partial V_{b'e}}{\partial I_e} \approx \frac{V_{th}}{I_e} \tag{2-6-8}$$

where $V_{th} = k_B T/q$ is the thermal voltage. Thus, from eqs. (2-4-11) and (2-6-8), we find

$$g_m = \frac{\partial I_c}{\partial V_{b'e}} = \alpha \frac{\partial I_e}{\partial V_{b'e}} \approx \frac{\alpha}{r_e} \approx \frac{I_c}{V_{th}} \tag{2-6-9}$$

The resistance $r_{b'e}$ in Fig. 2.6.3 may also be related to r_e or g_m. Indeed, assuming that the resistances r_{ce} and $r_{b'c}$ in Fig. 2.6.3 are very large and that the frequencies are sufficiently low such that the effects of the capacitances $C_{b'c}$ and $C_{b'e}$ can be neglected, we find

$$i_c \approx g_m v_{b'e} \qquad (2\text{-}6\text{-}10)$$

$$v_{b'e} \approx i_b r_{b'e} \qquad (2\text{-}6\text{-}11)$$

Substituting eq. (2-6-11) into eq. (2-6-10), we find

$$r_{b'e} \approx (i_c/i_b)/g_m \approx h_{fe}/g_m \qquad (2\text{-}6\text{-}12)$$

Another useful equivalent circuit is the T-equivalent shown in Fig. 2.6.4.

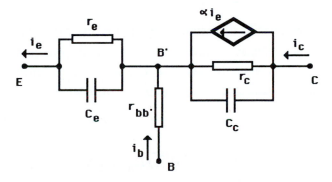

Fig. 2.6.4. T-equivalent circuit.

The emitter capacitance, C_e, in Fig. 2.6.4 may be estimated as the sum of the diffusion capacitance of the emitter-base junction, C_{edif}, and the depletion capacitance, C_{ed}, (see Section 1.8)

$$C_e = C_{edif} + C_{ed} \qquad (2\text{-}6\text{-}13)$$

where

$$C_{edif} = \left| dQ_b/dV_{b'e} \right| \qquad (2\text{-}6\text{-}14)$$

In the forward active mode, the excess minority carrier charge, Q_b, in the base (electrons in the case of *n-p-n* transistors) is given by (see Section 2.4)

$$|Q_b| = \frac{1}{2} S q W n_{be} \approx \frac{I_e W^2}{2 D_n} \tag{2-6-15}$$

(assuming that $W_{eff} \approx W$). Hence, from eqs. (2-6-8), (2-6-14) and (2-6-15), we obtain

$$C_{edif} \approx \frac{W^2}{2 D_n r_e} \tag{2-6-16}$$

leading to the base charging time constant

$$\tau_B = \left| \frac{Q_b}{I_e} \right| = C_{edif} r_e = \frac{W^2}{2 D_n} \tag{2-6-17}$$

τ_B is the characteristic diffusion time for the minority carriers across the base width, i.e., the effective transit time. In drift transistors, this transit time constant may be considerably reduced as a result of the built-in electric drift field caused by the non-uniformity of the doping profile. In this case, the transit time may be written as

$$\tau_B = \frac{W^2}{f D_n} \tag{2-6-18}$$

where $f = 2$ for a uniform doping profile, but may be much larger than 2 if the built-in electric field is sufficiently large. However, τ_B cannot be smaller than the minimum transit time, τ_{Bmin}, of carriers across the base determined by the effective saturation velocity, v_{sn}, of electrons in the base region, i.e.,

$$\tau_{Bmin} = W / v_{sn} \tag{2-6-19}$$

In very short bases, the effective saturation velocity may be substantially larger than the corresponding velocity found in long samples because of ballistic or overshoot effects.

The depletion capacitance of the emitter-base junction may be estimated as (see Section 1.8)

$$C_{ed} \approx S \sqrt{\frac{\varepsilon_s \, q N_{ab}}{2(V_{bieb} - V_{be})}} \tag{2-6-20}$$

Here, N_{ab} is the effective base doping concentration and V_{bieb} is the built-in voltage of the emitter-base junction.

The resistance $r_{b'c}$ and the capacitance $C_{b'c}$ in the hybrid-π equivalent circuit (Fig. 2.6.3) represent the dynamic (differential) resistance and the capacitance of the reverse-biased collector-base junction. (The collector-base capacitance $C_{b'c}$ is usually denoted as C_{ob} in manufacturer data sheets.) The resistance r_c in the T-equivalent circuit (Fig. 2.6.4) describes the Early effect. The resistance $r_{bb'}$ is the base spreading resistance.

Equations relating the parameters of the T- and the π-equivalent circuits to the h-parameters are given, for example, by Casasent (1973):

$$r_e = h_{re}/h_{oe} = h_{ib} - h_{rb}(1 + h_{fb})/h_{ob} \tag{2-6-21}$$

$$r_{bb'} = h_{ie} - h_{re}(1 + h_{fe})/h_{oe} = h_{rb}/h_{ob} \tag{2-6-22}$$

$$r_c = (1 + h_{fe})/h_{oe} = (1 - h_{rb})/h_{ob} \tag{2-6-23}$$

$$\alpha = h_{fe}/(1 + h_{fe}) = -h_{fb} \tag{2-6-24}$$

$$\beta = h_{fe} = -h_{fb}/(1 + h_{fb}) \tag{2-6-25}$$

Inversely, the h-parameters may also be related to the parameters of the hybrid-π equivalent circuit using circuit analysis. In particular, at low frequencies

$$h_{oe} = 1/r_{ce} + 1/r_{b'c} + g_m h_{re} \tag{2-6-26}$$

$$h_{ie} = r_{bb'} + r_{b'e} \tag{2-6-27}$$

$$h_{fe} = g_m r_{b'e} \tag{2-6-28}$$

Using the T-equivalent circuit for the common-base configuration shown in Fig. 2.6.4, we can evaluate the common base current gain α_ω at the frequency ω. The ac emitter current, i_e, is given by

$$i_e = v_{b'e}(1 + j\omega C_e r_e)/r_e = i_{eo}(1 + j\omega C_e r_e) \qquad (2\text{-}6\text{-}29)$$

where i_{eo} is the ac current through resistance r_e. Hence,

$$\alpha_\omega = i_c/i_e = i_c/[\,i_{eo}(1 + j\omega C_e r_e)] = \alpha/(1 + j\omega C_e r_e) \qquad (2\text{-}6\text{-}30)$$

where α is the common-base current gain at $\omega = 0$. Eq. (2-6-30) may be rewritten as

$$\alpha_\omega = \alpha/(1 + j\,\omega/\omega_\alpha) \qquad (2\text{-}6\text{-}31)$$

where

$$\omega_\alpha = 2\pi f_\alpha = 1/(C_e r_e) \qquad (2\text{-}6\text{-}32)$$

is called the alpha cutoff frequency. (Sometimes f_α is denoted as $f_{\alpha b}$.) At $f = f_\alpha$, the common-base current gain drops to 0.707 of its zero frequency value.

Let us now consider the common-emitter current gain. A simplified hybrid-π equivalent circuit used in the calculation of the short-circuit emitter current gain, β_ω, is shown in Fig. 2.6.5 (compare with Fig. 2.6.3).

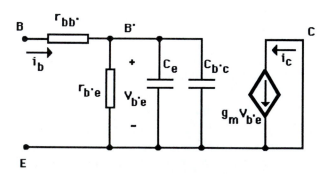

Fig. 2.6.5. Simplified hybrid-π equivalent circuit used in the calculation of β_ω.

The ac input (base) current can be expressed in terms of the voltage drop $v_{b'e}$ as

$$i_b = v_{b'e} \left[g_{b'e} + j\omega(C_e + C_{b'c}) \right] \qquad (2\text{-}6\text{-}33)$$

where $gb'e = 1/r_{b'e}$. The short circuit ac collector current, $i_c = g_m v_{b'e}$, is determined by the current source. Hence,

$$\beta_\omega = \frac{i_c}{i_b} = \frac{g_m}{g_{b'e} + j\omega(C_e + C_{b'c})} \qquad (2\text{-}6\text{-}34)$$

Eq. (2-6-34) can be rewritten as

$$\beta_\omega = \beta / \left(1 + j\omega/\omega_\beta \right) \qquad (2\text{-}6\text{-}35)$$

where

$$\omega_\beta = 2\pi f_\beta = g_{b'e} / (C_e + C_{b'c}) \qquad (2\text{-}6\text{-}36)$$

is called the beta cutoff frequency. At $f = f_\beta$, the common-emitter current gain drops to 0.707 of its zero frequency value. Eq. (2-6-36) may be rewritten as

$$f_\beta = g_m / [2\pi h_{fe} (C_e + C_{b'c})] \qquad (2\text{-}6\text{-}37)$$

In most cases, $C_e \gg C_{b'c}$ and, hence,

$$f_\beta \approx f_\alpha / h_{fe} \qquad (2\text{-}6\text{-}38)$$

Finally, we introduce the cutoff frequency, f_T, which is defined as the frequency at which the magnitude of the short-circuit common-emitter current gain equals unity, i.e., $|\beta_\omega| = 1$, which gives

$$f_T = f_\beta \sqrt{\beta^2 - 1} \approx f_\beta h_{fe} \approx g_m / [2\pi (C_e + C_{b'c})] \qquad (2\text{-}6\text{-}39)$$

where $\beta = h_{fe}$ is the low-frequency, short-circuit, common-emitter current gain (see eq. (2-6-25)). In many cases, the cutoff frequency is estimated from the measured current gain, β_ω, at frequencies much larger than f_β but much smaller than f_T. At such frequencies,

$$\beta_\omega \approx \beta/\left(j\omega/\omega_\beta\right) \qquad (2\text{-}6\text{-}40)$$

and, hence,

$$f_T \approx f_\beta h_{fe} \approx f|\beta_\omega| \qquad (2\text{-}6\text{-}41)$$

A more accurate expression for f_T may be obtained by including into eq. (2-6-39) additional time delays associated with the collector depletion layer transit time, τ_{cT}, the collector charging time, τ_c, and parasitic capacitance, C_p.

The collector transit time may be estimated as

$$\tau_{cT} \approx x_{dcb}/v_{sn} \qquad (2\text{-}6\text{-}42)$$

where x_{dcb} is the width of the collector-base depletion region and v_{sn} is the electron saturation velocity (for n-p-n transistors).

The collector charging time is

$$\tau_c = r_{cs}C_{b'c} \qquad (2\text{-}6\text{-}43)$$

where r_{cs} is the collector series resistance and $C_{b'c}$ is the collector capacitance.

The cutoff frequency may now be rewritten as

$$f_T \approx 1/\left(2\pi\tau_{eff}\right) \qquad (2\text{-}6\text{-}44)$$

where

$$\tau_{eff} = \tau_e + \tau_c + \tau_{cT} \qquad (2\text{-}6\text{-}45)$$

is the effective delay time and

$$\tau_e = \left(C_e + C_{b'c} + C_p\right)/g_m \approx \left(C_e + C_{b'c} + C_p\right)V_{th}/I_e \qquad (2\text{-}6\text{-}46)$$

In most practical transistors, τ_e is the dominant contribution to the total delay time. As can be seen from eq. (2-6-46), this time can be reduced by increasing the collector current. However, at very large collector currents, the displacement of the effective base-collector boundary into the collector region caused by the Kirk effect leads to an increase in the effective base width and to an increase in the emitter diffusion capacitance, C_{edif}. This may explain the decrease of f_T always observed at large collector currents (see eq. (2-6-39)).

An analysis of the above shows that in order to achieve a large cutoff frequency, narrow emitter stripes (small S), large emitter currents, very thin base regions, high base doping, and low parasitic capacitances are required. Very high cutoff frequencies (over 200 GHz) have been achieved in Heterojunction Bipolar Transistors where high doping in the base region may be combined with high emitter injection efficiency (see Section 2.7).

Another important characteristic of the transistor is the maximum oscillation frequency, f_{max}, which is defined as the frequency at which the power gain of the transistor is equal to unity under optimum matching conditions for the input and output impedances. Using a simplified π-equivalent circuit where we neglect $r_{b'c}$ and r_{ce}, we can obtain the following expression for f_{max} (see, for example, Sze (1981))

$$f_{max} = \sqrt{\frac{f_T}{8\pi\, r_{bb'}\, C_{b'c}}} \qquad (2\text{-}6\text{-}47)$$

2-7. HETEROJUNCTION BIPOLAR TRANSISTOR
(AIM-Spice MODEL HBTA1)

2.7.1. Principle of Operation

The idea of the Heterojunction Bipolar Transistor (HBT) was proposed by W. Shockley (1951) and was later developed by Kroemer (1957), (1982), (1983). This idea can be explained as follows. In a modern conventional bipolar junction transistor, the common-emitter current gain, β, is limited by the emitter injection efficiency. For the BJT, we have

$$\beta = \frac{I_c}{I_b} < \beta_{max} = \frac{D_n\, N_{de}\, X_e}{D_p\, N_{ab}\, W} \exp\left(-\frac{\Delta E_g}{k_B\, T}\right) \qquad (2\text{-}7\text{-}1)$$

where I_c and I_b are the collector and base currents, respectively, β_{max} is the maximum value of the common emitter current gain, D_n and D_p are hole and electron diffusion coefficients, respectively, N_{de} and N_{ab} are the donor concentration in the emitter region and acceptor concentration in the base region, respectively, X_e is the width of the emitter region (from the base boundary to the emitter contact), and W is the base width (from the emitter boundary to the collector boundary). The maximum gain is limited by the hole component of the emitter current (caused by the hole injection from the base into the emitter region). The exponential factor, $\exp(-\Delta E_g/k_B T)$, that reduces the gain is caused by the band gap narrowing in the highly doped emitter region (see, for example Sze (1981)). As pointed out by Kroemer (1982), a more realistic estimate of β_{max} is given by

$$\beta_{max} = \frac{v_{nb} N_{de}}{v_{pe} N_{ab}} \exp\left(-\frac{\Delta E_g}{k_B T}\right) \tag{2-7-2}$$

where v_{nb} and v_{pe} are effective velocities of electrons in the base and holes in the emitter, respectively, which include contributions from the carrier diffusion and drift.

In an HBT, the emitter region has a wider band gap than the base. Many HBTs are made using an AlGaAs/GaAs heterostructure because the lattice constants of these two materials are very close and, consequently, the heterointerface quality is excellent with very few dislocations and a low density of interface states. The band diagram of an n-p-n HBT biased in the active mode is shown in Fig. 2.7.1.

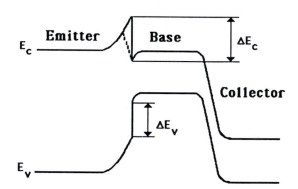

Fig. 2.7.1. Schematic HBT band structure.

With a total band discontinuity of ΔE_g, divided between ΔE_c in the conduction band and ΔE_v in the valence band, the expression for β_{max} becomes

$$\beta_{max} \approx \frac{v_{nb} N_{de}}{v_{pe} N_{ab}} \exp\left(\frac{\Delta E_v}{k_B T}\right) \tag{2-7-3}$$

where it is assumed that the notch in the conduction band of the base at the base-emitter interface is small compared to ΔE_c. Hence, a considerable value of β_{max} may be achieved even when N_{de} is much smaller than N_{ab}.

As pointed out by Kroemer (1982), electrons in the "spike-notch" region of the conduction band shown in Fig. 2.7.1 enter the base from the emitter with very large kinetic energies (close to ΔE_c) and, as a consequence, may have very high velocities (of the order of several times 10^5 m/s). Because of the directional nature of the dominant polar optical scattering in gallium arsenide – a typical material of choice for the narrow-gap semiconductor in an HBT – electrons may traverse the base region maintaining a very high velocity, even when they experience scattering. Kroemer described the conduction band spike as "a launching pad" for ballistic electrons. The magnitude of the spike can be reduced by grading the composition of the wide band gap emitter near the heterointerface (see Kroemer (1982)) as shown by the dashed line in Fig. 2.7.1. Such a grading plays an important role in optimizing the emitter injection efficiency of HBTs (see, for example, Grinberg et al. (1984)).

This discussion shows that the emitter injection efficiency of an HBT may be made very high. In fact, the transistor gain, β, is limited primarily by the recombination current caused by recombination at the emitter-base interface, I_{re}, and in the base, I_{rb}, i.e.,

$$\beta = \frac{I_{ne}}{I_{re} + I_{rb}} \qquad (2\text{-}7\text{-}4)$$

where I_{ne} is the electron component of the emitter current. In practical devices with narrow emitter stripes, an additional recombination current, I_{rs}, caused by surface recombination at the emitter edges becomes quite important (see Fig. 2.7.2). In fact, the surface recombination velocity in GaAs is quite high, on the order of 10^6 cm/s. Assuming an emitter area $S = LZ$, where L is the stripe length (long dimension) and Z is the stripe width (narrow dimension), eq. (2-7-4) can be written as

$$\beta = \frac{J_{ne}}{J_{re} + J_{rb} + j_{rs}/Z} \qquad (2\text{-}7\text{-}5)$$

where J_{ne}, J_{re}, and J_{rb} are the areal densities of the electron component of the emitter current, the recombination current caused by the recombination in the emitter-base space charge region, and the recombination current in the base region, respectively, and j_{rs} is the linear density of the surface recombination current. As can be seen from eq. (2-7-5), β is independent of Z for wide emitter stripes and is proportional to Z for narrow stripes.

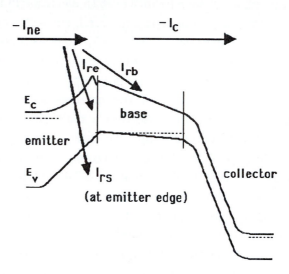

Fig. 2.7.2. Schematic diagram of current components in the HBT. The dashed lines show the position of quasi-Fermi levels in the emitter, base, and collector region.

A large recombination current may severely limit the maximum current gain of an HBT. However, as Kroemer (1982) pointed out, an even greater advantage of the HBT compared to the homojunction BJT, is the smaller base spreading resistance. Indeed, we may not need a very high current gain for many applications, but the wide band gap emitter allows us to dope the base region very strongly, drastically reducing the base resistance and the base width modulation.

In devices where the overall recombination current is small, the current gain could be as high as a few thousands or more. The recombination current, I_{rb}, caused by the recombination in the base, can be estimated as

$$I_{rb} \approx cqSn_p(0)W \, / \tau$$

(2-7-6)

where c is a numerical constant of the order of unity, $n_p(0)$ is the concentration of minority carriers (electrons) at the emitter end of the base, and τ is the minority carrier lifetime. The electron component of the emitter current can be estimated as follows

$$I_{ne} \approx qS n_p(0) v_{nb}$$ (2-7-7)

Thus, if I_{re} and I_{rs} can be neglected (which may be possible at a relatively high forward bias and with relatively wide emitter stripes), we obtain

$$\beta \approx \frac{I_{ne}}{I_{rb}} \approx \frac{\tau}{t_B}$$ (2-7-8)

where $t_B = W/v_{nb}$ is the electron transit time across the base. For a sufficiently short base (say, $W \sim 0.1$ μm), $\beta > 1000$ can be obtained even if the lifetime is only of the order of a nanosecond. Common emitter gains in access of 1600 have been achieved for AlGaAs/GaAs heterojunction bipolar junction transistors. For comparison, in a conventional homojunction silicon transistor, the dependence of the energy gap on the doping level leads to a shrinkage of the energy gap in the emitter region. (One possible mechanism of the energy gap narrowing is the formation of an impurity band when the wave functions of the donor electrons overlap.) The concentration of holes injected into the emitter region is proportional to n_{ie}^2/N_{de}, and the band gap narrowing leads to an exponential reduction of the maximum common emitter current gain (see eq. (2-7-2)), by up to a factor of 20 for $N_{de} = 10^{19}$ cm^{-3}. This energy gap shrinkage represents one of the dominant performance limitations for conventional Si BJTs (see, for example, Sze (1981)).

In addition to high injection efficiencies and correspondingly high common-emitter current gains, HBTs have a number of other advantages over conventional bipolar transistors. By allowing a higher base doping, the base spreading resistance or, alternatively, the base width could be reduced. Because of a relatively low doping of the emitter region, the emitter-base capacitance could be made quite small. All these factors result in a higher speed of operation. Recent advances in Molecular Beam Epitaxy (MBE) and Metal Organic Chemical Vapor Deposition (MOCVD) technology have made it possible to fabricate abrupt or graded heterojunctions with a high degree of reproducibility.

The simplest model for the descriptions of HBTs is the Ebers Moll model, which is identical to the model used for conventional BJTs. There are, however, many important effects which are specific for HBTs, such as surface recombination near the emitter edges (mentioned above, see Hiroaka et al.

(1987)) which greatly reduces the gain in devices with narrow emitter stripes. Hot electron, overshoot, and ballistic effects play a very important role both in the base and the collector region (see Maziar et al. (1986) and Morizuka et al. (1988)). Also, the very large doping density in the base region leads to an energy band narrowing in the base. According to Tiwari and Wright (1990), the energy gap shrinkage, ΔE_g, is approximately proportional to $N_a^{0.5}$ where N_a is the acceptor density in the base, i.e.,

$$\Delta E_g \text{ (eV)} = 2 \times 10^{-11} \, [N_a(\text{cm}^{-3})]^{0.5}$$

According to Klausmeier-Brown et al. (1989), the energy gap narrowing is

$$\Delta E_g \text{ (eV)} = 3 \times 10^{-8} \, [N_a(\text{cm}^{-3})]^{1/3}$$

For $N_a = 5 \times 10^{18}$ cm^{-3}, these equations yield $\Delta E_g = 45$ meV and 51 meV, respectively. Hence, there is some controversy regarding the band shrinkage. Also, the value of ΔE_g may depend on the type of the base dopants. In any case, these values of ΔE_g may substantially improve the emitter injection efficiency.

A more accurate analysis of the HBT operation should be based on more detailed models (see, for example, Tiwari (1986)). Such an analysis shows that both bulk recombination in the depletion regions and surface recombination at the emitter-base junction periphery limit the common-emitter short-circuit current gain. As a consequence, in heterojunction bipolar transistors optimized for high speed, the common-emitter short-circuit current gain can be as low as 10 or 20.

The propagation delay of an HBT logic gate is primarily limited by the time constant given by the product of the collector capacitance and the load resistance. The delay can be analyzed using the equation proposed by Dumke et al. (1972) for a typical HBT logic gate:

$$\tau_s = 2.5 R_B C_{BC} + R_B t_B / R_L + (3C_{BC} + C_L) R_L \qquad (2\text{-}7\text{-}9)$$

where t_B is the base transit time, R_B is the base resistance, R_L and C_L are the load resistance and the load capacitance, and C_{BC} is the collector-base capacitance.

2.7.2. Technological Aspects

As discussed in Section 2.3, there are two basic HBT technologies: Single HBT (SHBT) and Double HBT (DHBT). AlGaAs/GaAs/GaAs, InAlAs/InGaAs/InGaAs, and InP/InGaAs/InGaAs are typical material systems for SHBTs (the material order is emitter/base/collector). An SHBT does not have a barrier for electrons moving from the base into the collector. Thus, it has the advantage of a higher gain. However, the lack of symmetry between the base-emitter and base-collector junction current voltage characteristics may lead to a large collector-emitter offset voltage which is detrimental to low power operation.

Typical material systems for DHBTs are Si/SiGe/Si, AlGaAs/GaAs/AlGaAs, InAlAs/InGaAs/InP, InP/InGaAs/InP, and GaAs/GaInP/GaAs (again, the material order is emitter/base/collector). A DHBT may suffer from a low current gain because of the low base transport factor caused by the heterojunction barrier (spike) at the base-collector interface. However, this barrier can be reduced by a proper compositional grading. As a result of the symmetry between the base-emitter and the base-collector junction, the collector-emitter offset voltage is low. InP based HBTs (see Nottenburg et al. (1986)) are particularly attractive because a high conduction band discontinuity allows the injection of hot electrons with high initial velocities into the base region. These electrons may move near ballistically across the narrow base region resulting in subpicosecond transit times (see Chen et al. (1989)). In addition, this technology is compatible with long wavelength laser technology. Also, the InP substrate has a wider energy gap than the active base region, allowing back-side illumination and the use of emitter down structures. The surface recombination in InP based HBTs is less of a problem than in GaAs based devices. The contact resistance in this material system can be an order of magnitude smaller than for the AlGaAs/GaAs system. Especially intriguing is the possibility of using the InGaAs/InP emitter down structure for applications in I^2L logic (see Fig. 2.3.1). A lower collector capacitance and emitter reactance will lead to higher speed. InP based HBTs are expected to reach cut-off frequencies above 500 GHz.

Recently, Si/SiGe/Si DHBTs have drawn much attention because they can be fabricated using only a slight modification of the mature silicon bipolar technology (see Patton et al. (1990)). Cut-off frequencies exceeding 75 GHz at 298 K and up to 94 GHz at 85 K have already been obtained (see Crabbé et al. (1990)) and are expected to exceed 100 GHz in the near future.

A schematic structure of an AlGaAs/GaAs HBT is shown in Fig. 2.7.3 (after Chang et al. (1987)). This structure has a proton-implanted area that reduces the collector-base capacitance by as much as 60% leading to an increase in the cutoff frequency, f_T, and the maximum oscillation frequency, f_{max}.

A group at NTT (see Nakajima et al. (1986) and Nagata et al. (1987)) have reported a 5.5 ps/gate propagation delay for AlGaAs/GaAs HBTs with non-alloyed, compositionally graded InGaAs/GaAs emitter ohmic contacts. They reduced the specific emitter contact resistance to as low as 1.4×10^{-7} Ωcm^2. The emitter and collector dimensions were 2×5 and 4×7 μm^2, respectively. The structure also utilized proton implantation in order to reduce the collector-base capacitance. For a collector current of 12 mA, Nagata et al. reported $f_T = 80$ GHz and $f_{max} = 60$ GHz.

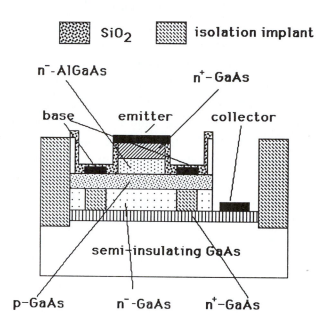

Fig. 2.7.3. A schematic structure of a high speed AlGaAs/GaAs HBT. (From Chang et al. (1987), © 1987 IEEE.)

Another approach to reducing the collector base capacitance was proposed by Kroemer (1982) who suggested the "collector-up" or "emitter-down" structure shown in Fig. 2.7.4. As can be seen from the figure, the area of the collector-base junction is limited to the active area of the device. The ion-implanted base contacts extend into the wide band gap emitter layer in order to minimize the injection of electrons directly from the emitter into the base contacts. (This injection is much less efficient inside the wide gap material than between the emitter and the narrow gap base.) Based on circuit simulations of HBTs, Akagi et al. (1987) predicted a higher speed for "collector-up" configurations than for non-self-aligned structures.

The first HBTs were implemented using the AlGaAs/GaAs material system.

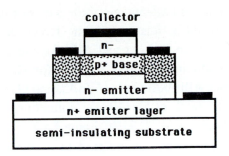

Fig. 2.7.4. Emitter-down HBT structure.

However, the use of $In_xGa_{1-x}As$ as base material has several advantages over GaAs (see, for example, Asbeck et al. (1986) and Asbeck et al. (1990)). The low-field mobility of InGaAs is higher than that of GaAs, and the energy separation between the central Γ minimum and the satellite L minima is higher than in GaAs (for example, for $In_{0.53}Ga_{0.47}As$, $\Delta E_{\Gamma\text{-}L} \approx 0.55$ eV compared to $\Delta E_{\Gamma\text{-}L} \approx 0.3$ eV in GaAs). This allows electrons to achieve higher velocities without being transferred to heavy mass satellite valleys, making ballistic and overshoot transport in the base much more efficient. Also, the composition of InGaAs in AlGaAs/InGaAs/GaAs HBTs can be graded with the percentage of In increasing from the emitter towards the base. This creates an additional built-in field assisting the electron transport across the base (see Section 1.10 and Asbeck et al. (1986)). A schematic band diagram of an HBT with a graded base is shown in Fig. 2.7.5.

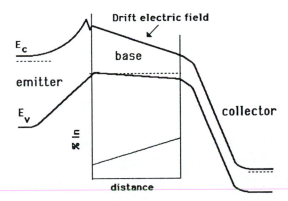

Fig. 2.7.5. Schematic band diagram of an HBT with a graded base.

In order to take full advantage of the superior properties of $In_xGa_{1-x}As$, the percentage of In should be relatively high. However, with a high In content, the lattice mismatch between InGaAs and AlGaAs and GaAs may become too large. Four other heterostructure systems suitable for HBTs are the InP/InGaAs/InP, AlInAs/InGaAs/InP, AlInAs/InGaAs/AlInAs, and GaAs/GaInP/GaAs systems (see Malik et al. (1983), Jalali et al. (1984), Nottenburg et al. (1986), Nottenburg et al. (1987), and Jalali et al. (1989), Lauterbach et al. (1991)). The potential advantages of these devices are related to a smaller built-in emitter junction potential, more pronounced ballistic and overshoot effects, a high electron drift velocity in InP, and low surface recombination velocity (see Nottenburg et al. (1986) and Jalali et al. (1989) and (1989a)). In addition, the quaternary InGaAsP emitter can be used with a band gap corresponding to the wavelength $\lambda = 1.3$ µm, compatible with fiber optic communication systems (see Su et al. (1985)). (Also, the AlInAs/InGaAs/AlInAs material system has the additional advantage of not using phosphorus in the MBE system.) Problems related to the relatively low breakdown voltage in these new material systems still remain to be solved.

Even though the base width in HBTs determines the transit time across the base and, hence, the device speed, the lateral dimensions of the emitter stripes are very important in practical transistors. Parasitic capacitances, resistances, and power dissipation are all dependent on the emitter dimensions. This was clearly demonstrated by Chang et al. (1987) who obtained a 14.5 ps gate propagation delay in HBT Common Mode Logic ring oscillators using devices with an emitter stripe width of only 1.2 µm. These devices exhibited a cutoff frequency $f_T = 67$ GHz and a maximum oscillation frequency $f_{max} = 105$ GHz. Another important feature of these devices was the very high base-region doping (up to 10^{20} cm^{-3}) which gave a small base spreading resistance.

Ishibashi and Yamaguchi (1987) reported an HBT with a new device structure. They replaced the n-type GaAs collector layer by a double layer that included a relatively thick i-GaAs layer (2000 Å in their devices) and a thin p^+-GaAs layer (200 Å thick doped at 2×10^{18} cm^{-3}). The p^+ layer is totally depleted and introduces a potential drop and an electric field in the i-layer resulting in a near ballistic collection of electrons in a certain voltage range. They obtained the very high cutoff frequency of 105 GHz.

Another type of heterojunction bipolar transistor – the Tunneling Emitter Bipolar Transistor (TEBT) – was proposed by Xu and Shur (1986) and fabricated by Najar et al. (1987). In this device, the wide band gap AlGaAs emitter is replaced by a conventional n^+-GaAs emitter, but a thin compositionally graded AlGaAs layer is inserted between the emitter and the base regions. This layer has vastly different tunneling rates for electrons and holes, so electrons can easily penetrate, but holes are prevented from being injected into the emitter region from the base. This is reminiscent of the "mass filtering" idea proposed by Capasso et al. (1985) for superlattice devices. The TEBT should have smaller emitter contact resistance, higher gain, fewer traps. Also, the TEBT is easier to fabricate than conventional HBTs.

We may conclude that as the technology matures, the HBTs will present a serious challenge to the GaAs MEtal-Semiconductor Field Effect Transistors (MESFETs) and Heterostructure Field Effect Transistors (HFETs) considered in Chapter 4.

2.7.3. HBT Theory

A. Thermionic-Emission-Diffusion model

The band structure of a single heterojunction n-p-n HBT is shown in Fig. 2.7.6.

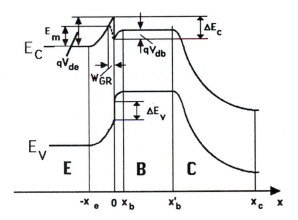

Fig. 2.7.6. Schematic single heterojunction n-p-n HBT band structure indicating important energies and positions used in the text.

As can be seen from this figure, the discontinuity in the conduction band edge at the emitter-base interface gives rise to a barrier which impedes the injection of electrons into the base region. Anderson (1962) proposed to describe the electron transport across this barrier using a thermionic model. However, as was shown, for example, by Grinberg et al. (1984), tunneling near the top of the barrier plays a very important role and must be accounted for in a realistic description of the HBT operation. Lauterbach (1992) considered a thermionic diffusion model which follows the work by Grinberg et al. (1984) but which has a more realistic description of several important factors determining the HBT operation, such as recombination current, emitter series resistance, base push out, etc. Here, we also account for tunneling in order to provide a complete analytical HBT model. We now consider the HBT structure shown in Fig. 2.7.6.

The electron current density across an abrupt heterojunction was discussed in Section 1.10. For the emitter-base interface in Fig 2.7.6, this can be written as

$$J_n'(x = 0) = J_{e \to b}(0) - J_{b \to e}(0)$$

$$= -qv_n \left[n(0^-) - n(0^+) \exp\left(-\frac{\Delta E_c}{qV_{th}} \right) \right] \qquad (2\text{-}7\text{-}10)$$

where $x = 0$ is the position of the heterointerface, $J_{e \to b}$ and $J_{b \to e}$ are the current densities associated with the flux of electrons from the emitter into the base and vice versa, respectively, ΔE_c is the conduction band discontinuity at the heterointerface, V_{th} is the thermal voltage, $v_n = [qV_{th}/(2\pi m_n)]^{1/2}$ is the mean electron thermal velocity in the x-direction (from emitter towards the base), m_n is the lighter of the electron effective masses in the emitter and base regions (see Grinberg (1986)), and $n(0^-)$ and $n(0^+)$ are the electron concentrations at the emitter and the base side of the heterointerface, respectively. These electron concentrations are related to those at the boundaries of the emitter-base space charge regions, $x = -x_e$ and $x = x_b$ (see Fig. 2.7.6), as follows:

$$n(0^-) = n(-x_e) \exp\left(-\frac{V_{de}}{V_{th}} \right) \qquad (2\text{-}7\text{-}11)$$

$$n(0^+) = n(x_b) \exp\left(\frac{V_{db}}{V_{th}} \right) \qquad (2\text{-}7\text{-}12)$$

where (see Section 1.10)

$$V_{db} = \left(V_{bi}^{eb} - V_{be} \right) \xi \qquad (2\text{-}7\text{-}13)$$

$$V_{de} = \left(V_{bi}^{eb} - V_{be}\right)(1 - \xi) \tag{2-7-14}$$

$$\xi = \left(1 + \frac{N_{ab}\,\varepsilon_b}{N_{de}\,\varepsilon_e}\right)^{-1} \tag{2-7-15}$$

Here, ε_b and ε_e are the dielectric constants and N_{ab} and N_{de} are the doping levels in the base and emitter regions, respectively, V_{bi}^{eb} is the built-in potential and V_{be} is the applied forward emitter-base bias voltage.

The electron current density given by eq. (2-7-10) is equal to the electron current density due to diffusion, $J_n(x_b)$, at the boundary of the space charge region in the base (neglecting interface recombination), which can be expressed as

$$J_n'(x_b) = -\frac{qD_{nb}}{L_{nb}} \frac{\left[n(x_b) - n_{bo}\right]\cosh\left(\dfrac{W_B}{L_{nb}}\right) - \left[n(x_b') - n_{bo}\right]}{\sinh\left(\dfrac{W_B}{L_{nb}}\right)} \tag{2-7-16}$$

Here, n_{bo}, D_{nb}, and L_{nb} are the equilibrium concentration, the diffusion coefficient and the diffusion length of the electrons in the base, respectively; W_B is the effective (neutral) base width, and x_b' is the location of the boundary of the base-collector space charge region inside the base. The excess electron concentration at the base side boundary of the base-emitter depletion region, $n(x_b) - n_{bo}$, can now be determined by equating eqs. (2-7-10) and (2-7-16), i.e.,

$$n(x_b) - n_{bo} = \frac{1}{\zeta_n}\left\{\eta_n\exp\left(\frac{\Delta E_n}{qV_{th}}\right)\left[n(x_b') - n_{bo}\right] + \right.$$

$$\left. + n(-x_e)\exp\left(\frac{-qV_{bi}^{eb} + qV_{be} + \Delta E_c}{qV_{th}}\right) - n_{bo}\right\} \tag{2-7-17}$$

where

$$\eta_n = \frac{D_{nb}}{v_n L_{nb}\,\sinh\left(W_B/L_{nb}\right)} \tag{2-7-18}$$

$$\zeta_n = 1 + \eta_n \cosh\left(\frac{W_B}{L_{nb}}\right) \exp\left(\frac{\Delta E_n}{qV_{th}}\right) \tag{2-7-19}$$

$$\Delta E_n = \Delta E_c - qV_{db} \tag{2-7-20}$$

Typically, in HBTs, the base is very highly doped. Hence, V_{db} is small so that $qV_{db} < \Delta E_c$ and $\Delta E_n > 0$, and the conduction band discontinuity creates an additional barrier for the electrons entering the neutral section of the base region. (Such a barrier is, of course, absent in homojunction BJTs.) In this case, ΔE_n is the additional energy barrier for electrons entering from the emitter into the neutral base. When $\Delta E_n < 0$, the conduction band discontinuity does not create an additional barrier because the conduction band spike at the heterointerface is below the bottom of the conduction band in the neutral section of the base. Using the following relationships:

$$n_{bo} = N_{de} \exp\left(\frac{\Delta E_c - qV_{bi}^{eb}}{qV_{th}}\right) \tag{2-7-21}$$

$$n(x_b') = n_{bo} \exp\left(\frac{V_{bc}}{V_{th}}\right) \tag{2-7-22}$$

where N_{de} is the equilibrium electron concentration in the emitter and V_{bc} is the base-collector voltage, we find from eq. (2-7-16):

$$J_n'(x_b) = -J_{ne}\left\{\left[\exp\left(\frac{V_{be}}{V_{th}}\right) - 1\right]\cosh\left(\frac{W_B}{L_{nb}}\right) - \left[\exp\left(\frac{V_{bc}}{V_{th}}\right) - 1\right]\right\} \tag{2-7-23}$$

Here,

$$J_{ne} = \frac{qD_{nb}n_{bo}}{\zeta_n L_{nb} \sinh\left(W_B/L_{nb}\right)} \tag{2-7-24}$$

is the electron current density with the base shorted to the emitter and the base-collector junction reverse biased. Since one of the advantages of the HBT is that the base doping level can be made as large as technically practical without degrading performance, the doping density in the base region is usually either much larger or at least comparable to the doping density in the emitter region. As a consequence, the concentration of the injected electrons in the base region of a typical HBT is smaller than the concentration of majority carriers (holes). Hence, we assume that $n(-x_e) = N_{de}$ and $p(x_b) = N_{ab}$.

Assuming that there is no spike in the valence band at the emitter-base heterointerface, the hole component of the current density can be treated as for a conventional BJT (see Section 1.10). In this case, the effect of the band gap discontinuity will enter only through the equilibrium minority carrier concentration p_{eo} in the expression for the hole component of the current density, i.e.,

$$J_p'(-x_e) = -J_{pe}\left[\exp\left(\frac{V_{be}}{V_{th}}\right) - 1\right]\cosh\left(\frac{X_E}{L_{pe}}\right) \tag{2-7-25}$$

where X_E is the width of the neutral part of the emitter region and

$$J_{pe} = \frac{q\,D_{pe}\,p_{eo}}{L_{pe}\,\sinh\left(X_E/L_{pe}\right)} \tag{2-7-26}$$

Here, L_{pe} and D_{pe} are the hole diffusion length and the diffusion coefficient in the emitter region, respectively.

Furthermore, based on the diffusion model for the base region, we can find the following expression for the electron current density at the base side edge of the base-collector depletion region (note that the current is defined as positive in the direction from collector towards base, see Fig. 2.7.7 and Section 2.4):

$$
\begin{aligned}
J_n'(x_b') &= \frac{qD_{nb}}{L_{nb}}\,\frac{[n(x_b') - n_{bo}] - [n(x_b') - n_{bo}]\cosh\left(W_B/L_{nb}\right)}{\sinh\left(W_B/L_{nb}\right)} \\
&= J_{ne}\left\{\left[\exp\left(\frac{V_{be}}{V_{th}}\right) - 1\right] - \left[\exp\left(\frac{V_{bc}}{V_{th}}\right) - 1\right]\frac{1 + \zeta_n \sinh^2\left(W_B/L_{nb}\right)}{\cosh\left(W_B/L_{nb}\right)}\right\}
\end{aligned} \tag{2-7-27}
$$

This is the electron component of the collector current density. The expression for the hole component of the collector current density is the same as that for the conventional diffusion model of a BJT:

$$J'_p(x_c) = -J_{pc}\left[\exp\left(\frac{V_{bc}}{V_{th}}\right) - 1\right]\cosh\left(\frac{X_C}{L_{pc}}\right) \tag{2-7-28}$$

where

$$J_{pc} = \frac{q\,D_{pc}\,p_{co}}{L_{pc}\,\sinh\left(X_C/L_{pc}\right)} \tag{2-7-29}$$

Here, L_{pc} and D_{pc} are the hole diffusion length and diffusion coefficient in the collector region, respectively, p_{co} is the equilibrium hole concentration in the collector, and X_C is the width of the neutral part of the collector.

Having solved for the current components at each junction, the equations for the intrinsic emitter and collector current densities may now be rewritten in a form similar to the Ebers-Moll equations for a homojunction BJT (see Section 2.4):

$$J'_e = A_{11}\left[\exp\left(\frac{V_{be}}{V_{th}}\right) - 1\right] + A_{12}\left[\exp\left(\frac{V_{bc}}{V_{th}}\right) - 1\right] \tag{2-7-30}$$

$$J'_c = A_{21}\left[\exp\left(\frac{V_{be}}{V_{th}}\right) - 1\right] + A_{22}\left[\exp\left(\frac{V_{bc}}{V_{th}}\right) - 1\right] \tag{2-7-31}$$

where

$$A_{11} = -J_{ne}\cosh\left(\frac{W_B}{L_{nb}}\right) - J_{pe}\cosh\left(\frac{X_E}{L_{pe}}\right) \tag{2-7-32}$$

$$A_{22} = -J_{ne}\frac{1 + \zeta_n\sinh^2(W_B/L_{nb})}{\cosh(W_B/L_{nb})} - J_{pc}\cosh\left(\frac{X_C}{L_{pc}}\right) \tag{2-7-33}$$

$$A_{12} = A_{21} = J_{ne} \tag{2-7-34}$$

Multiplying eqs. (2-7-30) and (2-7-31) with the base cross-section S and rearranging, the intrinsic emitter and collector currents can now be expressed as

$$I_e' = -\alpha_I \, I_c' - I_{eo}\left[\exp\left(\frac{V_{be}}{V_{th}}\right) - 1\right] \qquad (2\text{-}7\text{-}35)$$

$$I_c' = -\alpha_N \, I_e' - I_{co}\left[\exp\left(\frac{V_{bc}}{V_{th}}\right) - 1\right] \qquad (2\text{-}7\text{-}36)$$

where $I_{eo} = A_{12}A_{21}/A_{22} - A_{11}$ is the emitter reverse saturation current, $I_{co} = A_{12}A_{21}/A_{11} - A_{22}$ is the common-base collector reverse saturation current, $\alpha_N = -A_{21}/A_{11}$ is the normal (or forward) current gain, and $\alpha_I = -A_{12}/A_{22}$ is the inverse common-base current gain (see Section 2.4). We may also define the normal and inverse common-emitter gains as $\beta_N = \alpha_N/(1-\alpha_N)$ and $\beta_I = \alpha_I/(1-\alpha_I)$, respectively.

In a typical device, $L_{nb} \gg W_B$, $L_{pe} \gg X_E$, $L_{pc} \gg X_C$ and $\Delta E_n \gg qV_{th}$, which allow the coefficients of the Ebers-Moll equations to be simplified. Taking into account these inequalities, we have

$$\zeta_n \approx \frac{D_{nb}}{v_n W_B}\exp\left(\frac{\Delta E_n}{qV_{th}}\right) \qquad (2\text{-}7\text{-}37)$$

$$A_{12} = A_{21} = J_{ne} \approx qn_{bo}v_n\exp\left(-\frac{\Delta E_n}{qV_{th}}\right) \qquad (2\text{-}7\text{-}38)$$

$$A_{11} \approx -qn_{bo}v_n\exp\left(-\frac{\Delta E_n}{qV_{th}}\right) - \frac{qD_{pe}\,p_{eo}}{X_E} \qquad (2\text{-}7\text{-}39)$$

$$A_{22} \approx -\frac{qn_{bo}\,D_{nb}\,W_B}{L_{nb}^2} - \frac{qD_{pc}\,p_{co}}{X_C} \qquad (2\text{-}7\text{-}40)$$

We note that the term $\exp(-\Delta E_n/qV_{th})$ in A_{11}, A_{12} and A_{21} will result in a voltage offset in the I–V characteristics for both the common-base and the common-emitter configuration.

B. Thermionic-Field-Emission Diffusion Model

When electron tunneling is taken into account, eq. (2-7-10) should be rewritten as (see Grinberg et al. (1984))

$$J_n'(x=0) = -qv_n\left[n(0^-) - n(0^+)\exp\left(-\frac{\Delta E_c}{qV_{th}}\right)\right]\gamma_n \tag{2-7-41}$$

where

$$\gamma_n = 1 + \frac{\exp(V_{de}/V_{th})}{V_{th}}\int_{V^*}^{V_{de}} Tr\left(\frac{V_x}{V_{de}}\right)\exp\left(-\frac{V_x}{V_{th}}\right)dV_x \tag{2-7-42}$$

and

$$V^* = \begin{cases} V_{de} - \Delta E_c/q, & \text{for } V_{de} \geq \Delta E_c/q \\ 0, & \text{for } V_{de} < \Delta E_c/q \end{cases} \tag{2-7-43}$$

(see Fig. 2.7.6). The limits of the integral in eq. (2-7-42) are obtained by considering the range of barrier energies available for tunneling, and $Tr(V_x/V_{de})$ is the barrier transparency. Using an approach developed by Stratton (1969) and Padovani and Stratton (1966), we find for a triangular barrier

$$Tr(y) = \left[\sqrt{1-y} - y\ln\left(\frac{1+\sqrt{1-y}}{\sqrt{y}}\right)\right]\exp\left(-\frac{V_{de}}{V_{oo}}\right) \tag{2-7-44}$$

where V_{oo} is a constant voltage.

A comparison of the results obtained in this Section with those obtained above using the thermionic model shows that the effect of tunneling may be accounted for by replacing v_n by $\gamma_n v_n$ in the equations for the emitter and collector currents and in the expressions for the current gain.

C. Thermionic-Field-Diffusion Model for HBTs with Graded Emitter

A qualitative analysis of the effects of grading the AlGaAs composition at the heterointerface was given by Kroemer (1982). Also, a discussion of the role played by grading was given by Hayes et al. (1983). The analysis given below is based on the results obtained by Grinberg et al. (1984). The energy diagram of a graded emitter region is shown in Fig. 2.7.6 (dashed line). Here, we assume that the composition x of $Al_xGa_{1-x}As$ in the emitter region varies linearly over the grading length W_{GR}. As can be seen from the figure, the effect of grading is to make $\Delta E_n = \Delta E_c - qV_{db}$ smaller. For practical values of grading length, the spike is abrupt enough to justify the use of the thermionic-field emission model. Hence, the only necessary change in the theory given above is the replacement of ΔE_n by its value for the graded junction. For simplicity, we only consider the case of uniform doping profiles in the emitter and base regions. In this case, the potential distribution in the depletion layers is given by

$$V(x) = \begin{cases} -\dfrac{qN_{de}(x + x_e)^2}{2\varepsilon_e}, & \text{for } -x_e \le x \le 0 \\[4mm] -\dfrac{qN_{ab}(x - x_b)^2}{2\varepsilon_b} - V_{bi}^{eb} + V_{be}, & \text{for } 0 < x \le x_b \end{cases} \tag{2-7-45}$$

where the depletion lengths on the emitter and the base side of the interface, respectively, given by

$$x_e = \sqrt{\frac{2N_{ab}\,\varepsilon_e\varepsilon_b\left(V_{bi}^{eb} - V_{be}\right)}{qN_{de}\left(\varepsilon_b N_{ab} + \varepsilon_e N_{de}\right)}} \tag{2-7-46}$$

$$x_e = x_b N_{ab}/N_{de} \tag{2-7-47}$$

where

$$qV_{bi}^{eb} = \left(E_{gb} - \Delta E_{Fn} - \Delta E_{Fp} + \Delta E_c\right) \tag{2-7-48}$$

$$V_{bi}^{eb} - V_{be} = \frac{qN_{de}\,x_e}{2}\left(\frac{x_e}{\varepsilon_e} + \frac{x_b}{\varepsilon_b}\right) \tag{2-7-49}$$

Here, E_{gb} is the energy gap in the base region, ΔE_{Fn} is the energy difference between the bottom of the conduction band and the Fermi level in the emitter material far from any interface, and ΔE_{Fp} is the energy difference between the top of the valence band and the Fermi level in the base material far from any interface (see also Section 1.10). The variation of the conduction band edge in the graded region of the emitter is described by the following equation

$$E_c(x) = -qV(x) - \Delta E_c\left(1 + \frac{x}{W_{GR}}\right) \qquad (2\text{-}7\text{-}50)$$

which is valid for linear grading assuming that the electron affinity and the energy gap varies linearly with the composition. The barrier height on the emitter side of the interface, E_m, at $x = -W_{GR}$ (see Fig. 2.7.6) is given by

$$E_m = \frac{q^2 N_{de}\left(x_e - W_{GR}\right)^2}{2\varepsilon_e} \qquad (2\text{-}7\text{-}51)$$

Hence,

$$\Delta E_n = \Delta E_c + E_m - V_{bi}^{eb} + V_{be} \qquad (2\text{-}7\text{-}52)$$

This equation shows how the grading decreases the value of the additional barrier for the electrons entering the base from the emitter region. These equations are valid when the grading length is smaller than the width of the depletion layer in the emitter region. If this is not the case, or if $dE_c/dx > 0$ at $x = -W_{GR}$, then $E_m = 0$.

We should notice that the quality of the heterointerface may depend on the grading. In fact, grading may lead to an increase in the recombination current and actually cause a sharp decrease in gain (see Kusano et al. (1990)).

D. Non-Ideal Effects in HBTs
D1. Recombination currents

In order to achieve a more realistic HBT model, we have to consider some additional current contributions in the device. These are caused by recombination of minority carriers in the neutral base, surface recombination at the emitter periphery, and recombination in the emitter-base junction (see Fig. 2.7.7). Below, the corresponding contributions to emitter current are considered one by one as indicated in the equivalent circuit in Fig. 2.7.7.

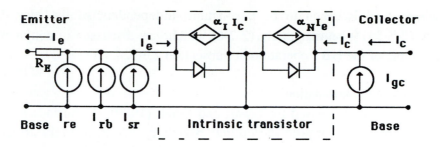

Fig. 2.7.7. Equivalent circuit of HBT accounting for recombination and generation currents and for emitter series resistance.

a) Recombination in the neutral base

For a high gain transistor, we may assume a linear variation of the electron concentration in the neutral base owing to diffusion transport. The base recombination current, I_{rb}, is then given by the average excess electron density, n_{bex}, divided by the minority electron life time in the base, τ_{eb}. Using eqs. (2-7-17) to (2-7-19), we find

$$I_{rb} = SW_B \frac{q n_{bex}}{\tau_{nb}} = \frac{SW_B \, q}{2 \tau_{nb}} \left\{ \left[n(x_b) - n_{bo} \right] + \left[n(x_b') - n_{bo} \right] \right\}$$

$$= \frac{SW_B \, q n_B}{2 \tau_{nb}} \left\{ \frac{1}{\zeta_n} \left[\exp\left(\frac{V_{be}}{V_{th}} \right) - 1 \right] \right.$$

$$\left. + \left[1 + \frac{\eta_n}{\zeta_n} \exp\left(\frac{\Delta E_n}{q V_{th}} \right) \right] \left[\exp\left(\frac{V_{bc}}{V_{th}} \right) - 1 \right] \right\} \qquad (2\text{-}7\text{-}53)$$

In our AIM-Spice model (see below), we will be using the following simplified version of eq. (2-7-53) which is approximately valid for the active forward mode of operation:

$$I_{rb} \approx I_{rb}^o \left[\exp\left(\frac{V_{be}}{V_{th}} \right) - 1 \right] \qquad (2\text{-}7\text{-}54)$$

where $I_{rb}{}^o$ is taken to be a constant, independent of the bias conditions. Eq. (2-7-54) is similar to eq. (2-7-6) which was discussed in conjunction of the principle of the HBT operation in Subsection 2.7.1.

b) Surface recombination

As was discussed by Tiwari and Frank (1989), the surface recombination primarily occurs around the emitter periphery. It is strongly affected by the Fermi potential pinning at the surface and by the conditions at the surface in general. Tiwari and Frank (1989) demonstrated that for collector currents above approximately 10^{-7} A, the surface recombination current, I_{rs}, is roughly proportional to $\exp(V/V_{th})$. At smaller collector currents, I_{rs} is roughly proportional to $\exp(V/(2V_{th}))$. Thus, the surface recombination current can be written as

$$I_{rs} = A_{rs1} P_e \left[\exp\left(\frac{V_{be}}{V_{th}} \right) - 1 \right] + A_{rs2} P_e \left[\exp\left(\frac{V_{be}}{2V_{th}} \right) - 1 \right] \qquad (2\text{-}7\text{-}55)$$

where A_{rs1} and A_{rs2} are constants which depend on the surface properties, and P_e is the emitter perimeter.

c) Interface recombination and recombination in the emitter-base space charge region.

These contributions can, in principle, also be expressed in terms of trap concentrations and capture cross sections. However, we restrict ourselves to an estimate of the interface recombination current of the form

$$I_{re} = I_{re}^o \left[\exp\left(\frac{V_{be}}{m_{re} V_{th}} \right) - 1 \right] \qquad (2\text{-}7\text{-}56)$$

where m_{re} is an ideality factor ($m_{re} \approx 2$ but may vary from device to device) and $I_{re}{}^o$ is the interface recombination "saturation" current which is proportional to the base-emitter interface cross section. A similar recombination/generation current contribution, I_{rc}, is associated with the base-collector interface.

D2. High current level effects

a) Emitter resistance voltage drop

The intrinsic base-emitter voltage, V_{be}, actually applied to the base-emitter junction is given by

$$V_{be} = V_{be}^{ext} - R_e I_e - R_b I_b \qquad (2\text{-}7\text{-}57)$$

where V_{be}^{ext} is the external emitter-base voltage, R_e and R_b are the emitter and base resistances, including contact resistances, and I_e and I_b are emitter and base terminal currents, respectively (see Fig. 2.7.7). In HBTs, the base is highly doped and, hence, R_b is usually less than or comparable to R_e. Furthermore, $I_b \approx I_e/\beta \ll I_e$ so that the voltage drop, $R_b I_b$, may be small. (This voltage drop may still be important because it may lead to emitter current crowding.)

b) Base push out (Kirk effect)

Base push out occurs when the collector current density, J_c, exceeds a "critical" value, J_{crit}, which is approximately equal to the value of the current corresponding to electron velocity saturation in the collector region, $qN_{dc}v_s$. Schrenk (1980) derived the following expression for the critical current density:

$$J_{crit} = qN_{dc} \, v_s + \frac{2v_s \, \varepsilon_c \left(V_{bi}^{cb} - V_{bc} \right)}{X_c^2} \qquad (2\text{-}7\text{-}58)$$

A further increase in the collector current is slowed down by the extension of an electron-hole plasma into the collector region, which increases the effective width of the base region. In eq. (2-7-58), v_s is the electron saturation velocity, V_{bi}^{cb} is the built-in voltage at the collector-base junction and X_c is the width of the collector region. Above this critical current density, the base width is enlarged by (see Schrenk (1978))

$$\Delta W_B = X_c \left(1 - \frac{J_{crit} - qv_s \, N_{dc}}{J_c - qv_s \, N_{dc}} \right) \qquad (2\text{-}7\text{-}59)$$

2.7.4. AIM-Spice HBT Model (HBTA1)

In Subsection 2.7.1, we discussed an idealized semi-analytical HBT model. However, practical devices often do not exhibit such an ideal behavior. Several possible non-ideal mechanisms associated with the heterointerface were discussed in conjunction with the heterojunction diode in Section 1.10. Above, we also mentioned that the effect of grading, for example, may be exactly opposite to what is expected from the ideal theory (see Kusano et al. (1990)). Therefore, for the purpose of HBT circuit simulation, it is more practical to rely on a simplified device model with parameters extracted from the experimental data.

Based on the physics discussed in Section 2.7.2, we can now write a simplified system of equations to model an HBT. This system of equations is nearly identical to the Ebers-Moll model (see Sections 2.4). However, the equation for the emitter current now also accounts for the important surface recombination term, I_{rs}:

$$I_e = I_{cc} + I_{be} + I_{re} + I_{rb} + I_{rs} \tag{2-7-60}$$

$$I_c = I_{cc} + I_{bc} + I_{rc} \tag{2-7-61}$$

$$I_b = I_e - I_c \tag{2-7-62}$$

$$I_{be} = \frac{I_F}{\beta_N} = \frac{I_i}{\beta_N}\left[\exp\left(\frac{V_{be}}{V_{th}}\right) - 1\right] \tag{2-7-63}$$

$$I_{bc} = \frac{I_R}{\beta_I} = \frac{I_i}{\beta_I}\left[\exp\left(\frac{V_{bc}}{V_{th}}\right) - 1\right] \tag{2-7-64}$$

$$I_{cc} = I_F - I_R = I_i\left[\exp\left(\frac{V_{be}}{V_{th}}\right) - \exp\left(\frac{V_{bc}}{V_{th}}\right)\right] \tag{2-7-65}$$

$$I_i = -\frac{\alpha_N I_{eo}}{1 - \alpha_N \alpha_I} \tag{2-7-66}$$

$$V_{be} = V_{be}^{ext} - R_e I_e - R_b I_b \tag{2-7-67}$$

$$V_{bc} = V_{bc}^{ext} - R_c I_c \tag{2-7-68}$$

$$I_{re} = I_{re}^o \left[\exp\left(\frac{V_{be}}{m_{re}V_{th}}\right) - 1 \right] \tag{2-7-69}$$

$$I_{rb} = I_{rb}^o \left[\exp\left(\frac{V_{be}}{V_{th}}\right) - 1 \right] \tag{2-7-70}$$

$$I_{rs} = A_{rs1} P_e \left[\exp\left(\frac{V_{be}}{V_{th}}\right) - 1 \right] + A_{rs2} P_e \left[\exp\left(\frac{V_{be}}{2V_{th}}\right) - 1 \right] \tag{2-7-71}$$

Let us now discuss the differences between HBT modeling and conventional BJT modeling in more detail.

1. Different recombination mechanisms play a significant role as described above. In particular, the surface recombination may be important (or even dominant).

2. The pre-exponential current densities A_{11} and A_{22} for the emitter-base and the collector-base characteristics, respectively, are much more different for a single heterojunction bipolar transistor than for a conventional BJT. When $\Delta E_n \gg V_{th}$, we have approximately $A_{11}/A_{22} \propto \exp(-\Delta E_n/qV_{th})$, leading to an offset voltage, $\Delta E_n/q$, in the output current-voltage characteristics.

3. High injection effects are not as important as in a conventional BJT because of a very high doping level in the base region. That is why we can use the Ebers-Moll model instead of the Gummel-Poon model.

4. The effects of base push out may be very important and may strongly affect the overall common emitter gain, β, at high collector currents.

The above equations may be further simplified in a first-order analysis by considering only the active forward mode of operation. In this case, we have

$$I_e \approx I_F \left(1 + \beta_N^{-1}\right) + I_{re} + I_{rb} + I_{rs} \tag{2-7-72}$$

$$I_c \approx I_F \tag{2-7-73}$$

and, hence,

$$\beta = \frac{I_c}{I_b} = \frac{\beta_o}{1 + \alpha_o \exp\left(-\dfrac{V_{be}}{2V_{th}}\right)} \tag{2-7-74}$$

where

$$\alpha_o = \frac{I_{re}^o + A_{rs2} P_e}{I_i/\beta_N + I_{rb}^o + A_{rs1} P_e} \tag{2-7-75}$$

$$\beta_o = \frac{I_i}{I_i/\beta_N + I_{rb}^o + A_{rs1} P_e} \tag{2-7-76}$$

At relatively small collector currents, just as in a conventional BJT, we expect the collector current to be proportional to $\exp(V_{be}/V_{th})$ and the overall recombination current to be proportional to $\exp(V_{be}/2V_{th})$. As a consequence, β should be proportional to $I_c^{1/2}$. At very high collector current, the common emitter gain should reach a plateau if the critical current for base push out, I_{crit}, is large enough. However, in practical devices where the collector is doped relatively low in order to obtain a higher breakdown voltage, I_{crit} is not very large, and β may start dropping at large collector currents owing to the increased effective base width, $W_{eff} = W_B + \Delta W_B$. The analysis of the different terms in eq. (2-7-74) shows that $\beta \sim 1/(W_{eff} + c_2 W_{eff}^2)$ where c_2 is a constant. However, for circuit modeling, it is more appropriate to account for the base push out by introducing an empirical term into eq. (2-7-74), keeping the effective base width constant, i.e.,

$$\beta \approx \frac{\beta_o}{\left[1 + (I_{crit}/I_c)^m\right]^{1/m}} \tag{2-7-77}$$

where m is a parameter which determines the shape of the β versus I_c relationship for I_c close to I_{crit}. (Usually, in the forward active mode, β_N is quite high and the first term in the denominators of eq. (2-7-76) can be neglected.)

Fig. 2.7.8 shows the dependence of the common-emitter current gain, β, of an AlGaAs/GaAs SHBT on the collector current calculated using this AIM-Spice model.

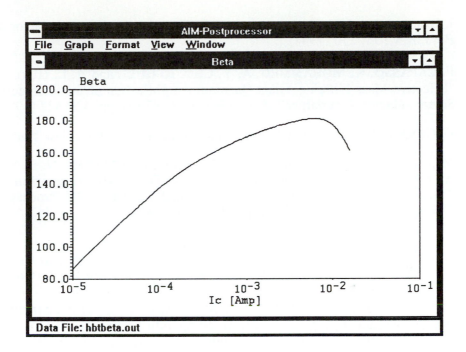

Fig. 2.7.8. Dependencies of the common emitter current gain on the collector current calculated for an AlGaAs/GaAs SHBT calculated using the AIM-Spice HBT model.

REFERENCES

J. AKAGI, J. YOSHIDA and M. KURATA, "A Model-Based Comparison of Switching Characteristics Between Collector-Top and Emitter-Top HBT's", *IEEE Trans. Electron Devices*, ED-34, p. 1413-1418 (1987)

R. A. ANDERSON, *Solid State Electron.*, 5, pp. 341-351 (1962)

P. M. ASBECK, M. F. CHANG, K. C. WANG, G. J. SULLIVAN and D. L. MILLER, "GaAlAs/GaInAs/GaAs Heterojunction Bipolar Technology for sub-35 ps Current-Mode Logic Circuits", *Proceedings of the 1986 Bipolar Circuits and Technology Meeting*, IEEE, Minneapolis, pp. 25-26 (1986)

P. M. ASBECK, M. F. CHANG, K. C. WANG and D. L. MILLER, "Heterojunction Bipolar Transistor Technology", in *Introduction to Semiconductor Technology. GaAs and Related Compounds*, ed. by Cheng T. Wang, John Wiley and Sons, New York (1990)

J. BARDEEN and W. H. BRATTAIN, "The Transistor, a Semiconductor Triode", *Phys. Rev.*, 74, 230 (1948).

F. CAPASSO, K. MOHAMMED, A. Y. CHO, R. HULL and A. L. HUTCHINSON, "Effective Mass Filtering: Giant Quantum Amplification of the Photocurrent in a Semiconductor Superlattice", *Appl. Phys. Lett.*, 47 (4), pp. 420-422 (1985)

D. CASASENT, *Electronic Circuits*, Quantum Publishers, Inc., New York (1973)

M. F. CHANG, P. M. ASBECK, K. C. WANG, G. J. SULLIVAN, N. H. SHENG, J. A. HIGGINS and D. L. MILLER, "AlGaAs/GaAs Heterojunction Bipolar Transistors Fabricated Using a Self-Aligned Dual-Lift-Off Process," *IEEE Electron Device Letters*, EDL-8, pp. 303-305 (1987)

Y. K. CHEN, R. N. NOTTENBURG, M. B. PANISH, R. A. HAMM and D.A. HUMPHREY, "Subpicosecond InP/InGaAs Heterostructure Bipolar Transistors", *IEEE Electron Device Letters*, EDL-10, No. 6, pp. 267-269, June (1989)

E. F. CRABBÉ, G. L. PATTON, G. M. C. STORK, G. H. COMFORT, B. S. MEYRSON and J. Y-C. SUN, "Low Temperature Operation of Si and SiGe Bipolar Transistors", *IEDM Technical Digest*, pp. 17-20, IEEE Catalog No. CH2865-4 (1990)

W. P. DUMKE, J. M. WOODALL and V. L. RIDEOUT, "GaAs-AlGaAs Heterojunction Transistor for High Frequency Operation," *Solid-State Electronics*, 15, pp. 1339-1343 (1972)

J. M. EARLY, "Effects of Space-Charge Layer Widening in Junction Transistors", *Proc. IRE*, 40, p. 1401 (1952)

J. J. EBERS and J. L. MOLL, "Large-Signal Behavior of Junction Transistors", *Proc. IRE*, 42, p. 1761 (1954)

D. ETIEMBLE and M. MOHSSINE, "Comparison of Performance of a Benchmark Circuit According to Various Design Styles for an Advanced Bipolar Technology", *Proc. IEEE 1987 Bipolar Circuits and Technology Meeting*, pp. 46-49, IEEE Catalog No. 87CH2509-8, Minneapolis (1987)

S. K. GHANDHI, *VLSI Fabrication Principles*, John Wiley & Sons, New York (1983)

A. A. GRINBERG, M. SHUR, R. J. FISHER and H. MORKOÇ, "Investigation of the Effect of Graded Layers and Tunneling on the Performance of AlGaAs/GaAs Heterojunction Bipolar Transistors", *IEEE Trans. Electron Devices*, ED-31, No 12, pp. 1758-1765 (1984)

A. A. GRINBERG, "Thermionic Emission in Heterosystems with Different Effective Electronic Masses", *Phys. Rev.*, B33, No. 10, pp. 7256-7258 (1986)

H. K. GUMMEL and H. C. POON, "An Integral Charge Control Model of Bipolar Transistors", *Bell Syst. Tech. J.*, 49, p. 827 (1970)

J. R. HAYES, F. CAPASSO, R. J. MALIK, A. C. GOSSARD and W. WIEGMANN, *Applied Physics Letters*, 43, p. 949 (1983)

Y. S. HIROAKA and J. YOSHIBA, "Two-Dimensional Analysis of the Surface Recombination Effect on Current Gain for GaAlAs/GaAs HBTs", *IEEE Trans. Electron Devices*, ED-35, no. 7, p. 857 (1987)

T. ISHIBASHI and Y. YAMAGUCHI, "A Novel AlGaAs/GaAs HBT Structure for Near Ballistic Collection", in *Program of 45th Annual Device Research Conference*, June 22-24, Santa Barbara, p. IV-A6 (1987)

B. JALALI, H. KANBE, J. C. VLCEK and C. G. FONSTAD, *IEEE Electron Device Letters*, EDL-5, p. 172 (1984)

B. JALALI, R. N. NOTTENBURG, Y. K. CHEN, A. F. J. LEVI, D. SIVCO, A. Y. CHO and D. A. HUMPHREY, *Applied Physics Letters*, 54, p. 2333 (1989)

B. JALALI, R. N. NOTTENBURG, J. C. BISCHOFF, W. S. HOBSON, Y. K. CHEN, T. FULLOWAN, S. J. PEARTON and A.S. JORDAN, "AlInAs/GaInAs Heterostructure Bipolar Transistors Grown by Metallorganic Chemical Vapor Deposition", *Electronics Letters*, 25, No. 22, p. 1496 (1989)

M. E. KLAUSMEIER-BROWN, M. S. LUNDSTROM AND M. R. MELLOCH, "The Effects of Heavy Impurity Doping on AlGaAs/GaAs Bipolar Transistors", *IEEE Trans. Electron Devices*, ED-36, No.10, pp. 2146-2155, Oct. (1989)

H. KROEMER, "Theory of a Wide-Gap Emitter for Transistors", *Proc. IRE*, 45, pp. 1535-1537 (1957)

H. KROEMER, "Heterostructure Bipolar Transistors and Integrated Circuits", *Proc. IEEE*, 70, pp. 13-25 (1982)

H. KROEMER, "Heterostructure Bipolar Transistors: What Should We Build?", *J. Vac. Sci. Technol.*, B1 (2), pp. 112-130 (1983)

C. KUSANO, H. MASUDA, K. MOCHIZUKI, M. KAWADA and K. MITANI, "The Effect of Turn-on Voltage (Vbe) of AlGaAs/GaAs HBT's Due to the Structure of the Emitter-Base Heterojunction", *Jap. J. Appl. Phys.*, 29, No. 8, pp. 1399-1402, Aug. (1990)

T. LAUTERBACH, M. SHUR, K. H. BACHEM AND W. PLETSCHEN, "Emitter-Base Electron Transport in GaInP/GaAs Heterojunction Bipolar Transistors and Tunnel Emitter Bipolar Transistors", in *Proceedings of The First International Semiconductor Device Research Symposium*, pp. 721-724, Charlottesville, VA, Dec. (1991)

T. LAUTERBACH, Doctoral Thesis, Erlangen-Nürnberg University (1992)

R. J. MALIK, J. R. HAYES, F. CAPASSO, K. ALAVI and A. Y. CHO, *IEEE Electron Device Letters*, EDL-4, p. 383 (1983)

C. M. MAZIAR, M. E. KLAUSMEIER-BROWN and M. S. LUNDSTROM, *IEEE Electron Device Letters*, EDL-7, no. 8, p. 483 (1986)

K. MORIZUKA et al., "Transit Time Reduction in AlGaAs/GaAs HBT's Utilizing Velocity Overshoot in p-Type Collectors", *IEEE Electron Device Letters*, EDL-9, no. 11, p. 585 (1988)

M. NAKAMAE, "Recent Progress and Future Prospects for VLSI Silicon Bipolar Transistors", *Proc. IEEE 1987 Bipolar Circuits and Technology Meeting*, pp. 5-6, IEEE Catalog No. 87CH2509-8, Minneapolis (1987)

F. E. NAJAR, D. C. RADULESCU, Y. K. CHEN, G. W. WICKS, P. J. TASKER and L. F. EASTMAN, "DC Characterization of the AlGaAs/GaAs Tunneling Emitter Bipolar Transistor", *Appl. Phys. Lett.*, 50, no. 26, p. 1915 (1987)

K. NAGATA, O. NAKAJIMA, Y. YAMAUCHI, H. ITO, T. NITTONO and T. ISHIBASHI, "High Speed Performance of AlGaAs/GaAs Heterojunction Bipolar Transistor with Non-Alloyed Emitter Contacts", in *Program of 45th Annual Device Research Conference*, June 22-24, Santa Barbara, p. IV-A2 (1987)

O. NAKAJIMA, K. NAGATA, Y. YAMAUCHI, H. ITO and T. ISHIBASHI, "High Speed AlGaAs/GaAs HBTs with Proton-implanted Buried Layers", *IEDM Technical Digest*, p. 416, Los Angeles, published by IEEE , Dec. (1986)

R. N. NOTTENBURG, J. C. BISCHOFF, J. H. ABELES, M. B. PANISH and H. TEMKIN, "Base doping effects in InGaAs/InP Double Heterostructure Bipolar Transistors", *IEDM Technical Digest*, p. 278, Los Angeles, published by IEEE, Dec. (1986)

R. N. NOTTENBURG, H. TEMKIN, M. B. PANISH, R. BHAT and J. C. BISCHOFF, "InGaAs/InP Double Heterostructure Bipolar Transistors with Near Ideal β versus I_c", *IEEE Electron Device Letters*, EDL-8, p. 282 (1987)

A. PADOVANI and R. STRATTON, *Solid State Electronics*, 9, p. 695 (1966)

G. L. PATTON, G. M. C. STORK, G. H. COMFORT, E. F. CRABBÉ, B. S. MEYRSON, D. L. HARAME and J. Y.-C. SUN, "Low Temperature Operation of Si and SiGe Bipolar Transistors", *IEDM Technical Digest*, pp. 13-16, IEEE Catalog No.CH2865-4 (1990)

H. SCHRENK, *Bipolare Transistoren*, Springer-Verlag, Berlin (1978)

W. SHOCKLEY, US Patent No. 2,569,347, issued 1951

M. SHUR, *Physics of Semiconductor Devices*, Series in Solid State Physical Electronics, Prentice Hall, New Jersey (1990)

R. STRATTON, "Theory of Field Emission from Semiconductors", *Phys. Rev.*, 125, pp. 67-82 (1969)

L. M. SU, N. GROTE, R. KAUMANNS and H. SHROETER, "An NpnN Double Heterojunction Bipolar Transistor on InGaAsP/InP", *Appl. Phys. Lett.*, 47, no. 1, pp. 28-30 (1985)

S. M. SZE, *Physics of Semiconductor Devices*, Second Edition, John Wiley and Sons, New York (1981)

S. M. SZE, *Semiconductor Devices. Physics and Technology*, John Wiley & Sons, New York (1985)

S. M. SZE, Editor, *VLSI Technology*, Second Edition, McGraw Hill, New York (1988)

C. G. THORTON and C. D. SIMMONS, "A New High Current Mode of Transistor Operation", *IRE Trans. Electron Devices*, ED-5, pp. 6-10 (1958)

S. TIWARI, "GaAlAs/GaAs Heterostructure Bipolar Transistors: Experiment and Theory", *IEDM Technical Digest*, p. 262, Los Angeles, published by IEEE , Dec. (1986)

S. TIWARI AND S. L. WRIGHT, "Material Properties of *p*-Type GaAs at Large Dopings", *J. Appl. Phys. Lett.*, 56, No. 6, pp. 563-565, Feb. (1990)

S. TIWARI and D. J. FRANK, "Analysis of the Operation of GaAlAs/GaAs HBT's", *IEEE Trans. Electron Devices*, ED-36, pp. 2105-2121, Oct. (1989)

F. F. VILLA, "Improved Second Breakdown of Integrated Bipolar Power Transistors", *IEEE Trans. Electron Devices*, ED-33, pp. 1971-1976, Dec. (1986)

R. M. WARNER and B. L. GRUNG, *Transistors Fundamentals for the Integrated-Circuit Engineer*, John Wiley & Sons, New York (1983)

J. XU and M. SHUR, "Tunneling Emitter Bipolar Junction Transistor," *IEEE Electron Device Letters*, EDL-7, pp. 416-418 (1986)

T. YAMAZAKI, I. NAMURA, H. GOTO, A. TAHARA and T. ITO, "A 11.7 GHz 1/8-Divider using 43 GHz Si High Speed Bipolar Transistor with Photoepitaxially Grown Ultra-Thin Base", *IEDM Technical Digest*, pp. 309-312, IEEE Catalog No. CH2865-4 (1990)

PROBLEMS

2-2-1. Calculate the temperature dependence of the collector current and transconductance of an *n-p-n* silicon transistor in the temperature range from 270 to 370 K. The base-emitter voltage is $V_{be} = 0.65$ V, the reverse collector-base voltage is $V_{bc} = -5$V, the base thickness is $W = 0.2$ μm, the acceptor concentration in the base region is $N_{ab} = 10^{16}$ cm^{-3}, the electron mobility is $\mu_n \approx 1000\,(300/T)^{2.4}$ cm^2/Vs, the intrinsic carrier concentration in silicon at 300 K is $n_i \approx 1.5 \times 10^{10}$ cm^{-3}, the silicon energy gap is $E_g \approx 1.12$ eV, and the emitter area is 50 μm^2.

*2-3-1. For the device shown in Fig. 2.3.1a, suggest appropriate doping levels and dimensions.

*2-3-2. Compare Si and GaAs BJTs. What are advantages and disadvantages of these materials for bipolar technology?

2-4-1. Consider a silicon *n-p-n* BJT with a graded base doping, $N_{ab}(x)$, biased in forward active mode at low injection. The position x is measured from the emitter-base junction. Show that in a typical case when the hole current contribution to the emitter current is negligible, the electric field F in the base is given by

$$F = V_{th} \frac{dN_{ab}/dx}{N_{ab}}$$

where V_{th} is the thermal voltage. Use this equation to show that F will be constant in regions of the base where the doping density varies exponentially with distance.

Assume that we have a constant field of -4 kV/cm in the base of such a transistor with a neutral base width of $W = 0.3$ mm and a doping density of 5×10^{17} cm^{-3} at the emitter-base edge. Find the doping density at the edge of the collector-base space-charge region.

From elementary transport equations and the above results, derive the following equation for the electron current density, j_n, in the base:

$$j_n \int_{x_1}^{x_2} \frac{p_b dx}{qD_n} = p_b(x_1)n_b(x_1) - p_b(x_2)n_b(x_2)$$

where $n_b(x)$ and $p_b(x)$ are the electron and hole concentrations, and x_1 and x_2 are two arbitrary positions in the neutral part of the base. Also show that for a uniform base area, S, the total collector current in the forward active mode can be written as:

$$I_c = \frac{qSD_nN_{ab}(x)n(x)}{W} \int_x N_{ab}(x)dx$$

(Note that the integral in the denominator on the right hand side of this expression is the Gummel number discussed in Subsection 2.4.2.).

Use this expression to plot the electron density profile in the base for the transistor specified above. Finally, calculate the total stored minority carrier charge in the base for this transistor.

2-4-2. Use the Ebers-Moll model to express the common-base collector saturation current, I_{cbo}, and the common emitter saturation current, I_{ceo}, in terms of the basic transistor and material parameters, such as electron and hole mobilities, base width, device temperature, doping levels, etc.

Hint: These currents are defined by

$$I_c = I_{cbo} + \alpha I_e$$

$$I_c = I_{ceo} + \beta I_b$$

Assume that recombination/generation current is negligible.

Use the following material and transistor parameters to calculate I_{cbo} and I_{ceo}: Emitter region stripe dimensions, $L = 10$ μm, $Z = 5$ μm, emitter region thickness, $X_e = 1$ μm, base region thickness, $W = 0.5$ μm, collector region thickness, $X_c = 5$ μm, emitter doping, $N_{de} = 5 \times 10^{18}$ cm^{-3}, base doping, $N_{ab} = 2 \times 10^{17}$ cm^{-3}, collector doping, $N_{dc} = 5 \times 10^{15}$ cm^{-3}, energy gap, $E_g = 1.12$ eV, effective density of states in the conduction band, $N_c = 3.22 \times 10^{19}$ cm^{-3}, effective density of states in the valence band, $N_v = 1.83 \times 10^{19}$ cm^{-3}, temperature, $T = 300$ K, dielectric permittivity of silicon, $\varepsilon_s = 1.05 \times 10^{-10}$ F/m, effective recombination time in the emitter-base depletion region, $t_{rec} = 1 \times 10^{-8}$ s, diffusion length of electrons in the base, $L_{nb} = 40$ μm, diffusion length of holes in the emitter, $L_{pe} = 25$ μm, electron mobility in the base region, $\mu_n = 0.1$ m^2/Vs, hole mobility in the emitter region, $\mu_p = 0.04$ m^2/Vs.

2-4-3. A typical circuit for measuring transistor switching characteristics is shown in the Fig. P2.4.3 along with the input voltage waveform. The short-circuit common-emitter gain is $\beta = 100$. Simulate the collector current waveform for $V_1 = 5$ V and $V_1 = 10$ V using AIM-Spice with the following BJT parameter listing of the Gummel-Poon model:

```
Is=14.34f Xti=3 Eg=1.11 Vaf=74.03 Bf=100 Ne=1.208
+ Ise=19.48f Ikf=.2385 Xtb=1.5 Br=9.715 Nc=2 Isc=0 Ikr=0
+ Rc=1 Cjc=9.393p Mjc=.3416 Vjc=.75 Fc=.5 Cje=22.01p
+ Mje=.377 Vje=.7+ Tr=10n Tf=408.8p Itf=.6 Vtf=1.7 Xtf=3
+ Rb=10
```

(The notation and units are explained in Appendix A9.)

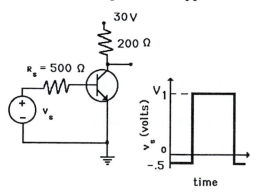

Fig. P2.4.3

2-4-4. Simulate with AIM-Spice the common-emitter current-voltage characteristics of a transistor with the following AIM-Spice model parameters: forward Early voltage, $v_{af} = 80$ V, corner for forward and reverse β high current roll-off, $i_{kf} = 0.24$ A and $i_{kr} = 0$ A, (with this choice of parameters the Gummel-Poon model is automatically selected, see Appendix A9). Perform the simulation for base currents of 5 µA, 10 µA, 15 µA, and 20 µA, and emitter-collector voltages varying from –0.5 V to 5 V.

2-5-1. Fig. P2.5.1 shows the dependence of the breakdown voltage in a Si p-n junction for $N_d = 10^{16}$ cm^{-3} (data from C. R. CROWELL and S.M. SZE, *Appl. Phys. Lett.*, 9, 242 (1966)). Find an approximate interpolation formula describing this temperature dependence. Assume that the common-emitter current gain in a silicon transistor is limited by the emitter band narrowing and given by

$$h_{fe} = 50 \times \exp\left[\frac{\Delta E_g}{k_B}\left(\frac{1}{T_o} - \frac{1}{T} \right) \right]$$

where the energy band narrowing is $\Delta E_g = 0.05$ eV, $T_o = 300$ K, and T is the transistor temperature.

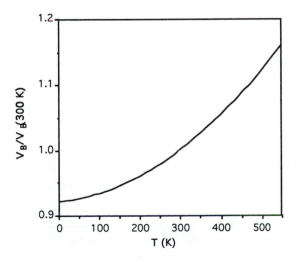

Fig. P2.5.1

For a collector doping of $N_d = 10^{16}$ cm^{-3} in an n-p-n BJT, calculate and plot the temperature dependence of the collector-base and collector-emitter breakdown voltages for a common-base configuration. Use the constant $m_b = 3$ in eq. (2-5-7).

2-6-1.   Calculate and plot the h-parameters for a silicon transistor for the common-emitter configuration as a function of the emitter current, I_e, for 0.1 mA $< I_e <$ 2 mA. (You can assume midband frequencies for this problem (i.e., you can neglect the capacitances while calculating the h-parameters)). The transistor design is shown in Fig. P2.6.1. Use the transistor parameters given in Problem 2-4-1. In addition, assume that the separation between the base contacts is $d = 10$ μm (the emitter is centered within d) and that the resistance of each base contact is $R_c = 10$ Ω (this includes the resistance of the metal interface and the material directly under the contact only). Neglect the base spreading resistance. The Early voltage is $V_A = 100$ V. (You can approximate the output resistance as $r_c \approx V_A/I_e$.)

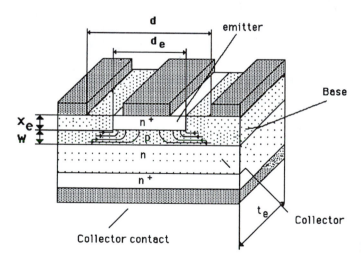

Fig. P2.6.1

Hint: Use the following equation (which can be derived using the Gummel-Poon model) to find the common-emitter current gain, β,

$$\beta^{-1} = \frac{D_p W^2 I_c}{q D_n^2 N_{de} X_e S} + \frac{D_p N_{ab} W}{D_n N_{de} X_e} + \frac{W^2}{2L_{nb}}$$

$$+ \frac{n_i W}{\mu_n F_{np} n_{bo} \tau_{rec} \left[W I_c / (q D_n n_{bo} S) \right]^{(1-1/m_{re})}}$$

Here, $m_{re} = 2$ is the ideality factor of the emitter base recombination current, the effective electric field in the emitter base junction is $F_{np} \approx$ 40 kV/cm, $\tau_{rec} = 10^{-7}$ s, and the emitter area is $S = 10^{-6}$ cm^2.

2-6-2. Calculate and plot the cutoff frequency, f_T, as a function of the emitter current, I_e, for 0.1 mA $< I_e <$ 2 mA for the silicon transistor described in Problem 2-6-1. Neglect the parasitic interconnect capacitance (C_p). The collector-base capacitance can be assumed to be entirely a depletion capacitance. The depletion capacitance between the base and emitter can be neglected. Assume a reverse base-collector bias voltage, $V_{bc} = -10$ V, a collector series resistance including all components of 5 Ω, and an electron saturation velocity, $v_{sn} = 10^5$ m/s.

2-6-3. What is the value of the emitter-base voltage at which the base spreading resistance, $r_{bb'}$, of an n-p-n silicon transistor will be reduced by 10% due to the injected electron charge? How will this change in the base spreading resistance affect the maximum oscillation frequency, f_{max}. The reverse collector-base voltage is $V_{bc} = -5$V, the base thickness is $W = 0.2$ μm, the transistor temperature is $T = 300$ K, the acceptor concentration in the base region is $N_{ab} = 10^{16}$ cm^{-3}, the ratio of electron and hole mobilities is $\mu_n/\mu_p = 3$, and the intrinsic carrier concentration in silicon at 300 K is $n_i \approx 1.5 \times 10^{10}$ cm^{-3}.

2-7-1. Using the AIM-Spice model HBTA1 (with default parameters), calculate the output HBT I-V characteristics for the common emitter configuration for values of the base current from 1 μA to 11 μA with 2 μA steps.

3

Metal Oxide Semiconductor Field Effect Transistors

3-1. INTRODUCTION

A Field Effect Transistor (FET) operates as a capacitor with one plate serving as a conducting channel between two ohmic contacts – the source and the drain contacts (see Fig. 3.1.1). The other plate – the gate – controls the charge induced into the channel. The carriers in the channel come from the source and move across the channel into the drain. Ideally, the gate contact is completely isolated from the conducting channel. Hence, the input impedance is infinitely high (at stationary conditions), a major advantage compared to Bipolar Junction Transistors. Another advantage is the relative ease of integrating FETs into very large integrated circuits because they are devices utilizing only the surface region of the semiconductor. The integration scale for FET technology is well into 10^7 devices on a single chip and continues to increase quite rapidly.

The most important FET is the silicon Metal Oxide Semiconductor Field Effect Transistor – MOSFET – (or the Metal Insulator Semiconductor Field Effect Transistor – MISFET). In a silicon MOSFET (see Fig. 3.1.1a), the gate contact is separated from the channel by a silicon dioxide layer. The channel is created by an electron gas (in n-channel devices) or by a hole gas (in p-channel devices) induced in the semiconductor at the silicon-insulator interface by the gate voltage. In an n-channel MOSFET, electrons enter the channel from an n^+ source, In a p-channel MOSFET, holes enter the channel from a p^+ source.

188

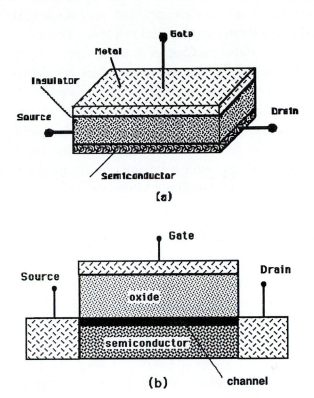

Fig. 3.1.1. Schematic illustration of the concept of a Field Effect Transistor (FET) (a) and schematic structure of a Metal Oxide Semiconductor Field Effect Transistor (MOSFET) (b). The device operates as a capacitor with one plate serving as a conducting channel between the source and drain contacts. The other plate – the gate – controls the charge induced into the channel.

Silicon technology is the most developed and advanced semiconductor technology thanks to the high purity that can be achieved for silicon (since it is an elemental semiconductor), and to the excellent properties of the "native" dielectric material – silicon dioxide – which can be thermally grown with elegance and ease on the surface of a silicon wafer. However, many other FETs have emerged and challenged the silicon MOSFET supremacy with a mixed success and great hopes, primarily at important fringes of silicon technology such as ultra high speed circuits, circuits operating at high temperatures and harsh environments, radiation hard circuits, and very large scale thin film integrated circuits used primarily in liquid crystal displays. These devices utilize compound semiconductor materials (see Chapter 4), solid state semiconductor solutions such as Si-Ge, hydrogenated amorphous silicon which is an alloy of amorphous silicon and hydrogen, or polycrystalline silicon (see Chapter 5).

In this chapter, we present device models for silicon MOSFETs which are used both as discrete devices and as components of monolithic Integrated Circuits (ICs). The minimum device size in such circuits has shrunk to 0.7 micrometers in commercial IC chips, with a commensurate increase in speed and in the integration scale (to many millions of transistors on a single chip for memory ICs). According to Sze (1988), the cost of a bit in a computer memory chip has dropped to about 6×10^{-3} cent. Sales of MOSFET digital integrated circuits in the United States alone reached six billion dollars per year in the 1980s and are projected to rise. (The total IC sales in the United States were $11 billion dollars in 1986 with approximately 3.5 billion in sales of digital bipolar ICs.) It is expected that IC chips will contain up to 100 million devices per chip by the year 2000 with the cost per bit of memory below 10^{-4} cent.

Complementary MOSFET (CMOS) technology that combines both *n*-channel and *p*-channel MOSFETs provides very low power consumption along with high speed. New silicon-on-insulator technology may help achieve three dimensional integration, i.e., packing of devices into many layers with a dramatic increase in integration density. New improved device structures and the combination of bipolar and field effect technologies may lead to further advances, yet unforeseen.

Because of the importance of CMOS technology, we present here device models for both *n*- and *p*- channel MOSFETs. These models are well suited for simulation and parameter extraction, even for submicron devices. They have been implemented in our circuit simulator – AIM-Spice, and we give examples of AIM-Spice simulations in Chapter 6 (see also Ytterdal et al. (1991)). Summaries of the basic relationships of the various MOSFET models are given in Appendix 8.

The emphasis of this Chapter is on MOSFET models suitable for circuit simulation. In Sections 3.2 to 3.8 we discuss background issues related to CMOS technology, the MOS capacitor and MOSFETs. In Section 3.9, we introduce a universal MOSFET model, MOSA1, for AIM-Spice. We use the term universal since the same basic formalism is applied to other FETs such as MESFETs (see Chapter 4) and HFETs (see Chapter 5). Sections 3.10 and 3.11 describe more detailed models for PMOS and NMOS. A unified capacitance model based on the quasistatic approximation is presented in Section 3.12. Finally, in Section 3.13, we briefly discuss the hot electron effect which is a very important mechanism in MOSFETs with submicrometer gate lengths.

Additional material on the interesting and complex device physics of MOSFETs can be found in books by Nicolian and Brews (1982), Milnes (1983), Pierret (1983), Schroder (1987), Tsividis (1987), and deGraaff and Klaassen (1990).

3-2. OVERVIEW OF CMOS TECHNOLOGY

In this Section, we briefly describe CMOS fabrication technology in order to provide the minimum information required for MOSFET device modeling. Such an overview may be useful to the readers who are already familiar with the basics of MOSFET operation. The readers who are not familiar with the subject may want to read this Section again after reviewing Sections 3.3 through 3.7.

We will discuss three different CMOS fabrication technologies: the so-called p-well, n-well, and twin well processes. More detailed information about CMOS technology can be found in books by Ghandhi (1983), Sze (1988), Shoji (1988), and Wolf (1990).

In the most common CMOS technologies, PMOS and NMOS are isolated using p-n junction isolation, i.e., utilizing depletion regions between n-type and p-type doped areas. (In contrast, NMOS devices are isolated from each other by a combination of a junction isolation and an oxide process called LOCOS (for LOCal Oxidation of Silicon).) Different CMOS technologies are distinguished by how the isolation between NMOS and PMOS is implemented. Figure 3.2.1 shows the cross section of a CMOS device fabricated using the p-well process.

In a p-well device, PMOS source and drain are isolated by the depletion regions between these contacts and the n-type substrate. We have to be careful to avoid an overlap of the source and drain depletion regions. The merging of the two depletion regions is called punch-through. Punch-through causes excessive leakage current and is detrimental to the device behavior. This requirement determines the minimum doping concentration in the bulk which depends on the drain-source separation. (The required doping is inversely proportional to the square root of the minimum feature size (see Sze (1981) for more details).)

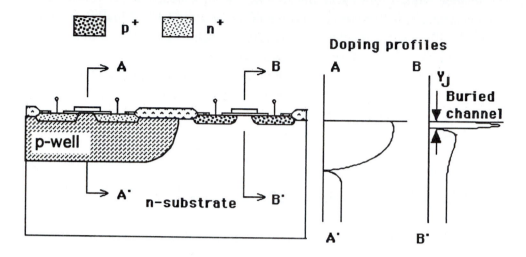

Fig. 3.2.1. Schematic cross-section of *p*-well CMOS. A-A' and B-B' indicate the doping profiles for NMOS and PMOS, respectively.

A *p*-well device is fabricated on a low *n*-doped substrate with a resistivity of the order of 5 ohm-cm, corresponding to doping densities of about 1×10^{15} cm^{-3}. Usually, the substrate doping density of the raw wafer received from the wafer manufacturer is not uniform and is too low to prevent punch-through of the PMOS. Hence, additional surface doping is done by ion implantation which yields a very good uniform doping with a lateral variation across the wafer of 2 to 3 percent. Typically, the required threshold voltage is approximately (0.7 ± 0.2) V at zero bulk bias. The doping of the *p*-well is usually taken to be 10 times larger than the substrate doping. For example, a gate oxide thickness of 500 Å and a *p*-well doping of 1×10^{16} cm^{-3} are typical for a 3 μm feature size technology.

As will be shown in Section 3.7, counter doping at the surface of PMOS is necessary to lower the absolute value of threshold voltage from 1.5 V to 0.7 V. The junction depth for this buried channel, Y_J, is very critical for PMOS scaling because of the leakage current path between source and drain along the *p-n* junction formed between the buried channel and the bulk. Since NMOS has a large bulk doping in *p*-well technology, the performance of NMOS fabricated using this process is not as good as for the other technologies because of the mobility degradation caused by the high doping in the well (see Section 3.8). However, PMOS determines the overall speed of CMOS circuits since it has about one third of the speed of NMOS due to the smaller mobility for holes than for electrons. PMOS fabricated using the p-well process has a better performance than for the n-well and twin well processes. This leads to a better overall performance of CMOS fabricated using the p-well technology. This technology is therefore especially suitable for static logic circuits utilizing equal numbers of NMOS and PMOS devices. However, as pointed out above, the p-well technology has a problem with PMOS scaling due to the leakage current.

Schematic doping profiles of an n-well device is shown in Fig. 3.2.2. These profiles are complementary to those of the p-well CMOS. The surface doping of NMOS should be as high as that in the p-well devices in order to have the same threshold voltage. A low bulk doping density near the depleted drain junction is very desirable in order to reduce the junction capacitance and to increase the switching speed.

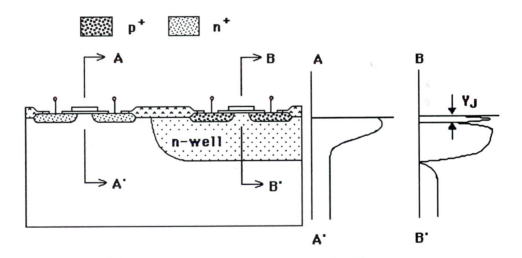

Fig. 3.2.2. Schematic cross-sections of n-well CMOS. A-A' and B-B' indicate the doping profiles for NMOS and PMOS, respectively.

However, if the bulk doping is too low, the depletion regions formed by the drain and source contacts with the substrate will merge and cause punch-through. These problems can be alleviated using a double implantation technology (with additional implantation costs). In this technology, a surface implant adjusts the threshold voltage and a relatively deep implant prevents punch-through. In the *n*-well process, the doping density of PMOS is much higher than for NMOS. This is very desirable for preventing leakage current but quite detrimental for the PMOS performance because of the large mobility degradation, as discussed in Section 3.8. Therefore, the *n*-well technology is appropriate for NMOS intensive circuits such as memories. Since the doping density of the *n*-well is very high, the leakage problems of PMOS are greatly reduced. Accordingly, this technology has less of a problem in PMOS scaling compared to the *p*-well process.

Most of the high performance circuits are fabricated using the twin well technology. This technology (see Fig. 3.2.3) allows us to optimize NMOS and PMOS independently of each other at the expense of additional processing costs.

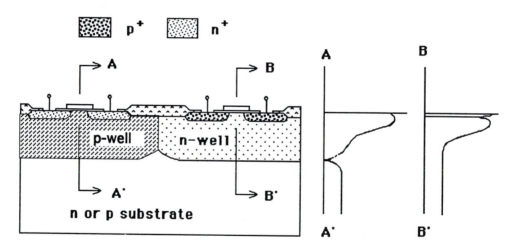

Fig. 3.2.3. Schematic cross sections of twin well CMOS. A-A' and B-B' indicate the doping profiles for NMOS and PMOS, respectively.

As can be seen from the above discussion, the CMOS technology shares may basic fabrication steps with bipolar technology. This can be used to a great advantage since BJTs and MOSFETs can be combined on the same chip. Such a technology (called BICMOS, see Alvarez (1989)) has emerged as one of the leading high speed VLSI technologies. At the same time, having *n*- and *p*-regions on the chip in close proximity leads to the formation of parasitic bipolar elements, schematically shown in Fig. 3.2.4. Two parasitic transistors are formed, one vertical and one lateral. Under normal operating conditions, all *p-n* junctions in these transistors are reverse biased. However, a spurious voltage discharge (due to an electrostatic charge, for example) may temporarily forward bias one of the junctions. As can be understood from the analysis of the circuit formed by these two parasitic transistors, this will result in a very high current, shorting the power supply provided that overall gain in the feedback loop formed by the two transistors is larger than 1 (i.e., $\beta_{vertical}\beta_{lateral} > 1$).

The vertical *n-p-n* transistor may have a large gain, greater than 50 or so. The lateral *p-n-p* transistor has a long base, and hence, a low gain of 0.01 to 1.

This phenomenon (called latch-up) was assumed to be an important limitation for CMOS technology. However, in modern CMOS technology, this phenomenon is primarily a concern for input/output circuits or for CMOS circuits operating in a radiation environment. To prevent latch-up, guard rings (p^+ or n^+ regions) may be diffused to separate the n^+ or p^+ drains from the edges of the tubs. This drastically decreases the lateral transistor gain. Alternatively, the distance between the drains and the tub can be increased. Similar techniques can be used for other CMOS processes (see Ong (1984) for details).

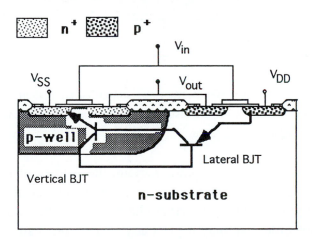

Fig. 3.2.4. Parasitic BJT elements in the CMOS structure.

This brief review of CMOS technology is intended as a background for understanding the requirements in realistic CMOS modeling. Such models must link the device characteristics to geometry and doping profiles, describe the subthreshold region and include the dependence of the carrier mobility on doping. These features are incorporated in the models considered in this Chapter.

3-3. SURFACE CHARGE AND THE METAL INSULATOR SEMICONDUCTOR CAPACITOR

3.3.1. Surface Charge

The Metal Insulator Semiconductor (MIS) capacitor consists of a metal layer (gate contact) separated from a semiconductor by a dielectric insulator. Such a capacitor is an integral part of any MOSFET (see Fig. 3.3.1). Understanding how an MIS capacitor operates is a prerequisite to understanding MOSFET physics and the modeling of MOSFETs.

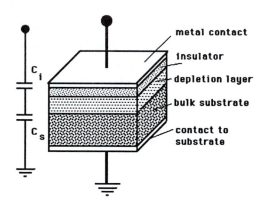

Fig. 3.3.1. Schematic diagram and simplified equivalent circuit of the MIS capacitor. C_i is the insulator capacitance and C_s is the equivalent semiconductor capacitance.

In the silicon-silicon dioxide system, the density of surface states at the dielectric-semiconductor interface is very low compared to the typical surface carrier density in a MOSFET. Also, the quality of the insulator is quite good. Hence, to first order, we may neglect any charges that can exist in the insulator layer and any possible surface states at the semiconductor-insulator interface. Indeed, a typical value of the surface free electron density in an MIS structure (or in a field effect transistor utilizing such a structure) is on the order of 10^{12} cm^{-2}. In a typical modern day Si-SiO$_2$ system, the density of the surface states is only 10^9 cm^{-2} (i.e., about 0.1 % of the free electron carrier concentration). However, in the GaAs-insulator system, for example, the density of the surface states is as high as 10^{13} cm^{-2}. This is why it is extremely difficult to realize metal-insulator-GaAs devices.

We may also assume that the insulator layer has infinite resistivity so that there is no charge carrier transport across the dielectric layer when a bias voltage is applied. These assumptions greatly simplify the analysis of the MIS capacitor, but they are not always valid in real devices.

At zero bias voltage, the bending of the energy bands in the semiconductor layer is determined by the difference in the work functions of the metal and the semiconductor (see Figs. 3.3.2 and 3.3.3).

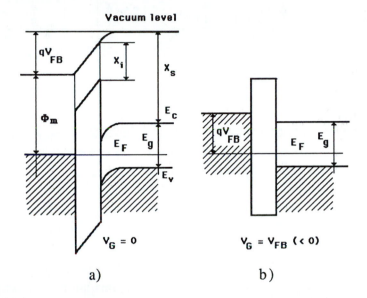

Fig. 3.3.2. Band diagrams of the MIS capacitor at zero applied voltage (a) and for an applied voltage equal to the flat band voltage (b). Zero voltage here corresponds to the depletion mode (i.e., the *p*-type semiconductor substrate is depleted near the insulator-semiconductor interface).

This band bending may be compensated by applying the flat band voltage, V_{FB}, equal to:

$$V_{FB} = \left(\Phi_m - X_s - E_c + E_F\right)/q \qquad (3\text{-}3\text{-}1)$$

where Φ_m is the work function of the metal, X_s is the electron affinity for the semiconductor, E_c is the energy of the conduction band edge, and E_F is the Fermi level. (In real devices, the flat band voltage may be affected by surface states at the interface and/or by fixed charges in the insulator layer.)

In typical devices, the gate metal of n-MOSFETs is substituted by a highly conductive n^+ polysilicon layer. In this case, the flat band voltage is approximately zero for MOSFETs with n-type doping in the channel (see Fig. 3.3.3).

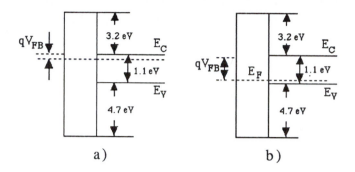

a)   b)

Fig. 3.3.3. Band diagrams of silicon MOSFETs with polysilicon gate at flat band. a) MOSFET with n-type doping in the bulk (p-channel), b) MOSFET with p-type doping in the bulk (n-channel).

When the voltage applied between the metal and the semiconductor, V_G, exceeds the flat band voltage, V_{FB}, the region close to the insulator-semiconductor interface becomes depleted of holes (the so-called depletion regime, see Fig. 3.3.2a). At even larger positive gate voltages, the band bending becomes so large that the Fermi level at the insulator-semiconductor interface becomes closer to the bottom of the conduction band than to the top of the valence band. In this case, the concentration of carriers near the interface actually corresponds to that of an n-type semiconductor. This is called <u>inversion</u>. According to the usual definition, <u>strong inversion</u> is reached when the difference between the Fermi level and the intrinsic Fermi level at the interface, $-q\varphi_s$, becomes equal (and opposite in sign) to this difference in the bulk of the semiconductor, $q\varphi_b$, far from the interface (see Fig. 3.3.4). Here,

$$\varphi_b = V_{th} \ln\left(\frac{N_a}{n_i}\right) \tag{3-3-2}$$

where N_a is the shallow acceptor density in the semiconductor, n_i is the intrinsic carrier concentration and $V_{th} = k_B T/q$ is the thermal voltage.

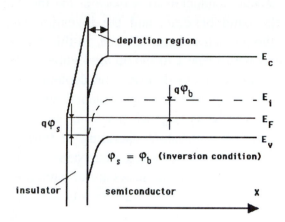

depletion region

E_c

$q\varphi_b$

E_i

$q\varphi_s$

E_F

E_v

$\varphi_s = \varphi_b$ (inversion condition)

insulator   semiconductor   X

Fig. 3.3.4. MIS capacitor band diagram at inversion threshold.

Values of φ_s such that $-\varphi_b < \varphi_s < \varphi_b$ correspond to the depletion and weak inversion regime, $\varphi_s = -\varphi_b$ is the flat band condition, and values of $\varphi_s < -\varphi_b$ correspond to the accumulation mode. Note that the surface potential, V_s, counted from the potential of the semiconductor far from the interface is equal to $\varphi_b + \varphi_s$.

Tsividis (1987) has given a more expanded definition of the different inversion regimes. His definition is based on the fact that the derivative $|d\varphi_s/dV_G|$ is relatively large in weak inversion and relatively small in strong inversion. He defines the condition $\varphi_s = \varphi_b$ as the onset of <u>moderate</u> inversion. According to Tsividis, strong inversion takes place when the band-bending is larger than φ_b by several V_{th}. In order to understand this definition, let us discuss the qualitative dependence of the surface potential, V_s, on the applied voltage. In depletion and weak inversion, an increase in the applied voltage is divided between the insulator and the semiconductor depletion region according to the relative size of the capacitances, C_i and C_s, of the two regions (see Fig. 3.3.1). Normally, in this regime, the change in band bending is a significant fraction of the change in the applied voltage. In the depletion regime, most of the charge is induced in the depletion region, although the free carrier charge, n_s, at the semiconductor-insulator interface increases by a factor $\exp(1) = 2.718$ for each increase in the band bending by V_{th} (≈ 25.8 meV at room temperature). In strong inversion, however, the inversion charge is dominant, almost all charge exchange in the semiconductor takes place at the interface with only very little variation in the band bending, and the inversion charge is nearly linearly dependent on the applied voltage. Hence, in this regime, the derivative of the surface potential with respect to the applied voltage, dV_s/dV_G, becomes small compared to the value of this derivative in depletion and weak inversion. The intermediate range of applied voltages, which Tsividis identifies as the moderate inversion regime, can be as large as 0.5 V or so. Hence, this regime is quite important for modern transistors operating at fairly low bias (see Tsividis (1987), p. 55 for a more detailed discussion). The importance of this intermediate regime highlights the necessity of having a model which is treating both the inversion and depletion modes in a unified way to allow an adequate description of this intermediate range of voltages. (Most conventional device models treat the depletion and strong inversion regimes separately.) Such a unified MIS model will be discussed in Subsection 3.3.3. It is based on the physics of the interface region discussed in this Subsection.

The surface concentrations of electrons and holes are given by

$$n_s = n_{po} \exp\left(V_s / V_{th}\right) \tag{3-3-3}$$

$$p_s = p_{po} \exp\left(- V_s / V_{th}\right) \tag{3-3-4}$$

where $p_{po} = N_a$, N_a is the concentration of shallow acceptors in the p-type semiconductor and $n_{po} = n_i^2/N_a$ is the equilibrium concentration of the minority carriers (electrons) in the bulk. (At the inversion threshold, $V_s = 2\varphi_b$). As can be seen from eqs. (3-3-3) and (3-3-4),

$$n_s p_s = n_i^2 \tag{3-3-5}$$

This corresponds to a quasi equilibrium situation since the Fermi level in the semiconductor is independent of position in the direction perpendicular to the insulator-semiconductor interface, a consequence of having zero current in the semiconductor.

The potential distribution in the semiconductor can be found from the solution of the one-dimensional Poisson's equation

$$\frac{d^2 V}{dx^2} = - \frac{\rho(x)}{\varepsilon_s} \tag{3-3-6}$$

where the space charge density $\rho(x)$ is given by

$$\rho(x) = q(p - n - N_a) \tag{3-3-7}$$

The electron and hole concentrations, n and p, may be expressed as

$$n = n_{po} \exp(V/V_{th}) \tag{3-3-8}$$

$$p = p_{po} \exp(-V/V_{th}) \tag{3-3-9}$$

where the potential V is taken to be zero far into the bulk of the semiconductor.

The neutrality condition for the bulk semiconductor region (far from the insulator-semiconductor interface) yields

$$N_a = p_{po} - n_{po} \tag{3-3-10}$$

The substitution of eqs. (3-3-7) to (3-3-10) into eq. (3-3-6) leads to the following equation:

$$\frac{d^2V}{dx^2} = \frac{q}{\varepsilon_s}\left\{ n_{po}\left[\exp\left(\frac{V}{V_{th}}\right) - 1\right] - p_{po}\left[\exp\left(-\frac{V}{V_{th}}\right) - 1\right]\right\} \tag{3-3-11}$$

Using the definition of the electric field,

$$F = -\,dV/dx \tag{3-3-12}$$

we can rewrite the left hand side of eq. (3-3-11) as $-dF/dx$ and then multiply eq. (3-3-11) by $F\,dx/dV$ to change the derivative with respect to x to a derivative with respect to V, i.e.,

$$F\frac{dF}{dV} = -\frac{q}{\varepsilon_s}\left\{ n_{po}\left[\exp\left(\frac{V}{V_{th}}\right) - 1\right] - p_{po}\left[\exp\left(-\frac{V}{V_{th}}\right) - 1\right]\right\} \tag{3-3-13}$$

Integrating both sides of this equation with respect to V, we obtain

$$F^2 = \frac{2qV_{th}}{\varepsilon_s}\left\{ p_{po}\left[\exp\left(-\frac{V}{V_{th}}\right) + \frac{V}{V_{th}} - 1\right] + n_{po}\left[\exp\left(\frac{V}{V_{th}}\right) - \frac{V}{V_{th}} - 1\right]\right\}$$

$$\tag{3-3-14}$$

Introducing the Debye length

$$L_{Dp} = \sqrt{\frac{\varepsilon_s V_{th}}{q N_a}} \qquad (3\text{-}3\text{-}15)$$

we can rewrite eq. (3-3-14) as (see, for example, Shur (1990))

$$F = \sqrt{2} \frac{V_{th}}{L_{Dp}} f(V) \qquad (3\text{-}3\text{-}16)$$

where

$$f(V) = \pm \left\{ \left[\exp\left(-\frac{V}{V_{th}}\right) + \frac{V}{V_{th}} - 1 \right] + \frac{n_{po}}{p_{po}} \left[\exp\left(\frac{V}{V_{th}}\right) - \frac{V}{V_{th}} - 1 \right] \right\}^{1/2} \qquad (3\text{-}3\text{-}17)$$

Here, the positive sign should be chosen for a positive potential and the negative sign for a negative potential. From eq. (3-3-17), we find for the electric field at the insulator-semiconductor interface

$$F_s = \sqrt{2} \frac{V_{th}}{L_{Dp}} f(V_s) \qquad (3\text{-}3\text{-}18)$$

where V_s is the value of V at the surface. Using Gauss' law, we can relate the total charge per unit area in the semiconductor, Q_s, to the surface electric field by

$$|Q_s| = \varepsilon_s |F_s| \qquad (3\text{-}3\text{-}19)$$

When $V_G = V_{FB}$, the surface charge is equal to zero. When $V_G < V_{FB}$, the surface charge is positive corresponding to the accumulation regime, and when $V_G > V_{FB}$, the surface charge is negative corresponding to the depletion and inversion regimes. It is easily shown from eqs. (3-3-17) to (3-3-19) that the surface charge is proportional to $\exp[|V_s|/(2V_{th})])$ in accumulation (when $|V_s|$ exceeds a few times V_{th}) and in strong inversion. In depletion and weak inversion, the surface charge varies as $V_s^{1/2}$.

The applied voltage can now be related to the surface potential, V_s. The electric field at the semiconductor surface can be found from eq. (3-3-18). Using the condition of continuity of the electric flux density

$$\varepsilon_s F_s = \varepsilon_i F_i \qquad (3\text{-}3\text{-}20)$$

(where ε_i is the dielectric permittivity of the insulator layer and F_i is the electric field in the insulator layer) we find the voltage drop $F_i d_i$ across the insulator, taking d_i is the insulator thickness. Hence, the applied voltage can be written as

$$V_G = V_{FB} + V_s + \varepsilon_s F_s / c_i \qquad (3\text{-}3\text{-}21)$$

where

$$c_i = \varepsilon_i / d_i \qquad (3\text{-}3\text{-}22)$$

is the insulator capacitance per unit area.

Eq. (3-3-21) allows us to calculate the threshold gate voltage, V_T, corresponding to the onset of the moderate inversion regime. As discussed above, the onset of moderate inversion occurs when the surface potential, V_s, measured with respect to the bulk of the semiconductor substrate far from the interface is equal to $2\varphi_b$. At this surface potential, the charge of the free carriers induced at the insulator-semiconductor interface is still small compared to the charge in the depletion layer which is given by

$$Q_{dT} = \sqrt{2\varepsilon_s q N_a V_s} = \sqrt{4\varepsilon_s q N_a \varphi_b} \qquad (3\text{-}3\text{-}23)$$

Hence, at the onset of moderate inversion, the electric field at the semiconductor-insulator interface is

$$F_{sT} = \sqrt{4 q N_a \varphi_b / \varepsilon_s} \qquad (3\text{-}3\text{-}24)$$

Using eq. (3-3-21), we find the following expression for the threshold voltage:

$$V_T = V_{FB} + 2\varphi_b + \sqrt{4\varepsilon_s q N_a \varphi_b} / c_i \qquad (3\text{-}3\text{-}25)$$

V_T is one of the most important parameters for metal-insulator-semiconductor devices. Typical calculated dependencies of V_T on doping level and dielectric thickness are shown in Fig. 3.3.5.

For the MIS structure shown in Fig. 3.3.1, the application of a substrate bias, V_{sub}, is simply equivalent to changing the applied voltage from V_G to $V_G - V_{sub}$. Hence, the threshold for moderate inversion is given by

$$V_T' = V_T - V_{sub} \qquad (3\text{-}3\text{-}26)$$

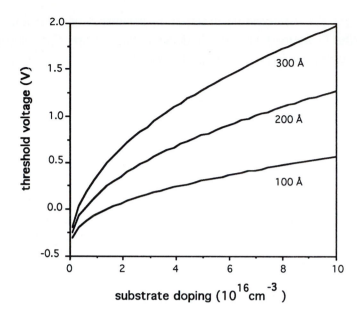

Fig. 3.3.5. Dependence of threshold voltage, V_T, on the substrate doping level for different gate dielectric thicknesses. Parameters used in calculation: energy gap: 1.12 eV, effective density of states in the conduction band: 3.22×10^{25} m^{-3}, effective density of states in the valence band: 1.83×10^{25} m^{-3}, semiconductor dielectric permittivity: 1.05×10^{-10} F/m, insulator dielectric permittivity: 3.45×10^{-11} F/m, flat band voltage: -1 V, temperature: 300 K, substrate bias: 0 V.

However, the situation will be different if a conducting layer of mobile electrons in the inversion layer is induced at the semiconductor-insulator interface under strong inversion conditions and maintained at some constant potential. This situation occurs in a MOSFET. Let us assume that the inversion layer is grounded. Then V_{sub} biases the effective junction between the inversion n-channel layer and the p-type substrate, changing the negative charge in the depletion layer. Under such conditions, the threshold voltage becomes

$$V_T = V_{FB} + 2\varphi_b + \sqrt{2\varepsilon_s\, qN_a\left(2\varphi_b - V_{sub}\right)}\,/c_i \qquad (3\text{-}3\text{-}27)$$

The threshold voltage may also be affected by so-called fast surface states at the insulator-semiconductor interface and by fixed charges in the insulator layer (see, for example, Shur (1990), p. 343).

The threshold voltage of moderate inversion, V_T, separates the subthreshold regime, where the mobile carrier charge increases exponentially with the increase in applied voltage, from the above-threshold regime, where the mobile carrier charge is proportional to the applied voltage. However, there is no clear distinction between the two regimes and different definitions and experimental techniques have been used to determine V_T. A detailed discussion of the experimental determination of threshold voltages will be given in Subsection 3.3.3 and in subsequent Sections of this Chapter.

3.3.2. MIS Capacitance

The MIS capacitance, C_{mis}, can be represented as a series combination of the insulator capacitance, C_i, and the capacitance of the semiconductor layer, C_s (see Fig. 3.3.1)

$$C_{mis} = \frac{C_i C_s}{C_i + C_s} \tag{3-3-28}$$

where

$$C_i = S \varepsilon_i / d_i \tag{3-3-29}$$

and S is the area of the MIS capacitor. The capacitance of the semiconductor part of the ideal MIS structure can be calculated as

$$C_s = S \left| \frac{dQ_s}{dV_s} \right| \tag{3-3-30}$$

where Q_s is the total charge density per unit area in the semiconductor and V_s is the surface potential. Using eqs. (3-3-18) and (3-3-19) and performing the differentiation in eq. (3-3-30), we obtain

$$C_s = \frac{C_{so}}{\sqrt{2} f(V_s)} \left\{ 1 - \exp\left(-\frac{V_s}{V_{th}} \right) + \frac{n_{po}}{p_{po}} \left[\exp\left(\frac{V_s}{V_{th}} \right) - 1 \right] \right\} \tag{3-3-31}$$

where

$$C_{so} = S \varepsilon_s / L_{Dp} \tag{3-3-32}$$

is the semiconductor capacitance under the flat band condition (i.e., for $V_s = 0$)

and L_{Dp} is the Debye length given by eq. (3-3-15).

The relationship between the surface potential, V_s, and the applied bias, V_G, is given by eq. (3-3-21) in combination with eqs. (3-3-18) and (3-3-17).

In depletion and inversion, the semiconductor capacitance, C_s, includes two components – the depletion layer capacitance, C_{dep}, and a capacitance C_{sc} related to the free minority carriers at the interface where

$$C_{dep} = S\varepsilon_s/d_{dep} \tag{3-3-33}$$

Here the depletion layer width is given by

$$d_{dep} = \sqrt{\frac{2\varepsilon_s V_s}{qN_a}} \tag{3-3-34}$$

for applied voltages smaller than the threshold voltage and

$$d_{dep} = 2\sqrt{\frac{\varepsilon_s \varphi_b}{qN_a}} \tag{3-3-35}$$

for applied voltages larger than V_T. The free carrier contribution is

$$C_{sc} = C_s - C_{dep} \tag{3-3-36}$$

The capacitance C_{sc} becomes important in the inversion regime, especially in strong inversion where most of the charge is induced at the semiconductor-insulator interface where the band bending is the largest. In an MIS structure, this minority carrier charge comes from electron-hole generation, primarily, owing to the recombination-generation centers in the semiconductor. In principle, the thermal generation of electron-hole pairs across the band gap can also contribute to the formation of the minority carrier charge in the inversion regime. Once an electron-hole pair is generated, the majority carrier (a hole in p-type material and an electron in n-type material) is swept from the space charge region by the electric field of this region into the substrate. The minority carrier is swept in the opposite direction towards the semiconductor-insulator interface. The build-up of minority carrier charge at the semiconductor-insulator interface in the inversion regime proceeds at a rate limited by the process of generation of electron-hole pairs. This is reflected in the equivalent circuit of the MIS structure shown in Fig. 3.3.6 (from Shur (1990)).

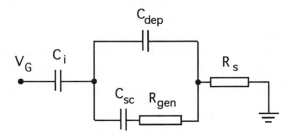

Fig. 3.3.6. Equivalent circuit of the MIS capacitor.

The resistance R_S in this equivalent circuit is the series resistance of the semiconductor layer. The differential resistance, R_{gen}, can be estimated as

$$R_{gen} = \frac{dV_S}{dI_{gen}}$$

(3-3-37)

Here, the generation current, I_{gen}, is approximately proportional to the volume of the depletion region (see for example Sze (1981)):

$$I_{gen} \approx \frac{qn_i S d_{dep}}{\tau_{gen}}$$

(3-3-38)

where τ_{gen} is an effective generation time constant. In the high frequency limit $(\omega \to \infty)$, the series capacitance of the equivalent circuit in Fig. 3.3.6, C_{eq}, becomes

$$C_{eq} = \frac{C_{dep} C_i}{C_{dep} + C_i}$$

(3-3-39)

In the limiting case of $\omega \to 0$, we find

$$C_{eq} = \frac{C_s C_i}{C_s + C_i}$$

(3-3-40)

(Here we have taken into account that $C_s = C_{dep} + C_{sc}$.) In strong inversion, we usually have $C_s \gg C_i$ and

$$C_{eq} \approx C_i \tag{3-3-41}$$

in the low-frequency limit.

The calculated dependence of C_{eq} on the applied voltage, V_G, for different frequencies is shown in Fig. 3.3.7. At negative applied voltages, C_{eq} is equal to C_i because the device is in the accumulation mode of operation. The calculated curves clearly show how the MIS capacitance in the inversion regime changes from $C_{dep}C_i/(C_{dep} + C_i)$ at high frequencies to C_i at low frequencies. It is important to notice that in a MOSFET where the carriers are supplied into the inversion layer from the highly doped source and drain regions, the time constant $R_{gen}C_{sc}$ must be substituted by a much smaller time constant equal to the transit time, t_{tr}, of the carriers under the gate. As a consequence, MOSFET gate-channel characteristics in the depletion and inversion regimes look like the zero frequency MIS characteristics even at very high frequencies close to $1/t_{tr}$.

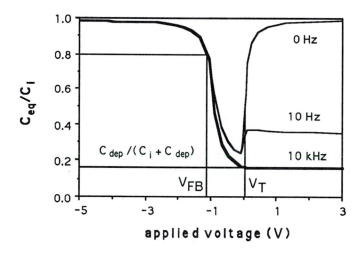

Fig. 3.3.7. Calculated dependence of C_{eq} on the applied voltage, V_G, for different frequencies. Parameters used: insulator thickness: 2×10^{-8} m, doping in the semiconductor: 10^{15} cm^{-3}, generation time: 10^{-8} s. Other parameters are the same as in Fig. 3.3.5. (From Shur (1990).)

Since the MIS capacitance in the strong inversion regime is constant, the inversion charge induced into the MOSFET channel can be approximated by

$$qn_s \approx c_i(V_G - V_T) \tag{3-3-42}$$

This equation serves as the basis of a simple charge control model which allows us to calculate MOSFET current-voltage characteristics in strong inversion.

Measured MIS $C-V$ characteristics allow us to determine important parameters of the MIS structure, including the flat band voltage, the gate insulator thickness and the doping of the semiconductor substrate. The maximum measured capacitance, C_{max}, (capacitance C_i in Fig. 3.3.6) yields the insulator thickness

$$d_i \approx S \varepsilon_i / C_{max} \tag{3-3-43}$$

The minimum measured capacitance, C_{min}, (at high frequency) allows us to find the doping concentration in the semiconductor substrate. First, we determine the depletion capacitance in the strong inversion regime from

$$1/C_{min} = 1/C_{dep} + 1/C_{max} \tag{3-3-44}$$

(see eq. (3-3-39)) and find the thickness of the depletion region

$$d_{dep} = S \varepsilon_s / C_{dep} \tag{3-3-45}$$

Then, we calculate the doping concentration, N_a, using eq. (3-3-35) for d_{dep} in strong inversion and eq. (3-3-2) for φ_b. This results in the following transcendental equation for N_a:

$$N_a = N_o \ln\left(\frac{N_a}{n_i}\right) \tag{3-3-46}$$

where

$$N_o = \frac{4\varepsilon_s V_{th}}{q d_{dep}^2} \tag{3-3-47}$$

(N_o corresponds to the acceptor doping density which would result in a depletion width d_{dep} using the surface potential $2V_{th}$). Eq. (3-3-46) can be solved by iteration. For example, we can first guess a reasonable value for N_a, apply this value on the right hand side of eq. (3-3-46) to find a more accurate estimate of N_a, and repeat the process until a desired accuracy has been reached. Alternatively, we observe that eq. (3-3-46) can be converted to the following normalized form of the non-ideal diode equation:

$$i = \exp(u - ri)$$

(3-3-48)

where, presently, $i = N_a/n_i$, $r = -n_i/N_o$ and $u = 0$. In Appendix A7, we give a very precise, approximate solution of i as a function of u for this equation (see also Fjeldly et al. (1991)).

Once d_i and N_a have been obtained, the device capacitance under flat band conditions, C_{FB}, can be determined using eqs. (3-3-29) and (3-3-32)

$$C_{FB} = \frac{C_i C_{so}}{C_i + C_{so}} = \frac{\varepsilon_i \varepsilon_s}{\varepsilon_i L_{Dp} + \varepsilon_s d_i}$$

(3-3-49)

The flat band voltage, V_{FB}, can then be determined as the applied voltage corresponding to this value of the device capacitance. This characterization process is illustrated by Fig. 3.3.8 (from Shur (1990)).

Strictly speaking, the above characterization technique applies to an ideal MIS structure. The capacitance-voltage characteristics of MIS structures may be strongly affected by mobile ions present in the gate dielectric layer and by surface states at the dielectric-semiconductor interface. Even for the widely studied silicon–silicon dioxide interface, we do not have a complete understanding of the nature of these states (see Sze (1981) for a detailed discussion). Their density depends very strongly on the oxide growth procedure and may vary from as high as 10^{15} cm^{-2} (which corresponds approximately to the total number of surface atoms) to as low as 10^9 cm^{-2}. The surface states at the insulator-semiconductor interface may be acceptor-like or donor-like (depending on their charge states when occupied by an electron and when empty). When the surface potential changes, the charge in the surface states changes as well, leading to a shift of the threshold voltage and to a change in the $C-V$ characteristics. In the equivalent circuit of an MIS structure, surface states may be represented by an additional series combination of an equivalent capacitance of the surface states, C_{ss}, and an additional resistance, R_{ss}. The time constant $C_{ss}R_{ss}$ represents the time response of the surface states.

Measurements of the frequency dependent MIS capacitance and MIS conductance can be interpreted using such an equivalent circuit yielding information about the density of surface states. This density is strongly dependent on the crystallographic orientation of the interface plane. A smaller density of states for the <100> orientation is correlated with the smaller density of available bonds on the silicon surface. Therefore, silicon MOSFETs are preferably fabricated on <100> substrates (see Sze (1981)). A more detailed discussion of the interface states for the silicon-silicon dioxide interface and of the fixed charges in the insulator layer is given by Nicolian and Brews (1982).

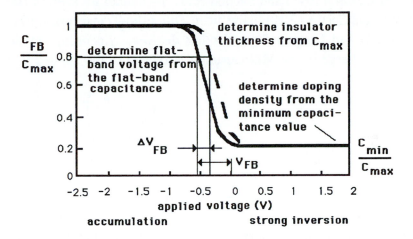

Fig. 3.3.8. Characterization procedure for the MIS capacitor using high frequency measurements.

The effect of the mobile ions in the gate dielectric layer is usually determined using a so-called high bias-temperature stress test. During such a test, the applied voltage corresponds to an electric field of the order of 10^6 V/cm in the dielectric and the temperature is chosen between 175 °C and 300 °C. The C–V characteristics are measured before and after the test (the application of a high bias and elevated temperatures usually lasts for approximately 3 minutes). The observed parallel shift of the C–V curves corresponds to the change, ΔV_{FB}, of the flat band voltage (see Fig. 3.3.8). For a good quality oxide, this change should be less than a few millivolts.

3.3.3. Unified Charge Control Model (UCCM) for MIS Capacitors

As discussed above, the standard charge control model postulates that the interface inversion charge of electrons, qn_s, is proportional to the applied voltage swing, $V_{GT} = V_G - V_T$, where V_G is the applied voltage and V_T is the threshold voltage, see eqs. (3-3-42) and (3-3-27). This model is an adequate description of the strong inversion regime of the MIS capacitor, but fails for applied voltages near and below V_T (i.e., in the weak inversion and depletion regime). Byun et al. (1990) proposed a new unified charge control model (UCCM) which they first applied to Heterostructure Field Effect Transistors (HFETs) (see Moon et al. (1990) and Section 4.7) and later to other FETs, including n-channel MOSFETs (see Park et al. (1991) and Section 3.11), p-channel MOSFETs (see Moon et al. (1991a) and Section 3.10), polysilicon Thin Film Transistors (TFTs) (see Byun et al. (1992) and Chapter 5), and other devices. According to this model, the inversion charge in the MIS capacitor is related to the applied voltage swing as follows:

$$V_{GT} = a(n_s - n_o) + \eta V_{th} \ln\left(\frac{n_s}{n_o}\right)$$
(3-3-50)

where $a \approx q/c_a$ and

$$c_a = \frac{\varepsilon_i}{d_i + \Delta d}$$
(3-3-51)

Here, ε_i is the dielectric permittivity of the gate insulator, d_i is the thickness of the gate insulator, η is an ideality factor, V_{th} is the thermal voltage, and Δd is a correction to the insulator thickness related to the shift in the Fermi level in the inversion layer with respect to the bottom of the conduction band. This correction is dependent on the interface electron density (see below). However, it can be approximately taken to be a constant for typical values of the interface carrier density. For Si-SiO$_2$ MOS capacitors, $\Delta d \approx 4$ Å which is normally much smaller than the insulator thickness (see Ohkura (1990) and Park et al. (1991)). Hence, we can usually assume that $a \approx q/c_i$ where $c_i = \varepsilon_i/d_i$ is the insulator capacitance per unit area. For the GaAs/AlGaAs interface, we have $\Delta d \approx 45$ Å or even larger (see Byun et al. (1990)). Hence, this term has to be taken into account for Heterostructure Field Effect Transistors (see Chapter 4).

We should notice that eq. (3-3-50) does not describe the mobile charge in the accumulation regime. However, this regime is not important in MOSFETs.

The ideality factor, η, reflects the gate voltage division between the insulator layer capacitance, C_i, and the depletion layer capacitance, C_{dep}. In the subthreshold regime, $C_{dep} \approx C_s$, see Subsection 3.3.2 and Fig. 3.3.1. At the onset of inversion ($V_{GT} = 0$), the interface potential, V_s, has the value $2\varphi_b$ (see Subsection 3.3.1). Below threshold, we have the following approximate relationship:

$$V_s = 2\varphi_b + V_{GT}/\eta \qquad (3\text{-}3\text{-}52)$$

where

$$\eta = 1 + C_{dep}/C_i \qquad (3\text{-}3\text{-}53)$$

The depletion capacitance and the insulator capacitance are given by eqs. (3-3-33) and (3-3-29), respectively. We note that, generally speaking, η is dependent on V_{GT}. At low substrate doping levels, η is close to unity near threshold where the gate depletion width is large (which corresponds to $C_{dep} \ll C_i$). Usually, we can estimate C_{dep} as follows (see Troutman (1974)):

$$C_{dep} = \frac{\varepsilon_s S}{\left(d_{dep}\right)_{av}} \qquad (3\text{-}3\text{-}54)$$

where

$$\left(d_{dep}\right)_{av} = \sqrt{\frac{2\varepsilon_s\left(1.5\varphi_b - V_{sub}\right)}{qN_a}} \qquad (3\text{-}3\text{-}55)$$

is an average width of the depletion region.

We note that eq. (3-3-50) can be rewritten in the form of the normalized non-ideal diode equation of eq. (3-3-48) using the dimensionless variables $i = n_s/n_o$, $u = (V_{GT} + an_o)/(\eta V_{th})$, and $r = an_o/(\eta V_{th})$. Hence, the very precise, approximate solution of eq. (3-3-48), discussed in Appendix A7, can also be used as an analytical solution for n_s in terms of V_{GT} (see also Fjeldly et al. (1991)).

Eq. (3-3-50) is an empirical equation which can be justified by comparing the calculation results with experiments and more precise calculations. In this Subsection, we will demonstrate an excellent agreement between eq. (3-3-50) and experimental data. In Subsection 3.3.4, we will demonstrate an equally good agreement with exact quantum mechanical calculations. Intuitively, the structure of the UCCM expression (eq. (3-3-50)) seems reasonable. In the strong inversion case, it reverts to the simple charge control model (see eq. (3-3-42)) while in the subthreshold regime, it predicts that the inversion charge is an exponential function of the applied voltage, as expected.

Since UCCM is an empirical model, it is especially important to have a clear and unambiguous procedure for extracting model parameters from experimental data. For the MIS structure, we will base the extraction on measured C–V characteristics. As can be seen from Fig. 3.3.7, the transition from the depletion to the strong inversion regime at low frequencies is accompanied by a sharp increase in the MIS capacitance. As shown below, the voltage at which the derivative of the MIS capacitance reaches its maximum is very close to the threshold voltage determined by eq. (3-3-27).

The first derivative of eq. (3-3-50) with respect to V_{GT} yields the following unified expression for the metal-channel capacitance per unit area, c_{gc}, valid for all values of applied bias voltage:

$$c_{gc} = q\frac{dn_s}{dV_{GT}} = \frac{qn_s}{\eta V_{th} + an_s} \tag{3-3-56}$$

The first derivative of this capacitance

$$\frac{dc_{gc}}{dV_{GT}} = \frac{qn_s \eta V_{th}}{\left(\eta V_{th} + an_s\right)^3} \tag{3-3-57}$$

reaches its maximum value for

$$n_s = \frac{\eta V_{th}}{2a} \tag{3-3-58}$$

Hence, we obtain the following sheet inversion charge density at threshold:

$$n_o \equiv n_s\left(V_{GT} = 0\right) = \frac{\eta V_{th}}{2a} \tag{3-3-59}$$

and the value for the unified capacitance per unit area at threshold becomes

$$c_{gcT} \equiv c_{gc}\left(V_{GT} = 0\right) = \frac{qn_o}{\eta V_{th} + an_o} = \frac{q}{3a} = \frac{c_{max}}{3} \tag{3-3-60}$$

Here, $c_{max} \approx c_i$ is the maximum value of c_{gc}. Eq. (3-3-60) serves as the basis for a very convenient and straightforward technique for determining the threshold voltage from experimental data.

Fig. 3.3.9 shows the measured gate-channel capacitance, $C_{gc} = Sc_{gc}$, versus the gate-source voltage, V_{GS}, of a large area (68,145 μm^2) n-channel

MOSFET (Fat-FET) (see Park et al. (1991)). The devices used in this experiment were fabricated by n-well technology using an n^+-doped polysilicon gate. The oxide thickness was 220 Å. The gate-channel capacitance was measured at 10 kHz using the split C–V technique proposed by Sodini et al. (1982). The measurements were done at three different substrate biases (V_{sub} = 0, –1, –4V). The threshold voltage was then determined from the measured C–V characteristics as described above.

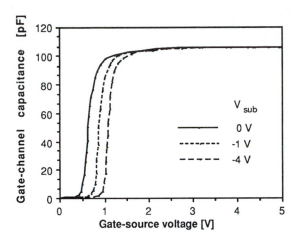

Fig. 3.3.9. Measured gate-channel capacitance versus gate-source voltage for a large area n-channel MOSFET for different values of substrate-channel bias. (From Park et al. (1991), © 1991 IEEE.)

From the experimentally determined gate-channel capacitance, we can calculate the inversion carrier sheet density as

$$qn_s = \int_{-\infty}^{V_{GS}} c_{gc}\, dV'_{GS} \tag{3-3-61}$$

According to UCCM, this should agree with eq. (3-3-56) which can be written as

$$\frac{1}{c_{gc}} = \frac{\eta V_{th}}{qn_s} + \frac{a}{q} \tag{3-3-62}$$

Hence, from a plot of $1/c_{gc}$ versus $1/n_s$, we can find η and a (see Fig. 3.3.10).

The slope of this plot gives η while the intercept with $1/n_s = 0$ yields a. The values of η obtained from the slopes in Fig. 3.3.10 agree very well with those determined directly from the subthreshold I–V characteristics. The value of d_i calculated from a (220 Å) is in excellent agreement with the gate oxide thickness measured by ellipsometry.

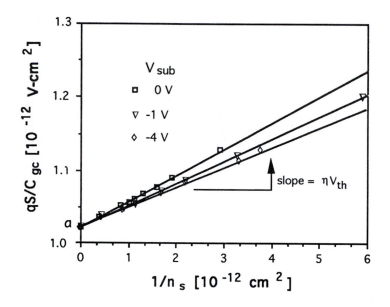

Fig. 3.3.10. Inverse gate-channel capacitance plotted as a function of the inverse mobile sheet charge density (see eq. (3.3.62)) for the Fat-FET n-channel MOSFET of Fig. 3.3.9. (From Park et al. (1991),© 1991 IEEE.)

The dC_{gc}/dV_{GS} versus V_{GS} characteristics from the MOSFET experiments are shown in Fig. 3.3.11. As was discussed above, the V_{GS} corresponding to the peak value of dC_{gc}/dV_{GS} should coincide with the V_{GS} at which the gate-channel capacitance has dropped to one third of its maximum value (see eqs. (3-3-55) to (3-3-58)).

In Fig. 3.3.12, we compare the measured n_s versus V_{GS} characteristics with UCCM using the parameters extracted above. The agreement is excellent for the entire range of the gate bias. The deviation in the measured curves found in the deep subthreshold region is due to C–V measurement errors. The errors are caused by the large series channel resistance at deep subthreshold compared with the reactance of the capacitance.

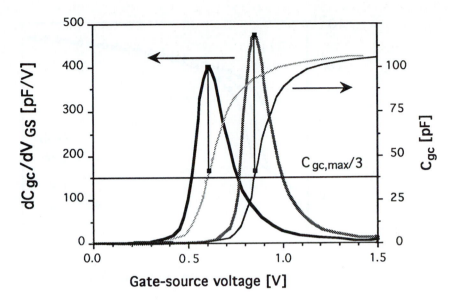

Fig. 3.3.11. Measured dependence of C_{gc} and dC_{gc}/dV_{GS} on V_{GS} for $V_{sub} = 0$ V (curves to the left) and -1 V (curves to the right). The threshold voltages determined by the two methods are indicated. (From Park et al. (1991),© 1991 IEEE.)

3.3.4. Analytical Unified MIS Capacitance Model

As mentioned above, the UCCM does not have an exact analytical solution for the inversion charge in terms of the applied voltage even though an accurate approximate solution can be obtained using the procedure described in Appendix A7. Here, we develop an analytical model for the MIS capacitance and the inversion charge which is in excellent agreement with UCCM, exact calculations, and experimental data. This model will be used in Section 3.9 and in Chapter 4 for Level 1 AIM-Spice models of different field effect transistors.

Above threshold, we use the expression for the differential channel capacitance per unit area, c_a, given by eq. (3-3-51). The corresponding sheet density of carriers in the inversion layer is

$$n_s = \frac{\varepsilon_i V_{GT}}{q(d_i + \Delta d)} \tag{3-3-63}$$

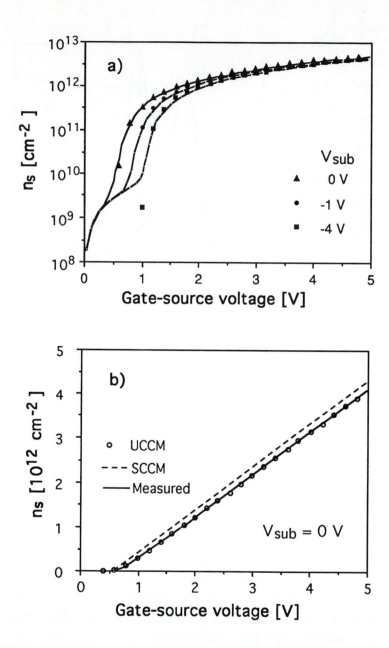

Fig. 3.3.12. Measured (solid lines) and calculated (UCCM, symbols) n_s versus V_{GS} characteristics in a semilog scale (a) and in a linear scale (b). In b) is also shown calculations using a simple charge control model (SCCM), see eq. (3-3-42). (From Park et al. (1991),© 1991 IEEE.)

Below threshold, the electron sheet density in the channel can be written as follows (see the discussion in Subsection 3.3.3)

$$n_s \approx n_o \exp\left(\frac{V_{GT}}{\eta V_{th}}\right) \tag{3-3-64}$$

where n_o is the carrier sheet density at the threshold point given by eq. (3-3-59), and η is the ideality factor. From eq. (3-3-64), the following expression is obtained for the subthreshold differential channel capacitance per unit area:

$$c_b = q\frac{dn_s}{dV_{GT}} = \frac{qn_o}{\eta V_{th}} \exp\left(\frac{V_{GT}}{\eta V_{th}}\right) \tag{3-3-65}$$

An approximate, unified expression for the effective differential metal-channel capacitance per unit area, c_{gc}, is obtained by representing c_{gc} as a series connection of the above-threshold and the subthreshold capacitances, i.e.,

$$c_{gc} = \frac{c_a c_b}{c_a + c_b} = \left[\frac{d_i + \Delta d}{\varepsilon_i} + \frac{\eta V_{th}}{qn_o} \exp\left(-\frac{V_{GT}}{\eta V_{th}}\right)\right]^{-1} \tag{3-3-66}$$

(see Jeng et al. (1988)). Hence, the unified carrier sheet charge density becomes (see eq. (3-3-61))

$$qn_s = \int_{-\infty}^{V_{GS}} c_{gc}\, dV'_{GS} = 2qn_o \ln\left[1 + \frac{1}{2}\exp\left(\frac{V_{GT}}{\eta V_{th}}\right)\right] \tag{3-3-67}$$

We notice that this expression is similar to an interpolation formula for n_S in heterostructure transistors given by Ruden (1989). Our calculations show that eq. (3-3-67) is in excellent agreement with UCCM.

3.3.5. Quantum Theory of the Two Dimensional Electron Gas

In Section 3.3.1, we implied that the electrons induced at the semiconductor-insulator interface of an MIS capacitor form a classical electron gas and behave essentially in the same way as electrons in a bulk semiconductor. This assumption is only correct if the thickness of the inversion layer is much larger than the deBroglie wavelength for electrons. For the classical electron gas considered above, this thickness, δd, can be estimated as

$$\delta d \approx \frac{V_{th}}{F_s} \tag{3-3-68}$$

where F_s is the surface field given by eq. (3-3-18) and V_{th} is the thermal voltage. Using Gauss' law, this field can be approximated as

$$F_s \approx \frac{\varepsilon_i \left(V_G - V_{FB}\right)}{d_i \varepsilon_s} \tag{3-3-69}$$

where V_{FB} is the flat band voltage. In this estimate we have used the condition that the electric displacement is continuous across the insulator-semiconductor interface and assumed that nearly all the applied voltage drops across the insulator. Hence,

$$\delta d \approx \frac{V_{th} \varepsilon_s d_i}{\left(V_G - V_{FB}\right)\varepsilon_i} \tag{3-3-70}$$

In modern MOSFETs, the oxide thickness can be well below 100 Å, and δd may become smaller than the deBroglie wavelength. For instance, for $d_i = 100$ Å, $V_G - V_{FB} = 1$ V and $T = 300$ K, we find $\delta d \approx 10$ Å. In this case, we must take into account the quantization of the energy levels in the potential well at the semiconductor-insulator interface in the direction perpendicular to the interface as illustrated by Fig. 3.3.13. (This was first pointed out by Schrieffer (1957).)

As discussed in Section 1.3, in the case of spherical energy surfaces, the dispersion relation in the direction parallel to the heterointerface is given by:

$$E_n = E_j + \frac{\hbar^2 \left(k_y^2 + k_z^2\right)}{2 m_n} \tag{3-3-71}$$

where E_n is the electron energy, E_j is the energy level of the j-th subband, k_y and k_z are the wave vector components parallel to the interface, and m_n is the effective electron mass. For a relatively thick electron gas layer, the number of subbands is large and the energy difference between the bottoms of the subbands is small. For a relatively thin electron gas layer, only the few lowest subbands are important and the energy difference between the bottoms of the subbands may become large compared to the thermal energy, k_BT. In this case, the electron gas is often referred to as a two-dimensional electron gas (2DEG). Pioneering work on the properties of the 2DEG was done by Stern (1972).

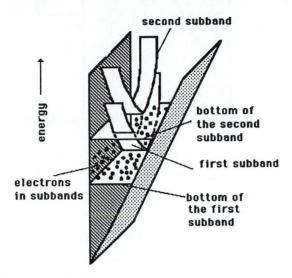

Fig. 3.3.13. Schematic diagram of energy subbands at the semiconductor-insulator interface (assuming constant effective field approximation).

The density of states, D, for each subband is given by

$$D = \frac{m_n}{\pi \, \hbar^2} \tag{3-3-72}$$

Notice that this density of states is independent of the subband energy, E_j. As a consequence, the overall density of states has a "staircase" dependence on energy, as shown in Fig. 3.3.14 for a triangular quantum well which is characteristic for the semiconductor-insulator interface of an MIS structure.

The number of electrons occupying a given subband, j, can be found by multiplying the density of states, D, for a single subband by the Fermi-Dirac distribution function and integrating from E_j to infinity:

$$n_j = D \int_{E_j}^{\infty} \left[1 + \exp\left(\frac{E - E_F}{k_B T} \right) \right]^{-1} dE \tag{3-3-73}$$

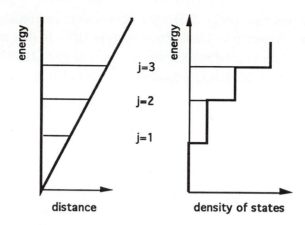

distance   density of states

Fig. 3.3.14. Energy levels (bottoms of subbands) and density of states for a triangular quantum well structure; $j = 1, 2,..$ correspond to the different subbands.

After evaluating this integral and adding the contribution from all subbands, we obtain

$$n_s = \sum_{j=0}^{\infty} \frac{m_n k_B T}{\pi \hbar^2} \ln\left[1 + \exp\left(\frac{E_F - E_j}{k_B T}\right)\right] \qquad (3\text{-}3\text{-}74)$$

These equations are only valid for a set of parabolic subbands corresponding to equivalent valleys in the first Brillouin zone and with equivalent orientation relative to the interface. As was discussed in Chapter 1, the conduction band structure of silicon is quite complicated (see Fig. 3.3.15).

The density of states, $D_{\parallel}$, for electron motion parallel to the interface is given by

$$D_{\parallel} = \frac{m_{\parallel}}{\pi \hbar^2} \qquad (3\text{-}3\text{-}75)$$

where $m_{\parallel}$ is the density of states effective mass for such motion. Usually MOSFETs are fabricated with the channel along the (100) surface (because of the low interface state density for this surface). The density of states effective mass is proportional to the area of the cross section of the energy ellipsoids by the plane of the inversion layer. Hence, as can be seen from Fig. 3.3.15, the effective mass $m_\parallel$ is m_t for two valleys and $(m_t m_l)^{1/2}$ for four valleys, where m_t is the transverse effective mass of electrons $(m_t = 0.190\ m_e)$ and m_l is the longitudinal effective mass $(m_l = 0.916\ m_e)$. Consequently, the relation between n_s and the Fermi energy, E_F, for silicon can be written as (compare with eq. (3-3-74)):

$$n_s = \sum_{i=1}^{6} \sum_{j=0}^{\infty} \frac{m_{\parallel i}\ k_B T}{\pi\ \hbar^2} \ln\left[1 + \exp\left(\frac{E_F - E_{ji}}{k_B T}\right)\right] \qquad (3\text{-}3\text{-}76)$$

where i is an integer from 1 to 6 corresponding to the six valleys, E_{ji} is the energy level of j-th subband in valley i, and $m_{\parallel i}$ is the appropriate effective mass for parallel motion.

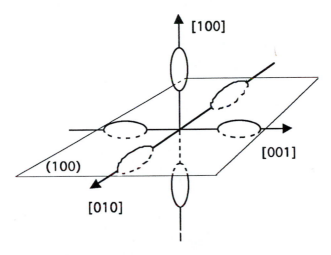

Fig. 3.3.15. Constant energy surfaces for the silicon conduction band relative to the plane of the inversion layer. The band minima correspond to the centers of the six ellipsoids oriented along [100] directions of momentum space, at approximately 85 % of the distance between the center and the boundary of the Brillouin zone. The long axis of the ellipsoids corresponds to the longitudinal effective mass of electrons in silicon, $m_l = 0.916\ m_e$. The short axes correspond to the transverse effective mass, $m_t = 0.190\ m_e$.

The quantized energy levels for the subbands can be found using a numerical self-consistent solution of the Schrödinger and Poisson's equations (see Ohkura (1990)). However, an excellent approximation for the exact solution can be found by assuming a linear potential profile (i.e., constant effective field, F_{eff}, see Fig. 3.3.13) in the semiconductor at and close to the semiconductor-insulator interface. In this case, the energy levels are given by

$$E_j - E_c(0) = \left(\frac{9\pi^2 \, \hbar^2 q^2 F_{eff}^2}{8m_\perp} \right)^{1/3} \left(j + \frac{3}{4} \right)^{2/3} \tag{3-3-77}$$

where $m_\perp$ is the effective mass for electron motion perpendicular to the (100) surface ($m_\perp = m_l$ for the two valleys where $m_\parallel = m_t$, and $m_\perp = m_t$ for the remaining four valleys) and $E_c(0)$ is the minimum conduction band energy at the Si-SiO$_2$ interface (see Stern and Sarma (1984), and Byun et al. (1990)). The effective field, F_{eff}, is expressed through the surface field, F_s, and the bulk field, F_B. For electrons, the relationship linking F_{eff}, F_B, and F_s, giving the best fit to the self-consistent solution of the Schrödinger and Poisson's equation is given by

$$F_s = \frac{q(n_s + n_B)}{\varepsilon_s} \tag{3-3-78}$$

$$F_B = \frac{qn_B}{\varepsilon_s} \tag{3-3-79}$$

$$F_{eff} = \frac{F_s + F_B}{2} \tag{3-3-80}$$

where qn_s is the interface electron sheet density,

$$qn_B = qN_a \, d_{dep} \tag{3-3-81}$$

is the sheet density of depletion charge and d_{dep} is the average depletion depth of eq. (3-3-34) (see Byun et al. (1990)). Correspondingly, we have for holes:

$$F_s = \frac{q(p_s + p_B)}{\varepsilon_s} \tag{3-3-82}$$

$$F_B = \frac{qp_B}{\varepsilon_s} \tag{3-3-83}$$

$$F_{eff} = \frac{F_s + F_B}{2} \tag{3-3-84}$$

where qp_s is the interface hole sheet density and

$$qp_B = q N_d \, d_{dep} \tag{3-3-85}$$

is the sheet density of depletion donor charge.

We should mention that even though the same effective electric field (compare eqs. (3-3-80) and (3-3-84)) determines energy levels for both electrons and holes, a detailed analysis given by Lee et al. (1991) shows that the effective thickness of the two dimensional electron and hole gas is determined by a different effective field:

$$F_{eff1} = \frac{F_s + 2 F_B}{3} \tag{3-3-84a}$$

in agreement with numerous experimental data on the electron and hole mobility of the two dimensional gases, as further discussed in Section 3.8.

Solving eqs. (3-3-76) to (3-3-80) iteratively, we can obtain the relation between n_s and the Fermi level, $E_F - E_c(0)$. Fig. 3.3.16 shows the calculated dependence of n_s on $E_F - E_c(0)$ at room temperature for different substrate doping levels, N_a (see Park et al. (1991)). In this calculation, it was assumed that the maximum value of n_B is given by

$$n_{B\,max} = \sqrt{2\varepsilon_s N_a \left(2\varphi_b + 5V_{th} - V_{sub} \right)/q} \tag{3-3-86}$$

where $q\varphi_b$ is the bulk Fermi energy relative to the intrinsic Fermi level (see eq. (3-3-2)), V_{th} is the thermal voltage, and V_{sub} is the substrate bias (see Lewyn and Meindl (1985)). In the subthreshold region, our calculation agrees reasonably well with the classical charge sheet model (CCSM) given by (see Brews (1978))

$$n_s = \sqrt{\frac{2\varepsilon_s N_c V_{th}}{q} \exp\left(\frac{E_F - E_c(0)}{k_B T} \right) + n_B^2} - n_B \tag{3-3-87}$$

especially at low levels of substrate doping (see Fig. 3.3.16).

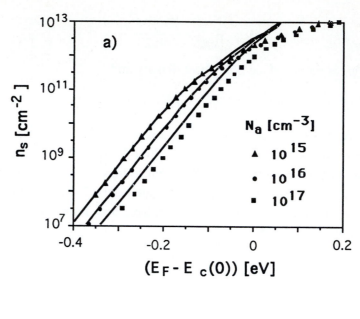

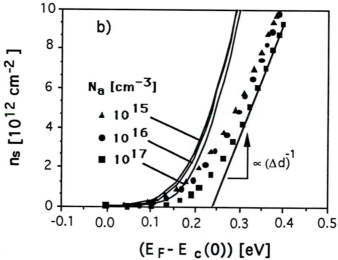

Fig. 3.3.16. Comparison of interface carrier density versus $E_F - E_C(0)$ characteristics for different substrate doping densities in a semilog plot (a) and a linear plot (b). Symbols: calculations based on a 2DEG formulation; solid lines: charge sheet model; straight line in b): linear approximation to 2DEG formulation, the slope gives $\Delta d \approx 4$ Å. (From Park et al. (1991), © 1991 IEEE.)

In eq. (3-3-87), N_c is the effective density of state of the conduction band in the three-dimensional formulation (see Section 1.3). The difference between the curves at high substrate doping levels is caused by the fact that the large bulk field quantizes the energy levels even in the subthreshold regime. However, at strong inversion, the difference between the charge sheet model and the 2DEG formulation is large.

Once the relation between n_s and $E_F - E_c(0)$ is found, we can calculate n_s as a function of V_{GS} for any given insulator thickness, see Fig. 3.3.17.

As can be seen from Fig. 3.3.16 (and first pointed out by Lee et al. (1983)), the dependence of n_s on E_F in the above-threshold regime can be approximated by a straight line

$$n_s = \frac{\varepsilon_i \left(E_F - E_{Fo} \right)}{q \Delta d} \tag{3-3-88}$$

where E_{Fo} is the intercept of this linear approximation with $n_s = 0$. This approximation means that a fraction of the applied voltage equal to $\Delta d/(d_i + \Delta d)$ is accommodated by a shift in the Fermi level with respect to the bottom of the conduction band.

The shift in the Fermi level with respect to the bottom of the conduction band changes the above-threshold capacitance from c_i to (see eq. (3-3-51))

$$c_a = \frac{\varepsilon_i}{d_i + \Delta d} \tag{3-3-89}$$

The parameter Δd can be interpreted as a correction to the insulator thickness. From the straight line approximation in Fig. 3.3.16b, we obtain $\Delta d = 4\text{Å}$ which is much smaller than that of GaAs ($\approx 45\text{Å}$ according to Byun et al. (1990)). This difference is caused not only by a much larger effective mass in the conduction band in silicon, which makes quantum effects much less pronounced, but also by the large difference in the dielectric constants between the insulator and the semiconductor for the MOS system.

In Fig. 3.3.17, we compare our numerical calculations with the UCCM (see Park et al. (1991) and Subsection 3.3.3). As can be seen, the agreement between the exact calculation and UCCM is excellent. The values of the ideality factor η calculated from the slope of the numerical subthreshold characteristics in Fig. 3.3.17a agree well with the expression discussed in Section 3.3.3.

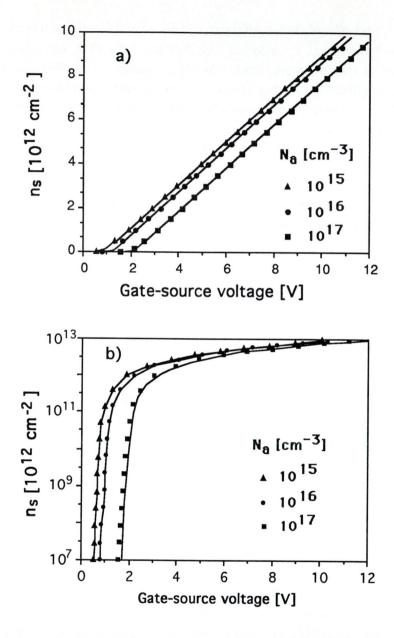

Fig. 3.3.17. MOSFET interface carrier density versus gate-source voltage in a linear scale (a) and in a semilog scale (b). Symbols correspond to numerical calculations, solid lines correspond to UCCM. Oxide thickness: 220Å. (From Park et al. (1991), © 1991 IEEE.)

3-4. BASIC MOSFET THEORY

3.4.1. Gradual Channel Approximation

A MOSFET essentially consists of an MIS capacitor and two diffused or implanted regions which serve as ohmic contacts to an inversion layer of free charge carriers at the semiconductor-insulator interface. The MIS capacitor is formed by a gate contact separated from the semiconductor substrate by an insulating layer (see Section 3.3). In an n-channel MOSFET, the substrate is p-type silicon and, accordingly, the inversion charge consists of electrons which form a conducting channel between the two n^+ ohmic contacts, called source and drain. The depletion regions between the n regions (source, drain and the channel) and the p-type substrate, provide isolation from other devices fabricated on the same substrate. A schematic cross-section of the n-channel MOSFET is shown in Fig. 3.4.1.

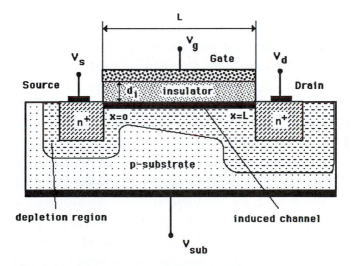

Fig. 3.4.1. Schematic cross-section of an n-channel MOSFET.

As discussed in Section 3.3, inversion charge can be induced in the MIS capacitor by applying a gate-source voltage, V_{GS}, above the threshold voltage, V_T. When a drain-source bias, V_{DS}, is applied with the MOSFET in the above-threshold state, induced charge carriers move in the channel inversion layer from source to drain. A variation in V_{GS} changes the electron concentration in the channel and, hence, the channel conductance and the device current.

The highly doped source and drain regions provide reservoirs from which carriers can enter the channel and to which they can escape from the channel. Hence, up to fairly high frequencies (comparable to the inverse transit time of carriers across the channel), the above-threshold C–V characteristics of the device (with source, drain, and substrate connected) look like the very low frequency C–V characteristics of an MIS structure (see Section 3.3).

Analytical or semianalytical modeling of MOSFET characteristics is usually based on the so-called Gradual Channel Approximation (GCA). In this approximation, the inversion and depletion charge densities under the gate are determined in terms of a one-dimensional electrostatic problem for the direction perpendicular to the channel, just as for the MIS structure. By inspection of the two-dimensional Poisson's equation for the semiconductor,

$$\frac{\partial F_x}{\partial x} + \frac{\partial F_y}{\partial y} = \frac{\rho}{\varepsilon_s} \tag{3-4-1}$$

we see that GCA is valid under the assumption that $|\partial F_x/\partial x| << |\partial F_y/\partial y| \approx |\rho|/\varepsilon_s$ where ε_s is the dielectric permittivity of the semiconductor, ρ is the space charge density, x is a space coordinate along the channel in the direction from source to drain, y is a space coordinate perpendicular to the channel, and F_x and F_y are the x and y components of the electric field in the channel, respectively.

The validity of GCA can be checked by making rough estimates of the variation in the longitudinal and vertical fields components. We consider a strong inversion situation where the inversion charge is much larger than the depletion charge. The voltage across the insulator is then approximately $V_{GS} - V_T$, where V_T is the threshold voltage (see Section 3.3), corresponding to a vertical insulator field strength of $F_{iy} = (V_{GS} - V_T)/d_i$, where d_i is the insulator thickness. From the requirement of continuity of the electric displacement across the semiconductor-insulator interface, we find

$$F_y \approx \frac{\varepsilon_i}{\varepsilon_s} \frac{V_{GS} - V_T}{d_i} \tag{3-4-2}$$

very close to the interface. Assuming an effective channel thickness, δd, we obtain

$$\frac{\partial F_y}{\partial y} \approx \frac{F_y}{\delta d} \tag{3-4-3}$$

The effective channel thickness may be estimated approximately as the distance from the interface where the vertical change in the channel potential reaches V_{th} = $k_B T/q$, i.e., $\delta d \approx V_{th}/F_y$. Hence,

$$\left| \frac{\partial F_y}{\partial y} \right| \approx \left(\frac{\varepsilon_i}{\varepsilon_s} \right)^2 \frac{(V_{GS} - V_T)^2}{d_i^2 V_{th}} \tag{3-4-4}$$

On the other hand, a very rough estimate of the variation of the longitudinal electric field gives

$$\left| \frac{\partial F_x}{\partial x} \right| \approx \frac{V_{DS}}{L^2} \tag{3-4-5}$$

where L is the gate length. (Here, we assumed that the field strength changes gradually from a small value near source to values of the order V_{DS}/L near drain.) Hence, the gradual channel approximation is valid when

$$\left(\frac{\varepsilon_i L}{\varepsilon_s d_i} \right)^2 \frac{(V_{GS} - V_T)^2}{V_{th} V_{DS}} \gg 1 \tag{3-4-6}$$

For $L = 1$ μm, $\varepsilon_i/\varepsilon_s \approx 0.33$, $V_{th} \approx 25.8$ mV (room temperature), $V_{GS} - V_T = 0.5$ V, $V_{DS} = 0.5$ V, and $d_i = 300$ Å, the left-hand side of inequality (3-4-6) is about 2300. This seems to imply that GCA is a very good approximation for a 1 μm device. However, some caution is warranted at this point. The inequality is weakened by a factor of 625 if, for example, both L and $V_{GS} - V_T$ are reduced by a factor of 5. But even without such a reduction, the longitudinal electric field in the channel is quite non-uniform at high drain-source biases. Indeed, based on a simple charge control model and GCA, we can estimate the inversion sheet charge density in the channel as follows:

$$qn_s \approx c_i (V_{GT} - V) \tag{3-4-7}$$

where c_i is the gate insulator capacitance per unit area, $V_{GT} = V_{GS} - V_T$, and V is the channel potential induced by the drain-source bias. Clearly, such an approximation may only be valid when qn_s is positive, corresponding to

$$V_{GT} \geq V \tag{3-4-8}$$

Hence, at large drain-source biases, GCA always becomes invalid at least in a certain section of the channel near the drain where the longitudinal field becomes comparable to or larger than the transverse field. When this takes place, the MOSFET is in "saturation" because the drain current becomes nearly independent of the drain bias (as will be shown below). Fig. 3.4.2 shows the calculated potential distribution under the gate in the saturation regime.

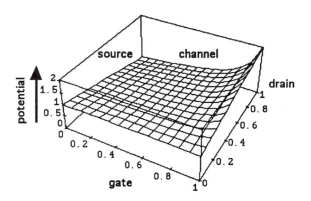

Fig. 3.4.2. Plot of the potential distribution in the gate insulator for a drain-source voltage above the saturation voltage.

In Fig. 3.4.2, we have used dimensionless units for the potential and for distances: $u = V/V_{GT}$ (the reduced channel potential) is shown along the vertical axis, $z = x/L$ is in the direction along the channel, and $w = y/y_o$ is in the direction perpendicular to the channel where $y_o = (\varepsilon_s V_{GT}/qN_a)^{1/2}$. The calculation was done using the charge control model described below. As can be seen from the figure, in the saturation regime, the derivative of the channel potential near the drain is very large. Closer to the source, the potential variation is relatively small and GCA is valid. Indeed, the estimate of eq. (3-3-6) seems to justify invoking GCA for a wide range of geometries and operating conditions in the section of the channel where the channel voltage satisfies eq. (3-4-8). Hence, with due caution, we may proceed on the basis of GCA and assume that the charges induced at a given position in the channel can be determined using the formulas derived for the MIS structure, but adding the effect of the channel voltage caused by the drain-source bias.

We now consider a situation when the device operates in the above-threshold regime, i.e., when the gate voltage is large enough to cause inversion everywhere in the channel. We will follow the derivation given, for example, by Shur (1990). The induced sheet charge density, Q_s, for an n-channel device can be found as follows (see Section 3.3):

$$Q_s = - c_i \left[V_{GS} - V_T - V(x) \right] \tag{3-4-9}$$

where the voltage difference in the bracket is the voltage drop across the insulator layer, and $\varphi_b = V_{th} \ln(N_a/n_i)$ where N_a is the acceptor concentration in the substrate and n_i is the intrinsic carrier density (see Section 3.3). For now, we assume that the source and the semiconductor substrate are both connected to ground.

The induced sheet density of free electrons, n_s, at each point of the channel can be determined by deducting the sheet charge density of acceptors in the depletion layer, $Q_{dep} = -qn_B$, from the total induced charge, Q_s,

$$n_s = c_i \left[V_{GS} - 2\varphi_b - V_{FB} - V(x) \right]/q - \left| Q_{dep} \right|/q \tag{3-4-10}$$

The depletion sheet charge density is given by

$$Q_{dep}(x) = - qN_a d_{dep} = - \sqrt{2\varepsilon_s qN_a \left[2\varphi_b + V(x) \right]} \tag{3-4-11}$$

where d_{dep} is the depletion width (see Section 3.3). Note that the total band bending between the bulk of the semiconductor and the interface in the above-threshold regime is $2\varphi_b + V(x)$ at an arbitrary position, x, since the n-channel – p-substrate junction is reverse-biased by the voltage $V(x)$. A simplified three-dimensional sketch of the potential distribution in the depletion region below the gate is shown in Fig. 3.4.3. Just as in Fig. 3.4.2, we use dimensionless units for the potential and for distances: $u = V/V_{GT}$ is along the vertical axis, $z = x/L$ is in the direction along the channel, and $w = y/y_o$ is in the direction perpendicular to the channel, where $y_o = (\varepsilon_s V_{GT}/qN_a)^{1/2}$. The potential distribution along the channel in Fig. 3.4.3 was calculated using a simple charge control model (described below). In the direction perpendicular to the channel a parabolic distribution was used between the interface and the edge of the depletion zone, corresponding to the depletion approximation of the GCA. We can clearly see how the depletion region increases towards the drain side of the channel because of an increasing potential difference between the substrate and the channel.

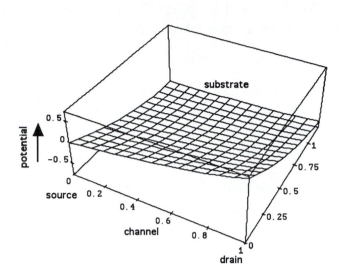

Fig. 3.4.3. Schematic representation of the potential distribution in the depletion region of a MOSFET biased above threshold.

As can be seen from Figs. 3.4.2 and 3.4.3, the potential distribution along the channel leads to a variable depletion region charge since the channel potential provides an additional reverse bias for the induced n-channel − p-substrate junction. The corresponding variation in the induced free electron charge in the channel is described by eqs. (3-4-10) and (3-4-11). Assuming a constant electron mobility, μ_n, and neglecting diffusion effects, the absolute value of the drain current, I_d, can be written as

$$I_d = W q \mu_n \frac{dV}{dx} n_s \tag{3-4-12}$$

where W is the device width. A constant mobility corresponds to an electron velocity, v_n, that is proportional to the longitudinal electric field $F = |F_x|$ in the channel, i.e.,

$$v_n = \mu_n F \tag{3-4-13}$$

In reality, however, velocity saturation sets in at sufficiently high fields and will strongly influence the electrical properties of the device at high drain-source bias.

This will be discussed further in Section 3.5.

Eq. (3-4-12) can be rewritten as

$$dx = \frac{W q \mu_n}{I_d} n_s \, dV_c \tag{3-4-14}$$

Integrating eq. (3-4-14) from zero to the channel length L (that corresponds to a change in V from zero at the source side of the gate to V_{DS} at the drain side of the gate), we obtain the following expression for the current-voltage characteristics:

$$I_d = \frac{W \, \mu_n \, c_i}{L} \left\{ \left(V_{GS} - V_{FB} - 2\varphi_b - \frac{V_{DS}}{2} \right) V_{DS} \right.$$
$$\left. - \frac{2\sqrt{2\varepsilon_s q N_a}}{3c_i} \left[\left(V_{DS} + 2\varphi_b \right)^{3/2} - \left(2\varphi_b \right)^{3/2} \right] \right\} \tag{3-4-15}$$

This equation is only valid for such values of V_{DS} that the inversion layer still exists even at the drain side of the gate, i.e.,

$$n_s \left(V = V_{DS} \right) \geq 0 \tag{3-4-16}$$

The condition $n_s = 0$ at the drain is called the pinch-off condition. As can be seen from eqs. (3-4-10) and (3-4-11), the pinch-off occurs at the drain side of the gate when

$$V_{DS} = V_{SAT}$$
$$= V_{GS} - 2\varphi_b - V_{FB} + \frac{\varepsilon_s q N_a}{c_i^2} \left[1 - \sqrt{1 + \frac{2c_i^2 (V_{GS} - V_{FB})}{\varepsilon_s q N_a}} \right] \tag{3-4-17}$$

where V_{SAT} is the saturation voltage. At low doping levels, we find that $V_{SAT} \to V_{GT}$ when the gate voltage approaches threshold. From Section 3.3, we have the following expression for the threshold voltage:

$$V_T = V_{FB} + 2\varphi_b + \sqrt{4\varepsilon_s q N_a \varphi_b} / c_i \tag{3-4-18}$$

Most MOSFETs have a fourth contact attached to the substrate (see Fig. 3.4.1) to allow the application of a substrate-source voltage, $V_{sub} = V_{BS}$. The difference between the interface potential and the substrate-source bias defines the potential drop across the gate depletion region and thereby the depletion charge. Hence, V_T in eq. (3-4-18) should be modified to account for the substrate bias:

$$V_T = V_{FB} + 2\varphi_b + \sqrt{2\varepsilon_s\, qN_a(2\varphi_b - V_{BS})}/c_i \qquad (3\text{-}4\text{-}19)$$

This expression is only valid for negative or slightly positive substrate-source bias when the interface – p-substrate junction is either reverse-biased or slightly forward biased, i.e., $V_{BS} < 2\varphi_b$. A larger positive substrate bias will lead to large leakage currents. The dependence of V_T on V_{BS} is discussed in Section 3.7.

The calculated dependence of the threshold voltage on substrate potential is shown in Fig. 3.4.4 for different values of gate insulator thickness.

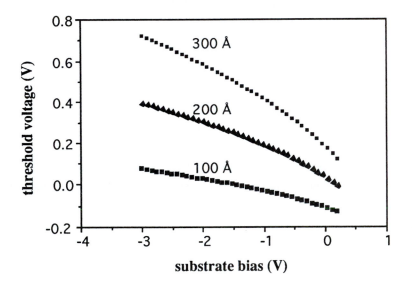

Fig. 3.4.4. Dependence of the threshold voltage on substrate potential in MOSFETs with different insulator thicknesses. Parameters used in the calculation: energy gap: 1.12 eV, density of states in the conduction band: 3.22×10^{25} m^{-3}, density of states in the valence band: 1.83×10^{25} m^{-3}, semiconductor dielectric permittivity: 1.05×10^{-10} F/m, insulator dielectric permittivity: 3.45×10^{-11} F/m, flat-band voltage: –1 V, substrate doping level: 10^{22} m^{-3}, generation time: 10^{-8} s, temperature: 300 K.

As can be seen from this figure and from eq. (3-4-19), the threshold voltage decreases with decreasing insulator thickness and is quite sensitive to the substrate bias. (This so-called body effect is important for the device characterization and threshold voltage engineering as will be discussed in more details in Section 3.7.)

The pinch-off condition implies that the carrier concentration at the drain side of the channel is equal to zero. Hence, at a first glance, one might think that the drain current should also vanish. However, this is not the case. Instead, the drain current saturates, i.e., becomes nearly independent of the drain-source bias at $V_{DS} > V_{SAT}$. What happens when the drain-source voltage approaches V_{SAT} may be understood by analyzing the electric field distributions under the gate.

The electric field in the channel in the direction parallel to the semiconductor-insulator interface, $F = |F_x| = dV/dx$, can be found from eq. (3-4-14):

$$F = \frac{I_d}{q\mu_n n_s(V)W} \tag{3-4-20}$$

When the drain-source voltage approaches the saturation voltage, $n_s \to 0$ and F tends to infinity (see Fig. 3.4.2). (Of course, in reality, at very large electric fields, the electron velocity is no longer proportional to the electric field, and velocity saturation has to be taken into account. However, if the device is very long or the saturation velocity is very large, we can ignore the velocity saturation effects for a while.) As we approach the saturation point, nearly all additional drain-source voltage will drop across a very narrow region near the drain. The field distribution in most of the channel varies very little and, as a consequence, the differential drain conductance

$$g_d = \left.\frac{dI_d}{dV_{DS}}\right|_{V_{GS}} \tag{3-4-21}$$

tends to zero when $V_{DS} \to V_{SAT}$. Hence, the current-voltage characteristics may be extrapolated in the voltage region $V_{DS} > V_{SAT}$ assuming a constant drain current, $I_d = I_{sat}$, (independent of the drain-source voltage). I_{sat} may be determined by substituting $V_{DS} = V_{SAT}$ from eq. (3-4-17) into eq. (3-4-15). In fact, neither does the electron concentration ever reach zero nor does the electric field become infinite. This is simply a preliminary remark indicating that GCA is not valid in this region and we need a more accurate and detailed analysis of the saturation regime. Such an analysis is given in Section 3.5.

3.4.2. A Simple Charge Control Model

For very small drain-source voltages, the terms in the curly brackets in eq. (3-4-15) may be expanded into a Taylor series leading to the following simplified expression for the current-voltage characteristics in the linear region

$$I_d = \frac{W \mu_n c_i}{L} V_{GT} V_{DS} \qquad (3\text{-}4\text{-}22)$$

This equation can be interpreted as follows: at very small drain-source bias, the charge induced into the channel does not depend on the channel potential and we have $qn_s = c_i V_{GT}$ (see eq. (3-4-7)). In this case, the electric field in the channel is nearly constant and given by

$$F = V_{DS} / L \qquad (3\text{-}4\text{-}23)$$

Multiplying the channel sheet charge density, qn_s, by the electron velocity, $v_n = \mu_n F$, and by the gate width, W, we obtain eq. (3-4-22).

A similar approach may be used for a simplified description of the current-voltage characteristics of a MOSFET based on the so-called charge control model. In this model, we assume that the concentration of free carriers induced into the channel is given by eq. (3-4-7). In other words, we neglect the effects of the charge in depletion layer, Q_{dep}, on the surface channel potential. The drain current, I_d, can then be written as

$$I_d = W \mu_n c_i \frac{dV}{dx} (V_{GT} - V) \qquad (3\text{-}4\text{-}24)$$

This equation can be rewritten as

$$dx = \frac{W \mu_n c_i}{I_d} (V_{GT} - V) \, dV \qquad (3\text{-}4\text{-}25)$$

Integrating eq. (3-4-25) from zero to the channel length L (which corresponds to a change in V from zero at the source side of the gate to V_{DS} at the drain side of the gate), we obtain the following expression for the current-voltage characteristics:

$$I_d = \frac{W \mu_n c_i}{L} \times \begin{cases} \left[(V_{GS} - V_T) V_{DS} - V_{DS}^2 / 2 \right], & \text{for } V_{DS} \leq V_{SAT} \\ (V_{GS} - V_T)^2 / 2, & \text{for } V_{DS} > V_{SAT} \end{cases} \qquad (3\text{-}4\text{-}26)$$

MOSFET current-voltage characteristics calculated using this simple charge control model are shown in Fig. 3.4.5.

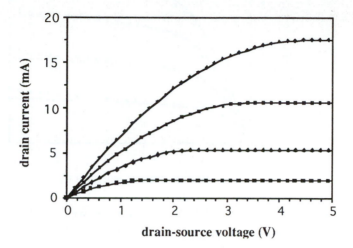

Fig. 3.4.5. MOSFET current-voltage characteristics calculated using the simple charge control model (see eq. (3-4-26). Parameters used in the calculation: μ_n = 0.06 m^2/Vs, L = 0.8 µm, W = 20 µm, V_T = 1 V, $\varepsilon_i/\varepsilon_o$ = 3.9, d_i = 300 Å. The top curve is for V_{GS} = 5 V, step: –1 V.

Fig. 3.4.6 shows measured I-V characteristics of a real silicon MOSFET (from Moon et al. (1991)). As can be seen from a comparison between Figs. 3.4.5 and 3.4.6, the simple charge control model gives a fairly crude description of the characteristics. It reproduces the qualitative trend of the current saturation at high drain voltages and the decrease of the current as the applied gate-source voltage approaches threshold. However, the experimental curves exhibit a much smaller drain current, much smaller saturation voltages, and a lack of complete saturation of the drain current at high drain voltages. At small gate voltages, the output conductance, dI_d/dV_{DS}, even increases somewhat in the deep saturation regime. The reason for these discrepancies is that we attempted to apply the simple charge control model to a short device (with L = 0.8 µm). In such devices, the saturation of electron velocity in the channel plays an important role, as explained in more detail in Section 3.5. As shown in Fig. 3.4.6, the experimental data can be accurately described by a model incorporating the velocity saturation effects, as will be discussed in Sections 3.9 - 3.12.

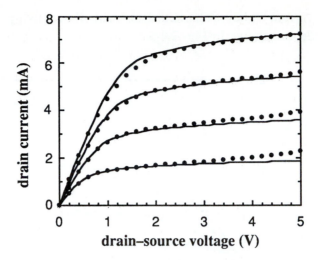

Fig. 3.4.6. Measured (solid dots) drain current–voltage characteristics of MOSFET with nominal gate length 0.8 μm. The top curve figure corresponds to V_{gs} = 5.V (extrinsic gate-source voltage). Gate voltage step: −1 V. Solid lines correspond to calculations based on a model considered in Section 3.11. (From Moon et al (1991), © 1991 IEEE.)

An important device characteristic is the transconductance defined as

$$g_m = \frac{dI_d}{dV_{GS}}\bigg|_{V_{DS}} \tag{3-4-27}$$

From eqs. (3-4-26) we find

$$g_m = \begin{cases} \beta V_{DS}, & \text{for } V_{DS} \leq V_{SAT} \\ \beta V_{GT}, & \text{for } V_{DS} > V_{SAT} \end{cases} \tag{3-4-28}$$

where $\beta = W\mu_n c_i/L$ is the transconductance parameter.

As can be seen from eqs. (3-4-28) and the expression for β, a high transconductance is obtained with high values of the low field electron mobility, thin gate insulator layers (i.e., larger gate insulator capacitance $c_i = \varepsilon_i/d_i$), and large W/L ratios. The dependence of the transconductance on the low field mobility and on the gate length, L, is, however, strongly affected by the velocity saturation effects in short channel devices (see Section 3.5).

3.4.3. Unified Charge Control Model for MOSFETs

As was discussed in Section 3.3, the capacitance-voltage characteristics of an MIS capacitor as well as the dependencies of the surface charge density in the MIS structure can be very accurately described using the new Unified Charge Control Model (UCCM). Moon et al. (1990), Park et al. (1991), Moon et al. (1991a), and Byun et al. (1990) applied this model to different field effect transistors including Heterostructure Field Effect Transistors, n-channel MOSFET, p-channel MOSFET, polysilicon Thin Film Transistors, and a-Si Thin Film Transistors. For all these devices, UCCM yielded excellent agreement with experimental data. In this Subsection, we discuss how to apply UCCM to FET modeling. The advanced version of UCCM which accounts for effects important in short channel devices will be considered in Sections 3.6 and 3.9 to 3.12.

For field effect transistors, the UCCM equation derived in Section 3.3 (see eq. (3-3-50)) has to be modified to account for the channel potential. According to this model, the inversion charge is related to the gate-source and channel potential as follows:

$$V_{GT} - \alpha V_F = \eta V_{th} \ln\left(\frac{n_s}{n_o}\right) + a(n_s - n_o) \qquad (3\text{-}4\text{-}29)$$

where V_F is the quasi-Fermi (electrochemical) potential measured relative to the Fermi potential at the source side of the channel, η is the ideality factor in the subthreshold region, V_{th} is the thermal voltage, and $a \approx q/c_a$ where

$$c_a = \frac{\varepsilon_i}{d_i + \Delta d} \qquad (3\text{-}4\text{-}30)$$

is the effective gate capacitance per unit area, ε_i is the dielectric constant of the gate insulator, d_i is the thickness of the gate insulator, and Δd is the correction to the insulator layer thickness (see Section 3.3).

The parameter α in eq. (3-4-29) accounts for the dependence of the threshold voltage on the channel potential in strong inversion, and, hence, on the position along the channel. In order to explain the meaning of this parameter, let us first consider the simplified version of the charge control model which was presented in eq. (3-4-7):

$$q n_s \approx c_i (V_{GT} - V) \qquad (3\text{-}4\text{-}31)$$

As was shown above, the threshold voltage depends on the depletion charge (see eq. (3-4-19)). Taking into account the dependence of this charge on the channel potential, we can write the corresponding position dependent threshold voltage, V_{TX}, to be used in eq. (3-4-31), as

$$V_{TX} = V_{FB} + 2\varphi_b + \sqrt{2\varepsilon_s\, qN_a(2\varphi_b + V - V_{BS})}/c_i \qquad (3\text{-}4\text{-}32)$$

This makes the charge control equation non-linear and difficult to use in device modeling. However, if we linearize eq. (3-4-32) with respect to V, we can write $V_{TX} \approx V_T + (\alpha - 1)V$ where now V_T is the value of the threshold voltage at the source side of the channel given by eq. (3-4-19). Substituting V_T by V_{TX} in eq. (3-4-31), we obtain

$$qn_s \approx c_i\left(V_{GS} - V_T - \alpha V\right) \qquad (3\text{-}4\text{-}33)$$

In order to make our model applicable to both the below and the above threshold regimes, we must take into account the diffusion current because, as will be discussed in Section 3.6, this current contribution is dominant in the subthreshold regime. Hence, instead of eq. (3-4-12) which accounts only for the drift current, we have to use the following, more general expression for the channel current:

$$I_d = qW\left|n_s\mu_n\frac{dV}{dx} - D_n\frac{dn_s}{dx}\right| = qn_sW\mu_n\frac{dV_F}{dx} \qquad (3\text{-}4\text{-}34)$$

(see Section 1.5 and Shur (1990), p. 92). In order to find the drain current from eq. (3-4-34), we determine n_s in terms of V_F from eq. (3-4-29) and integrate from $V_F = 0$ at the source to $V_F = V_{DS}$ at the drain. Hence, eqs. (3-4-29) and (3-4-34) describe both the subthreshold and the above threshold regimes. In Sections 3.10 to 3.12, we discuss advanced MOSFET models based on this formalism.

The dependence of n_s on $V_{GT} - V_F$ can be found by solving eq. (3-4-29) using the procedure described in Appendix A3. Alternatively, we can use a generalized form of eq. (3-3-67) for n_s, which is in excellent agreement with UCCM. Following the same chain of arguments as in the derivation of eq. (3-4-29), the generalized form of eq. (3-3-67) becomes

$$qn_s(x) = 2qn_o \ln\left[1 + \frac{1}{2}\exp\left(\frac{V_{GT} - \alpha V_F(x)}{\eta V_{th}}\right)\right] \tag{3-4-35}$$

This equation allows us to determine directly the carrier distribution along the channel as a function of $V_F(x)$. The applications of eq. (3-4-35) for FET analysis will be considered in Section 3.9 and in Chapter 4.

3.4.4. Effect of Source and Drain Series Resistances

So far we have considered an ideal device where the entire voltage drop between the source and drain is confined to the channel. In fact, both the drain and source parasitic series resistances, R_d and R_s, play an important role in limiting the device performance. These resistances may be accounted for by using the following expressions relating the "intrinsic" gate-source and drain-source voltages, V_{GS} and V_{DS}, considered so far, to the "extrinsic" (measured) gate-source and drain-source voltages, V_{gs} and V_{ds}, that include voltage drops across the series resistances:

$$V_{GS} = V_{gs} - I_d R_s \tag{3-4-36}$$

$$V_{DS} = V_{ds} - I_d(R_s + R_d) \tag{3-4-37}$$

The measured (extrinsic) transconductance of a field effect transistor is

$$g_m = \frac{dI_d}{dV_{gs}}\bigg|_{V_{ds}} \tag{3-4-38}$$

Using elementary circuit theory (see, for example, Sze (1985)), g_m can be related to the intrinsic transconductance, g_{mo}, and the intrinsic drain conductance, g_{do}, of the same device as follows:

$$g_m = \frac{g_{mo}}{1 + g_{mo}R_s + g_{do}(R_s + R_d)} \tag{3-4-39}$$

where

$$g_{mo} = \frac{dI_d}{dV_{GS}}\bigg|_{V_{DS}} \tag{3-4-40}$$

$$g_{do} = \frac{dI_d}{dV_{DS}}\bigg|_{V_{GS}} \tag{3-4-41}$$

Likewise, the measured ("extrinsic") drain conductance

$$g_d = \frac{dI_d}{dV_{ds}}\bigg|_{V_{gs}} \tag{3-4-42}$$

is related to the intrinsic conductances through

$$g_d = \frac{g_{do}}{1 + g_{mo}R_s + g_{do}(R_s + R_d)} \tag{3-4-43}$$

In the linear region, at very small drain-source voltages, g_{mo} is small and the term $g_{mo}R_s$ can be neglected in eqs. (3-4-39) and (3-4-43). In saturation, where g_{do} is very small in long channel transistors, the term $g_{do}(R_s + R_d)$ can be neglected. However, these approximations are not valid for very short devices (see Chou and Antoniadis (1987)).

In addition to its effect on the drain-source and gate-source voltages, the potential drop across the source series resistance also modifies the substrate-source voltage. The extrinsic substrate-source voltage, V_{bs}, of a MOSFET is related to its intrinsic counterpart, V_{BS}, as follows:

$$V_{BS} = V_{bs} - I_d R_s \tag{3-4-44}$$

In order to investigate the consequence of this shift on the extrinsic transconductance and the extrinsic drain conductance, it is convenient to define a new set of transconductances related to the substrate bias

$$g_{mb} = \frac{dI_d}{dV_{bs}}\bigg|_{V_{ds},V_{gs}} \tag{3-4-45}$$

$$g_{mbo} = \frac{dI_d}{dV_{BS}}\bigg|_{V_{DS},V_{GS}} \tag{3-4-46}$$

where g_{mb} and g_{mbo} are called the extrinsic and intrinsic substrate transconductances, respectively. Again, from simple circuit theory, we find the following generalized relationship between extrinsic and intrinsic conductances (this can also be seen directly from eqs. (3-4-39) and (3-4-43) by noting the symmetry between the definitions of gate transconductance and substrate transconductance):

$$\frac{g_m}{g_{mo}} = \frac{g_d}{g_{do}} = \frac{g_{mb}}{g_{mbo}} = \frac{1}{1 + R_s\left(g_{mo} + g_{mbo}\right) + g_{do}\left(R_s + R_d\right)} \quad (3\text{-}4\text{-}47)$$

(see Cserveny (1990)). Eq. (3-4-47) shows that all extrinsic conductances are reduced by the effect on the substrate-source voltage of the voltage drop across R_s, compared to the previous results where this effect was neglected. As noted by Cserveny (1990), the error incurred in saturation by neglecting the term $R_s g_{mbo}$ in eq. (3-4-47) can be quite important. In fact, for long-channel devices in saturation, this term will normally be more important than the term $g_{do}(R_s + R_d)$.

3-5. SATURATION REGIME

3.5.1. Introduction

When a MOSFET is biased in saturation, the gradual channel approximation (GCA) breaks down in a small region near the drain, as discussed in Section 3.4. This so-called saturated part of the channel is characterized by a significant degree of two-dimensionality in the electric field pattern, in contrast to the so-called linear region closer to the source where the field perpendicular to the channel is much larger than the longitudinal field. While the GCA still may offer a valid basis for the description of the device physics in the linear region, we need a different formulation to account for the two-dimensionality of the field pattern in the saturated part. Hence, it is natural to describe the channel in terms of a two region model, as indicated in Fig. 3.5.1.

In order to avoid discontinuities in principal physical variables such as the electric field and the average charge carrier velocity, we require that the description for the two regions merge smoothly at the boundary point. Otherwise, unwanted discontinuities may carry over to important modeled electrical characteristics of the device such as to the drain conductance and to the gate capacitance.

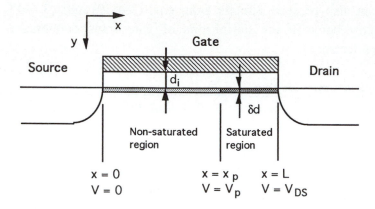

Fig. 3.5.1. Schematic representation of a MOSFET in saturation, where the channel is divided into a non-saturated region (where GCA is valid) and a saturated region. Carrier velocity and electric field are taken to be continuous at the boundary between the two regions.

In regions where the GCA is valid, analytical descriptions of field effect transistors are usually based on a velocity-field relationship and a model for charge control. The charge control model expresses the charge induced in the conducting channel as a result of the gate bias. A simple concept based on a parallel plate capacitor model for the induced channel charge was introduced in Section 3.3, along with the more comprehensive Unified Charge Control Model (UCCM) (see also Section 3.4). The velocity-field relationship expresses a local and instantaneous correspondence between the electric field and the average carrier drift velocity in a semiconductor. As pointed out in Section 1.5, such a correspondence does not exist at very high frequencies (comparable to the inverse energy relaxation time) or, conversely, in regions of space where the electric field changes rapidly. While high frequencies and rapid variations in the electric field are often the case in modern short-channel devices, velocity-field relationships are still needed in order to establish manageable models for use in circuit simulation. In fact, with a judicial choice of parameters and of the functional relationship between velocity and field, the fit to macroscopic experimental data, such as current-voltage characteristics, becomes quite acceptable, even for short-channel devices.

The simplest FET models use a linear velocity-field relationship (see Section 3.4). In these models, current saturation sets in when the conducting channel is pinched off at the drain side of the gate, i.e., when the charge carrier density at this point vanishes and, hence, the velocity becomes infinite (a consequence of current continuity). This description works reasonably well for long-channel devices, but the notion of an infinite carrier velocity is, of course, unphysical. Instead, current saturation is better described in terms of a saturation of the carrier velocity when the electric field near the drain becomes sufficiently high.

A very simple approximation to the velocity-field relationship is obtained by assuming that the carrier velocity, v, is proportional to the magnitude of the longitudinal electric field, $F = |F_x|$, until the saturation velocity, v_s, is reached. Thereafter, the velocity becomes constant, i.e.,

$$v(F) = \begin{cases} \mu F , & F < F_s \\ v_s , & F \geq F_s \end{cases} \tag{3-5-1}$$

Here, μ is the low-field mobility and $F_s = v_s/\mu$ is the saturation field. In this picture, current saturation in FETs occurs when the field at the drain side of the gate reaches the saturation field. A velocity saturation model for the MOSFET based on eq. (3-5-1) is discussed below, in Subsection 3.5.2. A more realistic velocity-field relationship, particularly for MOSFETs, is similar to that for bulk silicon (see Section 1.4 and Caugley and Thomas (1967)):

$$v(F) = \frac{\mu F}{\left[1 + \left(\mu F / v_s \right)^m \right]^{1/m}} \tag{3-5-2}$$

Here, $m = 2$ and $m = 1$ are reasonable choices for electrons and holes in MOSFETs, respectively, and the values of v_s and m are somewhat different from those for bulk silicon (see Cooper and Nelson (1983), Coen and Miller (1980), and Thornber (1980)). In particular, as discussed in Section 3.8, the values of μ are strongly dependent on the gate bias, especially for short channel devices. A more manageable velocity-field relationship, which is close to eq. (3-5-2) with $m = 2$, and which also describes electron velocity saturation quite well in compound semiconductor field effect transistors, is (see Sodini et al. (1984)):

$$v(F) = \begin{cases} \dfrac{\mu F}{1 + \dfrac{F}{2 F_s}} , & F < 2 F_s \\ \\ v_s , & F \geq 2 F_s \end{cases} \tag{3-5-3}$$

Fig. 3.5.2 shows different velocity-field models for electrons and holes in silicon MOSFETs where the electric field and the velocity have been normalized to F_S and v_S, respectively.

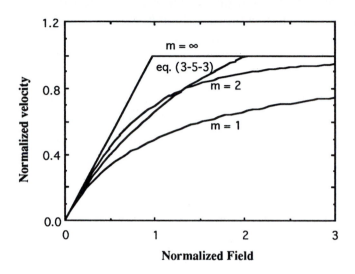

Fig. 3.5.2. Velocity-field relationships for electrons and holes in silicon MOSFETs. The electric field and the velocity are normalized to F_s and v_S, respectively. Two of the curves are calculated from eq. (3-5-2) with $m = 1$ (holes) and $m = 2$ (electrons), and one curve is calculated from eq. (3-5-3). The curve marked $m = \infty$ corresponds to the linear two-piece model in eq. (3-5-1).

3.5.2. A Simple Velocity Saturation Model

The MOSFET model discussed in Section 3.4 does not include the effects of velocity saturation in the channel at high drain-source bias. The simplest extension of this model, incorporating velocity saturation, is to use the two-piece linear approximation of eq. (3-5-1) for the carrier velocity.

In the strong inversion regime, the surface carrier concentration of electrons in the channel of an n-channel MOSFET can be found using the following charge control model (see Section 3.4, eq. (3-4-7)):

$$qn_s(x) \approx c_i\left[V_{GT} - V(x)\right]$$

(3-5-4)

where n_s is the sheet density of inversion charge, c_i is the gate insulator capacitance per unit area, $V_{GT} = V_{GS} - V_T$ is the gate-source voltage, V_{GS}, relative to the threshold voltage, V_T, $V(x)$ is the channel voltage induced by the drain-source bias, V_{DS}, and x is the space coordinate along the channel relative to the source. Eq. (3-5-4) should be solved together with the following expression for the absolute value of the drain current:

$$I_d = W q n_s v_n(F)$$

(3-5-5)

where W is the device width, and $v_n(F)$ is the electron velocity in the channel which is a function of the longitudinal electric field F. In the spirit of the gradual channel approximation, we consider only the component of the field that is parallel to the semiconductor-insulator interface. We also neglect the diffusion current (see Problem 3-5-1). Substituting eq. (3-5-4) into eq. (3-5-5), replacing $F\ (= |F_x|)$ in the resulting equation by dV/dx, and integrating with respect to x from 0 to L, where L is the gate length, yields the conventional charge control equation describing the current-voltage characteristics of a MOSFET

$$I_d = \beta\left(V_{GT} V_{DS} - \tfrac{1}{2}V_{DS}^2\right)$$

(3-5-6)

where

$$\beta = c_i \mu_n W / L$$

(3-5-7)

is the transconductance parameter and μ_n is the low-field mobility for electrons. Eq. (3-5-6) is valid only at relatively low drain-source voltages.

The drain saturation current, I_{sat}, is determined by assuming that saturation occurs when the electric field in the channel at the drain side of the gate exceeds the velocity saturation field, $F_s = v_s/\mu_n$. (This is a more realistic assumption than the pinch-off condition, $n_s(L) = 0$, used in the earlier model (see Section 3.4).) We define the saturation voltage, V_{SAT}, as the lowest drain-source voltage which gives $I_d = I_{sat}$ (or $F = F_s$ at the drain side of the gate).

In order to describe the longitudinal field distribution in the channel below saturation, we use the above equations assuming a constant electron mobility. The absolute value of the electric field in the channel, $F = dV/dx$, for $V_{DS} \leq V_{SAT}$ can be found from eqs. (3-5-4) and (3-5-5):

$$\frac{dV}{dx} = \frac{I_d}{\beta L\left(V_{GT} - V\right)}$$

(3-5-8)

Integrating this equation from 0 to x, we obtain an expression describing the variation of the channel potential, $V(x)$, along the channel below saturation:

$$V(x) = V_{GT} - \sqrt{V_{GT}^2 - \frac{2I_d\, x}{\beta L}}$$

(3-5-9)

Substituting this equation into equation (3-5-8) we find the electric field as a function of distance,

$$F = \frac{I_d}{\beta L \sqrt{V_{GT}^2 - \dfrac{2I_d\, x}{\beta L}}}$$

(3-5-10)

The maximum electric field occurs at the drain side of the gate and has the value

$$F(L) = \frac{I_d}{\beta L \sqrt{V_{GT}^2 - 2I_d/\beta}}$$

(3-5-11)

From the condition for the onset of saturation, $F(L) = F_s$, we now find the drain saturation current

$$I_{sat} = \beta V_L^2 \left[\sqrt{1 + \left(\frac{V_{GT}}{V_L}\right)^2} - 1 \right]$$

(3-5-12)

where

$$V_L = F_s L$$

(3-5-13)

The saturation voltage can be obtained by combining eqs. (3-5-9), (3-5-12) and (3-5-13)

$$V_{SAT} = V_{GT} - \frac{I_{sat}}{\beta V_L} = V_{GT} - V_L \left[\sqrt{1 + \left(\frac{V_{GT}}{V_L}\right)^2} - 1 \right]$$

(3-5-14)

At very large values of V_L such that $V_L \gg V_{GT}$, the square root on the right hand side of eqs. (3-5-12) and (3-5-14) may be expanded into a Taylor series yielding the following expressions for long-channel devices:

$$I_{sat} = \frac{\beta V_{GT}^2}{2}$$

(3-5-15)

$$V_{SAT} = V_{GT}\left(1 - \frac{V_{GT}}{2V_L}\right) \approx V_{GT}$$

(3-5-16)

coinciding with the result of the simple charge control model of Subsection 3.4.2 (see eq. (3-4-26)) which does not take into account the velocity saturation effects. Hence, the velocity saturation is not very important for long-channel FETs. Assuming that $V_{GT} \approx 3$ V, $\mu_n = 0.08$ m^2/Vs, and $v_s = 1\times10^5$ m/s, we find that velocity saturation effects on the drain saturation current may be neglected for $L \gg 2.4$ µm. However, typical gate lengths of modern MOSFETs may be less than one micrometer. Hence, velocity saturation is important in such devices.

For small V_L, i.e., $V_L \ll V_{GT}$, we obtain from eqs. (3-4-12) and (3-4-14)

$$I_{sat} = \beta V_L V_{GT}$$

(3-5-17)

$$V_{SAT} = V_L\left(1 - \frac{V_L}{2V_{GT}}\right) \approx V_L$$

(3-5-18)

Hence, the drain saturation current for the short channel device (i.e., small V_L) is reduced by a factor $2V_L/V_{GT}$ compared to that given by the constant mobility model. Since I_{sat} is proportional to V_{GS}^2 in a long channel device and to V_{GS} in a short channel device, this difference can be used to establish the importance of short channel effects in measured device characteristics (see Problem 3-5-2).

As was discussed in Subsection 3.4.4, the drain and source parasitic series resistances, R_d and R_s, may play an important role in limiting the device performance. These resistances may be accounted for by relating the intrinsic gate-source and drain-source voltages considered so far to the "extrinsic" (measured) gate-source and drain-source voltages, V_{gs} and V_{ds}, that include voltage drops across series resistances:

$$V_{GS} = V_{gs} - R_s I_d$$

(3-5-19)

$$V_{DS} = V_{ds} - (R_s + R_d)I_d$$

(3-5-20)

Substituting I_{sat} instead of I_d into eq. (3-5-19) and the resulting equation into eq. (3-5-14), we find for the saturation current

$$I_{sat} = \frac{\beta V_{gt}^2}{1 + \beta R_s V_{gt} + \sqrt{1 + 2\beta R_s V_{gt} + \left(V_{gt}/V_L\right)^2}} \qquad (3\text{-}5\text{-}21)$$

The extrinsic saturation voltage, V_{sat}, can be obtained from eq. (3-5-14) in combination with eqs. (3-5-19) to (3-5-21):

$$V_{sat} = V_{SAT} + (R_s + R_d) I_{sat}$$

$$= V_{gt} + \left(R_d - \frac{1}{\beta V_L}\right) I_{sat} \qquad (3\text{-}5\text{-}22)$$

3.5.3. The Region of the Channel with Velocity Saturation

Given that a satisfactory model exists for the non-saturated part of the channel (see above and Sections 3.4, 3.10 and 3.11), we now want to explore a model for the saturated part of the conducting channel in MOSFETs. The present model is based on previous work by El Mansy and Boothroyd (1977), Ko (1989), Moon et al. (1990), and by Steen et al. (1990) on Heterostructure Field Effect Transistors (HFETs). Here, we use a similar approach in order to find a solution for the longitudinal field in the MOSFET channel. The model relies on the fundamental assumption that the carrier velocity in the saturated part of the channel is constant and equal to the saturation velocity. This implies that the carrier sheet density in the saturated part of the channel is also constant. This assumption was earlier used by Pucel et al. (1975) to describe the saturation region in MESFETs (see also Shur (1978)). In addition, we limit ourselves to considering devices with a fairly low doping in the substrate. The above assumption clearly represents a great simplification of the true physics of the saturated region. However, a redeeming factor is that it leads to a manageable theory with qualitatively correct features, which gives a fairly good fit to experimental data with a judicious choice of parameters such as the saturation velocity and the effective channel thickness.

From the above discussion, the intrinsic saturation voltage, V_{SAT}, can be defined as the intrinsic drain-source voltage, V_{DS}, for which the longitudinal electric field at the drain end of the channel just reaches the saturation field, F_s.

For $V_{DS} > V_{SAT}$, the location in the channel where $|F_x| = F_s$ marks the boundary between the saturated and the non-saturated regions. The boundary point, $x = x_p$, moves closer to the source with increasing drain-source voltage. This effect is called channel length modulation. Another important parameter is the channel potential at the boundary point, $V_p = V(x_p)$. V_p and x_p depend on V_{DS} and on the intrinsic gate-source voltage, V_{GS}, and have to be determined self-consistently using the models for the two regions with the requirement that the potential, the electric field and the velocity be continuous at $x = x_p$.

For a description of the saturated region, it is necessary to consider a two-dimensional Poisson's equation of the form

$$\frac{\partial F_x}{\partial x} + \frac{\partial F_y}{\partial y} = \frac{\rho(x,y)}{\varepsilon_s} \tag{3-5-23}$$

where F_x and F_y are the longitudinal and transverse components of the electric field, respectively, ε_s is the semiconductor dielectric permittivity and $\rho(x,y)$ is the charge density in the semiconductor. The latter consists of a mobile charge density, qn, and a depletion charge density, qN_a, where N_a is the substrate doping density. Integrating eq. (3-5-23) with respect to y from the semiconductor-insulator interface through the effective channel thickness, δd, we obtain

$$\left\langle \frac{\partial F_x}{\partial x} \right\rangle \delta d + F_y(\delta d) - F_y(0) = -\frac{q}{\varepsilon_s}(n_s + N_a \delta d) \tag{3-5-24}$$

where $\langle \partial F_x/\partial x \rangle$ is the average of $\partial F_x/\partial x$ over the channel thickness and n_s is the electron sheet density in the channel. At low substrate doping and with the MOSFET biased in strong inversion such that $n_s \gg N_a \delta d$, the vertical electric field at $y = \delta d$ will be small compared to the vertical field at the interface, in which case $F_y(\delta d)$ can be neglected in eq. (3-5-24). Making the substitution $\langle \partial F_x/\partial x \rangle \rightarrow -\partial^2 V/\partial x^2$, where V is the average of the potential over the cross-section of the channel, eq. (3-5-24) can be written as

$$\frac{\partial^2 V}{\partial x^2} + \frac{F_y(0)}{\delta d} \approx \frac{qn_s}{\varepsilon_s \delta d} \tag{3-5-25}$$

The electric field, $F_y(0)$, at the interface is obtained by equating the electric displacement of the two sides of the semiconductor-insulator interface (see

Sections 3.3 and 3.4), leading to

$$F_y(0) = \frac{c_i}{\varepsilon_s}(V_{GS} - 2\varphi_b - V_{FB} - V) \qquad (3\text{-}5\text{-}26)$$

where $\varphi_b = V_{th}\ln(N_a/n_i)$ is the potential difference between the Fermi level and the intrinsic Fermi level in the bulk of the semiconductor far from the interface, V_{FB} is the flat-band voltage, $c_i = \varepsilon_i/d_i$ is insulator capacitance per unit area, ε_i is the dielectric permittivity and d_i is the thickness of the insulator.

From the conditions of velocity saturation and current continuity, we know that the electron sheet density is constant in the saturated region. Its value can therefore be determined at the boundary point, $x = x_p$, where the GCA is still valid. Thus,

$$n_s = \frac{c_i}{q}(V_{GS} - V_T - V_p) \qquad (3\text{-}5\text{-}27)$$

where V_T is the threshold gate-source voltage. We recall from Section 3.3 that V_T can be related to the flat-band voltage as follows:

$$V_T = V_{FB} + 2\varphi_b + \sqrt{4\varepsilon_s\,qN_a\varphi_b}\,/c_i \qquad (3\text{-}5\text{-}28)$$

The combination of eqs. (3-5-25) to (3-5-28) leads to the following second order differential equation for the channel potential in the saturated region:

$$\frac{\partial^2(V - V_p)}{\partial x^2} - \frac{V - V_p}{\lambda^2} = 0 \qquad (3\text{-}5\text{-}29)$$

where

$$\lambda = \sqrt{\frac{\varepsilon_s}{\varepsilon_i}\,d_i\,\delta d} \qquad (3\text{-}5\text{-}30)$$

It should be noted that the solution of eq. (3-5-29) is very sensitive to the magnitude of the characteristic length, λ, for the saturated region. In comparisons with experimental data, it is therefore convenient to treat λ as a fitting parameter rather than using eq. (3-5-30) which itself is a result of rough estimates and approximations.

The general solution of eq.(3-5-29) can be written on the following form:

$$V(x) = A \exp\left(\frac{x - x_p}{\lambda}\right) + B \exp\left(-\frac{x - x_p}{\lambda}\right) + V_p \qquad (3\text{-}5\text{-}31)$$

The coefficients A and B are determined from the boundary conditions, i.e., from the requirement that V and F_x be continuous at $x = x_p$ with the values V_p and $-F_s$, respectively, leading to

$$A = \tfrac{1}{2} V_\lambda \qquad (3\text{-}5\text{-}32)$$

$$B = -\tfrac{1}{2} V_\lambda \qquad (3\text{-}5\text{-}33)$$

where $V_\lambda = \lambda F_s$.

A relationship that links x_p and V_p to the drain-source voltage is obtained by considering eq. (3-5-31) at the drain side of the channel:

$$V_{DS} = V_p + V_\lambda \sinh\left(\frac{\Delta L}{\lambda}\right) \qquad (3\text{-}5\text{-}34)$$

where $\Delta L = L - x_p$ and L is the gate length. Eq. (3-5-34) can be solved with respect to ΔL, resulting in

$$\Delta L = \lambda \ln\left[\frac{V_{DS} - V_p}{V_\lambda} + \sqrt{\left(\frac{V_{DS} - V_p}{V_\lambda}\right)^2 + 1}\right] \qquad (3\text{-}5\text{-}35)$$

Combining eqs. (3-5-31) and (3-5-35), we find

$$V(x) = V_p + (V_{DS} - V_p) \cosh\left(\frac{L - x}{\lambda}\right)$$
$$- \sqrt{(V_{DS} - V_p)^2 + V_\lambda^2} \sinh\left(\frac{L - x}{\lambda}\right) \qquad (3\text{-}5\text{-}36)$$

A self-consistent determination of V_p is based on a model for the non-saturated part of the channel (see Subsection 3.5.2 and Sections 3.4, 3.10 and 3.11).

Owing to the complexity of eqs. (3-5-31) to (3-5-35), we cannot hope to be able to derive explicit, analytical expressions for important electrical properties of the MOSFET, such as I–V characteristics, using the present model for the saturation regime. However, a numerical solution can readily be obtained which may serve as a physically based reference for simpler, more empirical models (see Sections 3.9 to 3.11). Nonetheless, it is possible to simplify the equations somewhat in certain limiting cases. For $\Delta L \leq \lambda/3$, i.e., just beyond the onset of saturation, we can write, to first order in ΔL,

$$V_{DS} \approx V_p + F_s \Delta L \tag{3-5-37}$$

in which case $\Delta L \approx (V_{DS} - V_p)/F_s$, and

$$V(x) \approx V_{DS} - F_s(L - x) \tag{3-5-38}$$

For $\Delta L > 3\lambda$, i.e., in deep saturation, we have

$$V_{DS} \approx V_p + \frac{1}{2}V_\lambda \exp(\Delta L/\lambda) \tag{3-5-39}$$

From eq. (3-5-39), we obtain

$$\Delta L \approx \lambda \ln\left[\frac{2(V_{DS} - V_p)}{V_\lambda}\right] \tag{3-5-40}$$

The solutions obtained represent only an approximation of the actual potential distribution in the saturation region. However, they clearly show that the potential rises exponentially with distance inside this region.

Based on this result and on numerical simulations of the potential in the saturation region, Moon et al. (1990) proposed a simplified empirical expression linking the drain-source voltage to the length of the saturation region, ΔL,

$$V_{DS} \approx V_p + V_\alpha\left[\exp\left(\frac{\Delta L}{\lambda}\right) - 1\right] \tag{3-5-41}$$

where the constant V_α is determined from the condition of continuity in the drain conductance. In Fig. 3.5.3, we show the channel potential and field profiles for a 1 μm n-channel MOSFET calculated using the two-dimensional PISCES simulation program. The figure shows that the results of a simple exponential approximation for the electric potential in the saturated region is in excellent agreement with the more accurate two-dimensional calculation.

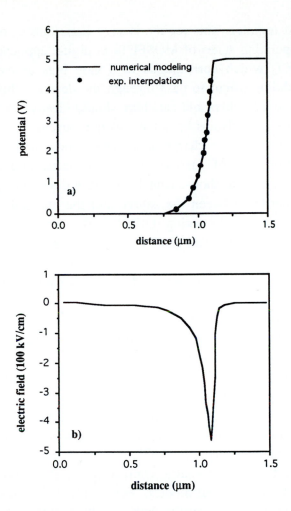

Fig. 3.5.3. Potential (a) and field (b) profiles in a 1 μm *n*-MOSFET calculated using the PISCES simulation program. Gate oxide 210 Å, substrate doping 2×10^{16} cm^{-3}, $V_{gs} = 0.2$ V (n^+ poly-gate), $V_{ds} = 5$ V, $V_{BS} = 0$. An exponential approximation for the saturated region is shown as open symbols in a).

3-6. SUBTHRESHOLD REGIME

When the gate electrode of a MOSFET is biased below the threshold voltage, such that the conducting channel is weakly inverted, the MOSFET is said to be in the subthreshold regime. This regime is very important since it is one of the two stationary operating states of MOSFETs in digital applications. The *on*-state of the MOSFET, which corresponds to a gate bias above threshold, allows a significant drain current to pass through the device, while the *off*-state which corresponds to a subthreshold gate bias, should ideally block all drain current. In practice, there will always be some leakage current in the *off*-state owing to a finite amount of mobile charge at the semiconductor-insulator interface and a finite injection rate of carriers from the source into the channel. In the subthreshold regime in short-channel devices, a drain voltage induces lowering of the energy barrier between the source and the channel. This effect is called Drain Induced Barrier Lowering (DIBL). DIBL causes excess injection of charge carriers into the channel and gives rise to an increased subthreshold current (see Troutman (1979)). This current has deleterious effects on the performance of digital circuits in terms of increased power dissipation and a possible shift in logical levels.

In the following analysis, based on the work by Fjeldly and Shur (1992), we first discuss the standard theory for subthreshold current of MOSFETs which works well for long-channel devices and can, with minor adjustments, even account for some short-channel effects. However, in order to explain important new features that occur in short-channel devices with gate lengths of the order of a micrometer or less, it is necessary to consider additional effects such as DIBL and the concomitant drain bias induced threshold voltage shift. (For a discussion of the short-channel concept for MOSFETs, see Sze (1981).) The results are used to generalize the unified charge control model (UCCM) to include important short-channel effects in our models. For simplicity, we restrict the discussion to *n*-channel devices, but the results apply equally well to *p*-channel transistors.

3.6.1. Subthreshold Current in Long Channel MOSFETs

A MOSFET in the subthreshold regime will normally carry only a small drain current, even at high drain-source bias. In fact, the source and drain contacts, together with the weakly inverted or depleted channel, act as a bipolar transistor where the channel acts as a long base, the n^+ source acts as the emitter and the drain plays the role of the collector. Most of the applied "collector-emitter" voltage will drop across the reverse biased junction which, in the MOSFET, corresponds to the drain-channel junction. Accordingly, away from the source and drain contacts, the channel itself will experience very little lateral potential variation. Instead, the potential distribution in the channel will be governed almost entirely by the gate electrode. This means that the electric field component perpendicular to the semiconductor-insulator interface is much larger than the longitudinal component, allowing the channel to be analyzed within the framework of the Gradual Channel Approximation (GCA, see Section 3.4). As a consequence of the small potential drop along the channel, the drain current is dominated by diffusion instead of drift for drain-source voltages above a few thermal voltages.

A standard theory for subthreshold current in MOSFETs can be derived by considering the channel current density as the sum of a drift term and a diffusion term. For an n-channel device, the current density, J_n, can be written as (see Section 1.5)

$$J_n = q\left(-n\mu_n \frac{d\psi}{dx} + D_n \frac{dn}{dx}\right) = qD_n\left(-\frac{n}{V_{th}}\frac{d\psi}{dx} + \frac{dn}{dx}\right) \qquad (3\text{-}6\text{-}1)$$

where x is the position in the channel measured from the source, n is the density of electrons in the channel, μ_n is the low-field electron mobility, D_n is the electron diffusion coefficient, and $\psi(x,y)$ is the potential of the channel region referred to the potential of the source electrode. In eq. (3-6-1), it is assumed that the longitudinal electric field in the channel, $F_x = -d\psi/dx$, is sufficiently small (except for the junction region near drain) such that velocity saturation can be neglected. Likewise, the diffusion coefficient can be expressed in terms of the Einstein relation as $D_n = \mu_n V_{th}$, where $V_{th} = k_B T/q$ is the thermal voltage.

Fig. 3.6.1 shows qualitatively the band diagram and the potential distribution at the interface in the channel, $\psi_s(x) = \psi(x,0)$, for two values of drain-source bias, $V_{DS} = 0$ and $V_{DS} > 0$. At the interface, the channel consists of three regions, the source-channel junction with length x_s, the drain-channel junction with length x_d, and the middle region of length $L - x_s - x_d$. At $V_{DS} = 0$, the interface potential in the middle of the channel, $\psi_s{}^o$, can be taken to be approximately constant. A drain-source bias gives rise to a positive contribution, $V(x)$, to the channel potential, as indicated in the figure. As a consequence, the minimum in the interface potential, $\psi_s(x_{min}) = \psi_s{}^o + V(x_{min})$, will be localized at the source side of the channel, at $x = x_{min} \approx x_s$. Associated with the shift in the potential minimum, we have a reduction in the interface energy barrier between the source and the channel by $-qV(x_{min})$. This is the so-called Drain Induced Barrier Lowering (DIBL). DIBL is a short-channel effect which causes a drain voltage induced shift in the threshold voltage. This will be discussed in more details below (see Subsections 3.6.2 to 3.6.5).

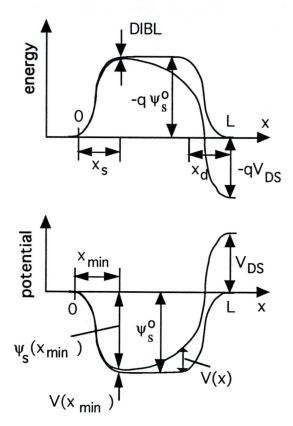

Fig. 3.6.1. Band diagram (top) and potential profile (bottom) at the semiconductor-insulator interface of an n-channel MOSFET. The symmetrical profiles correspond to $V_{DS} = 0$, and the asymmetrical profiles to $V_{DS} > 0$. The figure indicates the origin of the Drain Induced Barrier Lowering (DIBL). The various symbols are defined in the text. (From Fjeldly and Shur (1992), © 1992 IEEE.)

Multiplying eq. (3-6-1) by the integrating factor $\exp(-\psi/V_{th})$, the right hand side of this equation can be made into an exact derivative. A subsequent integration from source to drain yields (assuming that the current density remain independent of x)

$$J_n = qD_n \frac{n(L)\exp\left(-\dfrac{\psi(L,y)}{V_{th}}\right) - n(0)\exp\left(-\dfrac{\psi(0,y)}{V_{th}}\right)}{\displaystyle\int_0^L \exp\left(-\dfrac{\psi(x,y)}{V_{th}}\right)dx} \qquad (3\text{-}6\text{-}2)$$

where $n(L) = n(0)$ equals the source and drain contact doping density, N_c (neglecting degeneracy). With the source contact as a potential reference, we have $\psi(0,y) = 0$ at the source and $\psi(L,y) = V_{DS}$ at the drain, where V_{DS} is the intrinsic drain-source voltage.

When the device length is not too small, the channel potential can be taken to be independent of x over a portion of the channel length. We write the potential in this region as $\psi(x,y) = \psi^o(y)$. The integral in the denominator of eq. (3-6-2) is determined by the contribution from this portion of the channel. As can be seen from Fig. 3.6.1, the length of this section is approximately equal to $L - x_s - x_d$, and the current density can be expressed as

$$J_n = -\frac{qD_n N_c}{L - x_s - x_d}\exp\left(\frac{\psi^o(y)}{V_{th}}\right)\left[1 - \exp\left(-\frac{V_{DS}}{V_{th}}\right)\right] \qquad (3\text{-}6\text{-}3)$$

For long-channel devices, where $L \gg x_s - x_d$, we can write approximately

$$J_n \approx -\frac{qD_n N_c}{L}\exp\left(\frac{\psi^o(y)}{V_{th}}\right)\left[1 - \exp\left(-\frac{V_{DS}}{V_{th}}\right)\right] \qquad (3\text{-}6\text{-}4)$$

The drain current is obtained by integrating the current density over the cross section of the conducting channel. The absolute value of the drain current is

$$I_d = \frac{qW\,\delta d\,D_n N_c}{L}\exp\left(\frac{\psi_s^o}{V_{th}}\right)\left[1 - \exp\left(-\frac{V_{DS}}{V_{th}}\right)\right] \qquad (3\text{-}6\text{-}5)$$

where W is the width of the gate, δd is the effective channel thickness, and $\psi_s = \psi_s^o$ is the constant potential at the semiconductor-insulator interface. We again emphasize that ψ_s^o is defined relative to the source electrode. Hence, although the interface potential relative to the interior of the p-type substrate, $V_s = \psi_s^o + V_{bi}$, is positive, ψ_s^o will be negative for n-channel MOSFETs. Here, V_{bi} is the built-in potential between the source contact and the substrate. Note that eq. (3-6-5) could also be obtained directly by making the analogy between the MOSFET in the subthreshold regime and a long-base BJT, as discussed above (see Shur (1990)).

At threshold, the interface potential in the channel relative to the source can be expressed as

$$\psi_{sT}^{o} = V_{sT} - V_{bi} = 2\varphi_{b} - V_{bi} \qquad (3\text{-}6\text{-}6)$$

where V_{sT} is the potential relative to the interior of the substrate at threshold and φ_{b} is the position of the Fermi potential relative to midgap in the neutral region of the doped substrate (see Section 3.3). For simplicity, we have assumed that the substrate is shorted to the source. (The effects of a substrate-source bias, V_{BS}, are found simply by replacing V_{bi} by $V_{bi} - V_{BS}$. Of course, such a replacement is only valid for negative or small positive values of V_{BS}. A positive V_{BS} comparable to V_{bi} would lead to a large substrate leakage current.)

Below threshold, the interface potential can be written as

$$\psi_{s}^{o} = \psi_{sT}^{o} + \frac{C_{i}}{C_{i} + C_{dep}}V_{GT} = -V_{bi} + 2\varphi_{b} + V_{GT}/\eta \qquad (3\text{-}6\text{-}7)$$

where $V_{GT} = V_{GS} - V_{T}$, V_{T} is the threshold voltage, V_{GS} is the intrinsic gate-source voltage, C_{i} is the insulator capacitance, C_{dep} is the gate depletion capacitance of the semiconductor (see Section 3.3), and η is the ideality factor. The ideality factor reflects the gate voltage division between the insulator layer capacitance and the depletion layer capacitance. Generally speaking, η is dependent on V_{GT}, although at low substrate doping levels, η is close to unity near threshold where the gate depletion width is large (which corresponds to $C_{dep} \ll C_{i}$). Usually, however, we can estimate C_{dep} as discussed in Section 3.3 (see eqs (3-3-54) and (3-3-55)). Eqs. (3-6-5) and (3-6-7) show that the subthreshold drain current decreases nearly exponentially with decreasing V_{GT}. For $V_{DS} > 3V_{th}$, this current is practically independent of the drain-source voltage.

Since the mobile charge density in the channel has an exponential dependence on the interface potential, ψ_{s}^{o}, the effective channel thickness, δd, can be estimated as the distance from the interface where the potential has changed by V_{th}. For this estimate, we need to find the vertical component of the electric field at the interface. According to Gauss' law, this field is equal to

$|Q_{dep}/\varepsilon_s|$ in the subthreshold regime, where $Q_{dep} = -[2\varepsilon_s q N_a(\psi_s{}^o + V_{bi})]^{1/2}$ is the sheet depletion charge density under the gate (see Section 3.3), N_a is the acceptor doping density in the substrate and ε_s is the dielectric permittivity of the semiconductor. Hence, the effective channel thickness becomes

$$\delta d \approx V_{th}\sqrt{\frac{\varepsilon_s}{2qN_a(2\varphi_b + V_{GT}/\eta)}} \qquad (3\text{-}6\text{-}8)$$

We notice that this expression is only valid when $-2\varphi_b < V_{GT}/\eta < 0$, i.e., in the weak inversion and depletion regime. Still, this condition is fulfilled for values of the drain current that are many orders of magnitude smaller than the threshold current.

3.6.2. Short Channel Effects and the Charge Sharing Model

At short gate lengths, the depletion widths associated with the source-channel and the drain-channel junctions may represent a significant fraction of the total gate length. In this case, the integral in the denominator of eq. (3-6-2) should also reflect the variation in the channel potential in the junction depletion zones. The GCA, however, breaks down in these regions owing to the two-dimensionality of the field pattern. But, since the potential increases rapidly and the integrand becomes very small in the junction depletion zones, the contribution to the integral from these regions will be negligible. The net effect is, as discussed above, that the gate length L in eq. (3-6-5) should be replaced by

$$L \rightarrow L - x_s - x_d \qquad (3\text{-}6\text{-}9)$$

where x_s and x_d are the depletion widths of the source-channel and the drain-channel junctions, respectively. These widths can be expressed approximately in terms of conventional one-dimensional depletion widths for abrupt p-n junctions as (see Sze (1981))

$$x_s \approx \sqrt{\frac{2\varepsilon_s}{qN_a}(-\psi_s^o)} \qquad (3\text{-}6\text{-}10)$$

$$x_d \approx \sqrt{\frac{2\varepsilon_s}{qN_a}(V_{DS} - \psi_s^o)} \qquad (3\text{-}6\text{-}11)$$

Making this replacement in eq. (3-6-5) and using eqs. (3-6-7) and (3-6-8) for the interface potential and the effective channel thickness, respectively, leads to the following expression for the subthreshold drain current:

$$
I_d = \frac{W D_n V_{th}}{L - x_s - x_d} \sqrt{\frac{\varepsilon_s \, q N_a}{2(2\varphi_b + V_{GT}/\eta)}}
$$
$$
\times \exp\left(\frac{V_{GS} - V_T}{\eta V_{th}}\right)\left[1 - \exp\left(-\frac{V_{DS}}{V_{th}}\right)\right] \tag{3-6-12}
$$

Here, we have also used $V_{bi} = V_{th} \ln(N_c N_a/n_i^2)$ (disregarding degeneracy in the contacts) and $\varphi_b = V_{th}\ln(N_a/n_i)$, where N_c is the donor doping density in the source and drain contacts and n_i is the intrinsic carrier density. We note that this expression for the drain current has gained additional dependence on the drain-source bias through the dependence on V_{DS} of the drain-channel junction depletion width, x_d. This produces a finite output conductance in the subthreshold regime at all drain biases.

In the above analysis, it was assumed that the entire drain-source voltage drops across the drain-channel junction, and that the interface potential everywhere else in the channel is independent of V_{DS}. However, this assumption breaks down for short-channel devices. Experiments and detailed numerical calculations (see, for example, Fichtner and Potzl (1979)) show that MOSFETs with such short gate lengths will experience additional short-channel phenomena such as Drain Induced Barrier Lowering (DIBL), see Fig. 3.6.1, and a corresponding threshold voltage shift. Since the current in the subthreshold regime is exponentially dependent on the height of the source-channel barrier, it is clear that even a small lowering of this barrier will have dramatic effects on the charge transport properties of the device.

An explanation for the shift in the threshold voltage of short-channel devices was proposed in terms of the so-called charge sharing (or charge conservation) concept, first introduced by Yau (1974). The model considers the shared depletion charge in the regions where the gate depletion zone overlaps with the depletion zones of the source and drain contacts, as indicated in Fig. 3.6.2. The shared charge is balanced by counter charges distributed between the gate electrode and the source and drain contacts. In the long-channel case, the threshold voltage can be written as (see Section 3.3)

$$V_T = V_{FB} + 2\varphi_b + \frac{Q_{tot}}{C_i} \tag{3-6-13}$$

where V_{FB} is the flat-band voltage and $Q_{tot} = LWQ_{dep} = LW(4\varepsilon_s qN_a\varphi_b)^{1/2}$ is the total gate depletion sheet charge at threshold, neglecting charge sharing (and assuming zero drain-source and substrate-source bias).

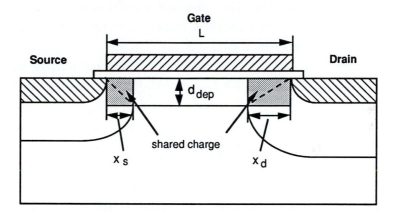

Fig. 3.6.2. Schematic illustration of the depletion zones associated with the source, the drain and the gate for an applied drain-source voltage. The regions of charge sharing are indicated by the shaded areas near source and drain. (From Fjeldly and Shur (1992), © 1992 IEEE.)

The charge sharing causes a reduction in Q_{tot} by a fraction of the shared charge (shaded regions in Fig. 3.6.2). According to eq. (3-6-13), this leads to an averaged shift in the threshold voltage which can be written as (see Fig. 3.6.2)

$$\Delta V_T = -\frac{2\kappa W\left(x_s + x_d\right)\sqrt{\varepsilon_s\, qN_a\varphi_b}}{C_i} \tag{3-6-14}$$

where κ is a constant (on the order of 0.5) which accounts for the charge sharing. In eq. (3-6-14), it was assumed that the source and drain depletion widths in the charge sharing region are x_s and x_d, respectively, and that the gate depletion width at threshold is $d_{dep} = (4\varepsilon_s\varphi_b/qN_a)^{1/2}$.

It should be emphasized that the above derivation of ΔV_T is equivalent to distributing the portion κ of the shared charge evenly under the gate to reduce uniformly the depletion charge density in the gate area. This may lead to a reasonable estimate of the threshold voltage shift at zero drain-source bias where the charge and potential distribution in the channel is quite symmetric. However, with an applied drain-source voltage, the charge distribution becomes asymmetric and eq. (3-6-14) becomes less reliable. Typically, at large V_{DS}, we have $x_d \gg x_s$ and eq. (3-6-14) predicts that the <u>additional</u> threshold voltage shift induced by the drain-source bias depends on V_{DS} as $[(V_{DS} - \psi_s{}^o)^{1/2} - (- \psi_s{}^o)^{1/2}]$ (see eqs. (3-6-10) and (3-6-11)). However, as will be shown in Subsection 3.6.3, a more accurate analysis predicts a linear variation of V_T with V_{DS}, in better agreement with experimental data.

By assuming that the ideality factor, η, does not change significantly with bias conditions (true if $C_i \gg C_s$), the shift in the interface potential can be evaluated from eq. (3-6-7) as

$$\Delta \psi_s = - \frac{\Delta V_T}{\eta} \qquad (3\text{-}6\text{-}15)$$

3.6.3. Drain Induced Barrier Lowering (DIBL)

Based on a simple consideration of charge sharing, the model of Subsection 3.6.2 gives averaged estimates of some important short-channel effects in the subthreshold regime of MOSFETs. This is achieved by distributing the effective gate depletion charge evenly along the channel in order to estimate the threshold voltage shift. While this may be a good approximation for $V_{DS} = 0$, it will fail to accurately predict the effect on V_T of an applied drain-source voltage. The reason is that a portion of the additional depletion charge induced by the drain-source bias will be distributed non-uniformly from source to drain, as indicated in Fig. 3.6.3. Likewise, the drain-source bias will induce a non-uniform shift, $V(x)$, in the interface potential along the channel which increases from $V(0) = 0$ at the source to $V(L) = V_{DS}$ at the drain.

Here, we proceed to develop a model for the distribution of the induced shift, $V(x)$, in the interface potential along the channel as a result of the applied drain-source bias. From such a model, it is possible to calculate the interface potential near its minimum, which defines the barrier for charge injection into the channel (see Fig. 3.6.1). An accurate estimate of the shift in the potential minimum is especially important since the channel current is exponentially dependent on the barrier height. In principle, this involves the solution of a two-dimensional Poisson's equation for the whole device, using proper boundary conditions. However, this requires extensive numerical calculations. Instead, we perform a simplified analytical calculation based on an approach similar to that used in our analysis of the velocity saturation region (see Section 3.5).

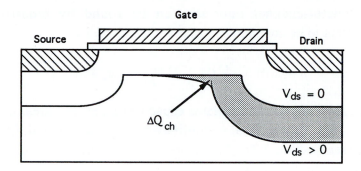

Fig. 3.6.3. Distribution of depletion charge induced by an applied drain-source bias, indicated by the shaded region. ΔQ_{ch} is the part of the induced charge located in the central channel region and which has its counter charge on the gate electrode. (From Fjeldly and Shur (1992), © 1992 IEEE.)

We start by considering the two-dimensional Poisson's equation for the depletion region under the gate, away from the source and drain contact depletion regions. In the subthreshold regime, the influence of the charge carriers on the electrostatics of the channel can be neglected, and we write the two-dimensional Poisson's equation as (see Section 3.5)

$$\frac{\partial F_x}{\partial x} + \frac{\partial F_y}{\partial y} = -\frac{qN_a}{\varepsilon_s} \tag{3-6-16}$$

where F_x and F_y are the longitudinal and perpendicular components of the electric field, respectively. Integrating this equation with respect to y from the semiconductor-insulator interface through the depletion region yields

$$\left\langle \frac{\partial F_x}{\partial x} \right\rangle d_{dep} - F_y(0) = -\frac{q}{\varepsilon_s} N_a d_{dep} \tag{3-6-17}$$

where $\langle \partial F_x/\partial x \rangle$ is the average of $\partial F_x/\partial x$ over the thickness of the depletion region. This thickness can be estimated approximately from a one-dimensional theory as

$$d_{dep} \approx \sqrt{\frac{2\varepsilon_s}{qN_a}(\psi_s + V_{bi})}$$

(3-6-18)

As discussed earlier, the vertical component of the electric field, $F_y(0)$, at the semiconductor-channel interface can be found by requiring the electric displacement to be continuous across the interface, i.e.,

$$F_y(0) = \frac{c_i}{\varepsilon_s}(V_{GS} - V_{FB} - \psi_s - V_{bi})$$

(3-6-19)

where $c_i = d_i/\varepsilon_i$ is the insulator capacitance per unit area, d_i is the insulator thickness, ε_i is the electrical permittivity of the insulator, and V_{FB} is the flat band voltage (see Section 3.3). In the presence of a drain-source bias, the interface potential can be written as

$$\psi_s = \psi_s^o + V(x)$$

(3-6-20)

where ψ_s^o is the constant interface potential of the middle part of the channel when $V_{DS} = 0$ and $V(x)$ is the addition to the channel potential caused by the applied drain-source voltage. Away from the source and drain contacts, we can assume that $\langle \partial F_x/\partial x \rangle = 0$ when $V_{DS} = 0$. We now consider eq. (3-6-17) with and without an applied drain-source bias and express the net effect of the drain-source bias by taking the difference, i.e.,

$$\frac{\partial^2 V}{\partial x^2}d_{dep}(x) - \frac{c_i}{\varepsilon_s}V(x) = \frac{q}{\varepsilon_s}N_a\left[d_{dep}(x) - d_{dep}^o\right]$$

$$= \frac{q}{\varepsilon_s}N_a d_{dep}^o\left(\sqrt{1 + \frac{V(x)}{\psi_s^o + V_{bi}}} - 1\right)$$

(3-6-21)

where d_{dep}^o is the depletion width for $V = 0$. In eq. (3-6-21), we have replaced $\langle \partial F_x/\partial x \rangle$ by $-\partial^2 V/\partial x^2$, assuming that $V(x)$ inside the gate depletion region is relatively weakly dependent on the distance from the interface. The second term on the left hand side of eq. (3-6-21), which is equal to the difference $F_y(0) - F_y^o(0)$, where $F_y^o(0)$ is the value of $F_y(0)$ for $V = 0$, is obtained by using eqs. (3-6-18) and (3-6-19). Since both $V(x)$ and its x-derivatives are small outside the depletion region of the drain contact, all terms in eq. (3-6-21) can be expanded to first order in V to give

$$\frac{\partial^2 V}{\partial x^2} - \frac{V}{\lambda^2} = 0 \tag{3-6-22}$$

where

$$\lambda = d_{dep}^{o} \left(1 + \frac{\varepsilon_i}{\varepsilon_s} \frac{d_{dep}^{o}}{d_i} \right)^{-1/2} \tag{3-6-23}$$

The general solution of eq.(3-6-22) can be written on the following form:

$$V(x) = A \exp\left(\frac{x}{\lambda}\right) + B \exp\left(-\frac{x}{\lambda}\right) \tag{3-6-24}$$

where the coefficients A and B are determined from the boundary conditions. Without much error, we can assume that eq. (3-6-24) is also valid through the source-channel junction region (i.e., $x < x_S$), in which case we have the boundary condition $V(x = 0) = 0$, which gives $A = -B = V_o/2$, such that eq. (3-4-24) can be written as

$$V(x) = V_o \sinh\left(\frac{x}{\lambda}\right) \tag{3-6-25}$$

Here, V_o is a constant that remains to be determined. We note that the shift, $-qV(x_{min}) \approx -qV(x_S)$ in the conduction band at the channel side of the source-channel junction is identical to the DIBL (see Fig. 3.6.1).

In order to find the voltage V_o, however, it is necessary investigate more closely the electrostatics of the entire channel. In particular, we have to consider the additional charges induced in the gate electrode and in the substrate as a result of the applied drain-source voltage. We start by taking an inventory of the induced charges and counter charges, as shown schematically in Fig. 3.6.4.

From the principle of charge sharing discussed previously, we can pair charges and counter charges as indicated in Fig. 3.6.4. According to this scheme, a fraction of the drain bias induced depletion charge, marked ΔQ_{ch} in the figure, will have its counter charge on the gate electrode. In order to be consistent with the potential variation along the channel, calculated above, the corresponding sheet charge distribution, $\Delta q(x)$, along the channel has to be as follows (see eq. (3.6.25)):

$$\Delta q(x) = \frac{V(x)}{c_{dep}(x)} \approx \frac{V_o \varepsilon_s}{d_{dep}^o} \sinh\left(\frac{x}{\lambda}\right) \qquad (3\text{-}6\text{-}26)$$

where we have invoked the gradual channel approximation and expanded to lowest order in $V(x)$; $c_{dep}(x)$ is the channel depletion capacitance per unit area.

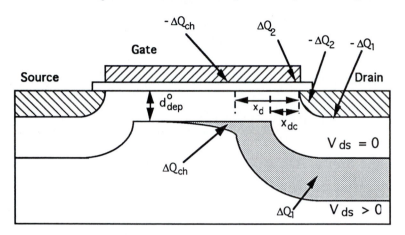

Fig. 3.6.4. Schematic overview of the drain bias induced charges and counter charges according to the principle of charge sharing. ΔQ_{ch} and $-\Delta Q_{ch}$ are the induced charges in the channel. The remaining charges and counter charges are between the drain and the substrate (ΔQ_1 and $-\Delta Q_1$) and between the drain and the gate (ΔQ_2 and $-\Delta Q_2$). (From Fjeldly and Shur (1992), © 1992 IEEE.)

Assuming, for simplicity, that eq. (3-6-26) is valid over the range $0 \leq x \leq L - x_d$, we obtain the following expression for V_o by requiring that the integral of $\Delta q(x)$ over this range equals ΔQ_{ch}:

$$V_o \approx \frac{\Delta Q_{ch}\, d_{dep}^o}{W\lambda \varepsilon_s} \left[\cosh\left(\frac{L - x_d}{\lambda}\right) - \cosh\left(\frac{x_s}{\lambda}\right)\right]^{-1} \qquad (3\text{-}6\text{-}27)$$

It now remains to determine the induced channel depletion charge, ΔQ_{ch}. A rough estimate of this charge can be obtained as indicated in Fig. 3.6.5. The shaded region in the substrate indicates roughly the amount of additional depletion charge, $\Delta Q \approx qWN_a(x_d^2 - x_{do}^2)$, induced under the gate by the drain-source bias. Here, $x_{do} \approx x_s$ is the depletion width of the drain-channel junction at zero drain-source voltage. From the concept of charge sharing, ΔQ_{ch} can be taken to be some fraction, χ, of ΔQ, i.e.,

$$\Delta Q_{ch} \approx \Delta Q \approx \chi W q N_a \left(x_d^2 - x_{do}^2 \right) = 2\chi W \, \varepsilon_s V_{DS} \qquad (3\text{-}6\text{-}28)$$

Typically, χ is of the order of 0.5. However, the value of this parameter and also the substrate doping density, N_a, can be adjusted to account for the shape and doping profiles in the drain junction (e.g., Low Doped Drain (LDD) MOSFETs) and substrate (e.g., ion-implantation). In other words, this fitting parameter is technology dependent.

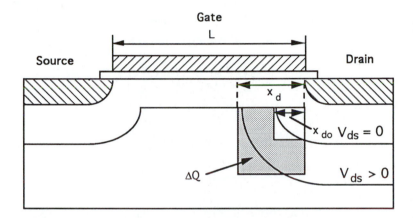

Fig. 3.6.5. Simplified model of the drain bias induced charge, ΔQ, in the substrate under the gate (shown as an estimate of the depletion charge under the gate between the depletion boundaries for $V_{DS} > 0$ and for $V_{DS} = 0$). The induced channel charge, ΔQ_{ch}, is a fraction of ΔQ, according to the charge sharing principle. (From Fjeldly and Shur (1992), © 1992 IEEE.)

The parameter V_O can be obtained by substituting eq. (3-6-28) into eq. (3-6-27), i.e.,

$$V_o \approx 2\chi V_{DS} \frac{d_{dep}^o}{\lambda} \left[\cosh\left(\frac{L - x_d}{\lambda} \right) - \cosh\left(\frac{x_s}{\lambda} \right) \right]^{-1} \qquad (3\text{-}6\text{-}29)$$

Substituting eq. (3-6-29) in eq. (3-6-25) and setting $x = x_{min} \approx x_s$, the lowering of the injection barrier is found to be

$$DIBL \approx qV(x_s) \approx qV_o \sinh\left(\frac{x_s}{\lambda}\right)$$

$$\approx 2\chi q V_{DS} \frac{d^o_{dep}}{\lambda} \frac{\sinh\left(\frac{x_s}{\lambda}\right)}{\cosh\left(\frac{L - x_d}{\lambda}\right) - \cosh\left(\frac{x_s}{\lambda}\right)} \qquad (3\text{-}6\text{-}30)$$

We note that the barrier lowering predicted by eq. (3-6-30) decreases exponentially with increasing gate length for $L - x_d > 3\lambda$. For sufficiently small gate lengths or sufficiently high drain-source bias such that $L \approx x_d + x_s$, the DIBL diverges and eq. (3-6-30) is no longer valid. This condition corresponds to severe punch-through in the device.

As a consequence of the barrier lowering, we have a drain bias induced shift in the threshold voltage. From eqs. (3-6-15) and (3-6-30), we find

$$\Delta V_T = -\eta \Delta \psi_s(x_s) = -\eta V(x_s) = -\sigma V_{DS} \qquad (3\text{-}6\text{-}31)$$

where

$$\sigma \approx \frac{2\eta \chi d^o_{dep}}{\lambda} \frac{\sinh\left(\frac{x_s}{\lambda}\right)}{\cosh\left(\frac{L - x_d}{\lambda}\right) - \cosh\left(\frac{x_s}{\lambda}\right)} \qquad (3\text{-}6\text{-}32)$$

In Fig. 3.6.6, we show experimental data of ΔV_T versus V_{DS} for two short channel NMOS devices (from Chung et al. (1991)). It is evident that the experimental data are in good agreement with the linear relationship of eq. (3-6-31), and the slopes of the straight lines give the parameter σ. A linear relationship was also observed experimentally by Liu et al. (1991), except for an accelerated shift in V_T at very small values of V_{DS}, typically $V_{DS} \leq 3V_{th}$. This deviation from linearity is expected since V_{DS} becomes comparable with or less than the subthreshold saturation voltage ($V_{SAT} = 2V_{th}$). Below saturation, drift current dominates over diffusion in the MOSFET channel and the applied drain-source voltage is distributed more evenly along the channel, resulting in a more pronounced DIBL than predicted above.

It is interesting to notice that the dependence of σ on N_a and L predicted by eq. (3-6-32) is similar to that obtained by de Graaff and Klaassen (1990) in cases where the hyperbolic functions can be expanded. Also, the present model expressions for the shift in threshold voltage with gate length is in good qualitative agreement with experimental data by Liu et al. (1991), as shown in Fig. 3.6.7. Like many others, Liu et al. observed that ΔV_T varies close to exponentially with L_{eff}. However, as indicated in the figure, an accelerated shift in the threshold voltage is obtained at very small values of L_{eff} (typically less than 0.3 μm for the technology used), in agreement with the present theory.

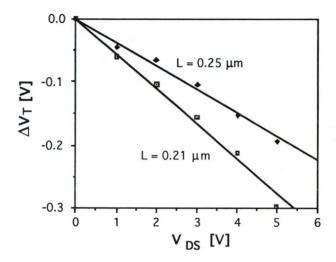

Fig. 3.6.6. Experimental threshold voltage shift versus drain-source voltage for two NMOS devices with effective gate lengths 0.21 μm and 0.25 μm. Eq. (3-6-31) is fitted to the two data sets yielding σ = 0.056 (L = 0.21 μm) and σ = 0.038 (L = 0.25 μm). (From Fjeldly and Shur (1992), © 1992 IEEE. Experimental data from Chung et al. (1991).)

The above estimates are simplified and partly empirical. They do not take directly into account, for example, the effect of the contact junction depth, r_j, on the short channel effects, which may be important. In fact, experimental studies show that a transition from long to short channel behavior takes place when

$$L < L_{min}(\mu m) = 0.4\left[r_j(\mu m)d_i(\text{Å})(W_d + W_s)^2(\mu m^2)\right]^{1/3} \quad (3\text{-}6\text{-}33)$$

where W_d and W_s are the drain-substrate and source-substrate depletion widths, respectively (see Brews et al. (1980)). This expression indicates the importance of the contact depth. Indirectly, however, this behavior may be accounted for by a judicial choice of the adjustable parameter χ.

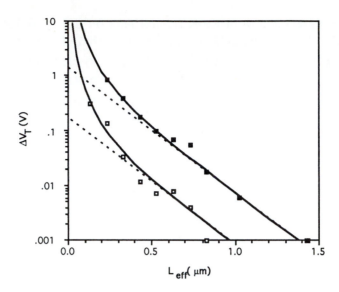

Fig. 3.6.7. Experimental values (symbols), fitted model calculations (solid lines) and exponential approximation (dotted lines) of shift in threshold voltage versus effective gate length, in a semilog plot. Upper curves: $T = 85$ K, lower curves: T = 300 K. (From Fjeldly and Shur (1992), © 1992 IEEE. Experimental data from Liu et al. (1991).)

3.6.4. Subthreshold Current in Short Channel MOSFETs

The current-voltage characteristics in the presence of DIBL can now be derived from the drift-diffusion expression in eq. (3-6-2). Alternatively, when the channel length becomes sufficiently small, we can obtain the characteristics in terms of the thermionic emission current across the energy barrier. In reality, however, pure thermionic emission current only takes place in a region of a few mean free paths of the maximum of the barrier, and drift-diffusion transport dominates in the rest of the channel. Hence, both processes are always present although one of them will normally dominate (see Sections 1.9 and 1.10 for comparison). Here, we calculate the drain current based on each of the two transport mechanisms separately, and propose a unified expression in which both are included, following the approach proposed by van der Ziel et al. (1983) for a short *n-i-n* diode.

In the drift-diffusion theory, only the region $x_s \leq x \leq L - x_d$ will contribute significantly to the integral in the denominator on the right hand side of eq. (3-6-2). Besides, since the integrand is exponentially dependent on the potential, we can safely expand the hyperbolic sine function in eq. (3-6-25) to first order in the distance, i.e.,

$$
\int_{x_s}^{L-x_d} \exp\left(-\frac{\psi(x)}{V_{th}}\right) dx \approx \exp\left(-\frac{\psi^o}{V_{th}}\right) \int_{x_s}^{L-x_d} \exp\left(-\frac{V_o\, x}{V_{th}\,\lambda}\right) dx
$$

$$
= \frac{\lambda V_{th}}{V_o} \exp\left(-\frac{\psi^o + V(x_s)}{V_{th}}\right)\left\{1 - \exp\left[-\frac{V_o\,(L - x_s - x_d)}{V_{th}\,\lambda}\right]\right\} \quad (3\text{-}6\text{-}34)
$$

Following the derivation used for the long-channel case, we obtain for the drift-diffusion drain current

$$
I_d(dd) = \frac{qN_c\, W\, \delta d D_n\, V_o}{\lambda V_{th}}\; \frac{\exp\left(\dfrac{\psi_s^o + V(x_s)}{V_{th}}\right)\left[1 - \exp\left(-\dfrac{V_{DS}}{V_{th}}\right)\right]}{1 - \exp\left[-\dfrac{V_o\,(L - x_s - x_d)}{V_{th}\,\lambda}\right]} \quad (3\text{-}6\text{-}35)
$$

This expression reduces to eq. (3-6-12) for small values of the applied V_{DS} and/or for large values of L, since in both cases V_o and $V(x_s)$ become very small.

We now instead assume that the drain current is governed by thermionic emission. At zero drain-source bias, the net drain current is zero, and the equilibrium thermionic emission current from source to drain, $I^o_{s \to d}$, and from drain to source, $I^o_{d \to s}$, are given by

$$
I^o_{s \to d} = I^o_{d \to s} = I_o \exp\left(\frac{\psi_s^o}{V_{th}}\right) \quad (3\text{-}6\text{-}36)
$$

where $I_o = W\delta d A^* T^2$, δd is the effective channel thickness (see eq. (3-6-8)), and A^* is Richardson's constant (see Section 1.9). With a finite drain-source voltage, the current contributions become (including the effect of DIBL)

$$
I_{s \to d} = I_o \exp\left[\frac{\psi_s(x_s)}{V_{th}}\right] = I_o \exp\left(\frac{\psi_s^o + V(x_s)}{V_{th}}\right) \quad (3\text{-}6\text{-}37)
$$

$$
I_{d \to s} = I_{s \to d} \exp\left(-\frac{V_{DS}}{V_{th}}\right) \quad (3\text{-}6\text{-}38)
$$

and the net thermionic emission current can be written as

$$I_d(te) = I_{s \to d} - I_{d \to s} = I_o \exp\left(\frac{\psi_s^o + V(x_s)}{V_{th}}\right)\left[1 - \exp\left(-\frac{V_{DS}}{V_{th}}\right)\right] \quad (3\text{-}6\text{-}39)$$

We now wish to find a unified expression for the I–V characteristics which includes both drift-diffusion and thermionic emission. A complete theory to this end would, however, be quite involved and impractical. But, as observed above, the two mechanisms act predominantly in a serial fashion, each representing a different part of the channel. Accordingly, the drain current will be dominated by the current limiting mechanism which will be the one that corresponds to the highest impedance. This is quite analogous to a series combination of two admittances where the current is dominated by smaller of the two. In line with this analogy, we propose a unified expression for the subthreshold I–V characteristics as the following combination of the two currents in eqs. (3-6-35) and (3-6-39):

$$I_d \approx \left[\frac{1}{I_d(dd)} + \frac{1}{I_d(te)}\right]^{-1} \quad (3\text{-}6\text{-}40)$$

It is clear from the above results that the current limiting mechanism will be thermionic emission at small gate lengths and drift-diffusion at larger gate lengths.

3.6.5. Generalized UCCM

According to the Unified Charge Control Model (UCCM), the inversion charge, n_s, in a MOSFET is related to the gate voltage swing, V_{GT}, and channel Fermi potential, V_F, as follows (see Sections 3.3 and 3.4):

$$V_{GT} - \alpha V_F \approx \eta V_{th} \ln\left(\frac{n_s}{n_o}\right) + a(n_s - n_o) \quad (3\text{-}6\text{-}41)$$

where η is the ideality factor in the subthreshold region, α is a parameter accounting for the body effect (i.e., the dependence of the threshold voltage of the strong inversion on the channel voltage), $a \approx q/c_i + q\Delta d/\varepsilon_s$, and Δd is a quantum correction to the effective insulator thickness. In eq. (3-6-41), we have not considered explicitly the effect of DIBL on the threshold voltage. However, by including the threshold voltage shift determined in eq. (3-6-31), UCCM can be generalized to

$$V_{GTo} + \sigma V_{DS} - \alpha V_F \approx \eta V_{th} \ln\left(\frac{n_s}{n_o}\right) + a(n_s - n_o) \qquad (3\text{-}6\text{-}42)$$

where V_{GTo} is the threshold voltage swing at zero drain-source bias.

In this Section, the drain bias induced threshold voltage shift was analyzed in terms of the lowering of the injection barrier between the source and the channel in the subthreshold regime. In strong inversion, however, the injection barrier is reduced owing to the effect of the gate-source bias, and will eventually disappear well above threshold. Hence, the importance of DIBL will be gradually less with increasing gate-source voltage and should gradually be phased out from the expression for V_T. For modeling purposes, σ should therefore include a dependence on V_{GT} as, for example, in the following empirical expression used in the unified MOSFET model discussed in Section 3.9:

$$\sigma = \frac{\sigma_o}{1 + \exp\left(\dfrac{V_{GTo} - V_{\sigma T}}{V_\sigma}\right)} \qquad (3\text{-}6\text{-}43)$$

which gives $\sigma \to \sigma_o$ for $V_{GTo} < V_{\sigma T}$ and $\sigma \to 0$ for $V_{GTo} > V_{\sigma T}$. The voltage V_σ determines the width of the transition between the two regimes. A qualitatively similar dependence of σ on the gate voltage was obtained by de Graaff and Klaassen (1990) who related this dependence to the interplay between two different physical mechanisms – DIBL (which is important in the subthreshold regime and just above the threshold) and static feedback (which is a capacitive coupling between the drain and the channel that may be important well above threshold).

3-7. THRESHOLD VOLTAGE ENGINEERING

The threshold voltage separates the *off*-state of the MOSFET operation, where the channel current is very low, from the *on*-state where the channel current is large. The *on*-to-*off* current ratio is very important for CMOS operation, especially for so-called dynamic circuits where the leakage current in the *off*-state determines the holding time (see, for example, Shoji (1988)). Hence, the threshold voltage, V_T, should be chosen in such a way that the difference between the threshold voltage and the gate voltage corresponding to the *off*-state is sufficiently large. For CMOS circuits, the gate-source voltage in the *off*-state must be equal to zero and the ratio of the threshold current to the leakage current in the *off*-state must be at least as large as 10^6.

As we discussed in Sections 3.4 and 3.6, the interface sheet carrier density in the MOSFET channel below threshold varies exponentially with the gate-source bias, V_{GS}. Since, the drain current, I_d, is proportional to the interface carrier concentration, the subthreshold current also varies exponentially with the gate-source bias. Hence, for NMOS, the subthreshold current can be written as

$$I_d(V_{GS}) \approx I_d(V_{TN}) \exp\left(-\frac{V_{GS} - V_{TN}}{\eta V_{th}}\right) \tag{3-7-1}$$

where V_{TN} is the NMOS threshold voltage, V_{th} is the thermal voltage, and η is the ideality factor. From eq. (3-7-1), we find that for an *on*-to-*off* ratio of 10^6, the threshold voltage has to satisfy the condition

$$V_{TN} > \eta V_{th} \ln\left(10^6\right) \approx 13.8 \, \eta V_{th} \tag{3-7-2}$$

For a typical operating temperature of 400 K and $\eta = 1.5$, we find that $V_{TN} > 0.72$ V. To this value, we typically add 0.2 V as a safety margin in order to account for process variation and DIBL (see Section 3.6). Hence, the target threshold voltage for NMOS should be $V_T = 0.92$ V. However, if in our design we use the expression for the threshold voltage derived for long channel devices (see Sections 3.3 and 3.4), as is commonly done, our target threshold voltage has to be even larger because the short channel effects will shift the NMOS threshold voltage towards more negative values. (However, as discussed below, narrow channel devices may experience an opposite shift of the threshold voltage, which also has to be taken into account.) On the other hand, the threshold voltage should not be chosen to be too high in practical circuits in order to retain a reasonable voltage swing at a relatively small power supply voltage (for example, $V_{DD} = 3$ V or less). All in all, in fast static CMOS digital circuits, which have a large noise margin, V_{TN} is normally chosen to be around 0.8 V.

One would expect that the threshold voltage of PMOS, V_{TP}, should then be chosen to be around -0.8 V in order to obtain symmetrical threshold characteristics. However, the absolute value of V_{TP} is often taken to be somewhat larger than that of NMOS in order to prevent a large leakage current in buried channel PMOS (see Section 3.2).

As was discussed in Sections 3.3 to 3.6, the threshold voltage depends on the substrate-source bias and on device parameters such as insulator thickness, doping profile, junction depth, channel length, and channel width.   In this Section, we will discuss some of these dependencies and their implications for the device properties.

In Sections 3.3 and 3.4, we showed that the threshold voltage for a long channel NMOS with a uniformly doped substrate is given by

$$V_{TN} = V_{FBN} + 2\varphi_b + \gamma_N \sqrt{(2\varphi_b - V_{BS})} \qquad (3\text{-}7\text{-}3)$$

where V_{FBN} is the flat-band voltage in the n-channel device, $\varphi_b = V_{th} \ln(N_a/n_i)$ is the absolute potential difference between the intrinsic Fermi level and the Fermi level in the substrate bulk, V_{BS} is the applied substrate-source bias,

$$\gamma_N = \sqrt{2\varepsilon_s q N_a}/c_i \qquad (3\text{-}7\text{-}4)$$

is called the body effect constant, and c_i is the insulator capacitance per unit area. If the surface states are important, this equation has to be modified as follows to account for the surface state charge density, q_{ss}:

$$
\begin{aligned}
V_{TN} &= V_{FBN} + q_{ss}/c_i + 2\varphi_b + \gamma_N \sqrt{(2\varphi_b - V_{BS})} \\
&= V_{TNo} + \gamma_N \left[\sqrt{(2\varphi_b - V_{BS})} - \sqrt{2\varphi_b} \right]
\end{aligned} \qquad (3\text{-}7\text{-}5)
$$

where

$$V_{TNo} = V_{FBN} + q_{ss}/c_i + 2\varphi_b + \gamma_N \sqrt{2\varphi_b} \qquad (3\text{-}7\text{-}6)$$

is the threshold voltage at zero substrate bias. In modern devices, the effect of the surface states is usually negligible. However, the value of q_{ss} may be determined from a comparison of eq. (3-7-5) with measurements of V_{TN} versus V_{BS} in very long and wide devices.

To account for short channel effects (see Section 3.6) and narrow channel effects (see the discussion below), eq. (3-7-6) has to be modified as follows (neglecting the term q_{ss}/c_i):

$$V_{TNo} = V_{FBN} + 2\varphi_b + \gamma_N \sqrt{2\varphi_b} + \Delta V_{TL} + \Delta V_{TW} \qquad (3\text{-}7\text{-}7)$$

where ΔV_{TL} is a negative, short channel correction and ΔV_{TW} is a positive, narrow channel correction (see eq. (3-6-14) and Figs. 3.7.1 and 3.7.2).

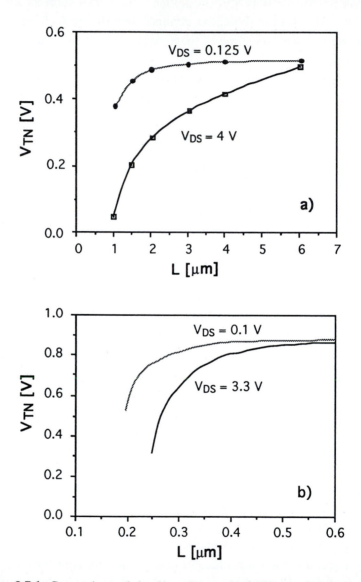

Fig. 3.7.1. Comparison of the dependence of NMOS threshold voltage on effective channel length for a relatively old fabrication technology (a) (see Fichtner and Potzl (1979)) and a modern state-of-the-art technology (b) (see Hayden et al. (1991)).

These corrections depend on the bias voltages and on device parameters such as channel length, channel width, and oxide thickness, and are also very much technology dependent. Nevertheless, to a first order approximation, the dependence of V_{TN} on the substrate bias can still be described by eq. (3-7-5).

Similarly, we have for PMOS,

$$V_{TP} = V_{TPo} - \gamma_P \left[\sqrt{(2\varphi_b + V_{BS})} - \sqrt{2\varphi_b} \right] \tag{3-7-8}$$

$$V_{TPo} = V_{FBP} - 2\varphi_b - \gamma_P \sqrt{2\varphi_b} + \Delta V_{TL} + \Delta V_{TW} \tag{3-7-9}$$

where now, $\varphi_b = V_{th} \ln(N_d/n_i)$, V_{FBP} is the PMOS flat-band voltage, and

$$\gamma_P = \sqrt{2\varepsilon_s q N_d} / c_i \tag{3-7-10}$$

(Note that the corrections ΔV_{TL} and ΔV_{TN} in PMOS have the opposite signs of those in NMOS.)

The correction of the threshold voltage related to short channel effects was discussed in detail in Section 3.6. The theory of Section 3.6 reproduces the correct trend of the dependence of V_T on the gate length, L. However, the numerical value of this correction is very sensitive to the fabrication process as illustrated in Fig. 3.7.1, where we compare measured values of V_T versus L for a relatively old fabrication technology and a modern state-of-the-art technology.

A shift in the opposite direction is observed in V_T with decreasing channel width, as shown in Fig. 3.7.2. This "narrow channel effect", is caused by the extra charge required to invert the surface under the field oxide next to the channel, as indicated in Fig. 3.7.3 where the device cross-section and the shape of the depletion region are shown. This effect also tends to increase with negative substrate bias (in NMOS). The threshold voltage shift in narrow channel devices usually exhibits a $1/W_{eff}$ dependence, where W_{eff} is the effective channel width, as shown in Fig. 3.7.3.

A calculated plot of V_T versus $g = (2\varphi_b - V_{BS})^{1/2} - (2\varphi_b)^{1/2}$ is shown in Fig. 3.7.4. This so-called "body plot" is usually determined experimentally for reverse or small forward substrate bias (in order to avoid substrate leakage current). The slope of the plot yields the body effect constant, γ_N (NMOS) or γ_P (PMOS). However, as discussed above, this analysis is less accurate for short channel devices. Nevertheless, body plots may still be useful for device characterization even for short channel MOSFETs.

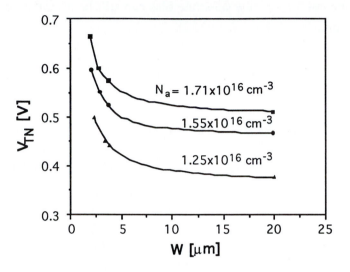

Fig. 3.7.2. Dependence of the threshold voltage on the channel width for different doping levels (see Akers et al. (1981)).

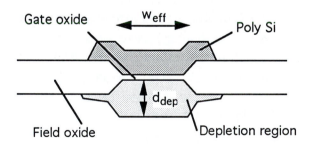

Fig. 3.7.3. MOSFET cross section perpendicular to the direction from source to drain.

Once the insulator capacitance is determined from $C-V$ measurements (see Section 3.3), the value of the substrate doping can be extracted from the body effect constant (see eqs. (3-7-4) and (3-7-10)). Then, after the substrate doping is determined, we can calculate $2\varphi_b$ (see above).

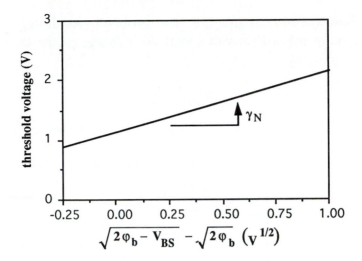

Fig. 3.7.4. A calculated body plot for NMOS with a uniformly doped substrate. The slope of the plot is equal to the body effect constant, γ_N.

The dependence of $2\varphi_b$ on substrate doping is shown in Fig. 3.7.5 for three different temperatures: 400 K (which may be close to a realistic operating temperature), 300 K, and 77 K (which may be of interest for cool CMOS applications).

As can be seen from Fig. 3.3.3 (Section 3.3), $V_{FBN} \approx -1$ V for NMOS with polysilicon gates so that $V_{FBN} + 2\phi_b \approx 0$. The desired threshold voltage of NMOS is approximately 1 V for a long channel device. In order to achieve a threshold voltage of 1 V for a long channel device at zero substrate bias, we need to have a body factor, γ_N, of approximately 1 $V^{1/2}$, as can seen from eq. (3-7-3). We obtain $\gamma_N \approx 1$ $V^{1/2}$ for typical parameter values such as $N_a = 4\times10^{16}$ cm^{-3} and oxide thickness $d_i = 300$ Å. For other values of d_i, the device parameters have to be scaled in such a way that the product $N_a d_i^2$ remains approximately constant so that the body effect constant remains close to 1 $V^{1/2}$. In short channel devices, this value can be reduced by 0.2 V or more, depending on the device parameters and on the technology used. In a typical design, the threshold voltage of short channel NMOS is approximately 0.8 V.

For PMOS with polysilicon gate, we have $V_{FBP} \approx -0.1$ V and the absolute value of V_{TP} would be larger than that of V_{TN} for the same fabrication process, using PMOS with a reasonable bulk n-type doping. To obtain a symmetrical CMOS, we need to have $-V_{TP}$ close to V_{TN}. This can be achieved by implanting boron ions (counter doping) at the surface. Assuming a delta function distribution of boron ions with a total sheet charge q_I, the shift of the threshold voltage is given by

$$\Delta V_{TP} = q_I / c_i \tag{3-7-11}$$

This approximation is acceptable for low dose and shallow counter doping.

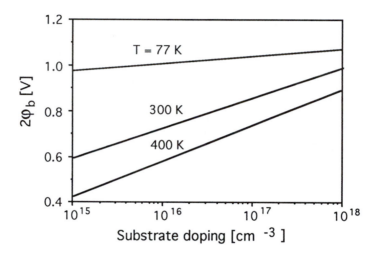

Fig. 3.7.5. The dependence of $2\varphi_b$ on substrate doping for three different temperatures: 400 K , 300 K, and 77 K. Parameters used in the calculation: energy gap: 1.12 eV, room temperature effective density of states in the conduction band: 3.22×10^{19} cm^{-3} and in the valence band: 1.83×10^{19} cm^{-3}.

Above, we discussed how to obtain a desired threshold voltage of 0.8 V for NMOS at zero substrate bias. However, the threshold voltage can also be adjusted by using a substrate bias. The latter approach is often preferred because a negative substrate bias $(V_{BS} < 0)$ increases the depletion widths of the source-substrate and drain-substrate junctions. The resulting reduction in the junction capacitances may give a considerable speed advantage. Also, in this case, the body effect constant can be chosen smaller than 1 V$^{1/2}$. The reduction of the body effect is advantageous for NMOS and CMOS circuits.

The negative substrate-source bias needed for NMOS can easily be generated by integrating special on-chip substrate bias generating circuits utilizing one positive power supply.

Using a typical value for the substrate bias, $V_{BS} \approx -2$ V, and a body effect constant, $\gamma_N \approx 0.5$ V$^{1/2}$, the shift in the threshold voltage is approximately 0.4 V. For an oxide thickness of 300 Å, $\gamma_N \approx 0.5$ V$^{1/2}$ corresponds to a doping density of about 10^{16} cm^{-3} which is higher than that needed to avoid punch-through. Punch-through (i.e., overlap of the source and drain depletion regions) occurs at zero drain-source bias for a substrate doping of approximately 10^{15} cm^{-3} for $L = 2$ µm. Hence, this value of doping is satisfactory for the NMOS design.

If the substrate doping is too high, the various depletion regions become quite narrow which result in high drain-substrate and source-substrate capacitances. Also, the substrate acts as a second "gate" (with the "gate"-channel spacing equal to the depletion region thickness) and a negative substrate potential causes a reduction in the channel carrier density. This body effect leads to a reduction in the drain saturation current. In order to avoid these negative consequences of high substrate doping and, at the same time, obtain the required threshold voltage, we can utilize a doping profile as shown in Fig. 3.7.6.

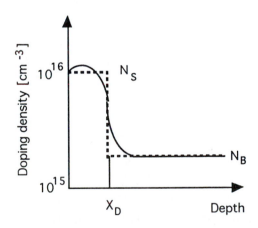

Fig. 3.7.6. Typical doping profile of ion implanted MOSFET. Solid
line: actual profile, dotted line: "box" approximation.

A body plot corresponding to this doping profile is shown in Fig. 3.7.7. This body plot has two different slopes, γ_{NS} and γ_{NB}, corresponding the doping density $N_S = 10^{16}$ cm^{-3} close to the interface and $N_B = 2 \times 10^{15}$ cm^{-3} in the bulk:

$$\gamma_{NS} = \sqrt{2\varepsilon_s\, qN_S}\, /c_i \qquad (3\text{-}7\text{-}12)$$

$$\gamma_{NB} = \sqrt{2\varepsilon_s\, qN_B}\, /c_i \qquad (3\text{-}7\text{-}13)$$

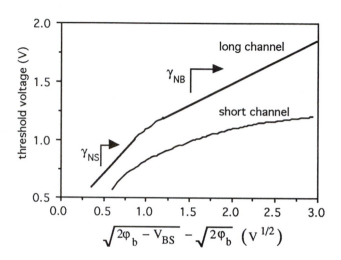

Fig. 3.7.7. Body plots: upper curve, measured threshold voltage for long channel device according to data from Ong (1984); lower curve: qualitative body plot for a short channel device. The doping profile corresponding to the upper curve is shown in Fig. 3.7.6 ($X_D = 0.35\ \mu$m).

When the substrate bias is so small that the depletion depth is less than X_D (0.35 μm in the example above), the slope is equal to γ_{NS}. When the depletion depth is larger than X_D, the slope is equal to γ_{NB}. As mentioned above, typically, γ_{NS} should be around 1 $V^{1/2}$ in order to obtain the desired threshold voltage of $V_{TN} \approx 1.0$ V. For these values of γ_{NS} and V_{TN}, the substrate bias corresponding to the transition point between the two regimes is

$$V_{BS} = 2\varphi_b - \left(\frac{V_{TN} - V_{TNo}}{\gamma_{NS}} + \sqrt{2\,\varphi_b}\right)^2$$

$$= 2\varphi_b - \frac{q\, N_S\, X_D^{\,2}}{2\,\varepsilon_s} \approx -2.2\ \text{V} \qquad (3\text{-}7\text{-}14)$$

Fig. 3.7.8 shows how the body plot changes for different doping profiles.

Using such body plots, we can optimize the doping profile in order to achieve the required device parameters.

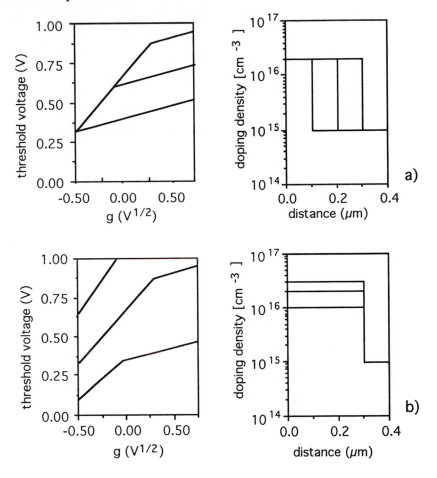

Fig. 3.7.8. Body plots for different doping profiles with constant N_S and different X_D (a), and constant X_D and different N_S (b); $g = (2\varphi_b - V_{BS})^{1/2} - (2\varphi_b)^{1/2}$. The oxide thickness is 300 Å for both cases.

Both NMOS and PMOS should be designed in such a way that both short and narrow channel effects are not appreciable at the target gate length and width. Otherwise, the threshold voltage would be too sensitive to the channel length and/or width variation caused by the process fluctuations.

Accurate threshold voltage modeling is vitally important for both device design and circuit simulation. However, in spite of numerous studies, simple analytical formulae do not always describe accurately the threshold voltage dependence on doping profile, channel length and width, source/drain junction depth, and substrate bias. The lower curve in Fig. 3.7.7 shows the qualitative effect of short channel effects on the body plot. As can be seen from the figure, the value of the body effect coefficient, γ_N, is smaller in short channel devices. This can be understood on the basis of the analysis given above (see eqs. (3-7-7) and (3-6-13)). Similar results were obtained by Sheu et al. (1987) and incorporated into the popular SPICE model BSIM (Berkeley Simulation). This model (which is also available in AIM-Spice) uses the following empirical expression:

$$V_{TNS} = V_{TNSo} + \gamma_{N\,eff}\sqrt{2\varphi_b - V_{BS}} - \Gamma_{Neff}\left(2\varphi_b - V_{BS}\right) \qquad (3\text{-}7\text{-}15)$$

In eq. (3-7-15), we have

$$V_{TNSo} = V_{FB} + 2\varphi_b \qquad (3\text{-}7\text{-}16)$$

$$\gamma_{N\,eff} = \gamma_{NL} + \gamma_{NW} \qquad (3\text{-}7\text{-}17)$$

$$\gamma_{NL} = \gamma_{NL1} + \Delta\gamma_{NL}\left(1 - \frac{L_o}{L}\right) \qquad (3\text{-}7\text{-}18)$$

$$\gamma_{NW} = \gamma_{NW1} + \Delta\gamma_{NW}\left(1 - \frac{W_o}{W}\right) \qquad (3\text{-}7\text{-}19)$$

$$\Gamma_{N\,eff} = \Gamma_{NL} + \Gamma_{NW} \qquad (3\text{-}7\text{-}20)$$

$$\Gamma_{NL} = \Gamma_{NL1} + \Delta\Gamma_{NL}\left(1 - \frac{L_o}{L}\right) \qquad (3\text{-}7\text{-}21)$$

$$\Gamma_{NW} = \Gamma_{NW1} + \Delta\Gamma_{NW}\left(1 - \frac{W_o}{W}\right) \qquad (3\text{-}7\text{-}22)$$

Here, γ_{NL1}, $\Delta\gamma_{NL}$, γ_{NW1}, $\Delta\gamma_{NW}$, Γ_{NL1}, $\Delta\Gamma_{NL}$, Γ_{NW1}, and $\Delta\Gamma_{NW}$ are empirical parameters extracted from the measured values of the threshold voltage and V_{TNo} is defined in eq. (3-7-6). This empirical model is used in designing CMOS devices and circuits. Nonetheless, simplified threshold voltage models provide valuable insight into dependencies on design and processing parameters.

Eq. (3-7-15) is valid for

$$2\varphi_b + V_F - V_{BS} \le \left(\frac{\gamma_{Neff}}{2\Gamma_{Neff}} \right)^2 \tag{3-7-23}$$

The limit of validity is obtained by equating the derivative of V_{TX} with respect to $(2\varphi_b + V_F - V_{BS})$ to zero. Beyond this limit, we assume that the threshold voltage remains constant at

$$V_{TM} = 2\varphi_b + V_{FB} + \Gamma_{Neff} \left(\frac{\gamma_{Neff}}{2\Gamma_{Neff}} \right)^2 \tag{3-7-24}$$

Fig. 3.7.9 shows a comparison of the variation of V_{TX} with the substrate-source bias between the present relationship and a precise two-dimensional calculation. As can be seen from the figure, the empirical equation is in excellent agreement with the two dimensional simulation. In particular, we notice a near constant value of V_{TX} when the substrate bias reaches a value close to that given by the limit of eq. (3-7-24).

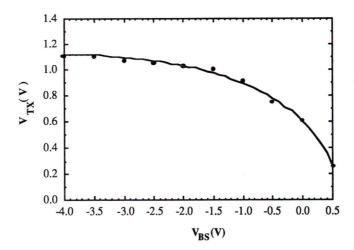

Fig. 3.7.9. Threshold voltage versus substrate-source bias of a short channel NMOS $(L = 1 \ \mu m, \ W = 20 \ \mu m, \ d_i = 220 \ \text{Å})$. Solid line – eq. (3-7-16), symbols – two dimensional device simulation. The parameter values used, $\gamma_{Neff} = 1.2 \ V^{1/2}$ and $\Gamma_{Neff} = 0.27$, are extracted as described in the text (from Rho et al. (1992)).

In the ideal case of a uniform bulk doping and a long channel device, the "effective" body effect constants reduce to: $\Gamma_{Neff} = 0$, and $\gamma_{Neff} = \gamma_N = (2\varepsilon_s qN_a)^{1/2}/c_i$. However, in the general case, γ_{Neff} and Γ_{Neff} can also be determined directly from measurements, and will include relevant information about non-uniform bulk doping and/or short channel effects. For the purpose of parameter extraction, eq. (3-7-15) can be re-written as follows:

$$\frac{V_{TX} - V_{FB} - 2\varphi_b}{\sqrt{2\varphi_b + V_F - V_{BS}}} = \gamma_{Neff} - \Gamma_{Neff}\sqrt{2\varphi_b + V_F - V_{BS}} \qquad (3\text{-}7\text{-}25)$$

Plotting the left hand side of this equation versus $(2\varphi_b + V_F - V_{BS})^{1/2}$, we find γ_{Neff} and Γ_{Neff} from the intercept and slope of this linear dependence, respectively.

3-8. CARRIER MOBILITY AND VELOCITY IN MOSFET CHANNELS

3.8.1. Velocity-Field Relationships

The carrier velocity in a MOSFET channel is a function of both the longitudinal electric field (in the direction from the drain to the source) and the transverse electric field (primarily determined by the gate-source bias). This velocity also depends on the temperature and the impurity concentration. In small devices, overshoot and ballistic effects make the velocity a non-local function of the electric field as well. Therefore, an accurate description of the carrier velocity in the channel is not an easy task. The simplest approach is to express the velocity in terms of parameters such as carrier mobility and saturation velocity which, in turn, may depend on the transverse electric field, the impurity concentration, the temperature, etc. A typical analytical approximation is given by (see Section 3.5)

$$v(F) = \frac{\mu F}{\left[1 + (F/F_s)^m\right]^{1/m}} \qquad (3\text{-}8\text{-}1)$$

Here, F is the absolute value of the longitudinal electric field in the channel, $F_s = v_s/\mu$ is the so-called saturation field, v_s is the saturation velocity and μ is the mobility. This dependence with $m = 2$ is in excellent agreement with experimental data for the electron velocity in bulk silicon for a wide range of impurity concentrations and device temperatures. It is also a good approximation for the electron velocity in n-channel MOSFETs provided that the gate length is not too small (greater than about 0.25 μm) and the electron mobility is taken to be a function of the effective transverse field. In eq. (3-8-1), μ is a function of the transverse electric field, but v_S is nearly independent of this field (see Thornber (1980) and Cooper and Nelson (1983)). However, it is difficult to obtain an analytical solution for MOSFET models using this expression. On the other hand, analytical solutions can easily be obtained using eq. (3-8-1) with $m = $ 1 (the so-called Trofimenkoff model). This approximation is reasonable for the velocity of holes in Si p-channel MOSFETs.

As was mentioned in Section 3.5, Sodini et al. (1984) proposed an approximation which yields results close to those obtained from eq. (3-8-1) with $m = 2$:

$$v(F) = \begin{cases} \dfrac{\mu F}{1 + \dfrac{F}{2F_s}} \, , & F < 2F_s \\[4mm] v_s \, , & F \geq 2F_s \end{cases}$$

$(3\text{-}8\text{-}2)$

Fig. 3.5.2 (Section 3.5) shows different velocity-field models for electrons and holes in silicon MOSFETs.

The approximation in eq. (3-8-2) is a widely used expression which allows us to obtain analytical model expressions for the current-voltage characteristics for many different types of FETs (see, for example, Sodini et al. (1984), Sheu et al. (1987), and Moon et al. (1990)). In Fig. 3.8.1, we compare calculated MOSFET I–V characteristics based on the different approximations for the velocity-field relationship shown in Fig. 3.5.2. As can be seen from the figure, Sodini's model agrees reasonably well, but Trofimenkov's model yields results which are very much different from a model using eq. (3-8-1) with $m = 2$. However, the comparison indicates that Sodini's model does not accurately reproduce the curvature of the NMOS I–V characteristics in the transition region (see Moon et al. (1991)). Our solution to this problem is described in Section 3.11.

As already mentioned, the hole velocity in the PMOS channel can be accurately described by eq. (3-8-1) with $m = 1$. This makes the modeling of p-channel MOSFETs distinctly different from the modeling of n-channel

MOSFETs, and somewhat easier because it is much more straightforward to find an analytical solution for such a velocity-field dependence (compare Section 3.10 (PMOS) and Section 3.11 (NMOS)).

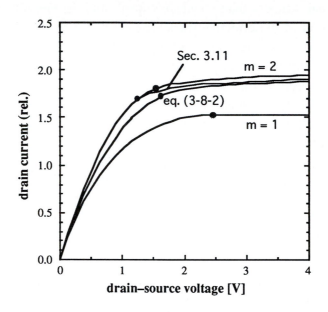

Fig.3.8.1. Comparison of I–V characteristics of a 1 μm NMOS based on the different velocity-field relationships shown in Fig. 3.5.2. The characteristic marked Sec. 3.11 is calculated by an analytical model discussed in Section 3.11. The open circles correspond to the saturation points. (From Moon et al. (1991), © 1991 IEEE.)

3.8.2. Mobility in the NMOS Channel

In existing Si MOSFET models, the dependence of the carrier mobility on the transverse electric field in the channel is usually described by the empirical relation

$$\mu = \frac{\mu_o}{1 + \theta(V_{GS} - V_T)} \tag{3-8-3}$$

where μ_o and θ are constants determined from experimental data, normally at zero substrate bias (see, for example, Klaassen (1976)), V_{GS} is the intrinsic gate-source voltage and V_T is the threshold voltage. However, as was first established by Sabnis and Clemens (1979), the electron channel mobility on the (100) surface, μ_n, is a universal function of the effective transverse field (see Section 3.3)

$$F_{eff} = \frac{F_s + F_B}{2} = \frac{q}{2\varepsilon_s}\left(2n_B + n_s\right) \tag{3-8-4}$$

defined in terms of the interface field

$$F_s = \frac{q(n_s + n_B)}{\varepsilon_s} \tag{3-8-5}$$

and the bulk field

$$F_B = \frac{qn_B}{\varepsilon_s} \tag{3-8-6}$$

where qn_B is the magnitude of the depletion sheet charge density of the bulk.

Eq. (3-8-4) can be understood as follows: The electric field at the semiconductor-insulator interface is equal to F_s while the field at the boundary of the inversion layer with the depletion layer is approximately equal to F_B. Hence, we can argue that the average effective field in the channel is close to $(F_s + F_B)/2$. Numerical calculations and experimental data prove this to be the case for electrons in the (100) inversion layers in Si MOSFETs (see Section 3.3).

The theoretical explanation of the electron and hole mobility dependencies on the effective transverse electric field was given by Lee et al. (1991). As they pointed out, the mobility determined by the surface scattering depends on i) the effective thickness of the inversion layer and ii) the relative population of the different valleys of the conduction band which may have different effective mobilities depending on their orientation with respect to the inversion layer. In the simplest case of the inversion layer in the (111) plane, the orientation of all six valleys is symmetrical with respect to this plane as can be seen from Fig. 3.8.2.

In this case, the surface scattering mobility is dependent only on the effective thickness of the inversion layer. As Lee et al. (1991) pointed out, the effective thickness of the two dimensional gas is dependent on the effective electric field given by

$$F_{eff1} \approx \frac{F_s + 2F_B}{3} \tag{3-8-7}$$

(A similar equation was derived by Stern (1972) based on a self-consistent solution of the Shrödinger and Poisson equations using a variational principle.) Hence, the field F_{eff1} represents the effective electric field determining the electron mobility, in good agreement with experimental data (see Lee et al. (1991)). The same arguments apply to holes in silicon which have only central valleys in the valence band.

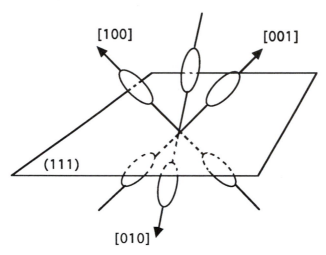

Fig. 3.8.2. Constant energy surfaces of the silicon conduction band relative to the (111) plane (see Fig. 3.3.15 for more details).

However, for the electrons on the (100) surface, the situation is different. As can be seen from Fig. 3.3.15 (Section 3.3), the inversion layer in this plane breaks the symmetry between the six lowest valleys of the silicon conduction band. In this case, the six valleys can be divided into two sets, one consisting of two equivalent valleys in the [100] direction, which is perpendicular to the inversion layer plane (we call these "transverse" valleys) and four valleys in the plane of the inversion layer (which we call "in-plane" valleys). The electrons in the transverse valleys have a transverse conductivity effective mass. This corresponds to a high mobility value for electrons in these valleys because the transverse effective mass is small.

At low transverse fields, electrons are equally divided between the six valleys. However, with a sufficiently high transverse electric field, the energy levels in the transverse valleys are much lower than those in the in-plane valleys because, as was discussed in Section 3.3, the effective mass determining the energy quantization is $m_\perp = m_l \gg m_t$ (see eq. (3-3-77)). Therefore, the relative population of electrons in a high transverse electric field is highest in the transverse valleys. This is the reason why the surface electron mobility along the (100) surface is higher than for other orientations.

As a consequence of having two "high mobility" valleys and four "low mobility" valleys, the electron mobility along the (100) surface is strongly dependent on the relative populations of these valleys which are determined by the energy levels given by eq. (3-3-77). Hence, to the first order, the effective field given by eq. (3-8-4) determines the universal mobility dependence for the electrons on the (100) surface because this effective field determines the subband energy levels, as was pointed out by Stern and Sarma (1984).

In the strong inversion regime, n_B is set equal to its maximum value

$$n_{B\,max} \approx \sqrt{2\varepsilon_s N_a (2\varphi_b - V_{BS})/q} \qquad (3\text{-}8\text{-}8)$$

where N_a is the acceptor density in the substrate, V_{BS} is the substrate-source bias, and $q\varphi_b$ is the bulk Fermi energy relative to the intrinsic Fermi level (see Section 3.3). In strong inversion, the electron sheet density in the channel is given by

$$n_s = c_a V_{GT} / q \qquad (3\text{-}8\text{-}9)$$

where $V_{GT} = V_{GS} - V_T$ and c_a is the gate-channel capacitance per unit area. Assuming zero surface charge, the threshold voltage can be written as (see Section 3.3)

$$V_T = V_{FB} + 2\varphi_b + \sqrt{2\varepsilon_s q N_a (2\varphi_b - V_{BS})}/c_i \qquad (3\text{-}8\text{-}10)$$

where V_{FB} is the flat band voltage and c_i is the insulator capacitance per unit area.

Substituting eqs. (3-8-5) to (3-8-10) into eq. (3-8-4), we obtain the following approximate expression for the effective transverse field in n-channel MOSFETs

$$F_{eff} \approx \frac{c_i}{2\varepsilon_s} (V_{GS} + V_T - V_{FB} - 2\varphi_b) \qquad (3\text{-}8\text{-}11)$$

In this derivation, we assumed that $c_a \approx c_i$, which is a good approximation except for devices with extremely thin gate oxides (less than 50 Å). Eq. (3-8-11) suggests that, to the first order, the mobility should be universally dependent on $V_{GS} + V_T$ because $V_{FB} + 2\varphi_b$ is only a logarithmic function of the substrate doping. (We also note that for modern n-channel Si MOSFETs with n^+ polysilicon gates, we have approximately $V_{FB} + 2\varphi_b \approx 0$ (see Section 3.2).)

The low-field electron mobility can be obtained by measuring the drain current, I_d, in the linear region, and determining the electron sheet density in the channel, n_s, from the measured gate-to-channel capacitance, C_{gc}, i.e., from

$$I_d (V_{DS} \to 0) = \frac{qW n_s \mu_n}{L} V_{DS} \tag{3-8-12}$$

and

$$q n_s = \int_{-\infty}^{V_{GS}} c_{gc} \, dV'_{GS} \tag{3-8-13}$$

Here, W and L are the gate width and the effective gate length, respectively, and V_{DS} is the drain-source voltage (typically about 10 mV for these measurements), see Section 3.3.

We now want to derive a universal expression for the electron mobility in terms of the gate–source voltage and the threshold voltage. By universality, we here mean that the dependence of mobility on device geometry and on substrate bias is implicitly included in the dependence of the threshold voltage on these parameters. This dependence was investigated experimentally by Park et al. (1991) for long channel NMOS and by Moon et al. (1991) for short channel NMOS. The measurements were performed in both the strong inversion and the subthreshold regimes, which helped to avoid the ambiguity associated with the definition and the determination of the threshold voltage. Their results are shown in Fig. 3.8.3 for devices with an oxide thickness of 220 Å.

As can be seen from this figure, the electron mobility is indeed nearly a universal function of $V_{GS} + V_T$. Furthermore, the following simple linear approximation yields excellent agreement with the experimental data:

$$\mu_n \approx \mu_{on} - \kappa_{1n} (V_{GS} + V_T) \tag{3-8-14}$$

The universal nature of the mobility shows that we can use the same parameters μ_{on} and κ_{1n} for different values of substrate bias. We also notice that the parameter values are fairly close even for devices with quite different gate lengths. All in all, this leads to a reduction in the number of parameters needed for accurate modeling of the MOSFET characteristics. The parameter values corresponding to the straight line in Fig. 3.8.3 are: $\mu_{on} = 625$ cm^2/Vs and $\kappa_{1n} = 28$ cm^2/V^{2}s.

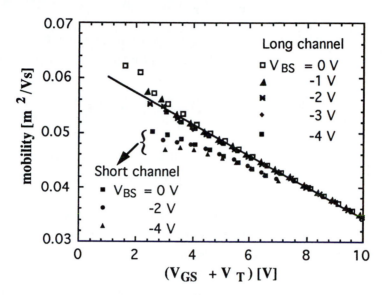

Fig.3.8.3. Electron mobility versus $V_{GS} + V_T$ for a long channel ($L = 20$ µm) and a short channel ($L = 1$ µm) NMOS for different values of substrate bias. The solid line indicates the linear approximation used in eq. (3-8-14). (From Park et al. (1991) and Moon et al. (1991), © 1991 IEEE.)

So far, we have only discussed the channel mobility at very low drain-source bias. However, it is clear that the transverse field, and thereby the mobility, will be influenced by the channel voltage, $V(x)$, caused by an applied drain-source voltage. As a first approximation, this can be taken into account by modifying eq. (3-8-14) as follows in the spirit of the gradual channel approximation (GCA):

$$\mu_n \approx \mu_{on} - \kappa_{1n} \left(V_{GS} + V_{TX} - V \right) \qquad (3\text{-}8\text{-}15)$$

In the frame of a simplistic charge control model, where we take $\alpha = 1$ (α is the parameter which enters in the unified charge control model (UCCM), see Section 3.4), the effective field is proportional to $V_{GS} + V_T - V$. However, in reality, the depletion charge depends on the channel potential which makes the threshold voltage a function of the electric potential as well. Hence, the effective field is, in fact, proportional to $V_{GS} + V_{TX} - V$ where the threshold voltage, V_{TX}, is dependent on the channel potential. As discussed in Section 3.4,

$$V_{TX} \approx V_T + (\alpha - 1)V \qquad (3\text{-}8\text{-}16)$$

where V_T is the threshold voltage close to the source and the parameter α accounts for the deviation from the gradual channel approximation. Hence, we obtain for the mobility

$$\mu_n = \mu_{on} - \kappa_{1n}(V_{GS} + V_T) + \kappa_{2n}V \qquad (3\text{-}8\text{-}17)$$

where $\kappa_{2n} \approx \kappa_{1n}(2-\alpha)$.

Usually, α is close to unity so that $\kappa_{2n} \approx \kappa_{1n}$. Hence, the effect of the drain-source bias is to increase the mobility in the channel. This is reasonable since the channel voltage counteracts the applied gate-source voltage and contributes to a reduction in the transverse electric field. Thereby, the average distance of the channel electrons from the interface increases, resulting in a reduced interface scattering rate and an increase in the mobility.

Finally, a note about the subthreshold regime. Below threshold, where diffusion current dominates, the channel voltage is small in most of the channel (except near drain, see the discussion in Section 3.6). Besides, the average distance of the electrons from the interface will be relatively large, also reducing the influence of the gate-source and the substrate-source bias voltages. Accordingly, in this regime, we can take the mobility to be a constant and equal to its value at threshold, i.e.,

$$\mu_{nsub} \approx \mu_{on} - 2\kappa_{1n}V_T \qquad (3\text{-}8\text{-}18)$$

(In reality, the mobility is known to decrease by about a factor of 2 or so in the subthreshold regime. However, this behavior is relatively unimportant since the drain current in this regime is primarily determined by the exponential dependence of the charge carrier density on the gate-source voltage.) In the practical implementation in the AIM-Spice we used eqs. (3-8-18) and (3-8-14) with a smoothing function to avoid possible problems with convergence.

3.8.3. Mobility in the PMOS Channel

As we discussed in the preceding Subsection, the hole mobility depends on the effective transverse electric field given by (see also Hairapetian et al. (1989))

$$F_{eff1} \approx \frac{F_s + 2F_B}{3} = -\frac{q}{\varepsilon_s}(n_B + p_s/3) \qquad (3\text{-}8\text{-}19)$$

The depletion charge sheet density, qn_B, includes a possible counter doping by boron implantation which is a usual procedure for PMOS (see Section 3.2).

Following the same line of argument as for the NMOS, we obtain the following approximate expression for the effective transverse field in p-channel MOSFETs (see Moon et al. (1991a))

$$|F_{eff1}| \approx \frac{c_i}{3\varepsilon_s}(V_{GS} + 2V_T - 6\varphi_b) \qquad (3\text{-}8\text{-}20)$$

which suggests a universal dependence of the hole mobility on $V_{GS} + 2V_T$. This dependence was investigated experimentally by Moon et al. (1991a) for long ($L = 20$ μm) and short channel ($L = 3$ μm and $L = 1.4$ μm) PMOS devices with an oxide thickness of 205 Å. The measurements were performed by the same technique as indicated in Subsection 3.8.2. A plot of the inverse mobility versus $V_{GS} + 2V_T$ is shown in Fig. 3.8.4.

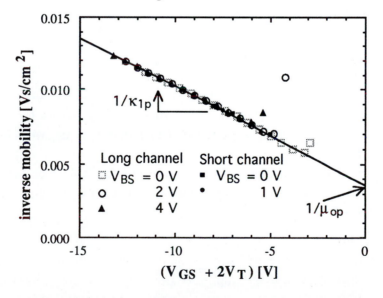

Fig.3.8.4. Inverse hole mobility versus $V_{GS} + 2V_T$ for long channel and short channel PMOS devices for different values of substrate bias. The solid line indicates the linear approximation used in eq. (3-8-21). (From Moon et al. (1991a), © 1991 IEEE.)

This figure indicates that the inverse hole mobility can be described approximately by the following linear relationship

$$\frac{1}{\mu_p} \approx \frac{1}{\mu_{op}} - \frac{1}{\kappa_{1p}}(V_{GS} + 2V_T)$$ (3-8-21)

The parameter values corresponding to the straight line in Fig. 3.8.4 are: $\mu_{op} = 279 \text{ cm}^2/\text{Vs}$ and $\kappa_{1p} = 1,508 \text{ cm}^2/\text{V}^2\text{s}$. Eq. (3-8-21) is equivalent to the dependence of mobility on F_{eff1} given by Aurora and Richardson (1989):

$$\mu_p = \frac{\mu^*}{1 + \beta F_{eff1}}$$ (3-8-22)

where μ^* is the zero field mobility and β is a constant.

As in the case of NMOS, the effect of the drain-source voltage on the PMOS mobility is obtained by invoking the gradual channel approximation and using the position dependent threshold voltage, V_{TX} (see eqs. (3-8-15) to (3-8-17)). The resulting expression for the hole mobility, including its dependence on the channel voltage, is as follows:

$$\frac{1}{\mu_p} = \frac{1}{\mu_{op}} - \frac{V_{GS} + 2V_T}{\kappa_{1p}} + \frac{V(x)}{\kappa_{2p}}$$ (3-8-23)

where $\kappa_{2p} \approx \kappa_{1p}/(3-2\alpha)$. Usually, α is close to unity such that $\kappa_{2p} \approx \kappa_{1p}$. Hence, the effect of the drain-source bias is the same as for NMOS, i.e., to increase the mobility in the channel (note that V is negative in a p-channel MOSFET).

As for the n-channel MOSFET, we take the hole mobility in the PMOS channel to be a constant in the subthreshold regime and equal to its value at threshold, i.e.,

$$\frac{1}{\mu_{psub}} = \frac{1}{\mu_{op}} - \frac{3V_T}{\kappa_{1p}}$$ (3-8-24)

3-9.   UNIVERSAL MODELING OF MOSFETs (AIM-Spice MODEL MOSA1)

As was mentioned in Section 3.4, the simple MOSFET models need improvements in several areas, especially for applications to modern submicron devices. Most importantly, the subthreshold current (unaccounted for in the basic models) must be considered. The subthreshold current is of great concern since it has consequences for the bias and logic levels in digital operations as well as for the holding time in dynamic circuits, as was discussed in Section 3.7. It also affects the power dissipation in logic circuits, especially at the VLSI scale.

In this Section, we upgrade the basic analytical model in order to describe both the below and above-threshold regimes of device operation. The description is based on a charge control model which uses one unified expression for the effective differential channel capacitance. Our approach follows the same basic philosophy used to derive the UCCM (Unified Charge Control Model) equations for MOSFETs (see Sections 3.3 and 3.4). The upgraded model also accounts for series drain and source resistances, velocity saturation in the channel, finite output conductance in the saturation regime, and for the threshold voltage shift due to drain bias induced lowering of the injection barrier between the source and the channel (DIBL). The model parameters, such as the effective channel mobility, the saturation velocity, the source and drain resistances, etc. are extractable from experimental data. This makes the model very suitable for incorporation into our circuit simulator, AIM-Spice (model MOSA1). It is also quite useful for MOSFET design and characterization. We apply the characterization procedure based on this model to a MOSFET with a quarter micron gate length and obtain excellent agreement with experimental data. The basic ideas used in developing this model are applied to the development of more accurate but also more complicated models described in Sections 3.10 and 3.11.

The approach developed in this Section can also be applied for modeling other field effect transistors, such as GaAs Metal Semiconductor Field Effect Transistors (see Section 4.4) and Heterostructure Field Effect Transistors (see Section 4.6). This is why we call this model "universal".

3.9.1. Unified Channel Capacitance

As mentioned above, the charge control in the MOSFETs can be described in terms of a unified expression for the channel capacitance, valid for all bias voltages. As was discussed in Section 3.3, the differential gate-channel capacitance per unit area can be written as (see also Jeng et al. (1988))

$$c_{gc} = \frac{c_a c_b}{c_a + c_b} \qquad (3\text{-}9\text{-}1)$$

where

$$c_a = \frac{\varepsilon_i}{d_i + \Delta d} \qquad (3\text{-}9\text{-}2)$$

is the contribution from the strong inversion regime,

$$c_b = \frac{qn_o}{\eta V_{th}} \exp\left(\frac{V_{GT}}{\eta V_{th}}\right) \qquad (3\text{-}9\text{-}3)$$

is the subthreshold contribution, and

$$n_o = \frac{\varepsilon_i \eta V_{th}}{2q(d_i + \Delta d)} \qquad (3\text{-}9\text{-}4)$$

is the sheet density of inversion charge at threshold.

A unified carrier sheet charge density can now be expressed in terms of the effective, unified differential channel capacitance. For an *n*-channel MOSFET at very low drain-source bias (well below saturation), we find (see Section 3.3):

$$qn_s = 2qn_o \ln\left[1 + \frac{1}{2}\exp\left(\frac{V_{GT}}{\eta V_{th}}\right)\right] \qquad (3\text{-}9\text{-}5)$$

Since the drain current is assumed to be small, the intrinsic gate voltage swing, V_{GT}, in this expression can be replaced by its extrinsic counterpart, $V_{gt} = V_{GT} + I_d R_s$, where I_d is the drain current and R_s is the source series resistance.

The capacitance model described above can be used for calculating the channel contribution to the gate-source capacitance, C_{gs}, and the gate-drain capacitance, C_{gd}, by using an approximation similar to that used in the model by Meyer (1971):

$$C_{gs} = C_f + \frac{2}{3}C_{gc}\left[1 - \left(\frac{V_{sate} - V_{dse}}{2V_{sate} - V_{dse}}\right)^2\right] \qquad (3\text{-}9\text{-}6)$$

$$C_{gd} = C_f + \frac{2}{3} C_{gc} \left[1 - \left(\frac{V_{sate}}{2V_{sate} - V_{dse}} \right)^2 \right] \tag{3-9-7}$$

Here, $C_{gc} = c_{gc}LW$, $V_{sate} = I_{sat}/g_{ch}$ is the effective extrinsic saturation voltage, I_{sat} is the saturation drain current, g_{ch} is the extrinsic channel conductance at low drain-source bias (see Subsection 3.9.2), and V_{dse} is an effective extrinsic drain-source voltage. V_{dse} is equal to V_{ds} for $V_{ds} < V_{sat}$ and is equal to V_{sate} for $V_{ds} > V_{sate}$. In order to obtain a smooth transition between the two regimes, we interpolate V_{dse} by the following equation:

$$V_{dse} = V_{ds} \left[1 + \left(\frac{V_{ds}}{V_{sate}} \right)^{m_c} \right]^{-1/m_c} \tag{3-9-8}$$

where m_c is a constant determining the width of the transition region between the linear and the saturation regime. The capacitance C_f in eqs. (3-9-6) to (3-9-7) is the side wall and fringing capacitance which can be estimated, in terms of a metal line of length W, as

$$C_f \approx \beta_c \, \varepsilon_s W \tag{3-9-9}$$

where β_c is on the order of 0.5 (see Gelmont et al. (1991)).

The source-substrate, drain-substrate, and gate-substrate capacitances also affect MOSFET performance in a circuit. The AIM-Spice model MOSA1 uses standard expressions for these capacitances which can be found, for example, in Shur (1990), p. 393.

3.9.2. Current-Voltage Characteristics

At drain-source voltages well below the saturation voltage, the drain current can always (i.e., at both above and below threshold) be expressed as

$$I_{ds} = g_{chi} V_{DS} = g_{ch} V_{ds} \tag{3-9-10}$$

where the intrinsic channel conduction in the linear region, g_{chi}, is related to its extrinsic counterpart, g_{ch}, as follows:

$$g_{ch} = \frac{g_{chi}}{1 + g_{chi} R_t} \tag{3-9-11}$$

Here, $R_t = R_s + R_d$ is the sum of the source and the drain series resistances. (In deriving eq. (3-9-11), we assumed that the drain current is so small that the transconductance is much less than the channel conductance, see Section 3.4). The intrinsic channel conductance can be written as

$$g_{chi} = \frac{q n_s W \mu_n}{L} \qquad (3\text{-}9\text{-}12)$$

where n_s is the unified carrier sheet density at low drain-source voltage given by eq. (3-9-5) and μ_n is the low-field carrier mobility. In eq. (3-9-10), V_{DS} and V_{ds} are the intrinsic and extrinsic drain-source voltages, respectively, which are related by $V_{ds} = V_{DS} + I_d R_t$. We note that, in the subthreshold regime, the channel resistance will normally be much larger than the parasitic source and drain resistances such that $g_{ch} \approx g_{chi}$, $V_{ds} \approx V_{DS}$ and $V_{gt} \approx V_{GT}$.

The MOSFET drain saturation current for the above-threshold regime was derived in Section 3.5 for the two-piece velocity saturation model. This current can be expressed as

$$I_{sat} = \frac{g_{chi} V_{gt}}{1 + g_{chi} R_s + \sqrt{1 + 2 g_{chi} R_s + \left(V_{gt} / V_L \right)^2}} \qquad (3\text{-}9\text{-}13)$$

Here, $V_L = F_s L$ where F_s is the saturation field and L is the gate length.

In order to link the linear and the saturation regions by one single expression, we propose to use the following basic interpolation formula for our model:

$$I_{ds} = \frac{g_{ch} V_{ds}}{\left[1 + \left(g_{ch} V_{ds} / I_{sat} \right)^m \right]^{1/m}} \qquad (3\text{-}9\text{-}14)$$

where m is a parameter that determines the shape of the characteristics in the knee region.

We notice that this model can utilize different approximations for I_{sat}. For example, we can use the model proposed by Sodini et al. (1984) and utilized in the popular MOS model BSIM (see Jeng et al. (1987)). Still another alternative is to use the empirical model proposed by Sakurai et al. (1991)   ($I_{sat} = BW(V_{gs} - V_T)^n / L$ where B and n are empirical fitting parameters). In this case, one can use the analytical equations for the propagation delays of logic gates by Sakurai et al. (1991).

We now want to improve this basic model to account for various non-ideal physical mechanisms such as subthreshold current, drain bias induced barrier lowering (DIBL), and other factors that contribute to a finite output conductance in saturation.

The subthreshold I–V characteristics of the MOSFET were discussed in Section 3.6. In the long channel limit, these characteristics can be expressed as

$$
\begin{aligned}
I_d &\approx qWD_n \frac{n_s(0) - n_s(L)}{L} \\
&= \frac{qWD_n n_o}{L} \exp\left(\frac{V_{gt}}{\eta V_{th}}\right)\left[1 - \exp\left(-\frac{V_{ds}}{V_{th}}\right)\right]
\end{aligned}
\tag{3-9-15}
$$

where $D_n = \mu_n V_{th}$ is the carrier diffusion coefficient. From eq. (3-9-15), we find the following subthreshold saturation current:

$$
I_{sat} = \frac{qW\mu_n V_{th} n_o}{L} \exp\left(\frac{V_{gt}}{\eta V_{th}}\right)
\tag{3-9-16}
$$

(The corresponding saturation voltage is of the order of the thermal voltage, V_{th}.)

A unified expression for the saturation current, valid both above and below threshold, can be obtained by substituting the gate voltage swing V_{gt} shown explicitly in eq. (3-9-13) by an equivalent gate voltage swing which coincides with V_{gt} well above threshold and is equal to $2V_{th}$ below threshold (note that the implicit dependence of g_{chi} on V_{gt} through eqs. (3-9-12) and (3-9-5) must be retained):

$$
V_{gte} = V_{th}\left[1 + \frac{V_{gt}}{2V_{th}} + \sqrt{\delta^2 + \left(\frac{V_{gt}}{2V_{th}} - 1\right)^2}\,\right]
\tag{3-9-17}
$$

Here, δ determines the width of the transition region. Typically, $\delta = 3$ is a good choice. The unified saturation current

$$
I_{sat} = \frac{g_{chi}\, V_{gte}}{1 + g_{chi}\, R_s + \sqrt{1 + 2g_{chi}\, R_s + \left(V_{gte}/V_L\right)^2}}
\tag{3-9-18}
$$

reverts to eq. (3-9-13) above threshold and reduces to eq. (3-9-16) below threshold. Hence, using eq. (3-9-18) for the saturation current, eq. (3-9-14) becomes a unified expression for the *I–V* characteristics with the correct limiting behavior in both the above and below threshold regimes. (We emphasize that, in contrast to Section 3.5, we are now expressing g_{chi} in terms of the unified electron sheet charge density, n_s (see eqs. (3-9-5) and (3-9-12)).)

An important effect which must be taken into account for an adequate description of the MOSFET behavior is the dependence of the threshold voltage on the drain-source voltage, as discussed in Section 3.6. This dependence can be fairly accurately described by the equation

$$V_T = V_{To} - \sigma V_{ds} \tag{3-9-19}$$

where V_{To} is the threshold voltage at zero drain-source voltage and σ is a coefficient which may depend on the gate voltage swing. This translates into the following shift in V_{gt}:

$$V_{gt} = V_{gto} + \sigma V_{ds} \tag{3-9-20}$$

where V_{gto} is the gate voltage referred to threshold at zero drain-source bias. The effects of the drain bias induced threshold voltage shift can be accounted for in the *I–V* characteristics of eq. (3-9-14) by using eq. (3-9-20) in the expressions for the saturation current and the linear channel conductance. This will result in a finite output conductance in the saturation region. However, as discussed in Section 3.6, σ will depend on the gate voltage swing since the effect of drain bias induced barrier lowering (DIBL) will vanish well above threshold. Hence, for modeling purposes, we adopted the following empirical expression for σ (see Section 3.6):

$$\sigma = \frac{\sigma_o}{1 + \exp\left(\dfrac{V_{gto} - V_{\sigma t}}{V_\sigma}\right)} \tag{3-9-21}$$

which gives $\sigma \to \sigma_o$ for $V_{gto} < V_{\sigma t}$ and $\sigma \to 0$ for $V_{gto} > V_{\sigma t}$. The voltage V_σ determines the width of the transition between the two regimes.

For a more accurate description of the saturation region, it is possible to include an additional empirical factor, $(1 + \lambda V_{ds})$, in the expression for the *I–V* characteristics (as usually done in SPICE FET models). The parameter λ is related to physical effects that are not explicitly included in our model such as, for example, gate length modulation. It is important to emphasize that λ is a strong function of the gate length, the aspect ratio, the contact depth, and other factors. As a rule of thumb (and based on the analysis given in Section 3.10), we can assume $\lambda \approx k_o/L$ where k_o is a constant. Hence, our model I–V characteristics can finally be expressed as

$$I_d = \frac{g_{ch} \, V_{ds} \, (1 + \lambda V_{ds})}{\left[1 + (V_{ds} / V_{sate})^m \right]^{1/m}}$$

(3-9-22)

where

$$V_{sate} \approx I_{sat}/g_{ch}$$

(3-9-23)

is the effective, extrinsic saturation voltage (the same as used in eqs. (3-9-6) to (3-9-8)). When the experimental I–V characteristics have a finite output conductance, g_{chs}, in saturation, the saturation current is conveniently defined, for the purpose of comparison with the present model, as the intersection of the linear asymptotic behavior at small drain-source voltages $(I_d = g_{ch}V_{ds})$ and in deep saturation $(I_d = g_{ch}V_{sate} + g_{chs}V_{ds})$. Hence, we can express V_{sate} in terms of the saturation current as: $V_{sate} = (1 - g_{chs}/g_{ch})I_{sat}/g_{ch}$. Normally, for well-behaved transistors, $g_{chs} \ll g_{ch}$ and we can use eq. (3-9-23). However, at sufficiently large gate voltage swing, i.e., for $V_{gto} > V_{\sigma t} + 3V_{\sigma}$, the effects of the drain voltage induced shift in the threshold voltage vanishes and we obtain $g_{chs} \approx \lambda g_{ch}V_{sate}$ in deep saturation. This can be used to determine λ as part of the device characterization and to obtain the corrected relationship $V_{sate} = (g_{ch}/I_{sat} + \lambda)^{-1}$ to be used in eq. (3-9-22) well into strong inversion.

The quality of the present unified MOSFET model is illustrated in Figs. 3.9.1 and 3.9.2 for a deep submicrometer NMOS. Fig. 3.9.1 shows the above-threshold I–V characteristics and Fig. 3.9.2 shows the subthreshold characteristics in a semilog plot. As can be seen from the figures, our model quite accurately reproduces the experimental data in the entire range of bias voltages, over several decades of the current variation. A residual drain current at large negative gate-source voltages is not accounted for in our model. This current is caused by a small leakage of the reverse biased p-n drain-substrate junction. The parameters used in this calculation were obtained using the parameter extraction procedure described below (see Subsection 3.9.3).

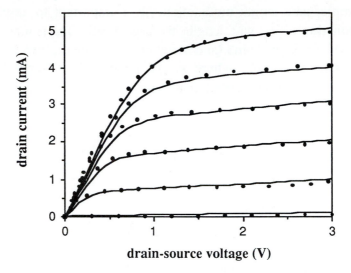

Fig. 3.9.1. Above-threshold experimental (symbols) and calculated (solid lines) *I–V* characteristics for a deep submicrometer NMOS with effective gate length $L = 0.25$ μm. Device parameters are: effective gate width $W = 10$ μm, oxide thickness $d_i = 5.6$ nm, mobility $\mu_n = 0.028$ m^2/Vs, saturation velocity $v_s = 4 \times 10^4$ m/s, knee parameter $m = 4.0$, source and drain series resistances $R_s = R_d = 75$ ohm, output conductance parameter $\lambda = 0.048$ V^{-1}, subthreshold ideality factor $\eta = 1.32$, threshold voltage at zero drain-source bias $V_{T_0} = 0.44$ V, threshold voltage parameters $\sigma_o = 0.048$, $V_\sigma = 0.2$ V, $V_{\sigma t} = 1.7$ V. Substrate-source bias $V_{bs} = 0$ V. Top curve: $V_{gs} = 3.0$ V, step: –0.5 V. (From Fjeldly et al. (1991a) and Shur et al. (1992). Data from Chung et al. (1991).)

The advantage of the model described in this Section is its simplicity and applicability to short channel devices. However, it is also important to point out the limitations of this simple model. As was shown in Section 3.4, the output conductance of long channel devices in the saturation regime is equal to zero (ideal saturation). The only way to approximate such an ideal saturation using eq. (3-9-14) (or eq. (3-9-22) with $\lambda = 0$) is to choose a large value of the shape parameter m. However, this leads to a very abrupt transition from the linear to the saturation regime in the computed current-voltage characteristics. In order to obtain a model which describes accurately devices with all channel lengths, we may choose to employ different equations for the linear and the saturation

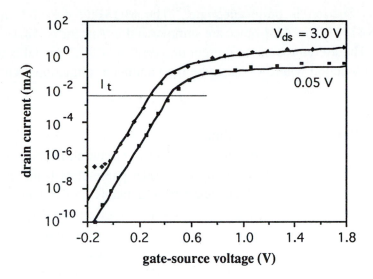

Fig. 3.9.2. Subthreshold experimental (symbols) and calculated (solid lines) *I–V* characteristics for a deep submicrometer NMOS (the same as shown in Fig. 3.9.1). I_t denotes the threshold current. (From Fjeldly et al. (1991a) and Shur et al. (1992).)

regimes while preserving the continuity of the drain current and its derivatives with respect to the terminal voltages. Such an approach is considered in Sections 3.10 and 3.11.

The dependence of the field effect mobility on the gate-source voltage can be accounted for within the framework of the model considered in this Section. However, the <u>effective</u> low field mobility may also depend on the drain-source voltage because the channel potential affects the surface electric field and, hence, the surface scattering, as was discussed in Section 3.8. (This will be considered in more detail in Sections 3.10 and 3.11.) To some extent, these effects can be indirectly accounted for in the present model by adjusting the values of the parameters λ and *m*. However, the ability of the model to reproduce these effects is limited. Here again, the more sophisticated models of Sections 3.10 and 3.11 become appropriate, especially since these models have already been implemented in AIM-Spice and, hence, the reader does not have to face the task of implementing the complex expressions in a circuit simulator. However, it is hard to compete with simplicity and with a set of equations that is easy to understand. Besides, for most practical modern day CMOS technologies, the model described in this Section is quite adequate. The MOSFET Level-1 model equations (MOSA1) used in AIM-Spice are summarized in Appendix A8.3.

The usefulness of the present universal MOSFET model is illustrated in the next subsection where we describe the parameter extraction technique using this model.

3.9.3. Parameter Extraction

In order to apply the present model in circuit simulation, it is necessary to find a practical way of extracting the model parameters from experimental data. The accuracy of the parameter extraction depends on the accuracy of the model and on the accuracy and completeness of the experimental data.

In real life, there are usually problems with both the model and the experimental data. Even though the analytical models developed in this Chapter reflect many important effects occurring in MOSFETs, such as velocity saturation, parasitic series resistances, the dependence of the threshold voltage on the drain bias, and finite output conductance, many other important phenomena are included indirectly, at best. For example, our model treats the dependence of the mobility on the bias conditions only through effective values and, indirectly, through the fitting parameter λ. Furthermore, many complicated phenomena associated with the threshold voltage dependence on the drain-source voltage are accounted for in a set of simple, empirical formulae (see eqs. (3-9-19) to (3-9-21)).

Hence, our approach to the device characterization of MOSFETs is to extract effective device parameters, such as the threshold voltage, V_{To}, the mobility, μ_n, the saturation velocity, v_s, the subthreshold ideality factor, η, the source and drain series resistances, R_s and R_d, the output conductance parameter, λ, the knee shape parameter, m, the threshold voltage coefficient, σ_o, and the voltages $V_{\sigma t}$ and V_σ characterizing the dependence of parameter σ on the gate voltage swing. In other words, we describe a parameter extraction procedure which gives fairly reasonable results without parameter adjustment and which can produce a nearly perfect fit with some parameter adjustment. We note that, even without a precise knowledge of the device geometry and process parameters, we can obtain a good fit to the device characteristics and, hence, predict the circuit performance fairly accurately using the AIM-Spice simulation program. Moreover, we still retain a limited ability to predict how device characteristics will scale with changes in the doping level and profile and device geometry. In other words, our model and parameter extraction procedure attempt to give the best possible answer based on the available information.

As an example, we perform the parameter extraction for the device with the characteristics shown in Figs. 3.9.1 and 3.9.2. We start by extracting the drain bias dependence of the threshold voltage from the measurements of the drain current versus gate-source voltage for different drain-source voltages. Plotting the results of these measurements in a semilog scale, as illustrated in Fig. 3.9.2, we can determine ηV_{th} from the slope of the characteristics in the subthreshold regime. We then estimate the threshold current in saturation, I_t, from eq. (3-9-16) using eq. (3-9-4) for electron sheet density at threshold, i.e.,

$$I_t = \frac{\varepsilon_i \mu_n W \eta V_{th}^2}{2(d_i + \Delta d) L} \qquad (3\text{-}9\text{-}24)$$

At this point, we do not need to know accurate values for the electron mobility, μ_n, the effective channel length, L, and the oxide thickness, d_i, because the threshold voltage is fairly insensitive to the exact value of I_t. (For our device, $I_t \approx 2 \times 10^{-6}$ A.) The intercept of the measured curves with the line $I_d = I_t$ yields the values of the threshold voltage versus drain-source voltage (see Fig. 3.9.2). A linear interpolation of this dependence yields the parameter σ_o, as illustrated earlier in Fig. 3.6.6 for devices from the same fabrication line as the one discussed here. For the present device, we find $V_{To} \approx 0.44$ V and $\sigma_o \approx 0.048$.

The next step is to find the shape parameter, m, for the knee region of the characteristics. For near ideal characteristics, with low output conductance in saturation, the extraction method is as indicated in Fig. 3.9.3. Each characteristic is approximated by two linear pieces, one describing correctly the linear region at small drain-source voltages (slope g_{ch}) and one approximating the characteristic in the saturated region (slope g_{chs}). The linear pieces intersect at the saturation current, I_{sat}, at the drain-source voltage $V_{ds} = I_{sat}/g_{ch} \approx V_{sate}$ (see the discussion in Subsection 3.9.2). According to eq. (3-9-22), the real current at this drain-source voltage is $I_d = I_{sat} - \Delta I \approx I_{sat}/2^{1/m}$. Hence, by locating the intersection in the two-piece linear model applied to the experimental characteristic, as shown in Fig. 3.9.3, and measuring the currents I_{sat} and $I_{sat} - \Delta I$, we can determine m as follows:

$$m \approx \frac{\ln(2)}{\ln[I_{sat}/(I_{sat} - \Delta I)]} \qquad (3\text{-}9\text{-}25)$$

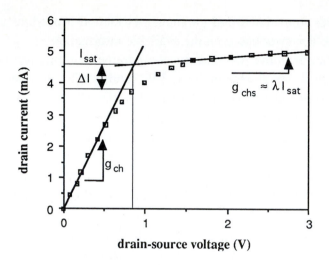

Fig. 3.9.3. Method for extracting the shape parameter m, the saturation current, I_{sat}, and the parameter λ. The I–V characteristic is taken from Fig. 3.9.1 ($V_{gs} = 3.0$ V). (From Shur et al. (1992).)

For the curve shown in Fig. 3.9.3, we obtain $m \approx 4$.

From the output conductance in deep saturation, g_{chs}, for the largest value of V_{gs}, we also determine the value of parameter λ from the expression

$$\lambda = \frac{g_{chs}}{g_{ch}V_{sate}} \approx \frac{g_{chs}}{I_{sat}} \tag{3-9-26}$$

(see the discussion in Subsection 3.9.2). From Fig. 3.9.3, we find $\lambda \approx 0.048$ V^{-1}.

The next step involves the determination of the total series resistance, $R_t = R_s + R_d$, and the transconductance parameter

$$\beta = \frac{\varepsilon_i \mu_n W}{L(d_i + \Delta d)} \tag{3-9-27}$$

(see Section 3.4). In the above-threshold regime, the intrinsic, linear channel conductance can be expressed as follows in terms of β: $g_{chi} \approx \beta V_{GT} \approx \beta V_{gt}$. Hence, we can write

$$\frac{1}{g_{ch}} = \frac{1}{g_{chi}} + R_t \approx \frac{1}{\beta V_{gt}} + R_t \tag{3-9-28}$$

This allows β and R_t to be determined from the slope and the intercept with $1/V_{gt} = 0$ by plotting $1/g_{ch}$ versus $1/V_{gt}$. Such a plot is shown in Fig. 3.9.4 for the NMOS device under investigation, resulting in the following parameter values: $R_t \approx 150$ ohm and $\beta \approx 0.011$ A/V^2. However, we note that since the series resistance in this short channel MOSFET dominates the overall device resistivity in the linear regime, the determination of β by this method is somewhat uncertain. Besides, this technique of extracting β emphasizes the properties of the *I–V* characteristics in the linear regime while, in fact, the properties of saturation region are of greater interest in digital applications. Therefore, we return below with alternative ways of estimating β.

For MOSFETs, we normally have $R_s \approx R_d \approx R_t/2$. If there is an asymmetry between the source and drain, an accurate estimate of the difference between R_s and R_d can be made by measuring the normal, extrinsic device transconductance in the saturation regime, g_m, and its counterpart, g_{mr}, with the source and drain interchanged (at constant drain current). Since the intrinsic transconductance is the same in both cases, we find

$$R_s - R_d = \frac{1}{g_m} - \frac{1}{g_{mr}} \tag{3-9-29}$$

In our case, however, we used $R_s = R_d = R_t/2 = 75$ ohm.

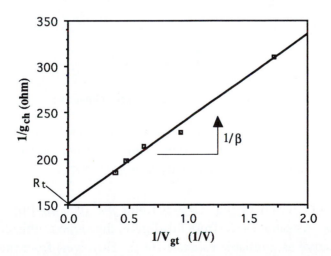

Fig. 3.9.4. Plot of inverse channel conductance versus inverse gate voltage swing, used for the determination of the parameters R_t and β. (From Shur et al. (1992).)

Fig. 3.9.5 shows the dependence of the above-threshold saturation current, I_{sat}, on the gate-source voltage. From a suitable interpolation formula representing the experimental data, we can determine the extrinsic transconductance in saturation, g_m, as a function of V_{gs} by differentiation. The intrinsic transconductance is subsequently obtained as $g_{mi} = g_m/(1 - g_m R_s)$. According to our model (see Section 3.5, eq. (3-5-12)), we can write

$$\frac{1}{g_{mi}^2} = \frac{1}{\beta^2 V_{GT}^2} + \frac{1}{\beta^2 V_L^2} \tag{3-9-30}$$

where $V_{GT} = V_{gt} - R_s I_{sat}$. Hence, by plotting $1/g_{mi}^2$ versus $1/V_{GT}^2$, we obtain β from the slope and βV_L from the intercept with the $1/g_{mi}^2$ axis (at $1/V_{GT}^2 = 0$).

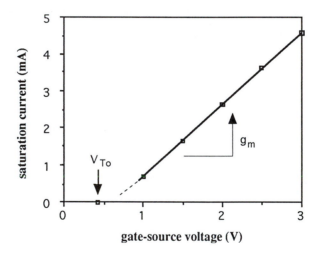

Fig. 3.9.5. Measured saturation current (defined in Fig. 3.9.3) versus V_{gs} for the same device as in Fig. 3.9.1. This plot is used to determine the extrinsic g_m in saturation. (From Shur et al. (1992).)

A plot based on eq. (3-9-30) and the data in Fig. 3.9.5 is shown in Fig. 3.9.6. As can be seen from this figure, the slope in this particular case is too small to serve as a reliable estimate for β. However, from the intercept, we find that $\beta V_L \approx 0.026$ A/V which corresponds to a saturation velocity $v_s \approx 4\times10^4$ m/s. This value for v_s seems to be too low compared to the well established measured value of $v_s \approx 8\times10^4$ m/s (see, for example, Chan et al. (1990)).

However, we checked that the values of v_s extracted from these experimental data by other models, such as the model presented in Section 3.11 and the model by Sodini et al. (1984), are quite similar. This may mean that there could be some uncertainty in the device parameters given by Chung et al. (1991).

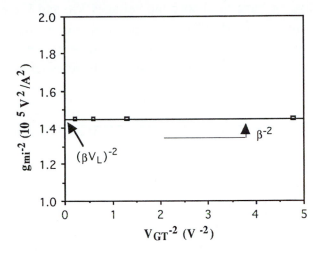

Fig. 3.9.6. Plot for the determination of the parameters β and V_L.
(From Shur et al. (1992).)

An alternative way to crudely estimate β from the experimental data (often used in literature) is to consider the saturation current slightly above threshold where $(V_{GT}/V_L)^2 \ll 1$. In this case, eq. (3-5-12) can be approximated by

$$I_{sat} = \beta V_L^2 \left[\sqrt{1 + \left(\frac{V_{GT}}{V_L}\right)^2} - 1 \right] \approx \frac{\beta}{2} V_{GT}^2 \qquad (3\text{-}9\text{-}31)$$

and $(\beta/2)^{1/2}$ is found as the slope of the plot of $(I_{sat})^{1/2}$ versus V_{GT} ($\approx V_{gt}$ in this regime). (It may also suffice to measure I_{sat} for one value of V_{gt} sufficiently close to threshold). However, this technique is not very accurate in short channel devices where the region $V_{GT} \ll V_L$ is very narrow and may be affected by subthreshold current caused by short channel effects.

Using the various techniques described above, we arrive at the following parameter values used to fit the experimental I–V characteristics in Figs. 3.9.1 and 3.9.2: $\beta \approx 0.0067$ A/V^2 and $V_L \approx 0.36$ V, which correspond to $v_s \approx 4 \times 10^4$ m/s and $\mu_n \approx 0.028$ m^2/Vs. However, we note that a higher value of β

(and μ_n), i.e., more in accordance with the values obtained from the slope in Fig. 3.9.4, would have given a better fit in the linear part of the $I-V$ characteristics, while the present values favor a good fit in the saturation region (which is of primary importance in digital applications).

Finally, we have to choose the voltages $V_{\sigma t}$ and V_σ which determine the dependence of the threshold voltage parameter σ on the gate-source voltage. Here, we have chosen $V_\sigma \approx 0.2$ V and $V_{\sigma t} = V_{To} + 0.5(V_{gt})_{max} \approx 1.7$ V. However, these parameter values can be slightly adjusted for a better fit.

3-10. CURRENT-VOLTAGE CHARACTERISTICS OF PMOS (AIM-Spice MODELS)

With this Section, we start a presentation of advanced CMOS models where we bring together some of the advanced features of MOSFET modeling discussed in the preceding Sections of this Chapter. These features include unification of the above-threshold and the subthreshold regimes through the Unified Charge Control Model (UCCM), modeling of the saturation regime, modeling of the velocity-field relationship in the non-saturated regime, modeling of the dependencies of the channel mobility and the threshold voltage on the bias voltages, incorporation of the effects of the source and drain series resistances, and last but not least, tractable characterization procedures for the extraction of the model parameters. PMOS is discussed first since it is easier to model than NMOS. After a short review of basic equations, we describe in succession several PMOS models with varying degrees of sophistication. This discussion of p-channel MOSFETs is partly based on the work by Moon et al. (1991a) and Park (1992), where additional information can be found.

3.10.1. Basic Assumptions and Equations

The Unified Charge Control Model for the p-channel MOSFET can be expressed as follows (in full analogy with the n-channel MOSFET discussed in Sections 3.4 and 3.6):

$$- V_{GT} + \alpha V_F \approx \eta V_{th} \ln\left(\frac{p_s}{p_o}\right) + a(p_s - p_o) \qquad (3\text{-}10\text{-}1)$$

Here, $V_{GT} = V_{GS} - V_T$ is the intrinsic gate voltage swing (gate-source voltage,

V_{GS}, relative to the threshold voltage, V_T), V_F is the Fermi potential relative to that at the source electrode, p_s is the hole sheet density in the channel and p_o is the same at threshold, V_{th} is the thermal voltage, η is the subthreshold ideality factor, α is a parameter accounting for the threshold voltage dependence on the channel potential, and $a = q(d_i + \Delta d)/\varepsilon_i = q/c_a$ where c_a is the gate-channel capacitance per unit area in strong inversion, d_i is the gate insulator thickness and Δd is a correction to account for the effective distance of the charge carriers from the interface, adjusted for the difference in dielectric constant between the oxide and the semiconductor. (PMOS usually has a buried channel (see Sections 3.9.7 and 3.9.8), in which case Δd accounts for the distance to this channel.). The hole sheet density at threshold is given by $p_o = \eta V_{th}/(2a)$.

In principle, eq. (3-10-1) has to be solved numerically in order to find p_s as a function of V_{GS} and V_F. However, an approximate, but very accurate, analytical solution can be obtained by the method discussed in Appendix A7 (see also Fjeldly et al. (1991)).

In short channel devices, the threshold voltage depends on both the channel length and the drain-source bias, as discussed in some detail in Section 3.6. Also, the lower mobility of holes makes short channel effects somewhat less important. Hence, in the devices discussed here, V_T remained nearly constant for gate lengths between 1 µm and 20 µm and within a normal range of drain-source biases. However, in deep submicrometer PMOS devices, the threshold voltage shift is quite important (see, for example, Hayden et al. (1991)), in which case we can use the expressions for the shift derived in Section 3.6 (see also Section 3.9).

As discussed in Section 3.8, the hole mobility, μ_p, at an arbitrary position, x, in the PMOS channel can be expressed in terms of the hole sheet concentration near source, p_S, and the channel voltage, $V(x)$, as

$$\frac{1}{\mu_p} = \frac{1}{\mu_{op}} - \frac{3V_T - ap_S}{\kappa_{1p}} + \frac{V(x)}{\kappa_{2p}} \qquad (3\text{-}10\text{-}2)$$

where μ_{op}, κ_{1p} and κ_{2p} are coefficients characterizing the mobility and its dependence on the bias conditions. Note that this expression is slightly modified in comparison with eq. (3-8-22) to prevent non-physical results at large positive values of V_{GS}. Regarding the structure of this expression, we note that a negative gate-source bias increases the transverse field in the channel and causes a reduction in the mobility, while an applied drain-source bias (negative value of

V) has the opposite effect. Moon et al. (1991a) showed that the parameter κ_{2p} can be related to κ_{1p} and to the coefficient α in eq. (3-10-1) as follows: $\kappa_{2p} \approx \kappa_{1p}/(3 - 2\alpha)$. (See Section 3.8 for a more detailed discussion of this expression and physical interpretations of the parameters used). For a given gate-source bias, the hole mobility at the source side of the channel, μ_S (which is also the mobility of the entire channel at low drain-source bias), is given by:

$$\frac{1}{\mu_p} = \frac{1}{\mu_{op}} - \frac{3 V_T - a p_S}{\kappa_{1p}} \tag{3-10-3}$$

It can be argued that the coefficients μ_{op} and κ_{1p} can be taken to be independent of the substrate bias and geometry (see Section 3.8 and Moon et al. (1991a)). Hence, all dependence of μ_S on substrate bias, device geometry and doping will be contained in the threshold voltage. This is in contrast to earlier expressions used for MOSFET mobility (see, for example, Arora and Richardson (1989)) which were explicitly dependent on device geometry and substrate bias conditions. The universality of eq. (3-10-3) was demonstrated in Section 3.8 for PMOS samples of different gate lengths and with different substrate biases. Excellent agreement was obtained between the measurements and the present, universal mobility model using the parameter values $\mu_{op} = 279$ cm^2/Vs and $\kappa_{1p} = 1510$ cm^2/V^{2}s.

Different velocity-field model relationships used for FET modeling were briefly discussed in Section 3.5. Based on previous experience, the following relationship is found to be suitable for the modeling of p-channel MOSFETs:

$$v(F) = \frac{F}{\dfrac{1}{\mu_p} + \dfrac{F}{v_s}} \tag{3-10-4}$$

where v_s is the hole saturation velocity and $F \equiv F_x = -dV/dx$ is the longitudinal electric field in the channel and μ_p is the low-field hole mobility given by eq. (3-10-2).

3.10.2. Intrinsic Unified Long Channel PMOS Model (PMOSA1)

In the long channel case, the effects of the source and drain resistances can be neglected, allowing us to relate experimental I–V characteristics directly to a model which expresses the drain current in terms of intrinsic bias voltages. Also, the threshold voltage shifts induced by the drain-source bias, which are typical for short channel devices, can be neglected.

Based on the velocity-field relationship of eq. (3-10-4), the drain current in a p-channel MOSFET can be approximated by

$$I_d = W q p_s v_p(F) = W q p_s \frac{F}{\frac{1}{\mu_p} + \frac{1}{v_s}F} \tag{3-10-5}$$

The sheet hole density, p_s, can be expressed in terms of the UCCM (see eq. (3-10-1), allowing eq. (3-10-5) to be integrated over the channel length, L, after multiplying the whole expression by the factor $(1/\mu_p + F/v_s)$. However, in performing this integration, we need to find an approximation for the average of $1/\mu_p + F/v_s$ over the channel length. This can be broken down to finding the channel averages of the potential, $<V>$, (associated with the term $1/\mu_p$) and the longitudinal field, $<F>$, (associated with the term F/v_s). Above threshold, $<V> \approx V_{DS}/2$ and $<F> \approx V_{DS}/L$ are reasonable average values. However, below threshold, $\mu_{psub} \approx \mu_S(V_{GS} = V_T)$ is taken to be a constant and F is small in most of the channel (except near the drain, see the discussion in Section 3.6). Hence, we can write the following approximate expression for the drain current:

$$I_d = \frac{q W v_s}{\alpha} \frac{\eta V_{th}(p_S - p_D) + \frac{a}{2}(p_S^2 - p_D^2)}{V_L - \left(1 - \frac{V_L \mu_S}{2\kappa_{2p}}\right) V_{DSe}} \tag{3-10-6}$$

where p_S and p_D are the sheet density of holes at the source end and the drain end of the channel, respectively, $V_L = v_s L/\mu_S$, and V_{DSe} is an effective drain-source voltage given by (noting that V_{DS} is negative for p-channel devices)

$$V_{DSe} = \frac{a(p_D - p_S)}{\alpha} \tag{3-10-7}$$

This effective voltage approaches V_{DS} above threshold ($V_{GT} < 0$) and vanishes below threshold ($V_{GT} > 0$), insuring the correct limiting behavior of the drain current in the two regimes and providing a reasonably good interpolation near threshold.

Furthermore, in the derivation of eq. (3-10-6), we have replaced the Fermi potential V_F (from the UCCM expression, see eq.(3-10-1)) by the channel voltage, V. This is a reasonable approximation in strong inversion where the drift current is dominant. But, at a first glance, this replacement does not appear to be feasible in the subthreshold regime where the diffusion transport mechanism dominates. However, the following argument shows that the replacement is also valid in the subthreshold regime:

When diffusion current dominates, the term F/v_s in the denominator of eq. (3-10-5) will be negligible compared to $1/\mu_p$ since the longitudinal channel field will be small (see the discussion above and in Section 3.6). Moreover, by replacing F in the numerator by $F_F = -dV_F/dx$, we obtain a formally correct expression for the current which contains both drift and diffusion contributions. Since both V_F and V vary by V_{DS} between source to drain, replacing V by V_F in eq. (3-10-5) or replacing V_F by V in the UCCM expression result in the same solution for the current.

One additional remark is necessary concerning eq. (3-10-6) for the drain current. The sheet hole densities, p_S and p_D, have to be evaluated in the channel near source and drain, respectively, using the UCCM expression of eq. (3-10-1). In the evaluation of the mobile charge density, it is important to use the Fermi potential and not replace it by the channel voltage, particularly in the subthreshold regime. Since the subthreshold current is dominated by diffusion, the total change V_{DS} in the Fermi potential is almost evenly distributed between source and drain, resulting in $V_F \approx V_{DS}$ at the drain end of the channel. On the other hand, the total change in the channel voltage is almost entirely confined to the drain-channel junction region, with little variation along the channel, which gives $V \approx 0$ at the drain end of the channel. Hence, it is necessary to use $V_F \approx V_{DS}$ in the evaluation of p_D.

In Figs. 3.10.1 and 3.10.2, we show a comparison of experimental I–V characteristics for a long channel PMOS and corresponding curves based on the present model. Fig. 3.10.1 shows above-threshold I_d versus V_{DS} characteristics, and Fig. 3.10.2 shows I_d versus V_{GS} characteristics in a semilog plot which emphasizes the behavior in the subthreshold region. Here, α, which is the only adjustable parameter, can be determined by fitting the present model in the saturation regime (see Subsection 3.11.2 for more details).

As can be seen from the figures, the agreement is excellent for the entire range of drain-source and gate-source bias voltages, spanning seven orders of magnitude in the drain current. In particular, we note the ability of the model to faithfully reproduce the various transition regions, such as the that between the linear and the saturation regimes (the knee region) in the above-threshold characteristics of Fig. 3.10.1 and the transition between the above-threshold and the subthreshold regimes shown in Fig. 3.10.2.

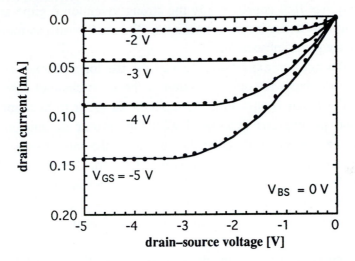

Fig. 3.10.1. Above-threshold experimental (dots) and calculated (solid lines) I–V characteristics for a long channel PMOS with $L = 19.5 \ \mu m$ and $W = 19.43 \ \mu m$. (From Moon et al. (1991a), [©] 1991 IEEE.)

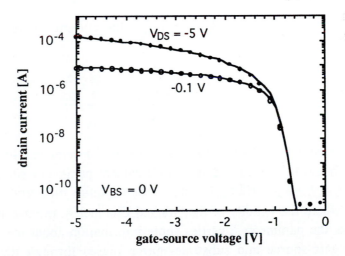

Fig. 3.10.2. Semilog plot of experimental (symbols) and calculated (solid lines) I–V characteristics for the same device as shown in Fig. 3.10.1. (From Moon et al. (1991a), [©] 1991 IEEE.)

3.10.3. Intrinsic Strong Inversion PMOS Model

We now want to incorporate short channel effect in the PMOS model, based on the insight gained in the above discussion of the long channel case. For simplicity, we first consider only the strong inversion regime and, again, restrict the discussion to the intrinsic transistor. We shall return to more general models in the subsequent Subsections.

Although the expression derived for the long channel I–V characteristics will form the basis for our discussion of the short channel case, we will apply it only to the linear (non-saturated) regime. In order to account properly for the additional phenomena associated with the shorter dimensions, we will use a different expression to describe the current in the saturation regime. In order to avoid discontinuities in important modeled electrical characteristics of the device, such as the drain conductance and the gate capacitance, we will require that the descriptions for the two regimes merge smoothly at the saturation point. Hence, a proper identification of the saturation voltage is an important objective of this model.

In strong inversion, the charge control model of eq. (3-10-1) simplifies to

$$p_s \approx c_a \left(-V_{GT} + \alpha V \right) / q = \frac{\varepsilon_i \left(-V_{GT} + \alpha V \right)}{d_i + \Delta d} \qquad (3\text{-}10\text{-}8)$$

Using this expression and $V_{DSe} \approx V_{DS}$, the current-voltage characteristics of eq. (3-10-6) reduce to

$$I_d = c_a W v_s \frac{V_{GT} V_{DS} - \dfrac{\alpha}{2} V_{DS}^2}{V_L - \zeta V_{DS}} \qquad (3\text{-}10\text{-}9)$$

where the source contact is used as a potential reference. In the short channel case, ζ should be considered as an independent parameter since its long channel value $[\zeta = 1 - V_L \mu_S / (2 \kappa_{2p})]$ may become quite inaccurate in short channel devices where the validity of the Gradual Channel Approximation is limited. In principle, the parameter ζ carries some information about the geometry and the applied gate-source and substrate-source biases through their effects on the mobility. But comparisons with experimental data indicate that ζ is always on the order of 0.5, that it can be taken to be independent of the gate-source bias, and that it has some residual dependence on substrate bias and geometry.

In the saturation regime, the channel is divided into two regions, as indicated in Fig. 3.10.3. In line with our discussion in Section 3.5, the Gradual Channel Approximation (GCA) is assumed to be valid in the non-saturated region, allowing the use of eq. (3-10-9) for that portion of the channel. However, in the saturated region, it is necessary to account for several short-channel effects, including channel length modulation and hot electron phenomena associated with charge injection from the channel into the drain contact, both of which contribute to a finite output conductance in saturation.

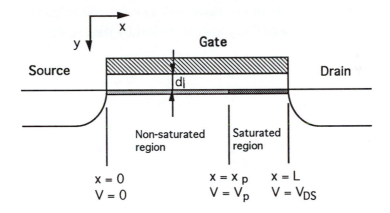

Fig. 3.10.3. Schematic representation of a MOSFET in saturation, where the channel is divided into a non-saturated region (where GCA is valid) and a saturated region.

Moon et al. (1991a) used the following expression for the differential output conductance in saturation to account for such effects (see also Ko (1989)):

$$\frac{dI_d}{dV_{DS}} \approx \frac{I_{sat} V_\lambda}{V_L - \zeta V_{SAT}} \left[\left(V_{DS} - V_{SAT} \right)^2 + \left(\theta V_\lambda \right)^2 \right]^{-1/2} \tag{3-10-10}$$

Here, V_{SAT} is the intrinsic saturation voltage, $V_\lambda = v_s \lambda / \mu_s$ is a characteristic voltage related to the characteristic length, λ, of the saturated region (see Section 3.5), and θ is a parameter which should be adjusted to secure a continuous output conductance at the onset of saturation.

In the deep saturation regime, eq. (3-10-10) can be rewritten as

$$I_{sat} \left. \frac{\partial V_{DS}}{\partial I_d} \right|_{\text{deep saturation}} \approx \frac{V_L - \zeta V_{SAT}}{V_\lambda} (V_{DS} - V_{SAT}) \qquad (3\text{-}10\text{-}11)$$

Hence, in deep saturation, the differential output resistance, $\partial V_{DS}/\partial I_d$, is approximately a linear function of the drain-source voltage, and V_{SAT} can be defined as the intercept of this linear dependence with $I_{sat}(\partial V_{DS}/\partial I_d) = 0$. However, Moon et al. (1991a) found that this value of the saturation voltage is numerically very close to the value of V_{SAT} defined in terms of the constant γ (which is independent of the gate-source bias) by the expression

$$\left[\frac{I_d}{V_{DS}} \frac{\partial V_{DS}}{\partial I_d} \right]_{SAT} = \gamma \qquad (3\text{-}10\text{-}12)$$

The extraction of V_{SAT}, I_{sat} and γ from experimental I–V characteristics is discussed in Subsection 3.10.5.

Combining eq. (3-10-9) at the onset of saturation with eq. (3-10-12), we obtain the following relationship between V_{SAT}, V_{GT} and I_{sat} (see Moon et al. (1991a)):

$$V_{SAT} = \frac{2(\gamma - 1)}{\alpha(2\gamma - 1)} V_{GT} + \frac{2\gamma\zeta}{\alpha(2\gamma - 1)} \frac{I_{sat}}{c_a W v_s} \qquad (3\text{-}10\text{-}13)$$

An equation for V_{SAT} in terms of V_{GT} can be now obtained by eliminating I_{sat} between eq. (3-10-9) (evaluated at the saturation point) and eq. (3-10-13). The constant θ is determined by equating the output conductances of eqs. (3-10-9) and (3-10-10) at the saturation point, resulting in

$$\theta = \frac{V_{SAT} \left(V_{GT} - \frac{\alpha}{2} V_{SAT} \right)}{\alpha V_L V_{SAT} - V_L V_{GT} - \frac{\alpha}{2} \zeta V_{SAT}^2} \qquad (3\text{-}10\text{-}14)$$

The drain current in saturation is obtained by integrating eq. (3-10-10), using the saturation point as the lower bound of the integration. Hence, we find for $V_{DS} > V_{SAT}$

$$I_d = I_{sat} \left\{ 1 + \left(\frac{V_\lambda}{V_L - \zeta V_{SAT}} \right) \right.$$

$$\left. \times \ln \left[\sqrt{1 + \left(\frac{V_{DS} - V_{SAT}}{\theta V_\lambda} \right)^2} - \frac{V_{DS} - V_{SAT}}{\theta V_\lambda} \right] \right\} \qquad (3\text{-}10\text{-}15)$$

3.10.4. Extrinsic Strong Inversion PMOS Model (PMOSA2)

In short channel MOSFETs with effective gate lengths of about one micrometer or less, the parasitic source and drain resistances play a significant role in the overall device behavior. It is therefore important to include the effects of these resistances into our model in order to improve the efficiency of the circuit simulation and to be able to extract model parameters directly from measured characteristics.

The intrinsic PMOS model described in the previous Subsection can be converted to an extrinsic model by expressing the intrinsic bias voltages (upper case subscripts) in terms of their extrinsic counterparts (lower case subscripts), i.e.,

$$V_{DS} = V_{ds} + I_d R_t \qquad (3\text{-}10\text{-}16)$$

$$V_{GT} = V_{gt} + I_d R_s \qquad (3\text{-}10\text{-}17)$$

R_s and R_d are the source and drain series resistance, respectively, and $R_t = R_s + R_d$. (Note that the signs of all voltages in these equations are reversed compared to the n-channel case).

Substituting eqs. (3-10-16) and (3-10-17) into eq. (3-10-9), we obtain the following analytical expression for the drain current in the linear regime in terms of extrinsic voltages:

$$I_d = \frac{2 k_1}{h_1 + \sqrt{h_1^2 - 4 g_1 k_1}} \qquad (3\text{-}10\text{-}18)$$

where

$$k_1 = V_{gt} V_{ds} - \alpha V_{ds}^2/2$$

$$h_1 = R_n V_L - R_t V_{gt} + (\alpha R_t - R_s - \zeta R_n) V_{ds}$$

$$g_1 = R_t (\zeta R_n + R_s - \alpha R_t/2) \qquad (3\text{-}10\text{-}19)$$

$$R_n = (c_a v_s W)^{-1}$$

Likewise, the saturation current is determined from eq. (3-10-9) evaluated at the saturation point, using eqs. (3-10-13) and (3-10-17),

$$I_{sat} = \frac{2 k_2}{h_2 + \sqrt{h_2^2 - 4 g_2 k_2}} \qquad (3\text{-}10\text{-}20)$$

where

$$k_2 = \frac{2\gamma(\gamma - 1)}{\alpha(2\gamma - 1)^2} V_{gt}^2$$

$$h_2 = R_n V_L - \frac{4\gamma(\gamma - 1)}{\alpha(2\gamma - 1)^2}\left[R_s + \frac{2\gamma^2 - 2\gamma + 1}{2\gamma(\gamma - 1)} \zeta R_n \right] V_{gt} \qquad (3\text{-}10\text{-}21)$$

$$g_2 = \frac{2[2\gamma(\alpha + 1) - \alpha]}{\alpha^2(2\gamma - 1)^2}[(\gamma - 1) R_s + \gamma \zeta R_n] \zeta R_n$$

The intrinsic and extrinsic saturation voltages are obtained from eqs. (3-10-13), (3-10-16) and (3-10-17):

$$V_{SAT} = \frac{2(\gamma - 1)}{\alpha(2\gamma - 1)} V_{gt} + \frac{2(\gamma - 1) R_s + 2\gamma \zeta R_n}{\alpha(2\gamma - 1)} I_{sat} \qquad (3\text{-}10\text{-}22)$$

$$V_{sat} = V_{SAT} - I_{sat} R_t \qquad (3\text{-}10\text{-}23)$$

In saturation, the drain current can be found from eq. (3-10-15) by substituting V_{DS} and V_{SAT} by extrinsic voltages inside the logarithm (as shown in Subsection 3.10.5, there is no need to substitute for V_{SAT} in the factor in front of the logarithm). The resulting expression cannot be solved analytically with respect to the current, but a first order expansion in terms of $(I_d - I_{sat})R_t/(\theta V_\lambda)$ gives the following approximation for the current

$$I_d \approx I_{sat} \left[1 + \frac{\dfrac{V_\lambda}{V_L - \zeta V_{SAT}} \ln\left(\sqrt{1+f^2} - f\right)}{1 + \dfrac{I_{sat} R_t}{\theta(V_L - \zeta V_{SAT})} \dfrac{1}{\sqrt{1+f^2}}} \right] \qquad (3\text{-}10\text{-}24)$$

where $f = (V_{ds} - V_{sat})/(\theta V_\lambda)$. Typically, the factor in front of the logarithm is smaller than λ/L, i.e., less than 0.1 for a one micrometer device (see Table 3.10.1). We also notice that, particularly in deep saturation,

$$\frac{I_{sat} R_t}{\theta(V_L - \zeta V_{SAT})} \frac{1}{\sqrt{1+f^2}} \ll 1 \qquad (3\text{-}10\text{-}25)$$

Hence, the drain current in saturation can be approximated by

$$I_d = I_{sat} \left\{ 1 + \left(\frac{V_\lambda}{V_L - \zeta V_{SAT}} \right) \right.$$
$$\left. \times \ln\left[\sqrt{1 + \left(\frac{V_{ds} - V_{sat}}{\theta V_\lambda}\right)^2} - \frac{V_{ds} - V_{sat}}{\theta V_\lambda} \right] \right\} \qquad (3\text{-}10\text{-}26)$$

3.10.5. Parameter Extraction Using Strong Inversion PMOS Model

We now describe a parameter extraction technique based on the present strong inversion model and demonstrate this technique on a short channel PMOS device with a nominal gate length of 1.4 μm and a gate oxide thickness of 205 Å. The various parameter values determined by this extraction technique are shown in Table 3.10.1.

We first consider the threshold voltage. The threshold voltage, V_T, can be defined by several techniques. Here, we use the technique which identifies V_T as the gate-source voltage at which the gate-channel capacitance, c_{gc}, drops to one third of its maximum value (see Section 3.3). However, since the gate-channel capacitance is proportional to the device transconductance, g_m, at low drain-source voltage, the threshold voltage may equally well be determined as the gate-source voltage where g_m reaches one third of its maximum value. A suitable drain-source voltage for this procedure is $V_{ds} \approx -10$ mV. Alternatively, V_T can be determined from the subthreshold characteristics as discussed in Section 3.9, by roughly estimating the threshold current. (For a comparison of various techniques of finding V_T, see Park et al. (1991).)

V_{bs} [V]	0	1	2	4
α	1.34	1.20	1.14	0.97
R_t [ohm]	162			
ΔL [μm]	0.57			
ζ	0.34	0.42	0.44	0.49
V_T [V]	−0.97	−1.20	−1.35	−1.65
λ [Å] (V_{gs}=−5V)	1,043	966	978	1,087
γ	2.35			

Table 3.10.1. Extracted device parameters for a PMOS with nominal gate length 1.4 μm and width 20 μm. (From Moon et al (1991a).)

The total parasitic resistance, R_t, is determined as the inverse of the linear channel conductance, g_{ch}, at large gate-source voltages. A good estimate of R_t is obtained by extrapolating a plot of $1/g_{ch}$ versus $1/V_{gt}$ to the intercept with the $1/g_{ch}$-axis (i.e., $1/V_{gt} = 0$, see Section 3.9). In general, such a plot may be non-linear since the hole mobility at small drain-source bias, μ_S, is a function of V_{GS} (see eq. (3-10-3)). On the other hand, the mobility for fixed gate-source biases can be obtained from plots of $1/g_{ch}$ versus the channel length. From eqs. (3-9-27) and (3-9-28) in Section 3.9, we see that the mobility is determined from the slopes of such plots, provided that μ_S does not vary much with L and that the different samples are fabricated by the same process. We also notice that such plots for different V_{gs} should intersect at $1/g_{ch} = R_t$ and $L = \Delta L$, where ΔL is the difference between the nominal and effective channel lengths (see Section 3.11 and Moon et al. (1991a)). For the present PMOS device, which has a nominal gate length of 1.4 μm, we find the effective gate length $L_{eff} \approx 0.83$ μm.

The determination of the saturation current and the saturation voltage is an important element of the parameter extraction. For $|V_{ds} - V_{sat}| \gg \theta V_\lambda$ (θV_λ is on the order of 0.4 V for $V_{gs} = -5$V), we obtain from eq. (3-10-25)

$$I_{sat} \left. \frac{\partial V_{ds}}{\partial I_d} \right|_{\text{deep saturation}} \approx \frac{V_L - \zeta V_{SAT}}{V_\lambda} (V_{ds} - V_{sat}) \qquad (3\text{-}10\text{-}27)$$

(compare with eq. (3-10-11)). Hence, the extrinsic saturation voltage, V_{sat}, can be found by plotting experimental data for the function on the left hand side of eq. (3-10-26) versus V_{ds}, and extrapolating to the intercept with $I_{sat}(\partial V_{ds}/\partial I_d) = 0$. The saturation current is subsequently determined from the experimental $I–V$ characteristics at $V_{ds} = V_{sat}$. This technique is demonstrated in Fig. 3.10.4 for the PMOS device under investigation. After determining the saturation point, the intrinsic saturation voltage is easily calculated from eq. (3-10-23).

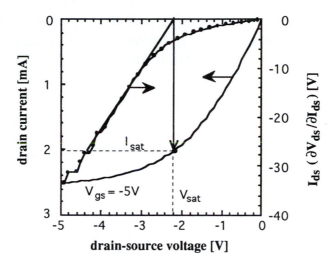

Fig. 3.10.4. Procedure for determination of the extrinsic saturation voltage, V_{sat}, and the saturation current, I_{sat}, from measured $I–V$ characteristics. The data are obtained for a PMOS with nominal gate length $L = 1.4\ \mu m$. (From Moon et al. (1991a), © 1991 IEEE.)

Previously, we also defined the saturation point in terms of the parameter γ, see eq. (3-10-12). This parameter is needed for the model description of the device. However, since the definition of γ was based on intrinsic bias voltages, a rigorous determination of γ from eq. (3-10-12) is done by first converting the extrinsic characteristics to intrinsic characteristics using eqs. (3-10-16) and (3-10-17). However, this is very inconvenient. Instead, Moon et al. (1991a) showed that only a small error is introduced if the derivative $\partial I_d/\partial V_{DS}$ in eq. (3-10-12) is replaced by $\partial I_d/\partial V_{ds}$, which can readily be obtained from the experimental characteristics. (Note that V_{SAT} and I_{sat} were determined above.) The value $\gamma \approx 2.35$ is obtained by averaging over a number of characteristics representing several values of gate-source and substrate source bias.

Next, we want to find the parameters α and ζ used in eq. (3-10-9) to describe the linear region of the PMOS *I–V* characteristics. For this purpose, it is useful to rewrite eq. (3-10-9) for the saturation point in the following way:

$$y \equiv \frac{V_{GT}}{V_{SAT}} - \frac{L_{eff} I_{sat}}{c_i \mu_S V_{SAT}^2} = \frac{\alpha}{2} - \zeta R_n \frac{I_{sat}}{V_{SAT}} \qquad (3\text{-}10\text{-}28)$$

Here, we have used $V_L = v_s L_{eff}/\mu_S$ and $R_n = (Wc_i v_s)^{-1}$. Fig. 3.10.5 shows plots for different substrate biases of *y* versus I_{sat}/V_{SAT}, based on the values of I_{sat} and V_{SAT} determined as described above. (V_{GT} is obtained from eq. (3-10-17).) We see from eq. (3-10-28) that such plots should be straight lines with slopes equal to ζR_n, intercepting the *y* -axis (i.e., $I_{sat}/V_{SAT} = 0$) at $y = \alpha/2$. We note that the current-voltage characteristics of *p*-channel MOSFETs are less sensitive to the value of the carrier saturation velocity than *n*-channel MOSFETs because of the much smaller mobility for holes. We therefore use a recognized value of the hole saturation velocity, $v_s = 6{\times}10^4$ m/s (see Hu et al. (1987)), in the expression for R_n. Hence, ζ can be determined from the slopes.

The values of α and ζ resulting from this extraction procedure are shown in Table 3.10.1 for different substrate biases. We notice that α decreases and ζ increases with increasing V_{bs}. This is plausible since the asymmetry of the channel created by the drain-source voltage is reduced with the application of a positive substrate bias, causing the importance of the third term on the right hand side of eq. (3-10-2) to diminish. This is reflected in an increased value of κ_2 and, hence, in the observed changes in α and ζ. (We recall that ζ in the long channel case depends explicitly on the parameter κ_{2p}, compare eqs. (3-10-6) and (3-10-9). It is reasonable to assume that a qualitatively similar dependence carries over to the short channel case.)

Finally, we extract the values of the characteristic voltage, V_λ, and the characteristic length, λ, of the saturated region of the channel. We observe from eq. (3-10-25) that the slope of the straight line $I_{sat}(\partial V_{ds}/\partial I_d)$ versus $V_{ds} - V_{sat}$, valid for deep saturation (see Fig. 3.10.4), is equal to $(V_L + \zeta V_{SAT})/V_\lambda$. Hence, since V_L, V_{SAT} and ζ have already been determined, V_λ is obtained from this slope and λ is calculated as $\mu_S V_\lambda/v_s$. The value of λ is near 0.1 µm for $V_{gs} = -5$ V and V_{bs} ranging from 0 V to 4 V (see Table 3.10.1). This is close to the value measured for HFETs by Moon et al. (1990) and suggested by Ko (1989).

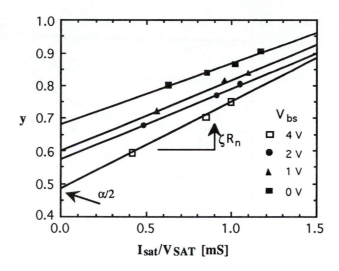

Fig. 3.10.5. Plots based on eq. (3-10-27) for determining the parameters α and ζ, shown for different values of substrate bias, V_{bs}. The plots are for a PMOS with a nominal gate length of 1.4 μm ($L_{eff} \approx 0.83$ μm). (From Moon et al. (1991a), © 1991 IEEE.)

In Figs. 3.10.6 and 3.10.7, we show comparisons of experimental *I–V* characteristics for the 1.4 μm PMOS considered above with curves based on the present model. The comparison is done for two values of substrate bias, using model calculations based on the extracted parameters given in Table 3.10.1, with no adjustment of parameters.

As can be seen in the figures, the agreement between the measured and the calculated curves is excellent. In particular, the transition region from the linear to the saturation regime is reproduced quite accurately. A correspondingly good agreement was also achieved for a device with shorter gate length (L_{eff} = 0.63 μm, see Moon et al. (1991a)).

3.10.6. Intrinsic Unified Short Channel PMOS Model (PMOSA3)

In this Subsection, we present an intrinsic unified short channel PMOS model based on UCCM and the velocity-field relationship for holes in silicon which were discussed in the preceding Subsections.

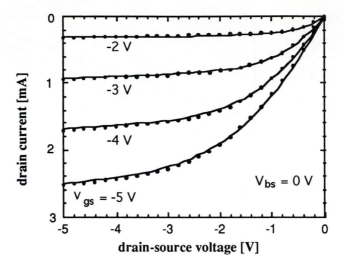

Fig. 3.10.6. Measured (symbols) and calculated (solid lines) current-voltage characteristics for a PMOS with gate length 1.4 μm (L_{eff} = 0.83 μm), bulk bias V_{bs} = 0 V. (From Moon et al. (1991a), © 1991 IEEE.)

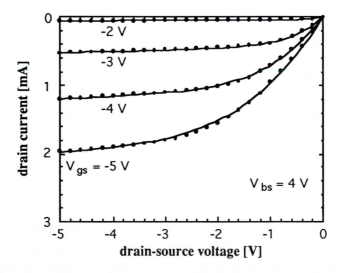

Fig. 3.10.7. Measured (symbols) and calculated (solid lines) current-voltage characteristics for a PMOS with gate length 1.4 μm (L_{eff} = 0.83 μm), bulk bias V_{bs} = 4 V. (From Moon et al. (1991a), © 1991 IEEE.)

Just as was done for the strong inversion model discussed in Subsections 3.10.3 and 3.10.4, we distinguish between the linear and the saturation regimes of operation. These two regimes should merge smoothly at the saturation point. We slightly modify UCCM for PMOS (see eq. (3-10-1)) in order to incorporate the effect of the Drain Induced Barrier Lowering (DIBL) effect (see Park et al. (1990) and Subsection 3.6.5):

$$- V_{GS} + \left(V_{To} - \sigma V_{DS} \right) + \alpha V_F \approx \eta V_{th} \ \ln\!\left(\frac{p_s}{p_o} \right) + a(p_s - p_o) \quad \text{(3-10-29)}$$

Here, V_{DS} is the intrinsic drain-source voltage and σV_{DS} represents the threshold voltage shift caused by DIBL.

We now want to find a unified expression for the drain current in the linear regime, covering both the subthreshold and the above-threshold biasing conditions. We start by rewriting the strong inversion expression for I_d, given by eq. (3-10-9), as follows (see Subsection 3.10.3):

$$I_d^{SI} = -\frac{1}{R_n} \frac{\left(a p_s + \dfrac{\alpha}{2} V_{DS} \right) V_{DS}}{V_L - \zeta V_{DS}} \quad \text{(3-10-30)}$$

where we have used $a p_S = -V_{GT}$ in strong inversion (SI) and

$$R_n = \frac{a}{q v_s W} \quad \text{(3-10-31)}$$

Eq. (3-10-30) can easily be converted to a unified expression with the correct limiting behavior in the linear regime both below and above threshold, by making the replacement $\alpha \rightarrow \alpha^*$ where

$$\alpha^* = \frac{\alpha}{1 + \dfrac{2\eta V_{th}}{a p_S}} \quad \text{(3-10-32)}$$

resulting in the following unified drain current:

$$I_d = -\frac{1}{R_n} \frac{\left(a p_s + \dfrac{\alpha^*}{2} V_{DS} \right) V_{DS}}{V_L - \zeta V_{DS}} \quad \text{(3-10-33)}$$

Here, ps is calculated from eq. (3-10-29) for all values of V_{DS}, and ζ should still be considered an independent parameter for the same reason as discussed for the strong inversion model (see Subsection 3.10.3). As was demonstrated by Byun et al. (1990), the introduction of the parameter α^* provides a smooth transition between the subthreshold and above threshold regimes, in excellent agreement with numerical calculations.

The unified saturation voltage, V_{SAT}, can be derived in full analogy with the strong inversion case (see Subsection 3.10.3), resulting in

$$V_{SAT} = \frac{-2(\gamma-1)}{\alpha^*(2\gamma-1)}aps + \frac{2\gamma\zeta}{\alpha^*(2\gamma-1)}R_n I_{sat} \qquad (3\text{-}10\text{-}34)$$

where the empirical constant γ relates to the properties of the I–V characteristics at the onset of saturation through eq. (3-10-12). I_{sat} in eq. (3-10-34) is obtained by replacing V_{DS} with V_{SAT} in eq. (3-10-33), i.e.,

$$I_{sat} = -\frac{1}{R_n}\frac{\left(aps + \dfrac{\alpha^*}{2}V_{SAT}\right)V_{SAT}}{V_L - \zeta V_{SAT}} \qquad (3\text{-}10\text{-}35)$$

Explicit expressions for V_{SAT} and I_{sat} can be found by combining eqs. (3-10-34) and (3-10-35):

$$V_{SAT} = \frac{1}{\alpha^*\zeta(\gamma-1)}\left[A^* - \sqrt{A^{*2} + 2\alpha^*\zeta(\gamma-1)^2 aps V_L}\right] \qquad (3\text{-}10\text{-}36)$$

where

$$A^* = \frac{\alpha^*(2\gamma-1)V_L}{2} + \zeta aps \qquad (3\text{-}10\text{-}37)$$

The drain current in the saturation regime is obtained by integrating eq. (3-10-10), using the saturation point as the lower bound of the integration. Hence, we find for $V_{DS} > V_{SAT}$

$$I_d = I_{sat} \left\{ 1 + \left(\frac{V_\lambda}{V_L - \zeta V_{SAT}} \right) \right.$$
$$\left. \times \ln \left[\sqrt{ 1 + \left(\frac{V_{DS} - V_{SAT}}{\theta V_\lambda} \right)^2 } - \frac{V_{DS} - V_{SAT}}{\theta V_\lambda} \right] \right\} \qquad (3\text{-}10\text{-}38)$$

Just as in Subsections 3.10.3, we determine the constant θ from the requirement of drain conductance continuity across the saturation point, i.e.,

$$\theta = \frac{ -\left(ap_S + \frac{\alpha^*}{2} V_{SAT} \right) V_{SAT} }{ \alpha^* V_L V_{SAT} + ap_S V_L - \frac{\alpha^*}{2} \zeta V_{SAT}^2 } \qquad (3\text{-}10\text{-}39)$$

Eqs. (3-10-29) to (3-10-39) form a complete set of equations for our unified intrinsic PMOS model implemented in the AIM-Spice circuit simulator (model PMOSA3).

3.10.7. Parameter Extraction for Unified Short Channel PMOS Model

The parameter extraction for the unified short channel PMOS model is almost identical to that for the strong inversion PMOS model (see Subsection 3.10.5). The parameters extracted in this way include the threshold voltage at zero drain-source bias (V_{To}), the total parasitic resistances (R_t), the hole mobility (μ_S), the difference between the nominal and effective channel length (ΔL), the drain saturation current (I_{sat}), and the drain-source saturation voltage (V_{SAT}). Then the ideality factor η and the DIBL parameter σ are extracted from subthreshold characteristics.

However, parameter values for α, ζ and λ should be extracted in a different way. As can be seen from Fig. 3.10.8 for a large area PMOS, the hole sheet density at large gate voltage swings can be approximated by

$$p_s \approx c_a' \left(-V_{GT} + \alpha V + \sigma V_{DS} \right) / q \qquad (3\text{-}10\text{-}40)$$

where

$$c_a' \approx 0.95 c_a \qquad (3\text{-}10\text{-}41)$$

In eq. (3-10-40), we dropped the term $V_{th}\ln(p_s/p_o)$ which is important not only below threshold but also in the inversion regime, and accounted for the contribution of this term by replacing c_a by c'_a. Eq. (3-10-40) leads to the following long channel I-V model:

$$I_d \approx \mu_S c'_a \frac{W}{L}\left(V_{GT}V_{DS} - \alpha'\frac{V_{DS}^2}{2}\right) \tag{3-10-42}$$

where $\alpha' = \alpha + 2\sigma$ (see Problem P3-10-1). Hence, ζ and λ can be extracted using eqs. (3-10-28) and (3-10-27), respectively, if we replace α with α' and c_a with c'_a in eq. (3-10-28) (see Fig. 3.10.5).

Fig. 3.10.9 through Fig.3.10.12 show a comparison between measured and calculated current-voltage characteristics using the extracted parameter values in Table 3.10.2 for a 1.4 µm PMOS (see Moon et al. (1991a)). As can be seen from these figures, the agreement between the measured and the calculated I–V characteristics is excellent for the entire range of gate and drain biases. In particular, the transition from the linear to the saturation regime is reproduced quite accurately.

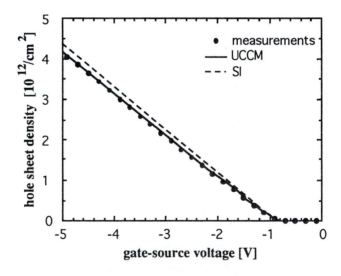

Fig. 3.10.8. Hole sheet density versus V_{GS} for a large area PMOS. Comparison of experimental data (symbols) with calculations based on UCCM (solid line) and the strong inversion (SI) model (from Moon et al. (1991a)). In strong inversion, the UCCM calculation can be approximated as $p_s = -c'_a V_{GT}$. (From Park (1992).)

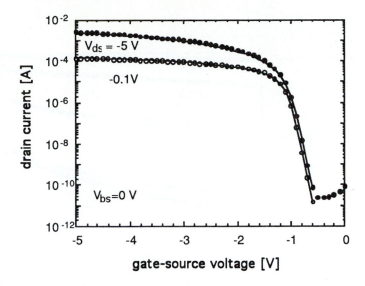

Fig. 3.10.9. Semilog plot of measured (symbols) and calculated (solid lines) *I–V* characteristics for a PMOS with gate length 1.4 μm (L_{eff} = 0.83 μm), substrate-source bias V_{bs} = 0 V, drain-source voltage V_{ds} = –0.1 V and –5 V. (From Park (1992) and Moon et al. (1991a), © 1991 IEEE.)

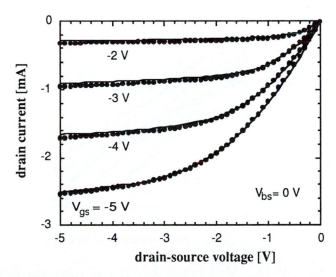

Fig. 3.10.10. Measured (symbols) and calculated (solid lines) *I-V* characteristics for a PMOS with gate length 1.4 μm (L_{eff} = 0.83 μm), substrate-source bias V_{bs} = 0 V. (From Park (1992).)

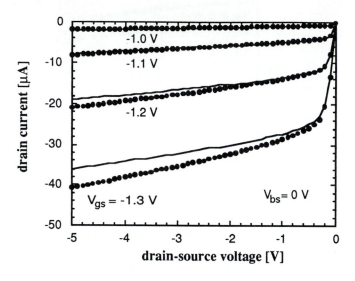

Fig. 3.10.11. Measured (symbols) and calculated (solid lines) *I–V* characteristics near threshold for a PMOS with gate length 1.4 μm (L_{eff} = 0.83 μm), V_{bs} = 0 V. (From Park (1992).)

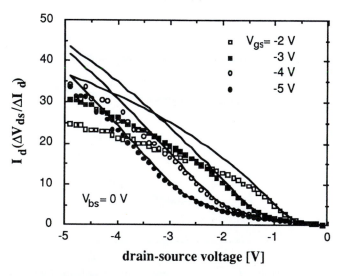

Fig. 3.10.12. Measured (symbols) and calculated (solid lines) $I_d(\partial V_{ds}/\partial I_d)$-voltage characteristics for a PMOS with gate length 1.4 μm (L_{eff} = 0.83 μm), V_{bs} = 0 V. (From Park (1992).)

V_{bs} [V]	0
α	1.3
R_t [ohm]	162
ΔL [μm]	0.57
ζ	0.28
V_{To} [V]	-0.99
λ [Å] (V_{gs}=-5V)	725
γ	2.35
σ	0.007
η	1.39

Table 3.10.2.   Extracted parameters using the unified short channel PMOS model for a device with nominal gate length 1.4 μm (from Park (1992)).

Fig. 3.10.9 clearly displays the DIBL effect.   Fig. 3.10.11 shows the measured and calculated *I–V* characteristics near threshold. Even in this regime, which is the most difficult to reproduce, our model yields a reasonable agreement with the experimental data.   Also the product of the drain current and the output resistance, $I_d(\partial V_{ds}/\partial I_d)$, is in quite satisfactory agreement with the measured data for the entire range of gate and drain biases, as shown in Fig. 3.10.12.   The continuity of the first derivative of the drain current is a feature of our model which helps to insure convergence of the AIM-Spice circuit simulator. Fig. 3.10.12 also clearly demonstrates the contributions of the channel length modulation and the DIBL to the output conductance in the saturation regime. At low gate bias (V_{gs} = –2V and –3V), the curves in Fig. 3.10.12 show that $I_d(\partial V_{ds}/\partial I_d)$ increases linearly with V_{ds} just after saturation. This is typical for the channel length modulation (see eq. (3-10-27)). However, in deep saturation, $I_d(\partial V_{ds}/\partial I_d)$ increases sublinearly, as can expected for the DIBL mechanism. The curves at higher gate biases (V_{gs} = –4V and –5V) are typical for the channel length modulation mechanism alone.

Here, we have used a constant value for the DIBL parameter σ which is different from the behavior discussed in Section 3.6 and used in Section 3.9.

The use of a gate voltage dependent σ (accounting for the DIBL at low gate bias and static feedback at higher gate biases), should give an even better fit to the measured data in Fig. 3.10.12. However, the accuracy of the present model is quite acceptable for digital circuits. But for analog circuits, a precise modeling of the drain resistance is more important. On the other hand, such circuits usually utilize transistors with larger gate lengths where DIBL and/or static feedback effects are less pronounced. Hence, our model can be used for the simulation of analog circuits as well.

3-11. CURRENT-VOLTAGE CHARACTERISTICS OF NMOS (AIM-Spice MODELS)

With this Section, we continue the presentation of advanced CMOS models by discussing NMOS along the same line as we discussed PMOS in Section 3.10. The present NMOS models are based on work by Park et al. (1991), Park (1992) and Moon et al. (1991), where additional information can be found.

3.11.1. Basic Assumptions and Equations
The Unified Charge Control Model for the n-channel MOSFET can be expressed as follows (see Sections 3.4 and 3.6):

$$V_{GT} - \alpha V_F \approx \eta V_{th} \ln\left(\frac{n_s}{n_o}\right) + a(n_s - n_o) \qquad (3\text{-}11\text{-}1)$$

Here, $V_{GT} = V_{GS} - V_T$ is the intrinsic gate voltage swing (gate-source voltage, V_{GS}, relative to the threshold voltage, V_T), V_F is the Fermi potential relative to that at the source electrode, n_s is the electron sheet density in the channel and n_o is the same at threshold, V_{th} is the thermal voltage, η is the subthreshold ideality factor, α is a parameter accounting for the threshold voltage dependence on the channel potential, and $a = q(d_i + \Delta d)/\varepsilon_i = q/c_a$ where c_a is the gate-channel capacitance per unit area in strong inversion, d_i is the gate insulator thickness and Δd is a correction to account for the effective distance of the charge carriers from the interface, adjusted for the difference in dielectric constant between the oxide and the semiconductor. (Δd is typically only about 4 Å in NMOS, see Section 3.3.) As was shown in Section 3.4, the electron sheet density at threshold is given by $n_o = \eta V_{th}/(2a)$.

In principle, eq. (3-11-1) has to be solved numerically in order to find n_S as a function of V_{GS} and V_F. However, an approximate, but very accurate, analytical solution can be obtained by the method discussed in Appendix A7 (see also Fjeldly et al. (1991)).

As discussed in Section 3.8, the electron mobility, μ_n, at an arbitrary position, x, in the NMOS channel can be expressed in terms of the electron sheet concentration near source, n_S, and the channel voltage, $V(x)$, as

$$\mu_n = \mu_{on} - \kappa_{1n}\left(an_S + 2V_T\right) + \kappa_{2n}V(x) \qquad (3\text{-}11\text{-}2)$$

were μ_{on}, κ_{1n} and κ_{2n} are coefficients characterizing the mobility and its dependence on the bias conditions. Note that this expression is slightly modified in comparison with eq. (3-8-16) to prevent non-physical results at large negative values of V_{GS}. Regarding the structure of this expression, we note that a positive gate-source bias increases the transverse field in the channel and causes a reduction in the mobility, while an applied drain-source bias (positive value of V) has the opposite effect. Park et al. (1991) showed that the parameter κ_{2n} can be related to κ_{1n} and the coefficient α in eq. (3-11-1) as follows: $\kappa_{2n} \approx \kappa_{1n}(2 - \alpha)$. (See Section 3.8 for a more detailed discussion of this expression and physical interpretations of the parameters used.) For a given gate-source bias, the electron mobility at the source side of the channel, μ_S (which is also the mobility of the entire channel at low drain-source bias), is given by:

$$\mu_n = \mu_{on} - \kappa_{1n}\left(an_S + 2V_T\right) \qquad (3\text{-}11\text{-}3)$$

Below threshold, the mobility is taken to be constant and equal to the value of μ_S at threshold, i.e., $\mu_{nsub} \approx \mu_S(V_{GS} = V_T) = \mu_{on} - 2\kappa_{1n}V_T$ (see Section 3.8).

It can be argued that the coefficients μ_{on} and κ_{1n} can be taken to be independent of the substrate bias and geometry (see Section 3.8 and Park et al. (1991)). Hence, all dependence of μ_S on substrate bias and geometry will be contained in the threshold voltage. This is in contrast to earlier expressions used for MOSFET mobility (see, for example, Klaassen (1976)) which were explicitly dependent on device geometry and substrate bias conditions. The universality of eq. (3-11-3) was demonstrated in Section 3.8 for NMOS samples of different gate lengths and with different substrate biases. Excellent agreement was obtained between the measurements and the present universal mobility model using the parameter values $\mu_o = 625$ cm^2/Vs and $\kappa_{1n} = 28$ cm^2/V^{2}s.

Among the different velocity-field model relationships for FET modeling discussed in Section 3.5, the following has been found to be a reasonable choice for *n*-channel MOSFETs:

$$v(F) = \frac{\mu_n F}{\left[1 + (\mu_n F / v_s)^2\right]^{1/2}} \tag{3-11-4}$$

where v_s is the electron saturation velocity and $F \equiv |F_x| = dV/dx$ is the absolute value of the longitudinal electric field in the channel and μ_n is the low-field hole mobility given by eq. (3-11-2). However, this velocity-field relationship makes it difficult to develop an analytical theory for the linear region of the NMOS *I–V* characteristics along the same line as was done for PMOS, and simplifications have to be introduced.

3.11.2. Intrinsic Unified Long Channel NMOS Model (NMOSA1)

As was done for PMOS, the effects of source and drain resistances can be neglected for long channel NMOS devices, allowing us to relate experimental *I–V* characteristics directly to a model which expresses the drain current in terms of intrinsic bias voltages. Also, the threshold voltage shift induced by the drain-source bias, typical for short channel devices, can be neglected in the long channel case. Furthermore, we can write the channel current in terms of the Gradual Channel Approximation (GCA) as follows, neglecting velocity saturation effects:

$$I_d = W q n_s \mu_n \frac{dV_F}{dx} \tag{3-11-5}$$

The electron sheet density, n_s, can be expressed in terms of the UCCM (see eq. (3-11-1)), allowing eq. (3-11-5) to be integrated over the channel length, *L*. In the above-threshold regime, after dividing both sides of eq. (3-11-5) by μ_n, the integral of the left hand side becomes

$$\int_0^L \frac{I_d}{\mu_n} dx = I_d \int_0^L \frac{dx}{\mu_s + \kappa_{2n} V(x)} \approx \frac{I_d}{\mu_s} \int_0^L \left(1 - \frac{\kappa_{2n}}{\mu_s} V(x)\right) dx \tag{3-11-6}$$

where we have assumed that $\kappa_{2n}V(x) \ll \mu_s$. The integral on the right hand side of eq. (3-11-6) involves an estimate, $<V>$, of the average channel potential over

the gate length. Above threshold, $<V> \approx V_{DS}/2$ is a reasonable average value.
Below threshold, $\mu_{nsub} \approx \mu_S(V_{GS} = V_T)$ is taken to be a constant. Hence, after
some manipulation, we can write the following approximate expression for the
drain current:

$$I_d = \frac{qW}{\alpha L}\left(\mu_S + \frac{\kappa_{2n}}{2}V_{DSe}\right)\left[\eta V_{th}\left(n_S - n_D\right) + \frac{a}{2}\left(n_S^2 - n_D^2\right)\right] \quad (3\text{-}11\text{-}7)$$

where n_S and n_D are the sheet density of electrons at the source end and the drain
end of the channel, respectively, and V_{DSe} is an effective drain-source voltage
given by

$$V_{DSe} = \frac{a\left(n_S - n_D\right)}{\alpha} \quad (3\text{-}11\text{-}8)$$

This effective voltage approaches V_{DS} above threshold $(V_{GT} > 0)$ and vanishes
below threshold $(V_{GT} < 0)$, insuring the correct limiting behavior of the drain
current in the two regimes and providing a reasonably good interpolation near
threshold. We note that, since V_{DS} is positive for an n-channel device, the drain-
source bias increases the effective distance of the carriers from the interface,
enhancing the average mobility in the channel. (In order to evaluate the mobile
charge density, n_D, we have to take $V_F \approx V_{DS}$.)

In Figs. 3.11.1 to 3.11.3, we show comparisons of experimental I–V
characteristics for a long channel NMOS with corresponding curves based on the
present model. I_d versus V_{GS} characteristics in the saturation regime $(V_{DS} = 5$ V$)$ for three values of substrate bias are shown Fig. 3.11.1. By fitting the
model to the experimental data, we obtain the values of the parameter α. We
notice that α decreases with increasing negative substrate bias, reflecting a
reduction in the asymmetry of the channel created by the drain-source voltage.
The reduced value of α also translates into an increase in the parameter κ_{2n}.

Fig. 3.11.2 shows the corresponding I–V characteristics for a relatively
small drain-source bias $(V_{DS} = 0.1$ V, which corresponds to the linear regime
above threshold and saturation below). The semilog plot of the subthreshold
current gives estimates of the ideality factor, η, and the threshold voltage, V_T.
We note that V_T shifts towards higher voltage with increased negative substrate
bias, as expected. A comparison of the measured and calculated above-threshold
I_d versus V_{DS} characteristics for the same long channel device is shown in
Fig. 3.11.3.

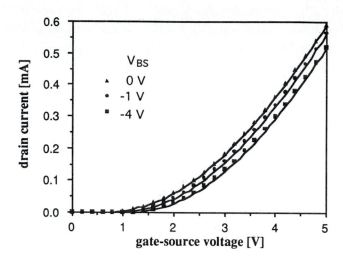

Fig. 3.11.1. Experimental (symbols) and calculated (solid lines) I_d versus V_{GS} characteristics in the saturation regime (V_{DS} = 5 V) for three values of substrate-source bias. The values obtained for the parameter α are: 1.164 (V_{BS} = 0 V), 1.094 (–1 V) and 1.059 (–4 V). $L = W = 20\,\mu$m. (From Park et al. (1991), © 1991 IEEE.)

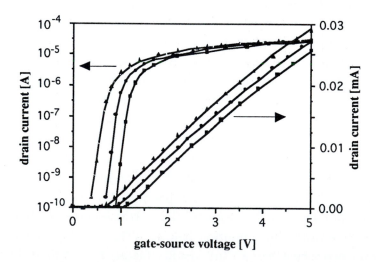

Fig. 3.11.2. Experimental (symbols) and calculated (solid lines) I_d versus V_{GS} characteristics for V_{DS} = 0.1 V and three values of substrate-source bias: 0 V (triangles), –1 V (circles) and –4 V (squares). Same device as in Fig. 3.11.1. (From Park et al. (1991), © 1991 IEEE.)

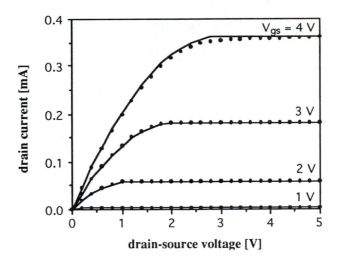

Fig. 3.11.3. Experimental (symbols) and calculated (solid lines) I_d versus V_{DS} characteristics for $V_{BS} = 0$ V. Same device as in Fig. 3.11.1. (From Park et al. (1991), © 1991 IEEE.)

As can be seen from these figures, the agreement is excellent for the entire range of drain-source, gate-source and substrate-source bias voltages. We notice the ability of the model to faithfully reproduce the various transition regions, i.e., between the linear and the saturation regimes (the knee region) in the above-threshold characteristics of Fig. 3.11.3 and between the above threshold and the subthreshold regimes shown in Fig. 3.11.2.

3.11.3. Intrinsic Strong Inversion NMOS Model

We now want to incorporate short channel effect in the NMOS model, based on the insight gained in the above discussion of the long channel case. For simplicity, we first consider only the strong inversion regime and, again, restrict the discussion to the intrinsic transistor. We shall return to more general models in the subsequent Subsections.

As mentioned in Subsection 3.11.1, it is difficult to obtain a manageable expression for the drain current based on the velocity-field relationship of eq. (3-11-4). Instead, Moon et al. (1991) used the following model by Grotjohn and Hoeflinger (1984) to approximate the linear, above-threshold regime of the NMOS I–V characteristics:

$$I_d \approx \frac{V_{GT}V_{DS} - \frac{\alpha}{2}V_{DS}^2}{R_n\sqrt{V_{DS}^2 + V_L^2}} \qquad (3\text{-}11\text{-}9)$$

Here, $R_n = (c_a W v_s)^{-1}$ and $V_L = v_s L_{eff}/\mu S$, where L_{eff} is the effective gate length.

The saturation region is described by the same type of expression as used for the PMOS (see Moon et al. (1991) and Section 3.10):

$$I_d = I_{sat}\left\{1 + \left(\frac{V_L V_\lambda}{V_L^2 + V_{SAT}^2}\right) \right. \\ \left. \times \ln\left[\sqrt{1 + \left(\frac{V_{DS} - V_{SAT}}{\theta V_\lambda}\right)^2} + \frac{V_{DS} - V_{SAT}}{\theta V_\lambda}\right]\right\} \qquad (3\text{-}11\text{-}10)$$

Here, V_{SAT} is the intrinsic saturation voltage, $V_\lambda = v_s\lambda/\mu S$ is a characteristic voltage related to the characteristic length, λ, of the saturated region (see Section 3.5), and θ is a parameter which should be adjusted to secure a continuous output conductance at the onset of saturation

The present models relies on the following empirical expression for the saturation voltage:

$$V_{SAT} = \frac{V_L V_{GT}}{\alpha V_L + \xi V_{GT}} \qquad (3\text{-}11\text{-}11)$$

This expression can be justified since it has the correct limiting behavior for long and short channel MOSFETs (see Section 3.5): For long channel devices, V_L becomes very large and V_{SAT} approaches the pinch-off voltage given by V_{GT}/α (α is the parameter used in UCCM (see eq. (3-11-1)). In short channel MOSFETs with velocity saturation, V_L becomes small and the saturation voltage approaches V_L/ξ. With the introduction of ξ as a modeling parameter, eq. (3-11-11) has been found to fit experimentally determined saturation voltages very well (see Moon et al. (1991)). Substituting eq. (3-11-11) into eq. (3-11-9), we find the following drain saturation current

$$I_d \approx \frac{V_{GT} V_{SAT} - \frac{\alpha}{2} V_{SAT}^2}{R_n \sqrt{V_{SAT}^2 + V_L^2}} \tag{3-11-12}$$

The constant θ is determined by equating the output conductances of eqs. (3-11-9) and (3-11-10) at the saturation point, resulting in

$$\theta = \frac{\left(\xi V_{GT} + \alpha V_L\right)\left(\xi V_{GT} + \frac{\alpha}{2} V_L\right)}{\xi\left(\xi V_{GT} + \alpha V_L\right)^2 - \frac{\alpha}{2}V_L V_{GT}} \tag{3-11-13}$$

In the parameter extraction, we also need the alternative definition of the saturation voltage presented in Section 3.10:

$$\left[\frac{I_d}{V_{DS}} \frac{\partial V_{DS}}{\partial I_d}\right]_{SAT} = \gamma \tag{3-11-14}$$

where the empirical parameter γ is obtained from the experimental current-voltage characteristics (see Subsection 3.11.5). Moon et al. (1991) showed that for short channel NMOS devices (note that Moon et al. defined γ differently),

$$V_{SAT} \approx \frac{2(\gamma - 1)}{\alpha(3\gamma - 2)} V_{GT} \left[1 - \frac{\gamma}{\gamma - 1}\left(\frac{R_n I_{sat}}{V_{GT}}\right)^2\right] \tag{3-11-15}$$

3.11.4. Extrinsic Strong Inversion NMOS Model (NMOSA2)

In short channel MOSFETs with effective gate lengths of about one micrometer or less, the parasitic source and drain resistances play a significant role in the overall device behavior. It is therefore important to include the effects of these resistances into our model in order to improve the efficiency of the circuit simulation and to be able to extract model parameters directly from measured characteristics.

The intrinsic NMOS model described in the previous Subsection can be converted to an extrinsic model by expressing the intrinsic bias voltages (upper case subscripts) in terms of their extrinsic counterparts (lower case subscripts), i.e.,

$$V_{DS} = V_{ds} - I_d R_t \tag{3-11-16}$$

$$V_{GT} = V_{gt} - I_d R_s \tag{3-11-17}$$

R_s and R_d are the source and drain series resistance, respectively, and $R_t = R_s + R_d$. Substituting eqs. (3-11-16) and (3-11-17) into eq. (3-11-9), we obtain the following analytical expression for the drain current in the linear regime in terms of extrinsic voltages:

$$I_d = \frac{2k_1}{h_1 + \sqrt{h_1^2 - 4g_1 k_1}} \tag{3-11-18}$$

where

$$k_1 = \left(V_{gt} V_{ds} - \alpha V_{ds}^2/2\right)\sqrt{V_{ds}^2 + V_L^2}$$

$$h_1 = \left[R_t V_{gt} - (\alpha R_t - R_s)V_{ds} + R_n\sqrt{V_{ds}^2 + V_L^2}\right]\sqrt{V_{ds}^2 + V_L^2} \tag{3-11-19}$$

$$g_1 = R_t R_n V_{ds} + R_t (R_s - \alpha R_t/2)\sqrt{V_{ds}^2 + V_L^2}$$

The saturation current is determined from eq. (3-11-9) evaluated at the saturation point, using eqs. (3-11-11) and (3-11-17),

$$I_{sat} = \frac{2k_2}{h_2 + \sqrt{h_2^2 - 4g_2 k_2}} \tag{3-11-20}$$

where

$$k_2 = \left(\xi V_{gt} + \alpha V_L/2\right) V_{gt}^2$$

$$h_2 = 3 R_s \xi V_{gt}^2 + \alpha R_s V_L V_{gt} + R_n \left(\xi V_{gt} + \alpha V_L\right)\sqrt{\left(\xi V_{gt} + \alpha V_L\right)^2 + V_{gt}^2}$$

$$g_2 = 3 R_s^2 \xi V_{gt} + \alpha R_s^2 V_L + 2R_s R_n \frac{\xi\left(\xi V_{gt} + \alpha V_L\right)^2 + \left(\xi V_{gt} + \alpha V_L/2\right)V_{gt}}{\sqrt{\left(\xi V_{gt} + \alpha V_L\right)^2 + V_{gt}^2}}$$

$$\tag{3-11-21}$$

The intrinsic and extrinsic saturation voltages are obtained from eqs. (3-11-12), (3-11-16), (3-11-17) and (3-11-20):

$$V_{SAT} = \left(\frac{\xi}{V_L} + \frac{\alpha}{V_{gt} - R_s I_{sat}} \right)^{-1} \tag{3-11-22}$$

$$V_{sat} = V_{SAT} + I_{sat} R_t \tag{3-11-23}$$

In saturation, the drain current can be found from eq. (3-11-10) by substituting V_{DS} and V_{SAT} with the corresponding extrinsic voltages inside the logarithm (there is no need to substitute for V_{SAT} in the factor in front of the logarithm). The resulting expression cannot be solved analytically with respect to the current, but using the same type of arguments as for the PMOS (see Section 3.10), we obtain the following approximate solution:

$$I_d = I_{sat} \left\{ 1 + \left(\frac{V_L V_\lambda}{V_L^2 + V_{SAT}^2} \right) \right.$$
$$\left. \times \ln \left[\sqrt{1 + \left(\frac{V_{ds} - V_{sat}}{\theta V_\lambda} \right)^2} + \frac{V_{ds} - V_{sat}}{\theta V_\lambda} \right] \right\} \tag{3-11-24}$$

3.11.5. Parameter Extraction Using Strong Inversion NMOS Model

The parameter extraction procedure for NMOS follows to a large extent that used for PMOS in the preceding Section (see Subsection 3.10.5). Here, we demonstrate the technique for NMOS devices with nominal gate lengths of 0.8 μm, 1.0 μm, and 1.2 μm. All samples have an oxide thickness of 220 Å and a nominal gate width of 20 μm.

We first consider the threshold voltage. The threshold voltage, V_T, can be defined by several techniques. Here, we use the technique which identifies V_T as the gate-source voltage at which the gate-channel capacitance, c_{gc}, drops to one third of its maximum value (see Section 3.3). However, since the gate-channel capacitance is proportional to the device transconductance, g_m, at low drain-source voltage, the threshold voltage may equally well be determined as the gate-source voltage where g_m reaches one third of its maximum value. A suitable drain-source voltage for this procedure is $V_{ds} \approx 10$ mV. Alternatively, V_T can be determined from the subthreshold characteristics as discussed in Section 3.9, by roughly estimating the threshold current. (For a comparison of various techniques of finding V_T, see Park et. al (1991).)

The total parasitic resistance, R_t, is determined as the inverse of the linear channel conductance, g_{ch}, at large gate-source voltages. A good estimate of R_t is obtained by extrapolating a plot of $1/g_{ch}$ versus $1/V_{gt}$ to the intercept with the $1/g_{ch}$-axis (i.e., $1/V_{gt} = 0$, see Section 3.9). In general, such a plot may be non-linear since the mobility at small drain-source bias, μ_S, is a function of V_{GS} (see eq. (3-11-3)). On the other hand, the mobility for fixed gate-source biases can be obtained from plots of $1/g_{ch}$ versus the channel length. From eqs. (3-9-27) and (3-9-28) in Section 3.9, we see that mobility is determined from the slopes of such plots, provided that μ_S does not vary much with L and that the different samples are fabricated by the same process. We also notice that such plots for different V_{gs} should intersect at $1/g_{ch} = R_t$ and $L = \Delta L$, where ΔL is the difference between the nominal and effective channel lengths. Fig. 3.11.4 shows such a plot based on a series of NMOS devices.

For the present series of NMOS devices, we find $R_t \approx 70$ ohm and $\Delta L \approx$ 0.13 μm.

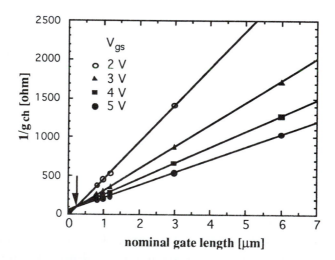

Fig. 3.11.4. Plot of inverse linear channel conductance versus nominal gate length for different gate-source voltages. The plots intersect at $1/g_{ch} = R_t$ and $L = \Delta L$ (arrow) The slopes are inversely proportional to the mobility. (From Moon et al. (1991), © 1991 IEEE.)

The determination of the saturation current and the saturation voltage is an important part of the parameter extraction procedure. For $V_{ds} - V_{sat} \gg \theta V_\lambda$, we obtain from eq. (3-11-24)

$$I_{sat} \frac{\partial V_{ds}}{\partial I_d}\bigg|_{\text{deep saturation}} \approx \frac{V_L^2 + V_{SAT}^2}{V_L V_\lambda}(V_{ds} - V_{sat}) \qquad (3\text{-}11\text{-}25)$$

Hence, the saturation voltage, V_{sat}, can be extracted by plotting experimental data for the function on the left hand side of eq. (3-11-25) versus V_{ds}, and extrapolating to the intercept with $I_{sat}(\partial V_{ds}/\partial I_d) = 0$. The saturation current is subsequently determined from the experimental I–V characteristics at $V_{ds} = V_{sat}$. This technique is demonstrated in Fig. 3.11.5 for the NMOS device with nominal gate length $L = 1.0\ \mu m$. After determining the saturation point, the intrinsic saturation voltage is easily calculated from eq. (3-11-23).

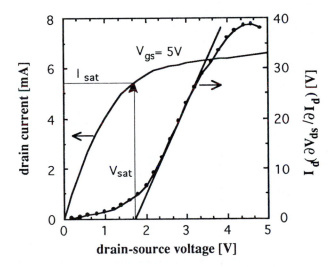

Fig. 3.11.5. Procedure for determination of the extrinsic saturation voltage, V_{sat}, and the saturation current, I_{sat}, from measured I–V characteristics. The data are obtained for an NMOS with nominal gate length $L = 1.0\ \mu m$. (From Moon et al. (1991), © 1991 IEEE.)

Previously, we also defined the saturation point in terms of the parameter γ, see eq. (3-11-14). Although the definition of γ was based on intrinsic bias voltages, we proceed as for PMOS (see Section 3.10) and replace the derivative $\partial I_d/\partial V_{DS}$ in eq. (3-11-14) with $\partial I_d/\partial V_{ds}$, which can readily be obtained from the experimental characteristics. (Note that V_{SAT} and I_{sat} were determined above.) The value $\gamma \approx 3.0$ is obtained for our devices by averaging over a number of characteristics representing several values of gate-source and substrate-source bias. (We note that the value of γ is technology dependent and should be independently extracted for each set of devices.)

Next, we want to find the parameters α and R_n used in eq. (3-11-9) which describes the linear region of the NMOS I–V characteristics. For this purpose, it is useful to rewrite eq. (3-11-15) as follows:

$$\frac{V_{GT}}{V_{SAT}} \approx \frac{\alpha(3\gamma - 2)}{2(\gamma - 1)} + \left(\frac{\gamma}{\gamma - 1}\right)\frac{R_n^2 I_{sat}^2}{V_{GT}\,V_{SAT}} \qquad (3\text{-}11\text{-}26)$$

Fig. 3.11.6 shows plots of V_{GT}/V_{SAT} versus $I_{sat}^2/(V_{GT}V_{SAT})$ for devices with different gate lengths, based on the values of I_{sat} and V_{SAT} determined as described above. (V_{GT} is obtained from eq. (3-11-17).)

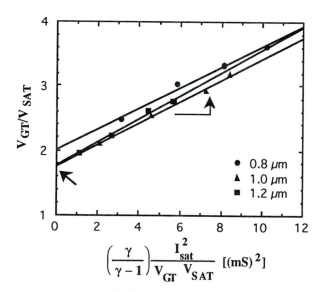

Fig. 3.11.6. Plots for determining the parameters α (from intercept with vertical axis) and R_n (from slope) shown for different values of nominal gate length. (From Moon et al. (1991), © 1991 IEEE.)

We see from eq. (3-11-26) that such plots should be straight lines with slopes equal to $\gamma R_n^2/(\gamma - 1)$, intercepting the V_{GT}/V_{SAT}-axis (i.e., $I_{sat}^2/(V_{GT}V_{SAT}) = 0$) at $V_{GT}/V_{SAT} = \alpha(3\gamma - 2)/(2(\gamma - 1))$. From the value of R_n, we can determine the saturation velocity. Extracted values of α and v_s for the different samples are shown in Table 3.11.1. This value of v_s is in good agreement with that reported by Chan et al. (1990).

Next, we extract the values of the characteristic voltage, V_λ, and the characteristic length, λ, of the saturated region of the channel. We observe from eq. (3-11-25) that the slope of the straight line $I_{sat}(\partial V_{ds}/\partial I_d)$ versus $V_{ds} - V_{sat}$, valid for deep saturation (see Fig. 3.11.5), is equal to $(V_L^2 + V_{SAT}^2)/(V_LV_\lambda)$. Hence, since V_L and V_{SAT} have already been determined, V_λ is obtained from this slope and λ is calculated as $\mu_S V_\lambda/v_s$. As shown in Table 3.11.1, the value of λ is near 0.1 μm for the devices investigated. This is close to the value obtained for PMOS (see Section 3.10) and suggested for HFETs by Moon et al. (1990). It is also in agreement with the value of λ proposed by Ko (1989).

Finally, we determine the saturation voltage parameter, ξ, from eqs. (3-11-22) and (3-11-23) using the experimentally obtained values of V_{SAT} and I_{sat} (see Fig. 3.11.5).

L [μm]	0.8	1.0	1.2
R_t [ohm]		70	
ΔL [μm]		0.13	
V_T [V]	0.60	0.65	0.62
α	1.15	0.99	1.00
v_s [10^4 m/s]	8.1	7.8	7.7
λ [Å]	957	994	1,080
ξ	0.82	0.79	0.84
γ		3.0	

Table 3.11.1. Extracted device parameters for NMOS devices with nominal gate lengths between 0.8 μm and 1.2 μm, and width 20 μm. (From Moon et al. (1991).)

In Fig. 3.11.7, we show comparisons of experimental I–V characteristics for the 1.0 μm NMOS with curves based on the present model. The comparison is performed using model calculations based on the extracted parameters given in

Table 3.11.1, with no adjustment of parameters. As can be seen in this figure, the agreement between the measured and the calculated curves is excellent. In particular, the transition region from the linear to the saturation regime is reproduced quite accurately. A correspondingly good agreement was also achieved for the other NMOS devices investigated (see Moon et al. (1991)).

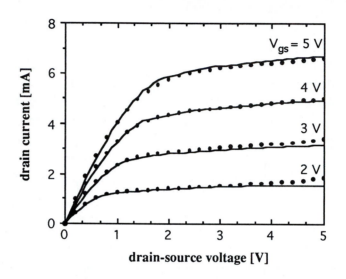

Fig. 3.11.7. Measured (symbols) and calculated (solid lines) I-V characteristics for an NMOS with nominal gate length 1.0 μm, substrate-source bias $V_{bs} = 0$ V. (From Moon et al. (1991), © 1991 IEEE.)

3.11.6. Intrinsic Unified Short Channel NMOS Model (NMOSA3)

Our unified NMOS model is similar to the unified PMOS model described in Section 3.10. In other words, we replace the parameter α with α^* and the gate voltage swing, V_{GT}, with an_S in the equations of the strong inversion model:

$$I_d = \frac{1}{R_n} \frac{\left(an_S - \frac{\alpha^*}{2} V_{DS}\right) V_{DS}}{\sqrt{V_{DS}^2 + V_L^2}} \tag{3-11-27}$$

In eq. (3-11-27),

$$R_n = \frac{a}{qv_s W} \tag{3-11-28}$$

and

$$\alpha^* = \frac{\alpha}{1 + \dfrac{2\eta V_{th}}{a n_S}} \qquad (3\text{-}11\text{-}29)$$

Here, n_S is the sheet density of electrons at the source, $V_L = v_S L_{eff}/\mu_S$, and L_{eff} is the effective gate length.

The drain current in the saturation regime is given by (see Subsection 3.11.3)

$$I_d = I_{sat} \left\{ 1 + \left(\frac{V_L V_\lambda}{V_L^2 + V_{SAT}^2} \right) \right.$$

$$\left. \times \ln \left[\sqrt{1 + \left(\frac{V_{DS} - V_{SAT}}{\theta V_\lambda} \right)^2} + \frac{V_{DS} - V_{SAT}}{\theta V_\lambda} \right] \right\} \qquad (3\text{-}11\text{-}30)$$

where V_{SAT} is the intrinsic saturation voltage, $V_\lambda = v_s \lambda / \mu_S$ is the characteristic voltage proportional to the characteristic saturation region length, λ (see Section 3.5), and θ is a parameter determined from the requirement of output conductance continuity at the onset of the saturation regime.

The expression for the saturation voltage, V_{SAT}, is given by (compare with eq. (3-11-11))

$$V_{SAT} = \left(\frac{\alpha^*}{a n_S} + \frac{\xi}{V_L} \right)^{-1} \qquad (3\text{-}11\text{-}31)$$

This expression yields the correct limiting behavior both above and below threshold. Substituting eq. (3-11-31) into eq. (3-11-27), we find the drain saturation current

$$I_{sat} = \frac{1}{R_n} \frac{\left(a n_S - \dfrac{\alpha^*}{2} V_{SAT} \right) V_{SAT}}{\sqrt{V_{SAT}^2 + V_L^2}} \qquad (3\text{-}11\text{-}32)$$

The constant θ is determined by equating the output conductances given by eqs. (3-11-27) and (3-11-30) at the saturation point, resulting in (compare with eq. (3-11-13))

$$\theta = \frac{\left(\xi a n_S + \alpha^* V_L\right)\left(\xi a n_S + \frac{\alpha^*}{2} V_L\right)}{\xi\left(\xi a n_S + \alpha^* V_L\right)^2 - \frac{\alpha^*}{2} V_L \, a n_S}$$

(3-11-33)

3.11.7. Parameter Extraction for Unified Short Channel NMOS Model

To a large extent, the parameter extraction for the present model follows the procedure for strong inversion NMOS described in Subsection 3.11.5, and is similar to that for the unified PMOS model (see Subsection 3.10.7). Here, we demonstrate the parameter extraction technique for a short channel NMOS device with a nominal gate length of 1.0 μm, a gate width of 20 μm, and a gate oxide thickness of 220 Å (the same device as described in Subsection 3.11.4). The parameter values determined by this extraction technique are shown in Table 3.11.2. Note that some parameter values have been changed from the strong inversion case (compare Tables 3.11.1 and 3.11.2). This is due to the difference in the charge control model and due to the incorporation of DIBL.

V_{bs} [V]	0
α	1.05
v_S [cm/sec]	8.6×10^6
ξ	0.76
ΔL [μm]	0.12
R_t [ohm]	70
V_{To} [V]	0.53
λ [Å] (Vgs=5V)	858
γ	3.0
σ	0.03
η	1.42

Table 3.11.2. Extracted device parameters for a unified short channel NMOS model for a device with nominal gate length 1.0 μm and gate width 20 μm. (From Park (1992).)

The measured and calculated current-voltage characteristics (using parameter values listed in Table 3.11.2) are shown in Figs. 3.11.8 to 3.11.11. As can be seen from these figures, the agreement between the measured and the calculated curves is quite satisfactory for the entire range of gate and drain biases, including the transition region from the linear to the saturation regime.

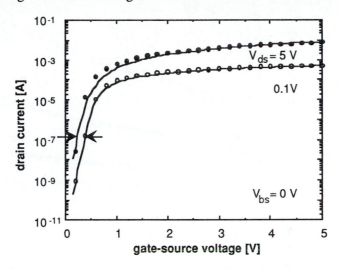

Fig. 3.11.8. Measured (symbols) and calculated (solid lines) near threshold $I–V$ characteristics for NMOS with gate length 1.0 μm ($L_{eff} = 0.88$ μm). (From Park (1992).)

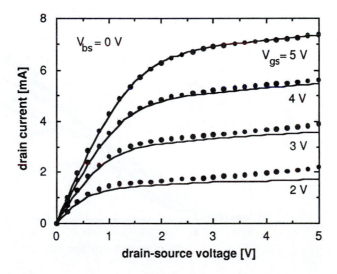

Fig. 3.11.9. Measured (symbols) and calculated (solid lines) $I–V$ characteristics for NMOS with gate length 1.0 μm ($L_{eff} = 0.88$ μm). (From Park (1992).)

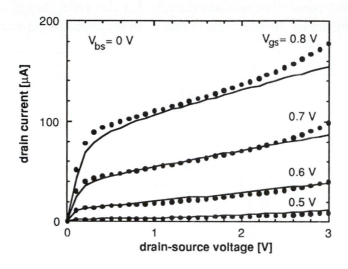

Fig. 3.11.10. Measured (symbols) and calculated (solid lines) $I-V$ characteristics near threshold for NMOS with gate length 1.0 μm (L_{eff} = 0.88 μm). (From Park (1992).)

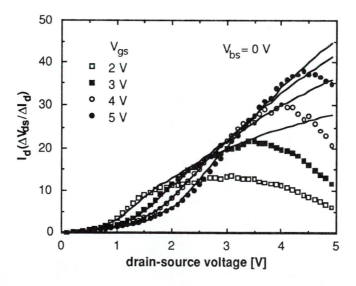

Fig. 3.11.11. Measured (symbols) and calculated (solid lines) $I_d (\partial V_{ds} / \partial I_d)$ versus V_{ds} characteristics for NMOS with gate length 1.0 μm (L_{eff} = 0.88 μm). (From Park (1992).)

3-12. UNIFIED QUASI-STATIC CAPACITANCE-VOLTAGE MODELING

In Section 3.9, we developed a semi-empirical C–V model based on the concept of a unified channel capacitance. This theory is an extension of Meyer's model (Meyer (1971)) which does not accurately take into account the bulk (depletion) charge. The problem of charge non-conservation related to the contribution of the bulk charge was considered by Ward and Dutton (1978). In this Section, we present the results of a more accurate model based on the Unified Charge Control Model (UCCM) and compare these results with the BSIM charge control model which is embedded into the popular circuit simulator SPICE3 (see Jeng et al. (1987)) and with our experimental data. Our new model follows the analysis given by Rho et al. (1992). This model includes the bulk charge and is more accurate in the near threshold regime. It is based on the Quasi-Static Approximation (QSA) and provides continuous C–V expressions for all gate, drain, and substrate biases, from subthreshold to above-threshold, and from the linear to the saturation regime. At the same time, our model is simple enough to be easily incorporated into circuit simulators.

Our calculation of the channel and bulk charges, and the partitioning of the channel charge are based on the QSA. This approximation assumes that the channel charges depend on the terminal voltages only and not on their history. This approach is only valid when the characteristic times of the signal variation are much larger than the carrier transit time which is approximately $(2\pi f_T)^{-1}$ where $f_T = g_m/(2\pi C_i)$ is the cutoff frequency, g_m is the transconductance and C_i is the gate insulator capacitance (see, for example, Shur (1990)). Hence, this approximation should be valid up to frequencies on the order of $f_T/2$. This corresponds to the most interesting range of frequencies of operation and time transients. A more detailed discussion of the validity and limitations of the QSA was given, for example, by Oh et al. (1980), Turchetti and Masetti (1983), Yang (1986), and Park et al. (1987).

An accurate calculation of MOSFET C–V characteristics has to account for the bulk charge which is affected by non-uniform doping profiles and short channel effects. In our approach, we relate the voltage dependencies of the bulk charge to the standard parameters of the body plots which are routinely measured during MOSFET characterization. This allows us to express the MOS capacitances in terms of the body effect parameters, γ_{Neff} and Γ_{Neff}, which can readily be obtained from the threshold voltage measurements (see Section 3.7).

3.12.1. The Inversion Sheet Charge Density

We find the sheet density of inversion electrons, n_s, in an n-channel MOSFET from the UCCM expression (see Sections 3.3 and 3.4)

$$V_{GS} - V_{TX} - V_F \approx \eta V_{th} \ln\left(\frac{n_s}{n_o}\right) + a\left(n_s - n_o\right) \qquad (3\text{-}12\text{-}1)$$

where V_F is the quasi-Fermi potential measured relative to that at the source side of the channel, V_{TX} is a threshold voltage which is dependent on the position, x, along the channel, V_{th} is the thermal voltage, η is the subthreshold ideality factor, n_o is the sheet density of inversion electrons at threshold (see Section 3.3), and $a = q(d_i + \Delta d)/\varepsilon_i \approx q/c_i$ where $c_i = d_i/\varepsilon_i$ is the gate insulator capacitance per unit area, d_i is the insulator thickness, Δd is the effective thickness of the two dimensional electron gas (which is typically much smaller than d_i for Si MOSFETs, see Section 3.3), and ε_i is the dielectric permittivity of the gate insulator. The parameter α, which was introduced in Section 3.4, is not used in eq. (3-12-1). Instead, in order to consider the depletion charge more carefully, the position dependent threshold voltage, V_{TX}, is adopted (see Subsection 3.4.3).

Since most modern MOSFETs have a non-uniform doping profile for threshold adjustment and show severe short channel effects, we cannot obtain a both simple and accurate expression for V_{TX}. Hence, as discussed in Section 3.7, the following empirical relationship can be expressed in terms of the body effect parameters, γ_{Neff} and Γ_{Neff}, which can readily be obtained from threshold voltage measurements (see, for example, Sheu et al. (1987) and Section 3.7):

$$V_{TX} = V_{FB} + 2\varphi_b + \gamma_{Neff}\sqrt{2\varphi_b + V_F - V_{BS}} - \Gamma_{Neff}(2\varphi_b + V_F - V_{BS}) \quad (3\text{-}12\text{-}2)$$

Here, V_{FB} is the flat-band voltage, V_{BS} is the substrate-source voltage, and $\varphi_b = V_{th}\ln(N_a/n_i)$ is the absolute potential difference between the intrinsic Fermi level and the Fermi level in the substrate bulk, N_a is the acceptor doping density in the substrate, and n_i is the intrinsic carrier density. The region of validity of eq. (3-12-2) was discussed in Section 3.7 (see eqs. (3-7-24) and (3-7-25)).

γ_{Neff} and Γ_{Neff} are empirical body effect constants determined from measurements and include relevant information about non-uniform bulk doping and/or short channel effects. In the ideal case of a uniform bulk doping and a long channel device, the "effective" body effect constants reduce to: $\Gamma_{Neff} = 0$, and $\gamma_{Neff} = \gamma_N = (2\varepsilon_s\, qN_a)^{1/2}/c_i$. (see Section 3.7).

3.12.1. The Bulk Sheet Charge Density

Just as was done for the inversion charge, we will express the bulk sheet charge density, Q_{dep}, in terms of the body plot parameters, γ_{Neff} and Γ_{Neff}. The bulk charge is a unique function of the band bending. Hence, once we determine the band bending for an arbitrary bias condition, we can equate Q_{dep} for this set of bias voltages to that of a different set which gives the same band bending. In particular, we are interested in identifying the conditions at which the channel just reaches threshold while retaining the initial values of the band bending and the quasi-Fermi potential. This threshold condition is reached for an "equivalent" substrate-source bias, $V_{BS\text{-}eq}$, from which the corresponding threshold voltage, $V_{TX}(V_{BS\text{-}eq})$, and Q_{dep} can be calculated.

We start our analysis by first considering a transistor with zero applied drain-source bias, i.e., with $V_F = 0$ in the channel. The resulting expression for Q_{dep} can then easily be generalized to account for an arbitrary V_F. Fig. 3.12.1a shows the energy diagram of an n-type MOSFET channel in the subthreshold regime for a given set of bias voltages, V_{GS}, V_{BS}, and $V_{DS} = 0$, corresponding to the band bending $q\psi_B$. Assuming the same band bending as in Fig. 3.12.1a, we obtain the energy band diagram at threshold as indicated in Fig. 3.12.1b. In this figure is shown the "equivalent" substrate-source voltage, $V_{BS\text{-}eq}$, and the corresponding gate-source voltage which is identical to the threshold voltage $V_{TX}(V_{BS\text{-}eq})$.

Considering Fig. 3.12.1a, we observe that the gate-substrate potential difference, V_{GB}, is the sum of voltage drops across the gate oxide and the silicon substrate, so that, accounting for the flat-band voltage, we obtain

$$V_{GB} - V_{FB} = \psi_B + \frac{1}{c_i}(qn_s - Q_{dep}) \tag{3-12-3}$$

(here, Q_{dep} is negative). By inspection of Fig. 3.12.1a, we find

$$|\psi_B| + |\psi_S| = |V_{FB}^*| + |V_{BS}| \tag{3-12-4}$$

where ψ_S is the difference between the surface potential and the quasi-Fermi level (here, $V_F = 0$). Taking into account the actual signs of the voltages in eq. (3-12-4), we obtain

$$\Psi_B = \Psi_S - V_{FB}^* - V_{BS} \tag{3-12-5}$$

where

$$V_{FB}^* = -\varphi_b - \frac{E_g}{2q} \tag{3-12-6}$$

and E_g is the band gap energy. (Note that for polysilicon n^+ gates, $V_{FB}^* \approx V_{FB}$ (see Fig. 3.3.3).)

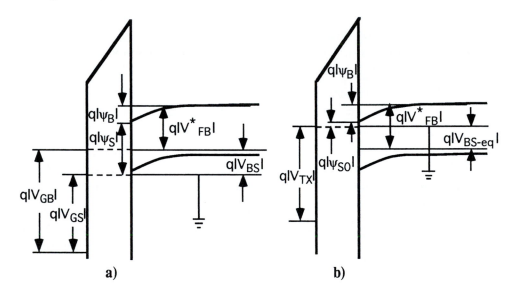

Fig. 3.12.1. Energy band diagrams of an n-channel MOSFET with applied bias voltages V_{GS} and V_{BS} (a), and at threshold with the "equivalent" biases $V_{GS} = V_{TX}$ and $V_{BS} = V_{BS\text{-}eq}$ (b). The source and drain terminals are grounded, i.e., $V_F = 0$ in the channel. The "equivalent" bias conditions in (b) are chosen such that the substrate band bending, $q\psi_B$, is the same as in (a).

Combining eqs. (3-12-3) and (3-12-5), we obtain

$$V_{GS} - V_{FB} + V_{FB}^* = \Psi_S + \frac{1}{c_i}(qn_s - Q_{dep}) \tag{3-12-7}$$

At the inversion threshold, $\psi_S = \psi_{SO}$ where

$$\psi_{SO} \approx -\left|V_{FB}^*\right| - 2\varphi_b \tag{3-12-8}$$

For the band diagram of Fig. 3.12.1b, we can relate the absolute value of the equivalent substrate bias, $|V_{BS\text{-}eq}|$, to the band bending as follows:

$$\left|\psi_B\right| + \left|\psi_{SO}\right| = \left|V_{FB}^*\right| - \left|V_{BS\text{-}eq}\right| \tag{3-12-9}$$

Taking into account the actual signs of the voltages in eq. (3-12-9), we obtain

$$\psi_B = \psi_{SO} - V_{FB}^* - V_{BS\text{-}eq} \tag{3-12-10}$$

Since the band bending in Figs. 3.12.1a and 3.12.1b are assumed to be identical, we can find an expression for $V_{BS\text{-}eq}$ by eliminating ψ_B between eqs. (3-12-5) and (3-12-10):

$$V_{BS\text{-}eq} - V_{BS} = -\psi_S + \psi_{SO} \tag{3-12-11}$$

The meaning of this equation is apparent: The same band bending, $q\psi_B$, can be obtained for two different surface potentials, ψ_S and ψ_{SO}, if the difference between them is exactly compensated via the substrate bias. Eliminating ψ_{SO} between eqs. (3-12-8) and (3-12-11), we can also express $V_{BS\text{-}eq}$ as follows:

$$V_{BS\text{-}eq} = V_{BS} - \psi_S + V_{FB}^* - 2\varphi_b \tag{3-12-12}$$

Now, an expression for the threshold voltage $V_{TX}(V_{BS\text{-}eq})$ can be derived by substituting $V_{BS\text{-}eq}$ for V_{BS} in eq. (3-12-2), recalling that $V_F = 0$. The bulk sheet charge density at threshold, denoted $Q_{dep}^{(o)}$, can be obtained in terms of this threshold voltage using a simple depletion charge relationship (see eqs. (3-3-23) and (3-3-25) in Section 3.3):

$$Q_{dep}^{(o)}(V_{BX\text{-}eq}) = -c_i\left[V_{TX}(V_{BX\text{-}eq}) - 2\varphi_b - V_{FB}\right]$$

$$= -c_i\left[\gamma_{Neff}\sqrt{-V_{BS} - V_{FB}^* + \psi_S} - \Gamma_{Neff}(-V_{BS} - V_{FB}^* + \psi_S)\right] \tag{3-12-13}$$

This result can now be generalized to an arbitrary value of the quasi-Fermi potential simply by making the substitution: $\psi_S \rightarrow \psi_S + V_F$, i.e.,

$$Q_{dep}^{(o)}(V_{BX-eq}) = -c_i\left[V_{TX}(V_{BX-eq}) - 2\varphi_b - V_{FB}\right]$$

$$= -c_i\left[\gamma_{Neff}\sqrt{V_F - V_{BS} - V_{FB}^* + \psi_S}\right.$$

$$\left. - \Gamma_{Neff}(V_F - V_{BS} - V_{FB}^* + \psi_S)\right] \qquad (3\text{-}12\text{-}14)$$

According to the discussion in Section 3.7, this result is valid up to $(V_F - V_{BS} - V_{FB}^* + \psi_S) = [\gamma_{Neff}/(2\Gamma_{Neff})]^2$. In the subthreshold regime (up to threshold), where the depletion approximation is accurate, Q_{dep} given by eq. (3-12-7) is identical to $Q_{dep}^{(o)}$. However, in the inversion regime, these two quantities do not quite coincide because of an additional small voltage drop in the inversion channel (see, for example, Sheu et al. (1987)). This can be accounted for by expanding the depletion charge in the inversion regime into a Taylor series in n_s, keeping only the first term in the expansion since Q_{dep} is only a weak function of n_s, i.e.,

$$Q_{dep} = Q_{dep}^{(o)} + \theta q n_s \qquad (3\text{-}12\text{-}15)$$

Substituting eq. (3-12-15) into eq. (3-12-7) and accounting for V_F, we obtain:

$$V_{GS} - V_{FB} + V_{FB}^* \approx \psi_S + V_F + \frac{q n_s}{c_b} - \frac{Q_{dep}^{(o)}}{c_i} \qquad (3\text{-}12\text{-}16)$$

where

$$c_b = \frac{c_i}{1-\theta} \qquad (3\text{-}12\text{-}17)$$

(Typically, $\theta \approx -0.05$ and $c_b \approx 0.95\, c_i$.) However, numerical calculations show that the effect of the difference between c_b and c_i on the computed $C-V$ characteristics is typically quite small (on the order of 1%). Hence, we can assume that $c_b \approx c_i$ and $Q_{dep} \approx Q_{dep}^{(o)}$.

The inversion sheet charge density, $q n_s$, was already calculated in Subsection 3.12.1. Hence, eliminating ψ_S between eqs. (3-12-14) and (3-12-16), we find:

$$Q_{dep} \equiv -q n_B = -c_i(\gamma_{Neff}\, y - \Gamma_{Neff}\, y^2) \qquad (3\text{-}12\text{-}18)$$

where

$$y = \frac{2}{\gamma_{Neff}} \frac{\left(V_{GS} - \frac{qn_s}{c_i} - V_{BS} - V_{FB}\right)}{1 + \sqrt{1 + \frac{4(1-\Gamma_{Neff})}{\gamma_{Neff}^2}\left(V_{GS} - \frac{qn_s}{c_i} - V_{BS} - V_{FB}\right)}} \qquad (3\text{-}12\text{-}19)$$

This result is valid when $V_{GS} - qn_s/c_i - V_{BS} - V_{FB} < (1+\Gamma_{Neff})[\gamma_{Neff}/(2\Gamma_{Neff})]^2$. Otherwise, Q_{dep} is saturated and has a constant value of $-c_i\Gamma_{Neff}[\gamma_{Neff}/(2\Gamma_{Neff})]^2$.

3.12.2. The Three Terminal MOS Capacitances

The three terminal MOS capacitances are defined for the MOSFET source and drain terminals tied together and biased to the potential V_{SB} with respect to the substrate. The gate capacitance, c_g, per unit area for this situation can be written as

$$c_g = c_{sg} + c_{bg} \equiv \left.\frac{q\partial n_s}{\partial V_{GB}}\right|_{V_{SB}} - \left.\frac{\partial Q_{dep}}{\partial V_{GB}}\right|_{V_{SB}} \qquad (3\text{-}12\text{-}20)$$

where c_g is composed of two components: the gate-channel capacitance, c_{sg}, and the gate substrate capacitance, c_{bg}. In the subthreshold regime, c_{sg} is negligible and c_{bg} decreases gradually with increasing V_{GB} due to the increase of the bulk depletion region. With an increase in V_{GB} above threshold, c_{sg} rapidly approaches c_i and c_{bg} tends to zero. Fig. 3.12.2 shows the calculated values of the three terminal MOS capacitances for different values of V_{SB} As can be seen from the figure, our unified model based on the calculation of qn_s and Q_{dep} accurately reflects the physics of the three terminal MOS structure (see Section 3.3).

3.12.3. The Four Terminal MOSFET Capacitances

The total channel charge, Q_N, and the total bulk charge, Q_B, are found by integrating the inversion sheet charge density, qn_s, and the depletion sheet charge density, Q_{dep}, along the channel. This integration requires the knowledge of the potential distribution in the channel. This distribution depends on the device structure, the velocity-field relationship, and the bias conditions and cannot be evaluated analytically.

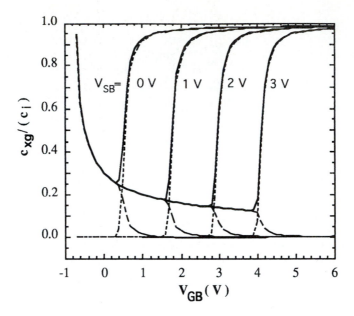

Fig. 3.12.2. Normalized three terminal MOS gate capacitances for different substrate-source biases, calculated using unified expressions for the inversion charge and the depletion charge (x = s (channel), b (substrate)). c_{sg} – dotted lines, c_{bg} – dashed lines, and c_g – solid lines. The threshold voltage data in Fig. 3.7.9 are used for the calculations. (From Rho et al. (1992), © 1992 IEEE.)

Therefore, we use a reasonable and fairly accurate approximation for these charges based on a three point Simpson's integration formula:

$$Q_N = -qW \int_0^L n_s \, dx \approx -\frac{qWL}{6} \left(n_{ss} + 4n_{sm} + n_{sd} \right) \tag{3.12.21}$$

$$Q_B = W \int_0^L Q_{dep} \, dx \approx -\frac{qWL}{6} \left(n_{Bs} + 4n_{Bm} + n_{Bd} \right) \tag{3.12.22}$$

where n_{ss}, n_{sd}, n_{sm} and n_{Bs}, n_{Bd}, n_{Bm} are the channel and bulk sheet densities, respectively, evaluated at the source, the drain, and in the middle of the channel. This approach is acceptable since n_s and n_B are slowly varying functions of the coordinate along the channel in the region that determines the integrals. The calculation of n_{ss}, n_{Bs} n_{sd} and n_{Bd} from eqs. (3.12.1) and (3.12.18) is straightforward since we know V_F at these points.

In order to find n_{sm} and n_{Bm}, we must know $V_{F(L/2)} \equiv V_F(x = L/2)$. Above threshold, $V_{F(L/2)}$ is found by combining eqs. (3-3-31) and (3-3-34), and integrating the resulting equation from 0 to $L/2$ (see Problem P3-12-2). This results in the following quadratic expression in $V_{F(L/2)}$:

$$I_d = \frac{2qc_iW\mu}{L}\left[V_{GT}V_{F(L/2)} - \frac{V_{F(L/2)}^2}{2} \right] \qquad (3.12.23)$$

For short channel devices, we can use the same approach if we substitute the drain-source bias, V_{ds}, with the effective drain-source bias, V_{dse}, which is equal to V_{ds} in the linear region and approaches asymptotically the saturation value, V_{sat}, for $V_{ds} > V_{sat}$ (see Section 3.9 for details).

Under the quasi-static approximation, the channel inversion charge, Q_N, can be partitioned into two parts, $Q_S = F_pQ_N$ and $Q_D = (1 - F_p)Q_N$, assigned to the source and the drain, respectively. Partitioning theories by Ward and Dutton (1978) and Tsividis (1987) suggest the value $F_p \approx 0.6$ for the partitioning factor. This factor is about 0.5 in the linear regime. Tsividis (1988) suggested the following expression which satisfies both limiting cases:

$$F_p = \frac{3 + 6\beta + 4\beta^2 + 2\beta^3}{5(1 + 2\beta + 2\beta^2 + \beta^3)} \qquad (3.12.24)$$

where

$$\beta = \begin{cases} 1 - \dfrac{V_{DS}}{V_{SAT}}, & \text{for } V_{DS} \leq V_{SAT} \\[2ex] 0, & \text{for } V_{DS} > V_{SAT} \end{cases} \qquad (3.12.25)$$

However, just as in the BSIM model, we have the option of using a different charge partitioning since the value $F_p \approx 0.6$ may not be realistic for certain circuit simulations such as those of switched capacitor filters.

All independent device capacitances are derivatives of the various charges, Q_S, Q_D, Q_B, with respect to the terminal voltages, i.e.,

$$C_{ij} = \chi_{ij}\frac{\partial Q_i}{\partial V_j}\bigg|_{V_{k \neq j} = 0} \qquad (3\text{-}12\text{-}26)$$

where the indices i, j, k represent either of the four terminals (gate, source, drain or substrate), and $\chi_{ij} = -1$ for $i \neq j$ and $\chi_{ij} = +1$ for $i = j$. The equations for the charges were derived above. The differentiation of these expressions should normally be done numerically since the analytical expressions are too cumbersome.

Measured data and model calculations for the gate and bulk capacitances are compared in Figs. 3.12.3 and Fig. 3.12.4, respectively. The measurements were done on long channel devices using the technique proposed by Weng and Yang (1985).

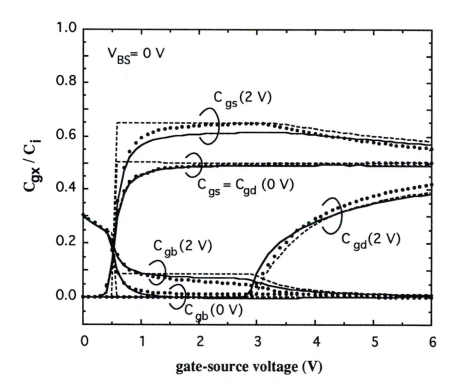

Fig. 3.12.3. Comparison of normalized gate capacitances (x = s (source), d (drain), b (substrate)) for different drain-source voltages (shown in parentheses) with $V_{BS} = 0$ V. Symbols – measurements, solid lines – unified $C–V$ calculations, dashed lines – BSIM charge model calculations. (From Rho et al. (1992), © 1992 IEEE.)

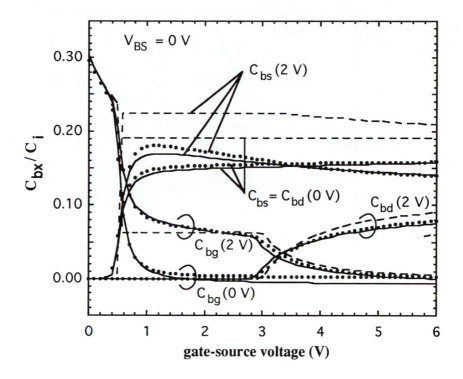

Fig. 3.12.4. Comparison of normalized substrate capacitances (x = s (source), d (drain), g (gate)) for different drain-source voltages (shown in parentheses) with $V_{BS} = 0$ V. Symbols – measurements, solid lines – unified C–V calculations, dashed lines – BSIM charge model calculations. (From Rho et al. (1992), © 1992 IEEE.)

Fig. 3.12.5 provides a further comparison of our unified capacitance model with BSIM calculations. This comparison is done for the capacitance related to the channel charge which is very difficult to measure. The agreement is quite good for the entire range of bias voltages. As can be seen from Fig. 3.12.3, the calculated capacitances C_{gs}, C_{gd}, and C_{gb} related to the gate charge are in good agreement with the BSIM simulations in the strong inversion regime. The agreement with the experimental data is also quite good for the capacitances related to the bulk charge (see Fig. 3.12.4) where the BSIM simulations are considerably less accurate. In addition, our model reproduces quite well the subthreshold behavior of the measured data. The BSIM simulation exhibits a discontinuity at threshold which can cause errors in small signal ac analyses.

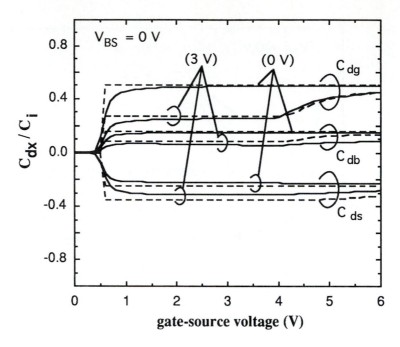

Fig. 3.12.5. Comparison of normalized drain capacitances (x = s (source), g (gate), b (substrate)) for different drain-source voltages (shown in parentheses) with substrate-source bias $V_{BS} = 0$ V. Solid lines – unified C-V calculations, dashed lines – BSIM charge model calculations. (From Rho et al. (1992), © 1992 IEEE.)

3-13. HOT ELECTRON EFFECTS

As characteristic device sizes are scaled down, the electric field in the MOSFET channel increases and, in the saturation regime, the high field region near the drain occupies a large fraction of the device channel. This leads to so-called hot electron effects which manifest themselves in a superlinear increase of the drain current in the saturation regime (the "kink" effect) and in the degradation of device parameters with time. These effects represent a major obstacle to a further down scaling of characteristic MOSFET feature sizes (see Hu et al. (1985) and Chan et al. (1985)).

The physics of the hot electron effects can be described as follows: The high electric field near the drain leads to electron heating. Some "lucky" electrons acquire so much energy from the electric field that they can cause generation of electron-hole pairs. The generated holes lead to a substrate current whereas the generated electrons increase the drain current in the saturation regime, causing the kink. The process of electron hole pair generation can be described by a generation rate which is an exponential function of the maximum electric field in the channel, F_{max}, which is reached at the drain:

$$G = AI_d \exp\left(-\frac{F_o}{F_{max}} \right) \qquad (3\text{-}13\text{-}1)$$

Here A is a constant, I_d is the drain current, and F_o is the characteristic field of the impact ionization, given by

$$F_o = \frac{\phi_i}{q\lambda_i} \qquad (3\text{-}13\text{-}2)$$

where ϕ_i is the energy required for an ionization event and λ_i is the mean free path for the ionization process. According to Hu et al. (1985), $F_o \approx 1.7$ mV/cm for Si n-channel MOSFETs. The generation rate is proportional to the drain current since it ought to be proportional to the product of the electron sheet density in the channel and the electron velocity. The maximum electric field is given by

$$F_{max} = \frac{V_{DS} - V_{SAT}}{\Delta L} \qquad (3\text{-}13\text{-}3)$$

Here, V_{DS} is the intrinsic drain-source voltage, V_{SAT} is the intrinsic drain-source saturation voltage, and ΔL is the length of the pinch-off region,

$$\Delta L \approx \lambda \ln\left[\frac{2\left(V_{DS} - V_p\right)}{V_\lambda} \right] \qquad (3\text{-}13\text{-}4)$$

(see eq. (3-5-39). Here, $V_p \approx V_{SAT}$, and $V_1 = \lambda F_s$ where λ is a characteristic length of the electric field variation in the high field region near the drain and F_s is the velocity saturation field.

The substrate current, I_{sub}, is proportional to the generation rate. Hence, substituting eqs. (3-13-2) to (3-13-4) into eq. (3-13-1), we obtain

$$\frac{I_{sub}}{I_d} = B\left(V_{DS} - V_{SAT}\right)\exp\left(-\frac{F_o\lambda}{V_{DS} - V_{SAT}}\right)$$ (3-13-5)

where B is a constant. Based on a two dimensional simulation and a quasi-two-dimensional analysis by Hu et al. (1985) and Chan et al. (1985), we find that $B \approx 1.2$ V^{-1}.

Eq. (3-13-5) can be re-written as

$$Y = \frac{V_{DS} - V_{SAT}}{F_o\lambda}$$ (3-13-6)

where

$$Y = \frac{1}{\ln\left[B\left(V_{DS} - V_{SAT}\right)\right] + \ln\left(I_d / I_{sub}\right)}$$ (3-13-7)

Our analysis showed that only one iteration is sufficient to accurately solve eq. (3-13-6) by iteration if we substitute V_{SAT} with V_{GT} in eq. (3-13-7). As can be seen from Fig. 3.13.1, the measured values of Y depend linearly on the drain-source voltage in the kink region. This means that eq. (3-13-6) can be used for the extraction of the saturation voltage, V_{SAT}, from experimental data.

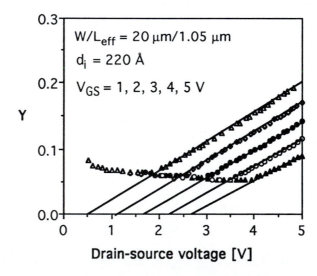

Fig. 3.13.1. Measured Y versus V_{DS} curves. (From Kim et al. (1992)).

The accuracy of such an extraction is illustrated in Fig. 3.13.2 where we compare two different parameter extraction techniques – from the drain resistance measurements as described in Section 3.10 and from the measured values of Y. The excellent agreement between the two techniques and the linearity of the Y versus V_{DS} curves shown in Fig. 3.13.1 prove the validity of the hot electron model considered in this Section.

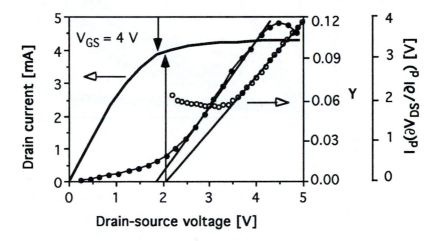

Fig. 3.13.2. Determination of saturation voltage by two independent techniques (compare the intercept with the V_{DS} axis of the two straight lines shown in the figure). (From Kim et al. (1992).)

Hot electrons can also tunnel into traps in the gate oxide near the drain. The negative charge in the oxide causes a partial channel depletion near the drain leading to an increase in the channel resistance and a decrease in the threshold voltage in this region. Hence, the device characteristics change with time when the drain voltage is high enough to cause significant electron heating (i.e., under "voltage stress"). The increase in the channel resistance should lead to a shift in the drain-source saturation voltage. This effect is clearly confirmed by the stress measurements reported by Kim et al. (1992), see Fig. 3.13.3. As can be seen from the figure, the Y versus V_{SAT} curves experience a parallel shift as a results of the voltage stress.

As was shown by Kim et al. (1992), this electron trapping cause a change, ΔI, in the drain current given by

$$\frac{\Delta I}{I_d} = S_o \ln\left(\frac{t}{\tau} + 1\right) \tag{3-13-8}$$

where S_O is a constant (see Fig. 3.13.4).

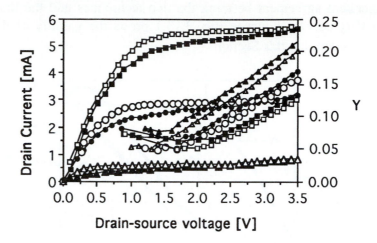

Fig. 3.13.3. Measured *I–V* characteristics and Y-functions for *n*-channel Si MOSFETs. Open symbols – data before stress, dark symbols – data after stress at $V_{DS} = 2$ V and $V_{GS} = 4.5$ V for 104 s. (From Kim et al. (1992).)

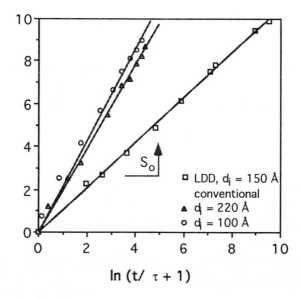

Fig. 3.13.4. Measured values of $\Delta I/I_d$ (in percent) under stress versus time (after Kim et al. (1992).)

The model of hot electron effects described in this Section is simple and accurate enough to be implemented in circuit simulators. We are planning to incorporate hot electron models in future releases of AIM Spice.

REFERENCES

L. A. AKERS, M. M. E. BEGUWALA and F. Z. CUSTODE, "A Model of a Narrow-Width MOSFET Including Tapered Oxide and Doping Encroachment", *IEEE Trans. Electron Devices*, ED-28, No. 12, pp. 1490-1494 (1981)

A. R. ALVAREZ, *BiCMOS Technology and Applications* , Kluwer Academic Publishers, Boston (1989)

N. D. ARORA AND L. M. RICHARDSON, "MOSFET Modeling for Circuit Simulation", in *Advanced MOS Device Physics,* N. Einspruch and G. Gindelblat, Eds., Academic Press, New York, p. 250 (1989)

J. R. BREWS, W. FICHTNER, E. H. NICOLLIAN and S. M. SZE, "Generalized Guide for MOSFET Miniaturization", *IEEE Electron Device Letters*, EDL-1, p. 2 (1980)

Y. BYUN, K. LEE and M. SHUR, "Unified Charge Control Model and Subthreshold Current in Heterostructure Field Effect Transistors", *IEEE Electron Device Letters*, EDL-11, No. 1, pp. 50-53, Jan. 1990 (see erratum *IEEE Electron Device Letters*, EDL-11, no. 6, p. 273, June (1990))

Y. BYUN, M. SHUR, M. HACK and K. LEE, "New Analytical Poly-Silicon Thin-Film Transistor Model for CAD and Parameter Characterization", *Solid-State Electron.*, 35, No. 5, pp. 655-663, May (1992)

D. M. CAUGLEY and R. E. THOMAS, *Proc. IEEE*, 55, pp. 2192-2193 (1967)

T. Y. CHAN, P. K. KO AND C. HU, "Dependence of Channel Electric Field on Device Scaling", *IEEE Electron Device Letters,* EDL-6, pp. 551-553 (1985)

T-Y. CHAN, S-W. LEE and H. GAW, "Experimental Characterization and Modeling of Electron Saturation Velocity in MOSFET's Inversion Layer from 90 to 350 K", *IEEE Electron Device Letters*, EDL-11, pp. 466-467, Oct. (1990)

S. Y. CHOU and D. A. ANTONIADIS, *IEEE Trans. Electron Devices*, ED-34, p. 448 (1987)

J. E. CHUNG, M. C. JENG, J. E. MOON, P.K. KO and C. HU, "Performance and Reliability Design Issues for Deep-Submicrometer MOSFET's", *IEEE Trans. Electron Devices*, ED-38, No. 3, pp. 545-554 (1991)

R. COEN and R. S. MILLER, "Velocity of surface carriers in the inversion layer on silicon", *Solid–State Electron.*, 23, pp. 35–40, (1980)

S. CSERVENY, "Relationship Between Measured and Intrinsic Conductances of MOSFETs", *IEEE Trans. Electron Devices*, ED-37, No. 11, pp. 2413-2414, Nov. (1990)

J. A. COOPER, Jr. and D. F. NELSON, "High–field drift velocity of electrons at the Si–SiO$_2$ interface as determined by time–of–flight technique", *J. Appl. Phys.*, 54, pp. 1445–1456 (1983)

H. C. de GRAAFF and F. M. KLAASSEN, *Compact Transistor Modeling for Circuit Design*, Computational Microelectronics Series, ed. S. Selberherr, Springer-Verlag, Wien, New York (1990)

Y. A. EL-MANSY and A. R. BOOTHROYD, "A Simple Two–Dimensional Model for IGFET Operation in the Saturation Region", *IEEE Trans. Electron Devices*, ED-24, No.3, pp. 254-267, (1977)

W. FICHTNER and H. W. POTZL, "MOS Modeling by Analytical Approximations, I. Subthreshold Current and Threshold Voltage", *Int. J. Electron.*, 46, p. 33 (1979)

T. A. FJELDLY, B. MOON and M. SHUR, "Analytical Solution of Generalized Diode Equation", *IEEE Trans. Electron Devices*, ED-38, No. 8, pp. 1976-1977, August (1991)

T. A. FJELDLY, M.. SHUR, T. YTTERDAL and K. LEE, "Unified CMOS Model for Circuit Simulation", *Proc. Int. Semicond. Device Research Symp., ISDRS'91*, Charlottesville, Virginia, pp. 407 - 411, Dec. (1991a)

T. A. FJELDLY and M. SHUR, "Threshold Voltage Modeling and the Subthreshold Regime of Operation of Short-Channel MOSFETs",*IEEE Trans. Electron Devices* , 40, No. 1, (1992)

B. GELMONT, M. SHUR and R. J. MATTAUCH, "Capacitance-Voltage Characteristics of Microwave Schottky Diodes", *IEEE Trans. Microwave Theory and Technique*, 39, No. 5, pp. 857-863, May (1991)

S. K. GHANDHI, *VLSI Fabrication Principles, Silicon and Gallium Arsenide*, John Wiley and Sons, New York (1983)

T. GROTJOHN and B. HÖFFLINGER, "A Parametric Short-Channel MOS Transistor Model for Subthreshold and Strong Inversion Current", *IEEE J. Solid-State Circuits*, SC-19, pp. 100–112 (1984)

A. HAIRAPETIAN, D. GITLIN and C. R. VISWANATHAN, "Low–temperature mobility measurement on CMOS devices", *IEEE Trans. Electron Devices*, ED-36, pp. 1448–1455 (1989)

J. D. HAYDEN, F. K. BAKER, S. A. ERNST, R. E. JONES, J. KLEIN, M. LIEN, T. F. MCNELLY, T. C. MELE, H. MENDEZ, L. C. PARRILLO, W. PAULSON, J. R. PFIESTER, F. PINTCHOVSKI, Y. C. SEE, R. D. SIVAN, B. M. SOMERO and E. O. TRAVIS, "A High-Performance Half-Micrometer Generation CMOS Technology for Fast SRAM's", *IEEE Trans. Electron Devices*, ED-38, No. 4, pp. 876-886 (1991)

C. HU, C. TAM, F. C. HSU, P. K. KO, T.Y. CHAN AND K. W. TERRIL, "Hot Electron Induced MOSFET Degradation", *IEEE Trans. Electron Devices,* ED-32, pp. 375-385 (1985)

G. J. HU, C. CHANG and Y. T. CHIA, "Gate–Voltage–Dependent Effective Channel Length and Series Resistance of LDD MOSFET's", *IEEE Trans. Electron Devices,* ED-34, pp. 2469–2475 (1987)

M. C. JENG, P. M. LEE, M. M. KUO, P. K. KO and C. HU, *Theory, algorithms, and user's guide for BSIM and SCALP*, Memorandum No. UCB/ERL M87/35, University of California, Berkeley (1987)

M. C. JENG, P. K. KO and C. HU, "A Deep-Submicrometer MOSFET Model for Analog/Digital Circuit Simulations", in *IEDM Technical Digest*, pp. 114-117 (1988)

M. C. JENG, R. S. MULLER and C. HU, "A Unified Model for Hot-Electron Currents in MOSFETs", in *IEDM Technical Digest*, pp. 600 (1981)

S. H. KIM, K. S. MIN, S. W. HONG, K. R. LEE, M. SHUR and T. A. FJELDLY, "AIM-spice FET Device Models: New Device Modeling in VLSI Era", in *Proc. of ISSSE-92*, Paris, Sept. (1992)

F. M. KLAASSEN, "A Model for Computer–Aided Design", *Philips Res. Repts.*, 31, pp. 71–83 (1976)

P. K. KO, "Approach to Scaling", in *Advanced MOS Device Physics*, p. 18, N. Einspruch and G. Gildenblat, editors, Academic Press (1989)

K. LEE, M. SHUR, T. J. DRUMMOND and H. MORKOÇ, "Electron Density of the Two-Dimensional Electron Gas in Modulation Doped Layers", *J. Appl. Phys.*, 54, No. 4, pp. 2093-2096, April (1983)

K. LEE, J. S. CHOI, S. P. SIM and C. K. KIM, "Physical Understanding of Low Field Carrier Mobility in Silicon Inversion Layer", *IEEE Trans. Electron Devices*, ED-38, No. 8, pp. 1905-1912, August (1991)

L. L. LEWYN and J. D. MEINDL, "An IGFET Inversion Charge Model for VLSI Systems", *IEEE Trans. Electron Devices*, ED-32, no. 2, pp. 434-440 (1985)

Z. H. LIU, J. H. HUANG, J. DUSTER, P. K. KO, C. HU, M. C. JENG and Y. C. CHEN, "Threshold Voltage Modeling for Deep-Submicrometer Conventional and LDD MOSFETs at 300 K and 85 K", *Proc. Int. Semicond. Device Research Symp., ISDRS'91*, Charlottesville, Virginia, pp. 411-414, Dec. (1991)

J. E. MEYER, "MOS Models and Circuit Simulation", *RCA Review*, 32, pp. 42-63, March (1971)

A. D. MILNES, Editor, *MOS Devices, Design and Manufacture*, Edinburgh University Press, Edinburgh (1983)

B. MOON, Y. BYUN, K. LEE and M. SHUR, "New Continuous Heterostructure Field Effect Transistor Model and Unified Parameter Extraction Technique", *IEEE Trans. Electron Devices*, ED-37, No. 4, pp. 908-918, April (1990)

B. J. MOON, C. K. PARK, K. LEE and M. SHUR, "New Short Channel *n*–MOSFET Current-Voltage Model in Strong Inversion and Unified Parameter Extraction Method", *IEEE Trans. Electron Devices*, ED-38, No. 3, pp. 592-602, March (1991)

B. MOON, C. PARK, K. RHO, K. LEE, M. SHUR and T. A. FJELDLY, "Analytical Model for *p*-Channel MOSFETs", *IEEE Trans. Electron Devices*, ED-38, pp. 2632-2646 (1991a)

E. H. NICOLLIAN and J. R. BREWS, *MOS Physics and Technology*, John Wiley, New York (1982)

S. Y. OH, D. E. WARD and R. W. DUTTON, "Transient Analysis of MOS Transistors", *IEEE J. Solid State Circuits*, SC-15, No. 4, August (1980)

Y. OHKURA, "Quantum Effects in *n*-MOS Inversion Layers", *Solid-State Electronics*, 33, No. 12, pp. 1581-1585, Dec. (1990)

D. G. ONG, *Modern MOS Technology, Processes, Devices, & Design*, McGraw-Hill, Inc., New York (1984)

C. K. PARK, C. Y. LEE, K. R. LEE, B. J. MOON, Y. BYUN and M. SHUR, "A Unified Charge Control Model for Long Channel *n*-MOSFETs", *IEEE Trans. Electron Devices*, ED-38, pp. 399-406, Feb. (1991)

C. K. PARK, *A Unified Current-Voltage Modeling for Deep Submicron CMOS FETs*, Ph. D. Thesis, Korea Advanced Institute of Science and Technology (1992)

H. J. PARK, P. K. KO and C. Hu, "A Non-Quasi-static MOSFET Model for SPICE", in *IEDM Technical Digest*, pp. 652-655 (1987)

R. F. PIERRET, *Field Effect Devices*, Addison Wesley Modular Series on Solid State Devices, vol. 4, Reading, Massachusetts (1983)

R. A. PUCEL, H. HAUS and H. STATZ, in *Advances in Electronics and Electron Physics*, vol. 38, Academic Press, New York, pp. 195-205 (1975)

K. RHO, K. LEE, M. SHUR and T. A. FJELDLY, "Unified Quasi-Static Capacitance Model", *IEEE Trans. Electron Devices* , 40, No. 1, Jan. (1992)

P. P. RUDEN, "Heterostructure FET Model Including Gate Leakage", *IEEE Trans. Electron Devices*, ED-37, No. 10, pp. 2267-2270, Oct. (1989)

A. G. SABNIS and J. T. CLEMENS, "Characterization of the Electron Mobility in the Inverted (100) Si Surface", *IEDM Tech. Digest*, pp. 18-21 (1979)

T. SAKURAI and A. R. NEWTON, "A Simple MOSFET Model for Circuit Analysis", *IEEE Trans. Electron Devices*, ED-38, No. 4, pp. 887-894, April (1991)

J. R. SCHRIEFFER, "Mobility in Inversion Layers: Theory and Experiments", in *Semiconductor Surface Physics*, edited by R.H. Kingstone, Philadelphia, Univ. of Pennsylvania Press, pp. 55–69 (1957)

D. K. SCHRODER, *Advanced MOS Devices*, Addison Wesley Modular Series on Solid State Devices, Reading, Massachusetts (1987)

B. J. SHEU, D. L. SCHARFETTER, P. K. KO and M. C. JENG, "BSIM, Berkeley Short-Channel IGFET Model," *IEEE J. Solid State Circuits*, SC–22, pp. 558–566 (1987)

M. SHOJI, *CMOS Digital Circuit Technology*, Prentice Hall, New Jersey (1988)

M. S. SHUR, "Analytical Model of GaAs MESFET", *IEEE Trans. Electron. Devices*, ED-25, No. 6, pp. 612-618, June (1978)

M. SHUR, *Physics of Semiconductor Devices*, Prentice Hall, New Jersey (1990)

M. SHUR, T. A. FJELDLY, T. YTTERDAL and K. LEE, "Unified MOSFET Model", in *Solid-State Electron.*, 35, No. 12, pp. 1795-1802, Dec. (1992)

C. G. SODINI, T. W. EKSTEDT and J. L. MOLL, "Charge Accumulation and Mobility in Thin Dielectric MOS Transistors", *Solid State Electron.*, 25, No. 9, pp. 833-841, Sep. (1982)

C. G. SODINI, P. K. KO and J. L. MOLL, "The Effect of High Fields on MOS device and Circuit Performance", *IEEE Trans. Electron Devices*, ED–31, pp.1386–1393 (1984)

T. STEEN, T. A. FJELDLY and M. SHUR, "Analytical Modeling of Heterostructure Field Effect Transistors", in *Proc. 14th Nordic Semiconductor Meeting*, Aarhus, Denmark, June (1990)

F. STERN, "Self-Consistent Results for n-type Si Inversion Layers", *Phys. Rev.*, B-5, No. 12, pp. 4891–4899 (1972)

F. STERN and S. SARMA, "Electron Energy Levels in GaAs-Ga$_{1-x}$Al$_x$As Heterojunctions", *Phys. Rev.*, B-30, No. 2, pp. 840–848 (1984)

S. M. SZE, *Physics of Semiconductor Devices*, John Wiley & Sons, New York (1981)

S. M. SZE, *Semiconductor Devices. Physics and Technology*, John Wiley & Sons, New York (1985)

S. M. SZE, Editor, *VLSI Technology*, Second Edition, McGraw-Hill, New York (1988)

K. K. THORNBER, "Relation of Drift Velocity to Low-Field Mobility and High-Field Saturation Velocity", *J. Appl. Phys.*, 51, pp. 2127-2136 (1980)

R. TROUTMAN, *IEEE J. Solid State Circuits*, SC-9, p. 55 (1974)

R. TROUTMAN, "VLSI Limitations from Drain-Induced Barrier Lowering", *IEEE Trans. Electron Devices*, ED-26, p. 461 (1979)

Y. P. TSIVIDIS, *Operation and Modeling of the MOS Transistor*, McGraw-Hill, New York (1987)

C. TURCHETTI and G. MASETTI, "On the Small-Signal Behavior of the MOS Transistor in the Quasi-Static Operation", *Solid-State Electron.*, 26, pp. 941-949 (1983)

P. W. TUINENGA, *SPICE. A Guide to Circuit Simulation & Analysis Using PSpice*, Prentice Hall, New Jersey (1988)

D. E. WARD and R.W. DUTTON, "A Charge-Oriented Model for MOS Transistor Capacitances", *IEEE J. Solid-State Circuits*, SC-13, No. 5, Oct. (1978)

K. C.-K. WENG and P. YANG, " A Direct Measurement Technique for Small Geometry MOS Transistor Capacitances", *IEEE Trans. Electron Devices Lett.*, 6, No. 1, Jan. (1985)

P. YANG, "Capacitance Modeling for MOSFET", in *Advances in CAD for VLSI*, vol. 3, Part 1, A. E. Ruehli, Editor, North Holland, Amsterdam, pp. 107-130 (1986)

L. D. YAU, "A Simple Theory to Predict the Threshold Voltage of Short-Channel IGFET's", *Solid State Electron.*, 22, p. 701 (1974)

A. VAN DER ZIEL, M. S. SHUR, K. LEE, T. H. CHEN and K. AMBERIADIS, "Carrier Distribution and Low-Field Resistance in Short n^+-n-n^+ Structures", *IEEE Trans. Electron Devices*, ED-30, No. 2, pp. 128-137, Feb. (1983)

T. YTTERDAL, K. LEE, M. SHUR AND T.A. FJELDLY, "AIM-Spice, a New Circuit Simulator Based on a Unified Charge Control Model", *Proc. Int. Semicond. Device Research Symp., ISDRS'91*, Charlottesville, Virginia, pp. 481-485, Dec. (1991)

PROBLEMS

3-3-1. Using Fig. 3.3.3 as a guide, sketch the band diagrams of *n*-channel and *p*-channel silicon MOSFET channels with n^+ polysilicon gate at zero gate-channel bias.

3-3-2. Sketch small-signal capacitance-voltage curves (C_{xg}/C_i versus V_G) for the MOS structures shown in Fig. P3.3.2 (a)-(d). In all cases, the gate dc bias (V_G) is varied slowly and a small signal voltage source (v_g) is applied to the gate terminal at 100 kHz. Label each region on the curves (e.g., accumulation, depletion, and inversion). Assume a substrate doping density of 10^{16} cm^{-3}, an oxide thickness of 500 Å, and $V_{FB} + 2\varphi_b = 0$ V. Calculate numerical values of the capacitances and gate voltages at flat band and threshold conditions. In the figures, A represent the small signal current meter to measure capacitance by measuring the imaginary part of the admittance.

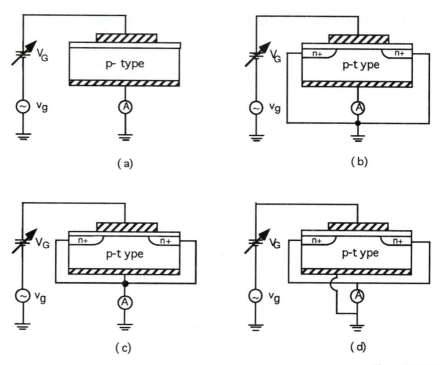

Fig. P3.3.2

3-3-3. Using Fig. 3.3.7, extract the transistor parameters using the procedure described in Subsection 3.3.2. Compare the extracted parameter values with the values of parameters given in the figure caption to Fig. 3.3.7. c_i is measured to be 18.5 nF/cm^2.

3-3-4. The unified charge control model (see eq. (3-3-50)) is not applicable at very small surface electron concentrations, n_s, because in practical devices n_s usually saturates at large negative values of the gate voltage swing, V_{GT}, reaching a certain small value, $n_s = n_{leak}$. Use eq. (3-3-50) to estimate the value of $V_{GT} = V_{leak}$ at which n_s reaches the saturation value n_{leak}. How should you modify eq. (3-3-50) to account for this "leakage" carrier concentration?

3-3-5. Using UCCM (the Unified Charge Control Model), derive eqs. (3-3-59) and eq.(3-3-60), and show that the derivative of the gate-channel capacitance reaches its maximum value at threshold.

3-3-6. Using eq.(3-3-77), calculate the lowest energy levels for all conduction band valleys shown in Fig. 3.3.15 for $F_{eff} = 10^5$ V/cm and 10^6 V/cm.

3-3-7. In the classical limit, the separation of the energy subbands in a 2D electron gas is small compared to the thermal energy, $k_B T$. In this case, the sheet density of the 2D gas is given by the classical charge sheet model (see eq. (3-3-87)) which is derived using a conventional 3D electron gas approach. Show that in this limit $(E_j - E_{j-1} << k_B T)$, eq. (3-3-76) reduces to eq. (3-3-87).

Do it first for a single valley with an effective mass m_n and show that N_c in eq.(3-3-87) is the same as that in eq. (1-3-7). Repeat for the (100) silicon surface and show that in this case N_c reduces to that in eq. (1-3-7) with the effective mass defined by eq. (1-3-19) with $m_x = m_t$ and $m_y = m_z = m_l$.

Hint: Change the summation over j in eq. (3-3-76) to an integration. To do this, solve eq. (3-3-77) for j and differentiate the resulting equation. Find dE_j by putting $dj =1$, and substitute dE_j into eq. (3-3-76).

3-3-8. In the frame of the classical charge sheet model of eq. (3-3-87), derive the expressions for n_s for the limiting cases of very small and very large bulk charge, qn_B.

3-4-1. Consider a field effect transistor with a gate length of 0.1 μm, a gate width of 0.25 μm, and a gate silicon oxide thickness of 20 Å. The silicon dioxide dielectric constant is 3.9. Estimate the gate capacitance and gate voltage swing required to induce 1 electron into the channel. At what temperature will this gate voltage swing equal the thermal voltage? Calculate the drain current assuming that the electron velocity in the channel is saturated at 10^5m/s. Estimate the transconductance per mm gate width assuming velocity saturation.

(This problem is relevant to the so-called Single Electronics. See, for example, K. K. LIKHAREV and T. CLAESON, "Single Electronics", *Scientific American*, pp. 80-85, June (1992).)

3-4-2. Assume that the breakdown field of a *p*-type Si NMOS substrate is 300 kV/cm. The silicon dielectric constant is 11.7. Calculate the substrate breakdown voltage as a function of the substrate doping for 10^{15} cm^{-3} $\leq N_a \leq 10^{17}$ cm^{-3}.

3-4-3. Sketch the minimum substrate doping required to avoid the punch-through of the drain and source depletion regions in a Si MOSFET at zero source-drain bias as a function of the gate length. Assume that the one-dimensional depletion approximation is valid and that the built-in voltage between the source and drain contacts and the substrate is equal to 0.7 V.

3-4-4. Consider a MOSFET with a substrate thickness d. Assume that the depletion width, d_{dep}, reaches d before the onset of strong inversion. Derive an equation for the threshold voltage for this case and compare with eq. (3-4-19). Sketch for typical MOSFET parameters the dependence of the threshold voltage on the substrate bias for the device with substrate thickness d and compare with the conventional dependence for a thick substrate.

3-4-5. Calculate constant α in eq. (3-4-33) as a function of the substrate doping for 10^{14} cm^{-3} $< N_a < 10^{16}$ cm^{-3} 10^{14} cm^{-3} for a silicon *n*-channel MOSFET with an oxide thickness of 200 Å ($V_{FB} = -1.0$ V).

3-4-6. Show graphically how the *I–V* characteristics of a MOSFET change as
(a) R_s increases and
(b) R_d increases.

What are the implications of an increase in R_s and R_d for the noise margin of an inverter?

3-4-7. Derive eqs. (3-4-39) and (3-4-43).

3-4-8. Derive equations for the simple charge control model (see Subsection 3.4.2) which take into account the parasitic source and drain series resistances, R_s and R_d, and compare these equations with eq. (3-4-26).

3-5-1. Estimate the ratio of the diffusion and drift currents in the linear, strong inversion regime of MOSFET operation. Assume a device temperature of $T = 300$ K, and a drain-source voltage of 0.5 V.

3-5-2. Fig. P3.5.2 shows measured transfer characteristics (I_{sat} versus V_{gs}) for two different MOSFETs. Which device is likely to have the longer gate? Why?

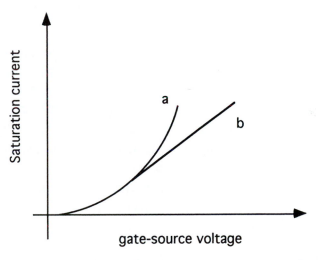

gate-source voltage   **Fig. P3.5.2**

3-5-3. Calculate and plot the ratio of the saturation currents predicted by the velocity saturation model of Section 3.5 and the constant mobility model of Section 3.4 for a silicon n-channel MOSFET with an oxide thickness of 200 Å and a gate width of 20 μm, for the gate lengths 0.5 μm $< L <$ 10 μm. The electron mobility and the saturation velocity are 700 cm^2/Vs and 6x10^4 m/s, respectively. Do the calculation for

$V_{GT} = 4$ V.

(a) Neglect the parasitic source and drain resistances, R_s and R_d.

(b) Repeat the calculation for $R_s = R_d = 1$ Ωmm, compare and comment.

3-5-4. Plot the length of the saturated region for 1 V $< V_{DS} - V_p < 10$ V for a silicon n-channel MOSFET with an oxide thickness of 200 Å. Assume that the effective channel thickness near the drain is $1,000$ Å and the saturation field is $F_s = 15$ kV/cm.

*3-6-1. Using eq. (3-6-32), choose the design parameters of a Si MOSFET with a gate length of 0.25 μm in such a way that short channel effects will be suppressed.

*3-6-2. Show from eq. (3-5-31) that the maximum lateral channel electric field in a MOSFET can be expressed as

$$F_{max} = \left(V_{DS} - V_p\right)/\lambda$$

where λ is given by eq. (3-5-30). Large values of F_{max} may cause long-term device instability due to the so-called hot carrier effect. In other words, the higher F_{max}, the worse is the device reliability. Therefore larger values of λ are desirable in order to reduce F_{max}. However, this leads to larger short channel effects. In light of this trade-off, discuss why devices with Low Doped Drain (LDD) region should have better reliability and less pronounced short channel effects. **Hint:** Sketch and compare qualitative distributions of the longitudinal electric field along the channel for a conventional device and for an LDD device.

3-6-3. An NMOS has a subthreshold ideality factor of 1.5 and a threshold voltage of 0.8 V at $V_{ds} = 0$ V. The DIBL constant is $\sigma = 0.02$. Calculate I_d as a function of V_{ds} at $V_{gs} = 0$ V for 77 K, 300K, and 400K. Use default AIM-Spice values for the other parameters. **Hint:** Use UCCM.

3-7-1. (a) Use the data shown in Fig. 3.7.4 to estimate the substrate doping density.

(b) From the data shown represented by the top curve in Fig. 3.7.7, estimate the oxide thickness using the substrate doping data in the text.

3-7-2. Sketch the body plot for a PMOS device with n-type ploysilicon gate and the doping profile shown in Fig. P3.7.2. $N_s = 5 \times 10^{16}$ cm^{-3}, $X_d = 0.1$ μm, $N_B = 10^{16}$ cm^{-3}. The oxide thickness is 300 Å. How will the body plot change as N_s and/or X_d increase?

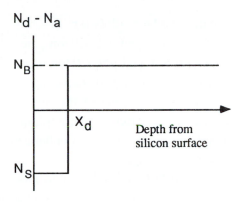

Fig. P3.7.2

*3-7-3. Propose doping profiles for a long channel NMOS device with a threshold voltage of 1 V for a back bias of −3 V. How should these profile change for 0.3 μm gate NMOS devices? What other considerations should affect your choice of the doping profile? The optimum thickness of the gate oxide is typically around 500 Å for a 2 μm gate device. How would you scale the oxide thickness with the gate length? Why?

Hint: Use Fig. 3.7.1b for estimating the change in the threshold voltage with the gate length.

3-7-4. Calculate the ideality factor for the subthreshold slope defined in eq. (3-3-53)

(a) as a function of the substrate bias for the uniformly doped case with a body effect constant of $\gamma_N = 0.5 \, (N_a(\text{cm}^{-3})/10^{16})^{1/2} V^{1/2}$

(b) for the doping profile corresponding to the top curve in Fig. 3.7.7. The oxide thickness is 300 Å.

3-7-5. Using a charge sharing model similar to that described by eq. (3-6-13) for short channel devices, show that the threshold voltage increase shown in Fig. 3.7.2 for a narrow channel device can be expressed as

$$\Delta V_{TW} = K(2\varphi_b - V_{BS})/W_{eff}$$

where K is a constant.

Hint: Replace ΔQ_{tot} in eq. (3-6-13) with a fringing charge.

*3-7-6. Sketch qualitative body plots for

(a) NMOS with n-doped polysilicon gate.

(b) PMOS with p-doped polysilicon gate.

(c) PMOS with n-doped polysilicon gate.

What is a reasonable value of the threshold voltage for such a PMOS used in a CMOS inverter? Also explain why we need an additional p-type surface doping to obtain an enhancement device with such a value of the threshold voltage.

(d) NMOS with p-doped polysilicon gate. Assume a uniformly doped substrate and neglect the surface state charge density (see eq. (3-7-5)).

3-7-7. Derive inequality (3-7-24) and eq. (3-7-25).

3-8-1. Using the data given in Fig. 1.4.1, propose reasonable values of parameters μ, m, and v_s for Si, GaAs, InP, and $In_{0.53}Ga_{0.47}As$ in eq. (3-8-1).

3-8-2. Comment on the substrate bias dependence of the electron channel mobility predicted by eq. (3-8-3). (Assume θ = constant). Why is the predicted dependence unphysical?

*3-8-3. How is the mobility parameter κ_{1n} in eq. (3-8-13) related to the oxide thickness? Using a long channel MOSFET model, find the oxide thickness which gives maximum drain current for a given gate-source voltage and channel length. Find and plot this optimum value as a function of the gate length for $V_{gs} = 1.5$ V, 3.3 V, and 5 V.

The transistor speed is roughly proportional to the transconductance and inversely proportional to the oxide capacitance. Therefore, for a given channel length and power supply voltage, there will be an optimum oxide thickness for maximum speed. Discuss what happens when we include a constant parasitic capacitance in addition to the intrinsic oxide capacitance. Also consider the case of a short channel NMOS with velocity saturation. Assume a threshold voltage of 1 V.

3-8-4. From eq. (3-8-13), calculate the dependence of the channel electron mobility on the gate-source voltage for a silicon n-channel MOSFET with an oxide thickness of 100 Å and a substrate doping of $N_a = 10^{17}$ cm^{-3}. Use the data in Fig. 3.8.3 to estimate the parameters needed.

3-8-5. Repeat the calculation in Problem 3-8-4 for holes in a p-channel MOSFET using the same oxide thickness and substrate doping level. In this case, use the data in Fig. 3.8.4 to estimate the parameters needed.

3-9-1. Use the AIM-Spice model MOSA1 with default parameters to calculate $C-V$ characteristics of an n-channel Si MOSFET for three different temperatures: 77 K, 300 K, and 425 K.

*3-9-2. Consider the p-channel MOSFET characteristics shown in Figs. 3.10.6 and 3.10.7. The device width is 20 μm. Use the parameter extraction techniques described in Section 3.9 to estimate device parameters. Assume typical values for device parameters which cannot be extracted this way (i.e., subthreshold parameters). Calculate the device $I-V$ characteristics using the extracted and/or typical values of the parameters with the AIM-Spice model of Section 3.9 (MOSA1) and compare the results with the experimental data shown in Figs. 3.10.6 and 3.10.7.

3-10-1. From the long channel strong inversion PMOS model (see Subsection 3.10.3), derive eq. (3-10-43). How should we change eq. (3-10-43) for a short channel device?
Hint: Refer to eq. (3-10-9) and explain.

3-10-2. Plot and compare the current-voltage characteristics (I_d versus V_{ds}) for

three long channel NMOS models:
(a) eq. (3-4-15) with $\gamma_N = 0$ (see also Section 3.7)
(b) eq. (3-4-15) with $\gamma_N = 0.4$ V$^{1/2}$
(c) the BSIM model with $\gamma_N = 0.4$ V$^{1/2}$, which states that

$$I_d = \mu_n c_i \frac{W}{L}\left[\left(V_{gs} - V_T\right)V_{ds} - \frac{\alpha}{2}V_{ds}\right]$$

where

$$\alpha = 1 + \frac{\gamma_N}{2\sqrt{2\varphi_b - V_{BS}}}\left[1 - \frac{1}{1.744 + 0.8364\left(2\varphi_b - V_{BS}\right)}\right]$$

Vary the drain-source bias from 0 to 5 V for gate-source voltages of
1, 2, 3, 4, and 5 V. Assume $\mu_n c_i = 40$ µA/V^2, $W/L = 20$, and $V_{To} = 1$ V. Repeat the problem for two values of the substrate-source bias,
$V_{BS} = 0$ V and –5 V.
Explain why the BSIM expression for α is inaccurate for non-uniformly doped and/or for short channel MOSFETs.
Hint: See Section 3.7.

*3-10-3. Use the AIM-Spice PMOS model PMOSA2 to calculate the maximum
p-channel MOSFET transconductance per mm gate width as a function
of the gate length. Start from the default parameters given in the AIM-
Spice program. Scale the gate length down to 0.2 µm. Scale the oxide
thickness and gate width proportionally to the gate length.

*3-11-1. Repeat Problem 3-10-3 for the NMOS model NMOSA2.

*3-11-2. Use the AIM-Spice PMOS and NMOS models in Problems 3-10-3 and
3-11-1 (PMOSA2 and NMOSA2) to calculate the propagation delay of a
CMOS inverter as a function of the gate length, L. Start from the
default parameters given in the AIM-Spice program. Scale L down to
0.2 µm. Scale the device width, W, and the gate oxide thickness
proportionally to L. Estimate the effect of interconnects by adding an
equivalent interconnect capacitance, C_{ic}, between the output node and
ground. Assume that C_{ic} (fF)$= 0.2 \cdot W$(µm). Repeat the calculation for
two values of the power supply voltage, 3 V and 1.5 V.

*3-11-3.Consider inverter circuits shown in Fig. P3.11.3 consisting of enhancement NMOS drivers with (i) resistive, (ii) enhancement, (iii) depletion, and (iv) PMOS enhancement loads, respectively.

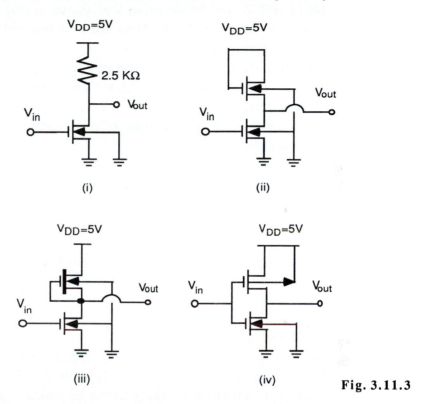

Fig. 3.11.3

Assume the gain factors μc_i of 40, 40, 15 μA/V^2 and threshold voltages at zero substrate bias of 1 V, -4V, -1V, for enhancement NMOS, depletion NMOS, and enhancement PMOS, respectively. The W/L ratio of all enhancement NMOS drivers is (20μm/1μm). Ignore the channel length modulation and use a long channel strong inversion current-voltage model (see eq. (3-4-15)). Assume the same body effect constants ($\gamma_N = \gamma_P$) for all MOSFETs, with the values 0 and 0.4 V$^{1/2}$.

(a) Construct a set of I_d–V_{ds} characteristics for the enhancement driver MOSFET for gate-source voltages equal to 0, 1, 2, 3, 4, and 5V for the drain-source voltage varying from 0 to 5V.

(b) Draw a load line for the resistive load shown in (i). Draw the inverter circuit transfer characteristic.

(c) Find W/L for the enhancement load shown in (ii) such that the logic inversion voltage is 2.5 V. Draw the load line and the inverter circuit transfer characteristic.

(d) Find W/L for the depletion load shown in (iii) such that the logic inversion voltage is 2.5 V. Draw the load line and the inverter circuit transfer characteristic.

(e) Find W/L for the PMOS enhancement load shown in (iv) such that the logic inversion voltage is 2.5 V. Draw the inverter circuit transfer characteristic.

Discuss the effect of the body effect constant on each of the four technologies.

3-11-4. Both the subthreshold ideality factor η and the UCCM parameter α are related to the body effect parameters γ_{Neff} and Γ_{Neff}. Explain qualitatively how they are related.

3-12-1. Draw energy diagrams similar to those in Fig. 3.12.2 assuming a finite quasi-Fermi potential, V_F, relative to the source. Indicate the signs of the various potentials used in the diagram.

3-12-2. (a) Derive eq. (3-12-24).

(b) Use eqs. (3-4-29) and (3-3-34) to express the drain current as a function of the electron sheet densities at the source, n_{ss}, and at the drain, n_{sd}. From the resulting strong inversion expression, show that the electron sheet density in the middle of the channel, n_{sm}, can be written as:

$$n_{sm} \equiv n_s\left(x = \frac{L}{2}\right) = \sqrt{\frac{n_{ss}^2 + n_{sd}^2}{2}}$$

(c) Comment on how much error will result in the C–V characteristics from using the same equation for $V_{F(L/2)}$ in the subthreshold regime as in the above threshold regime.

Hint: Compare the inversion charge in the subthreshold regime with typical fringing and overlap charges.

4

Compound Semiconductor Field Effect Transistors

4-1. INTRODUCTION

Compound semiconductor field effect transistors occupy an important niche in the electronics industry. GaAs FET amplifiers, oscillators, mixers, switches, attenuators, modulators, and current limiters are widely used, and high speed integrated circuits based on GaAs FETs and Heterostructure FETs (HFETs) have been developed.

By combining elements from columns 3 and 5 of the Periodic Table (such as Ga and As, for example), one creates compound semiconductors with four valence electrons per atom, the same as in silicon. Compounds such as GaAs, InP, InAs, InSb, AlAs and others have semiconducting properties and band structures somewhat similar to "classic" elemental semiconductors such as silicon or germanium. At the same time, they provide a gamut of materials with different band gaps (direct and indirect), different lattice constants, and other physical properties for the semiconductor device designer. Moreover, some of these compounds can be combined to form solid state solutions, such as $Al_xGa_{1-x}As$ where the composition, x, may vary continuously from 0 to 1 with a corresponding change in physical properties from those of GaAs into those of AlAs.

Gallium arsenide is the most studied and understood compound semiconductor material. It has been proven indispensable for many device applications – from ultra-high speed transistors to lasers and solar cells. Its room temperature lattice constant (5.653 Å) is very close to that of AlAs (5.661 Å) and the heterointerface between the two materials has a very small density of interface states. New technologies, such as Molecular Beam Epitaxy (MBE) and Metal Organic Chemical Vapor Deposition (MOCVD) allow us to grow these materials with very sharp and clean heterointerfaces and to have very precise control over doping and composition profiles, literally (in case of MBE) changing these parameters within an atomic distance.

Other compound semiconductors, important for applications in ultra-high speed submicron devices, include $In_xGa_{1-x}As$, $Al_xIn_{1-x}As$, $In_xGa_{1-x}P$, GaP, InP, AlN, etc.

There are several advantages of GaAs and related compound semiconductors for applications in submicrometer devices. First of all, the effective mass of electrons in GaAs is much smaller than in Si ($0.067m_e$ in GaAs compared to $0.98m_e$ longitudinal effective mass and $0.19m_e$ transverse effective mass in Si, where m_e is the free electron mass). This leads to a much higher electron mobility in GaAs – approximately 8500 cm^2/Vs in pure GaAs at room temperature compared to 1500 cm^2/Vs in Si. Moreover, in high electric fields, the electron velocity in GaAs is also larger than in Si (see Section 1.4), which is even more important in submicrometer devices where the electric fields are high. The light electrons in GaAs are much more likely to experience so-called "ballistic transport", i.e., to move through the short active region of a high speed device without having any collisions (with lattice imperfections or lattice vibration quanta). This may boost their velocity far beyond the values expected for long devices. Such ballistic transport may become important in very short devices with sizes on the order of 0.1 μm or less. This was first predicted by Shur and Eastman (1979) and was later observed by Heiblum et al. (1985) and Levi et al. (1985). In somewhat longer GaAs devices (with dimensions between 0.1 and 1.5 μm), so-called "overshoot" effects are important. These effects (first predicted in computer simulations by Ruch (1972)), are related to the finite time that it takes for an electron to relax its energy. They may also result in boosting the electron velocity to considerably higher levels than the stationary values shown in Fig. 1.4.1 (Chapter 1). As shown in Fig. 4.1.1, electrons very close to the injecting contact are moving ballistically and the electron velocity is proportional with time. Further from the contact, the velocity reaches a peak value and then decreases. However, due to overshoot effects, the peak value of the velocity is much higher than the stationary value reached as the distance increases further.

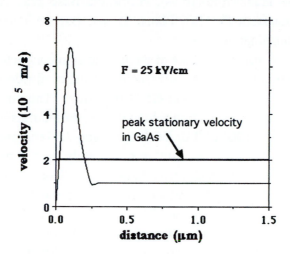

Fig. 4.1.1. Electron velocity versus distance for electrons injected into a region of constant electric field (from Shur (1976)).

In silicon, ballistic and overshoot effects may also be important. However, they are much less pronounced because of the larger electron effective mass. Another important advantage of materials such as GaAs and InP, for applications in high speed submicron devices, is the availability of semi-insulating substrates that eliminate parasitic capacitances related to junction isolation (present in silicon circuits) and allow us to fabricate microstrip lines with small losses (the latter being especially important for applications in Microwave Monolithic Integrated Circuits (MMICs)).

We should also mention that GaAs is a direct gap material that is widely used in optoelectronic applications. The direct gap makes possible a monolithic integration of ultra high speed submicron transistors together with lasers or Light Emitting Diodes (LEDs) on the same chip for use in optical communication. The direct gap also results in high electron-hole recombination rates which may lead to a better radiation hardness.

The availability of excellent heterostructure systems, such as AlGaAs/GaAs, GaInAs/InP, InGaAs/AlGaAs, etc., and new technologies, such as MBE and MOCVD, open up unlimited opportunities for experimentation with new devices, such as Heterostructure Field Effect Transistors (HFETs), Heterojunction Bipolar Transistors (HBTs), Hot Electron Transistors (HETs), Induced Base Transistors (IBTs), Permeable Base Transistors (PBTs), Vertical Ballistic Transistors (VBTs), Planar Doped Barrier Transistors (PDBTs), and many novel quantum devices.

There are also drawbacks of compound semiconductor technology. Silicon is blessed with an excellent native oxide. Silicon nitride is also used as a high quality insulator in silicon field effect transistor. The poor quality of the oxide on GaAs and the correspondingly high density of interface states make it difficult to fabricate GaAs MOSFETs. Wide band gap AlGaAs and, more recently, AlN may substitute for such an insulator, but only in a limited way. Only very recently has a new approach of oxidizing thin silicon layers grown by Molecular Beam Epitaxy on the GaAs surface offered hope for the development of a viable GaAs MOSFET technology (see Tiwari et al. (1988)). Schottky barrier MEtal Semiconductor Field Effect Transistors (MESFETs), Junction Field Effect Transistors (JFETs), and Heterostructure Field Effect Transistors (HFETs) are the most commonly used GaAs devices. In many cases, GaAs MESFETs and JFETs are fabricated by direct ion implantation into the GaAs semi-insulating substrate.

Compound semiconductor transistors are mostly used for microwave and ultra-high speed applications where their high speed properties are the most important. Hence, scaling down the device size in order to enhance the electron velocity and reduce the transit time by ballistic and/or overshoot effects is especially important for such transistors. Other possible uses include optoelectronics, radiation-hard electronics (because of the direct band gap), high temperature electronics (because of the relatively large energy gap in some of the materials, including GaAs), power devices (because of the high breakdown field and the ability to speed up their turn-on by light).

The development of accurate device models is a prerequisite for the commercialization of compound semiconductor technology. However, this technology is much less developed, and reliable device and circuit modeling is therefore especially important, even more so than for its silicon counterpart.

Accurate device models have to be based on insight into the physics of the devices. This type of insight may be obtained from numerical simulations such as self-consistent two-dimensional Monte Carlo modeling (see, for example, Hess and Kizilyalli (1986), and Jensen et al. (1991) and (1991a)). Clearly, numerical device simulations are not directly applicable to circuit design involving tens or hundreds of devices interacting with each other and with other circuit elements, nor in device design where numerous dependencies of device characteristics on the design parameters have to be optimized, nor in device characterization where the device and process parameters must be extracted from experimental data. All these tasks require accurate analytical or semi-analytical device models. These models must be based on physical device and material parameters (rather than using look-up tables and simple interpolations of the measured device characteristics) in order to provide the necessary feed-back between the fabrication process and the device and circuit design.

In this chapter, we will describe such models for GaAs MESFETs and HFETs. We will discuss both long and short channel devices and so-called non-ideal effects which interfere with ideal device behavior. The models considered relate the device *I–V* and *C–V* characteristics to device parameters such as doping density, doping profile, and active layer thickness. They have been incorporated into our integrated circuit simulator, AIM-Spice.

4-2. DEVICE TECHNOLOGY

Here, we briefly review the basics of compound semiconductor FET fabrication with emphasis on different device structures. Compound semiconductor field effect transistors can be fabricated utilizing epitaxially grown wafers or by direct ion implantation into semi-insulating substrates. Epitaxial layers are usually grown by Molecular Beam Epitaxy (MBE) or Metal Organic Chemical Vapor Deposition (MOCVD) on GaAs semi-insulating substrates grown by the Czochralski method. A brief review of material growth, characterization, ion implantation, and ohmic and Schottky contact fabrication is given by Shur (1987).

As was mentioned in Section 4.1, compound semiconductor materials do not have a good quality natural oxide. Therefore, even though GaAs and other compound semiconductor MOSFETs have been demonstrated, the most important compound semiconductor FETs include MEtal Semiconductor Field Effect Transistors (MESFETs) and Heterostructure Field Effect Transistors (HFETs) which rely on the depletion layer between the gate electrode and the channel for gate-channel isolation. Junction Field Effect Transistors (JFETs) may also be important but we will not consider them separately because their modeling is quite similar to that of MESFETs.

Depending on the desired application, MESFETs may differ in device design, gate material, doping profile, and layer composition. Two examples of ion implanted MESFET structures are shown in Fig. 4.2.1.

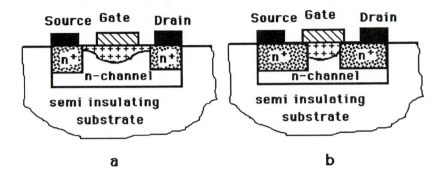

Fig. 4.2.1 Ion implanted GaAs MESFETs.

The MESFET schematically shown in Fig. 4.2.1a has depleted surface regions between the gate and the source and drain contacts. This increases parasitic resistances but usually leads to a relatively small carrier injection from the n^+ source contact into the substrate and, hence, to a small output conductance in the saturation regime. In the structure shown in Fig. 4.2.1b, the gate itself serves as a mask for implanting the source and drain contacts. Such "self-aligned" MESFETs have smaller parasitic contact resistances but may be more susceptible to carrier injection into the substrate and high gate leakage current caused by the n^+ implant straggle into the region under the gate. A useful modification of this structure utilizes thin silicon nitride "side walls" (~ 0.1 μm) on the gate sides which prevent the n^+ implant straggle into the region under the gate. Both structures can be further improved by utilizing a relatively deep, low dose p-type implant. The charge of the depleted acceptors of this implant provides an energy barrier for injection into the substrate (see, for example, Jensen et al. (1991)). A similar effect can be obtained by utilizing an AlGaAs buffer layer or superlattice buffers providing barriers for charge injection into the buffer owing to the conduction band discontinuities (see, for example, Eastman and Shur (1979)).

Fig. 4.2.2 shows typical recessed gate MESFETs. Recessing is a technique for adjusting the device threshold voltage by reducing the thickness of the active layer under the gate while maintaining a relatively low series resistance between the gate region and the source and drain contacts.

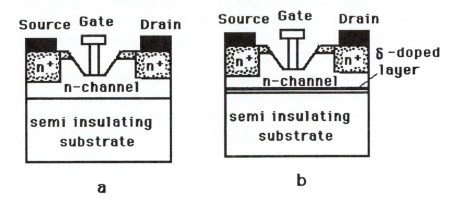

Fig. 4.2.2. Recessed gate MESFETs.

The device layers typically grown by MBE allow us to obtain desired doping and/or composition profiles, making these structures very appropriate for microwave devices. In the structures shown in Fig. 4.2.2, the top n^+ layer allows us to decrease the contact resistance and compensate the depletion caused by the surface states. Figs. 4.2.2a and b show a uniformly doped and a delta doped (or pulse doped) MESFET, respectively. However, a variety of other doping profiles are utilized as well. Both ion-implanted and epitaxially grown MESFETs are usually passivated by a layer of silicon nitride (Si_3N_4). This passivation layer may affect surface states (and, hence, the surface depletion) and may also lead to stress effects which in turn cause a shift in the threshold voltage and in overall changes in the current-voltage characteristics. The devices exhibited in Fig. 4.2.2 have "T"-gates in order to minimize the gate series resistance, which is important in microwave devices. Other gate shapes are also possible (see, for example, Shur (1987) for more details).

The shape of the recess is a very important design feature. In power MESFETs, for example, the recess is often asymmetric with a larger gate-drain than gate-source distance in order to encompass the high electric field region at the drain side of the gate, preventing it from reaching the drain contact. Such a design (see, for example, Eastman et al. (1980)) may allow us to increase the drain-source breakdown voltage. The asymmetry in favor of a reduced source series resistance, R_s, is generally beneficial since it helps reduce the erosion of the extrinsic device transconductance caused by R_S (see Section 3.4).

In the simplest MESFET structures, the source, drain, and gate electrodes are simply parallel metal stripes. However, in power devices, with a large gate width, such a design would yield unacceptably high values of the gate series resistance. Interdigitated designs with several gate "fingers" can alleviate this problem (see Shur (1987), Chapters 7 and 8).

As can be seen from this discussion, a large variety of MESFET designs are used for different applications. In our models, all these designs are treated on equal footing. Model parameters are extracted from measured device characteristics or assigned "typical values". This approach is very effective for circuit simulation but gives only a limited insight into the device physics. A full blown two dimensional, self-consistent Monte Carlo simulation is one of the best tools for gaining insight into the device physics (see, for example, Jacoboni and Lugli (1989) and Jensen et al. (1991) and (1991a)). In the not so distant future, when more computer resources become available, even three dimensional Monte Carlo simulation may be used to simulate GaAs MESFETs.

These arguments equally apply to other advanced device structure, such as HFETs (see Fig. 4.2.3). Different heterostructure systems utilized in HFETs were briefly discussed in Chapter 1 and in Section 4.1.

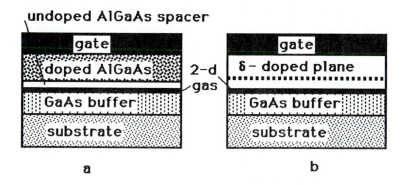

Fig. 4.2.3. Examples of modulation doped layers for HFETs:
a) conventional modulation doped layer, b) delta-doped layer.

The first HFETs (called HEMTs – High Electron Mobility Transistors) were made using the $Al_xGaAs_{1-x}/GaAs$ material system (see Fig. 4.2.3). The advantages of this system include a nearly perfect lattice match between $Al_xGa_{1-x}As$ and GaAs. However, quantum well HFETs utilizing InGaAs quantum wells sandwiched between the AlGaAs layer and the GaAs substrate have demonstrated a superior performance because of the larger band gap discontinuity between InGaAs and AlGaAs and the superior transport properties of InGaAs compared to GaAs. Other important variations of HFET structures include a p^+-type gate in order to increase the gate current turn-on voltage and a p-type or p-i-p^+ buffer in order to limit the carrier injection into the substrate (see Lee and Shur (1990)). Other variations include structures with no dopants in the AlGaAs layer for both n-channel and p-channel devices, called Heterostructure Insulated Gate Field Effect transistors (HIGFETs), a doped superlattice in the AlGaAs layer, a superlattice in the GaAs buffer layer, and HFETs with multiple quantum well channels (for a larger current swing). More details can be found in books by Shur (1987), Wang (1990), Sze (1990), and Morkoç et al. (1991).

The device design determines the basic device parameters such as the field effect mobility, the effective saturation velocity, the threshold voltage, the turn-on voltage for the gate leakage current, the subthreshold slope (i.e., the ideality factor), the output conductance in the saturation regime, the device transconductance, and the source and drain series resistances. Typically, the room temperature field effect mobility in GaAs MESFETs varies from 2000 to 5000 cm^2/Vs depending on the doping level and the material quality. The effective saturation velocity varies from 6×10^4 to 2.5×10^5 m/s (it may increase somewhat in submicrometer devices due to overshoot and ballistic effects). The turn-on voltage of the gate leakage current varies from $0.7 - 0.8$ V for GaAs MESFETs to $1 - 1.1$ V in HFETs with Schottky gates and to $1.4 - 1.5$ V in Dipole HFETs or π-HFETs utilizing p-type dopants to enhance the gate barrier height (see Akinwande et al. (1990) and Lee and Shur (1990)). The series resistances vary within $0.1 - 1$ ohm-mm or even more. Threshold voltages for MESFETs vary from 0.2 to 0.4 V for devices used in digital applications to -10 V for power MESFETs. HFET threshold voltages do not vary as much because of the limited voltage swing related to the material limitations imposed on the maximum density of the two dimensional gas. They may vary from 1 V or so (for undoped devices) to -2 V. These numbers crudely outline a typical parameter space for the model parameters of compound semiconductor FETs.

Following Shur (1987), we now briefly describe integrated circuit fabrication processes for compound semiconductor FETs. Initially, GaAs ICs were fabricated using epitaxial n-type layers. The depletion mode MESFETs were isolated by mesa-etching through the doped layer. However, now most MESFET ICs are fabricated using selective ion-implantation into semi-insulating substrates. Ion implantation allows us to obtain excellent reproducibility and threshold voltage uniformity which are difficult to achieve with epitaxial devices. A second layer of metal and a dielectric layer are used in order to connect the circuit elements. Alternatively, plated air bridges can be used to isolate the first and second level metals at the crossings.

In a typical fabrication process, a GaAs semi insulating substrate is coated with a thin Si_3N_4 film which remains almost intact throughout all subsequent processing steps. Localized implantation steps are carried out by implanting through this layer. The first implant is optimized for the FET channels, the second, deeper implant is for ohmic contacts and for Schottky barrier switching diodes (if required). Additional implants may be incorporated into the process if needed. After the implantation steps, an additional insulator is deposited and the implants are annealed. A typical annealing temperature is around 850 °C. This approach provides a good threshold voltage uniformity with a threshold voltage standard deviation between 20 and 40 mV.

Interface states on the GaAs surface lead to a pinning of the Fermi level at the surface at approximately 0.6 to 0.8 eV below the bottom of the conduction band. This causes a surface depletion of the active channel which may substantially increase the parasitic series resistances (see Fig.4.2.1a). A self-aligned gate fabrication process (see Fig.4.2.1b) solves this problem, reducing the series resistances and making the fabrication process more planar. A self-aligned gate MESFET IC process using refractory metal-silicon gates was developed for LSI circuits designed for gigabit/s speeds (see, for example, Vu et al. (1988)). The cross section of a self-aligned gate IC at different stages of its fabrication process is shown in Fig. 4.2.4.

The circuits are fabricated by selective ion-implantation into three inch Liquid Encapsulated Czochralski (LEC) wafers. The FET channel and load implants are annealed using a Si_3N_4 cap. Following the anneal, the cap is stripped and the refractory metal-silicon gate is sputter deposited and patterned using reactive ion etching. This metal serves as a mask for the source and drain implants. The n^+ implant is also annealed using a Si_3N_4 cap. The devices are completed using AuGe ohmic contacts formed by liftoff. Interconnect metallization begins with the deposition of a Si_3N_4 layer which assists the liftoff. A Au based metal (approximately 6000 Å thick) is deposited and lifted off using the plasma etched dielectric. A SiO$_2$ interlevel dielectric is then deposited. The formation of the second interconnect level is quite similar. Both interconnect levels have sheet resistances of typically less than 0.07 ohms/square. This process yields excellent threshold voltage uniformity (see Vu et al. (1988)).

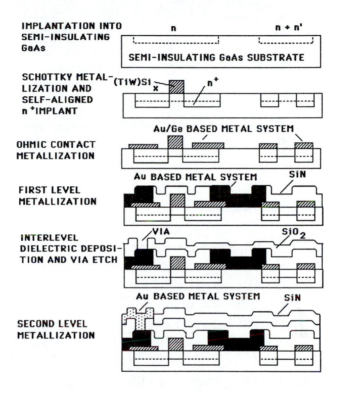

Fig. 4.2.4. Cross section of self-aligned gate GaAs IC at various stages of the fabrication process.

A similar process can be used to fabricate HFET self-aligned ICs (see Cirillo et al. (1984)). The fabrication process starts from growing a modulation doped heterostructure layer on a semi-insulating GaAs substrate using MBE. The cross sectional view of a self-aligned gate HFET IC is shown in Fig. 4.2.5 where the fabrication steps are explained.

A more detailed description of fabrication processes can be found in books by Shur (1987), Wang (1990), Sze (1990), and Morkoç et al. (1991).

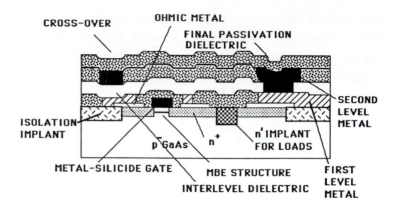

Fig. 4.2.5. Cross sectional view of self-aligned gate HFET IC. (From Cirillo et al. (1984), © 1984 IEEE).

4-3. BASIC MESFET MODELS

GaAs MESFETs (MEtal Semiconductor Field Effect Transistors) are widely used in both analog and digital applications. Their microwave performance challenges that of Heterostructure Field Effect Transistors (see Feng et al. (1990)) and the integration scale of GaAs MESFET Integrated Circuits approaches 100,000 transistors per chip. With thin, highly doped channels and low parasitic resistances, GaAs MESFETs can obtain high transconductances and currents.

A schematic representation of a MESFET is shown in Fig. 4.3.1. The gate electrode forms a Schottky barrier contact (see Section 1.9) with the conducting channel between the source and drain ohmic contacts. The gate bias modulates the depletion region under the gate and, hence, governs the drain current being passed through the channel. Since the charge carriers of the channel are effectively removed from the gate-semiconductor interface by the gate depletion layer, problems related to interface traps are largely avoided. On the other hand, the MESFET has a limited voltage swing in the forward direction of the gate Schottky barrier (limited by the built-in voltage of the contact), which is a drawback in the application of MESFETs in enhancement (normally *off*) logic. However, this limitation is less important in low power circuits operating with a low supply voltage.

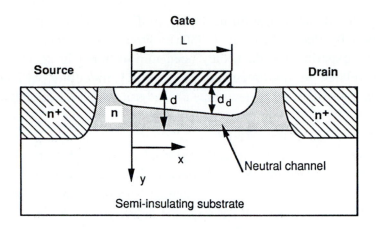

4.3.1. Schematic representation of a MESFET.

The MESFET was first discussed in terms of the Shockley model (see below), where velocity saturation is neglected. In this model, current saturation at high drain-source bias takes place as a result of channel pinch-off at the drain-side of the channel. The Shockley model may be applicable for very long channel devices, but gives a poor description of modern day devices with gate lengths of the order of one micrometer or less. A deeper insight into MESFET device physics can be obtained from detailed two-dimensional Monte Carlo simulation (see, for example, Warriner (1977)). However, simple analytical or semi-analytical models based on the device physics are still required for circuit simulators, such as AIM-Spice (see Ytterdal et al. (1991)) and UM-SPICE (see Peczalski et al. (1986), and Jenkins (1989)).

The first such model valid for MESFETs with very high pinch-off voltages was proposed by Williams and Shaw (1978) and by Shur (1978). The model was based on the assumption of complete velocity saturation in the channel at high drain-source bias and allowed one to calculate the device saturation current for an arbitrary doping profile in the channel. An empirical model for interpolating MESFET current-voltage characteristics was proposed by Curtice (1980). For MESFETs with relatively low pinch-off voltages, Shur (1982) developed a "square law" model, which is qualitatively valid for arbitrary pinch-off voltages. Statz et al. (1987) proposed a more accurate model (by introducing an additional empirical parameter) which is widely known as the "Raytheon model" and is implemented in recent releases of the popular circuit simulator, PSpice® (see Tuinenga (1988)). An extension of the analytical model that takes into account the effects of the non-uniform doping profile on the channel conductance was proposed by Peczalski et al. (1986a). Even though these models adequately describe experimental data, they need improvement in several areas. Several issues related to improvements in MESFET modeling will be discussed in Section 4.4. Here, we review some of the basic MESFET models.

4.3.1. The Shockley Model

We consider first the gate region of a MESFET (intrinsic device) with a uniform channel doping N_d, a channel thickness d, and a built-in voltage V_{bi} for the gate contact. With a channel potential V (relative to the intrinsic source) and an intrinsic gate-source voltage V_{GS}, the depletion width, d_d, can be expressed as follows, using the gradual channel approximation (GCA):

$$d_d = \sqrt{\frac{2\varepsilon_s}{qN_d}(V_{bi} - V_{GS} + V)} \qquad (4\text{-}3\text{-}1)$$

where ε_s is the semiconductor dielectric permittivity. The threshold voltage, V_T, corresponds to the gate-source voltage at which the depletion width at zero drain-source bias $(V = 0)$ equals the channel width, or in terms of eq. (4-3-1)

$$V_T = V_{bi} - V_{po} \qquad (4\text{-}3\text{-}2)$$

where V_{po} is the so-called the pinch-off voltage which, for a uniformly doped channel, is given by

$$V_{po} = \frac{qN_d d^2}{2\varepsilon_s} \qquad (4\text{-}3\text{-}3)$$

At gate-source voltages above the threshold voltage, a finite neutral region exists in the channel allowing significant drain current to pass when applying a drain-source bias. For $V_{GS} < V_T$, this current drops to a low value characteristic of the subthreshold regime.

From eq. (4-3-1), it is obvious that the depletion width under the gate increases from source to drain when a positive drain-source voltage is applied. The depletion width at the drain side of the gate, $d_d(L)$, is obtained by replacing the channel potential by the intrinsic drain-source voltage, V_{DS}, in eq. (4-3-1). Without velocity saturation, $d_d(L)$ increases with increasing drain-source voltage until the channel is pinched off. This occurs when $d_d(L) = d$, corresponding to $V_{DS} = V_{GS} - V_T \equiv V_{GT}$.

In the Shockley model, it is assumed that the electron drift velocity, v_n, is proportional to the absolute value of the longitudinal electric field $F = |dV/dx|$, i.e.,

$$v_n = \mu_n F \tag{4-3-4}$$

where μ_n is the low-field electron mobility.   In the gradual channel approximation, the potential drop, dV, in a small section, dx, at the lateral position x in the channel (see Fig. 4.3.1) can be written as

$$dV = I_d \, dR = \frac{I_d \, dx}{q N_d \mu_n W \left[d - d_d(x) \right]} \tag{4-3-5}$$

where I_d is the drain current, dR is the channel resistance of the small section of length dx, and W is the gate width. In eq. (4-3-5) it was assumed that $d_d(L) < d$ (no pinch-off). The depletion width, $d_d(x)$, at position x in terms of the channel potential, $V(x)$, is given by eq. (4-3-1). Integrating eq. (4-3-5) over the entire gate length, L, leads to the following expression for the drain current

$$I_d = g_o \left\{ V_{DS} - \frac{2 \left[\left(V_{DS} + V_{bi} - V_{GS} \right)^{3/2} - \left(V_{bi} - V_{GS} \right)^{3/2} \right]}{3 V_{po}^{1/2}} \right\} \tag{4-3-6}$$

where

$$g_o = \frac{q N_d \mu_n W d}{L} \tag{4-3-7}$$

is the conductance of the undepleted channel.   Eq. (4-3-6) is called the fundamental equation of field-effect transistors, and is valid only for $V_{DS} \leq V_{GT}$. It is easily shown that the channel conductance, dI_d/dV_{DS}, becomes zero when $V_{DS} = V_{GT}$. Hence, it can be argued the drain current saturates at this drain-source voltage, which is called the saturation voltage, V_{SAT}. Accordingly, we have for the Shockley model

$$V_{SAT} = V_{GT} \tag{4-3-8}$$

and the corresponding saturation drain current becomes

$$I_{sat} = g_0 \left[\frac{V_{po}}{3} + \frac{2(V_{bi} - V_{GS})^{3/2}}{3 V_{po}^{1/2}} - V_{bi} + V_{GS} \right] \tag{4-3-9}$$

From eq. (4-3-6), we find that the intrinsic device transconductance below saturation is given by

$$g_{mi} = \frac{\partial I_d}{\partial V_{GS}}\bigg|_{V_{DS}} = g_0 \frac{\sqrt{V_{DS} + V_{bi} - V_{GS}} - \sqrt{V_{bi} - V_{GS}}}{\sqrt{V_{po}}} \tag{4-3-10}$$

For small drain-source voltages, i.e., when $V_{DS} \ll V_{bi} - V_{GS}$, eqs. (4-3-6) and (4-3-10) can be simplified to

$$I_d \approx g_0 \left(1 - \sqrt{\frac{V_{bi} - V_{GS}}{V_{po}}} \right) V_{DS} \tag{4-3-11}$$

$$g_{mi} \approx \frac{g_0 V_{DS}}{2 \sqrt{V_{po} (V_{bi} - V_{GS})}} \tag{4-3-12}$$

From eq. (4-3-9), we find the intrinsic transconductance in the saturation region:

$$g_{msi} = g_0 \left(1 - \sqrt{\frac{V_{bi} - V_{GS}}{V_{po}}} \right) \tag{4-3-13}$$

4.3.2. Velocity Saturation Model

In the Shockley model, it was assumed that the drift velocity increases linearly with the electric field. From current continuity, it follows that the drift velocity at the drain side of the gate increases to infinity as we approach the pinch-off condition, $d_d(L) = d$. This is, of course, unphysical. Instead, velocity saturation will occur at sufficiently high electric fields (see Section 1.4), which gives an alternative mechanism for current saturation in the device. A simple way of dealing with velocity saturation is to assume a two-piece linear velocity-field relationship of the form (see Section 3.5)

$$v(F) = \begin{cases} \mu F, & F < F_s \\ v_s, & F \geq F_s \end{cases} \qquad (4\text{-}3\text{-}14)$$

where v_s is the saturation velocity and $F_s = v_s/\mu$ is the saturation field. When $F(L) \leq F_s$, the results from the Shockley model are still valid. Hence, the saturation voltage, defined as the drain-source voltage at the onset of velocity saturation, can be determined from eq. (4-3-5) in combination with eqs. (4-3-1) and (4-3-6), resulting in the expression

$$\frac{\dfrac{V_{SAT}}{V_{po}} - \dfrac{2}{3}\left(1 - \dfrac{V_{GT} - V_{SAT}}{V_{po}}\right)^{3/2} + \dfrac{2}{3}\left(1 - \dfrac{V_{GT}}{V_{po}}\right)^{3/2}}{1 - \sqrt{1 - \dfrac{V_{GT} - V_{SAT}}{V_{po}}}} = \frac{V_L}{V_{po}} \qquad (4\text{-}3\text{-}15)$$

where $V_L = F_s L$. The corresponding saturation current is

$$I_{sat} = g_o V_L \left[1 - \sqrt{1 - \frac{V_{GT} - V_{SAT}}{V_{po}}}\right] \qquad (4\text{-}3\text{-}16)$$

From eq. (4-3-15), we see that the Shockley saturation voltage, $V_{SAT} = V_{GT}$, is recovered when $V_L/V_{po} \gg 1$. However, in the opposite limit, when $V_L/V_{po} \ll 1$, which corresponds to near velocity saturation in the entire channel, we find

$$V_{SAT} \approx V_L \qquad (4\text{-}3\text{-}17)$$

In this expression, it is also assumed that $V_L/V_{po} \ll z(1 - z)$ where $z = (1 + V_{GT}/V_{po})^{1/2}$. For intermediate cases, V_{SAT} can be found either by solving eq. (4-3-15) as a third order equation in $[1 - (V_{GT} - V_{SAT})/V_{po}]^{1/2}$, or by solving the equation numerically. However, a simple interpolation formula for the saturation voltage can be established by combining the results for the two limiting cases as proposed by Shur (1982):

$$V_{SAT} \approx \left(\frac{1}{V_L} + \frac{1}{V_{GT}}\right)^{-1} \qquad (4\text{-}3\text{-}18)$$

Likewise, an interpolation formula, valid for devices with relatively low pinch-off voltages, can be found for the saturation current (see Shur (1982)):

$$I_{sat} \approx \beta V_{GT}^{2} \tag{4-3-19}$$

where

$$\beta = \frac{2\varepsilon_s v_s W}{d\left(V_{po} + 3V_L\right)} \tag{4-3-20}$$

We note that the square law of eq. (4-3-19) is of the same form as that used in SPICE modeling of the saturation current in JFETs, and it has also been used to describe the saturation characteristics in SPICE simulation of GaAs MESFETs. Statz et al. (1987) proposed a more general version of eq. (4-3-19) to also cover devices with higher pinch-off voltages:

$$I_{sat} \approx \frac{\beta V_{GT}^{2}}{1 + \alpha V_{GT}} \tag{4-3-21}$$

where α is an additional parameter. Eq. (4-3-20) can be used to determine the dependence of I_{sat} and the device transconductance on channel doping, gate length, electron mobility, and saturation velocity.

The velocity saturation model now allows us to make a rough estimate of the intrinsic high speed performance of the MESFET. From eq. (4-3-19), we can calculate the transconductance at the saturation point:

$$g_{msi} = \frac{dI_{sat}}{dV_{GT}} \approx \frac{4\varepsilon_s v_s W}{d\left(V_{po} + 3V_L\right)} V_{GT} \tag{4-3-22}$$

Furthermore, we may argue that the gate-source capacitance, C_{gs}, at saturation will be on the order of $\varepsilon_s LW/d$. Hence, the cut-off frequency can be written approximately as

$$f_T = \frac{g_{msi}}{2\pi C_{gs}} \propto \frac{v_s V_{GT}}{L\left(V_{po} + 3V_L\right)} \tag{4-3-23}$$

From this expression, it is obvious that a high cut-off frequency can be obtained using a small gate length and a small pinch-off voltage (see eq. (4-3-3)). Normally, it is also desirable to have a device with a high current level, which favors a large doping sheet density (N_dd). The best tradeoff is therefore to use a thin and highly doped channel.

Source and drain series resistances, R_s and R_d, may play an important role in determining the current-voltage characteristics of GaAs MESFETs. These resistances can be taken into account by using the following relationships between the extrinsic (lower case subscripts) and intrinsic (upper case subscripts) drain-source and gate-source bias voltages:

$$V_{ds} = V_{DS} + R_t I_d \tag{4-3-24}$$

$$V_{gs} = V_{GS} + R_s I_d \tag{4-3-25}$$

where $R_t = R_s + R_d$. The saturation current in terms of the extrinsic gate voltage swing, V_{gt}, is readily obtained by combining eqs. (4-3-19) and (4-3-25):

$$I_{sat} = \frac{2\beta V_{gt}^2}{1 + 2\beta V_{gt} R_s + \sqrt{1 + 4\beta V_{gt} R_s}} \tag{4-3-26}$$

For device modeling suitable for Computer Aided Design (CAD), one has to model the current-voltage characteristics in the entire range of drain-source voltages, not only in the saturation regime. An empirical interpolation expression for the full, extrinsic MESFET *I–V* characteristics was proposed by Curtice (1980) using a hyperbolic tangent function:

$$I_d = I_{sat} \left(1 + \lambda V_{ds}\right) \tanh\left(\frac{g_{ch} V_{ds}}{I_{sat}}\right) \tag{4-3-27}$$

Here, λ is an empirical constant that accounts for the finite output conductance in saturation, and g_{ch} is the extrinsic channel conductance of the linear region given by

$$g_{ch} = \frac{g_{chi}}{1 + g_{chi}\left(R_s + R_d\right)} \tag{4-3-28}$$

where g_{chi} is the intrinsic channel conductance at very low drain-source voltage. For a uniformly doped channel, we have from eq. (4-3-11)

$$g_{chi} = \left. \frac{\partial I_d(V_{DS} \to 0)}{\partial V_{DS}} \right|_{V_{GS}} = g_0\left(1 - \sqrt{1 - \frac{V_{GT}}{V_{po}}}\right) \qquad (4\text{-}3\text{-}29)$$

The finite output conductance in saturation, described in terms of the constant λ in eq. (4-3-27), may be related to the short channel effects (see Pucel et al. (1975)) and to parasitic currents in the substrate, such as space charge limited current. Hence, the output conductance may be greatly reduced by using a heterojunction buffer to prevent carrier injection into the substrate (see, for example, Eastman and Shur (1979)).

The analytical models discussed above are suitable for Computer Aided Design of GaAs MESFETs and GaAs MESFET circuits. However, these models do not explicitly take into account important effects such as subthreshold current and drain voltage induced shift in the threshold voltage. However, these effects will be addressed in the next Section where a complete, unified model for the MESFET is developed along the same line as was done in Section 3.9 for MOSFETs. Many other complicated effects which are not explicitly included are, for example, deviation from the gradual channel approximation (which may be especially important at the drain side of the channel, see Pucel et al. (1975)), possible formation of a high field region (i.e., a dipole layer) at the drain side of the channel (Engelmann and Liehti (1977), Shur and Eastman (1978), Fjeldly (1986), Fjeldly et al. (1989)), inclusion of diffusion and incomplete depletion at the boundary between the depletion region and conducting channel (Yamaguchi and Kodera (1976)), ballistic or overshoot effects (Ruch (1972), Maloney and Frey (1975), Shur (1976), Warriner (1977), Cappy et al. (1980), Fjeldly (1988)), effects of donor diffusion from the n^+ contact regions into the channel (Chen et al. (1987)), effects of the passivating silicon nitride layer (Asbeck et al. (1984), Chen et al. (1987)), and effects of traps (Chen et al. (1986)). These factors may still be included indirectly by adjusting model parameters such as mobility, saturation velocity, pinch-off voltage, etc. In a rigorous way, they can be only treated using numerical solutions. However, for practical circuit simulators used in circuit design, analytical or very simple numerical models are still a necessity.

4-4. UNIVERSAL MESFET MODEL (AIM-Spice MODEL)

As mentioned in Section 4.3, the simple MESFET models, such as the Shockley model and the model based on the two-piece velocity-field relationship, need improvements in several areas. First of all, it is important to account for the effects related to non-uniform doping in the channel. Non-uniformities are caused, for example, by ion implanted channel doping. Also, a non-uniform doping profile may be tailored as a part of the device design, which is the case for so-called delta-doped GaAs MESFETs considered in Subsection 4.4.4. Second, the subthreshold current (unaccounted for in basic models) is of great concern since it is has consequences for the bias and logic levels in digital operations, which again affect the power dissipation in logic circuits. In addition, the accuracy of the basic models needs to be improved. In particular, we need a better fit of the current-voltage characteristics in the knee region, which is especially important for logic devices, and a more accurate model for the output conductance in the saturation region, which is crucial for microwave devices and also plays a role in digital applications.

Here, we develop a new, universal model for MESFETs which covers all ranges of operation, including the subthreshold regime. The current-voltage (*I–V*) and capacitance-voltage (*C–V*) characteristics are described by continuous, analytical expressions with relatively few, physically based parameters. Our approach follows the same basic philosophy used for MOSFETs (see Sections 3.3, 3.4, and 3.9), starting from one unified expression for the effective differential channel capacitance. The model accounts for the series drain and source resistances, the velocity saturation in the channel, the finite output conductance in the saturation regime, and the drain bias induced threshold voltage shift. The model parameters, such as the effective channel mobility, the saturation velocity, the source and drain resistances, etc. are extractable from experimental data. In the case of a uniform or delta-doped profile, the model is analytical. This makes the model very suitable for incorporation into circuit simulators and for the design of microwave GaAs MESFETs. We apply our new characterization procedure to an ion-implanted MESFET and a delta-doped MESFET and obtain good agreement with the experimental data.

The MESFET models described in this Sections are implemented in our circuit simulator AIM-Spice. Examples of the simulation of a MESFET ring oscillator are presented in Chapter 6.

4.4.1. Unified Channel Capacitance

In a MESFET biased above threshold, the differential capacitance per unit area associated with the gate depletion region can be written as (see Sections 1.9 and 4.3)

$$c_a = \frac{\varepsilon_s}{d_d} \tag{4-4-1}$$

where ε_s is the dielectric permittivity of the semiconductor and d_d is the depletion width. For an arbitrary doping profile, d_d can be expressed as follows using the gradual channel approximation:

$$V_{bi} - V_{GS} + V = \frac{q}{\varepsilon_s} \int_0^{d_d} y\, N(y)\, dy \tag{4-4-2}$$

Here, V_{GS} is the intrinsic gate-source voltage, V is the channel voltage, V_{bi} is the built-in voltage, y is the distance into the semiconductor from the gate electrode and $N(y)$ is the doping density profile of the conducting channel. In eq. (4-4-2) we have assumed that the doping profile does not vary too abruptly, so that the extent of the partially depleted boundary layer separating the neutral channel from the depletion region is small compared to the characteristic length of the doping profile variation. (The special but important case of delta-doped MESFETs will be addressed separately, see Subsection 4.4.4.) Above threshold, the sheet density of electrons in the channel is

$$n_s = \int_{d_d}^{d} N(y)\, dy \tag{4-4-3}$$

where d is the effective channel depth.

The threshold voltage, V_T, corresponding to the gate-drain voltage where the channel is fully depleted, is given by eq. (4-3-2), where the pinch-off voltage can be expressed as

$$V_{po} = \frac{q}{\varepsilon_s} \int_0^{d} y\, N(y)\, dy \tag{4-4-4}$$

Below threshold, the electron sheet density in the channel can be written as follows, assuming a Boltzmann energy distribution (see the discussion in Sections 3.3 and 3.4):

$$n_s \approx n_o \exp\left(\frac{V_{GT} - V_F}{\eta V_{th}}\right) \tag{4-4-5}$$

Here, n_o is the electron sheet density at the threshold point, V_F is the Fermi potential relative to the source, $V_{th} = k_B T/q$ is the thermal voltage and η is the ideality factor. Note that for $V_{DS} > V_{th}$ in the subthreshold regime, most of the applied drain-source voltage will drop across a depletion zone extending from the gate towards the drain, and the current in the channel will be dominated by diffusion. Hence, we have $V \approx 0$ in the channel while the Fermi potential varies from $V_F(0) = 0$ at the source side to $V_F(L) = V_{DS}$ at the drain side of the gate. Hence, the subthreshold differential capacitance per unit area, c_b, becomes

$$c_b = q\frac{dn_s}{dV_{GS}} = \frac{qn_o}{\eta V_{th}} \exp\left(\frac{V_{GT} - V_F}{\eta V_{th}}\right) \tag{4-4-6}$$

An approximate, unified expression for the effective differential gate-channel capacitance per unit area, c_{gc}, is obtained by representing c_{gc} as a series connection of the depletion capacitance and the subthreshold capacitance, extending the validity of both beyond threshold, i.e.,

$$c_{gc} = \frac{c_a c_b}{c_a + c_b} \tag{4-4-7}$$

Note that this approach is identical to that used in Sections 3.3, 3.4, and 3.9.

The electron sheet density, n_o, in eq. (4-4-6) is found by considering the effective differential capacitance at threshold and zero drain-source voltage (i.e., $V = V_F = 0$). For MOSFETs, the gate-channel capacitance at threshold, c_T, is very close to one third of the maximum capacitance (which for a MOSFET corresponds to the oxide capacitance, see Section 3.3). However, contrary to the situation in the MOSFET, the gate-channel capacitance in a MESFET with an arbitrary doping profile does not reach a distinct maximum value because the thickness of the depletion region decreases monotonically with increasing gate voltage. Instead, we define the threshold capacitance based on the following consideration: Above threshold, we have $c_a \ll c_b$ and below threshold, $c_b \ll c_a$. Hence, it is natural to associate the threshold point with the condition that $c_b = c_a$. Also, at threshold, the channel is totally depleted so that $c_b = c_a = \varepsilon_s/d$. We can therefore write $c_T = c_a/2 = c_b/2 = \varepsilon_s/(2d)$. (We note, however, that the value of the threshold voltage is relatively insensitive to the precise numerical value of c_a and c_b at threshold.) Hence, we find from eq. (4-4-6)

$$n_o = \frac{\varepsilon_s \eta V_{th}}{qd} \tag{4-4-8}$$

A unified sheet charge density of electrons in the channel can be expressed in terms of the effective differential capacitance as

$$qn_s = \int_{-\infty}^{V_{GS}} c_{gc}\left(V'_{GS}\right) dV'_{GS} \tag{4-4-9}$$

The capacitance model described above can be used for calculating the gate-source capacitance, C_{gs}, and the gate-drain capacitance, C_{gd}, by using an approximation similar that used in a model by Meyer (1971). In terms of the extrinsic gate bias voltage, $V_{gt} = V_{GT} + I_d R_s$, where I_d is the drain current and R_s is the source series resistance, the capacitances can be written as

$$C_{gs} = C_f + \frac{2}{3} C_{gc}\left[1 - \left(\frac{V_{sate} - V_{dse}}{2V_{sate} - V_{dse}}\right)^2\right] \tag{4-4-10}$$

$$C_{gd} = C_f + \frac{2}{3} C_{gc}\left[1 - \left(\frac{V_{sate}}{2V_{sate} - V_{dse}}\right)^2\right] \tag{4-4-11}$$

Here, $C_{gc} = c_{gc}LW$, $V_{sate} = I_{sat}/g_{ch}$ is the effective extrinsic saturation voltage, I_{sat} is the saturation drain current and g_{ch} is the extrinsic channel conductance at low drain-source bias (see Subsection 4.4.2). In eqs. (4-4-10) and (4-4-11), V_{dse} is an effective extrinsic drain-source voltage which is equal to V_{ds} for $V_{ds} < V_{sate}$ and is equal to V_{sate} for $V_{ds} > V_{sate}$. In order to obtain a smooth transition between the two regimes, we interpolate V_{dse} as follows:

$$V_{dse} = V_{ds}\left[1 + \left(\frac{V_{ds}}{V_{sate}}\right)^{m_c}\right]^{-1/m_c}$$ (4-4-12)

where m_c is a constant determining the width of the transition region between the linear and the saturation regime. The capacitance C_f in eqs. (4-4-11) and (4-4-12) is the side wall and fringing capacitance which can be estimated, in terms of a metal line of length W, as

$$C_f \approx \beta_c \, \varepsilon_s W$$ (4-4-13)

where β_c is about 0.5 (see Gelmont et al. (1991)). A more detailed analysis of parasitic capacitances in MESFET was given by Anholt and Swirhun (1991) and (1991a).

Fig. 4.4.1 shows calculated values of C_{gs} and C_{gd} for the MESFET considered in Subsections 4.4.3 and 4.4.5.

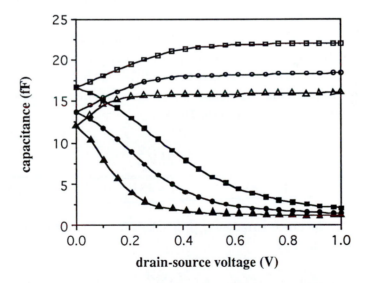

Fig. 4.4.1.   Gate-source capacitance (upper curves) and gate-drain capacitance (upper curves) versus drain-source voltage for $V_{GS} = 0$ V (squares), –0.4 V (circles) and –0.8 V (triangles). The device parameters are the same as for Fig. 4.4.2. In addition, we use $m_c = 3$. (From Shur et al. (1992).)

4.4.2. Unified Current-Voltage Characteristics

At drain-source voltages well below the saturation voltage, the drain current can always (i.e., both below and above threshold) be expressed as

$$I_d = g_{chi} V_{DS} = g_{ch} V_{ds} \qquad (4\text{-}4\text{-}14)$$

where g_{chi} and g_{ch} are the intrinsic and extrinsic channel conductances of the linear region, and V_{DS} and V_{ds} are the intrinsic and extrinsic drain-source voltages, respectively. The intrinsic channel conductance can be written as

$$g_{chi} = \frac{q n_s W \mu_n}{L} \qquad (4\text{-}4\text{-}15)$$

where μ_n is the low-field mobility and n_s is the unified electron sheet density at low drain-source voltage given by eq. (4-4-9) with $V_F = 0$ (see eq. (4-4-6)). In the linear region, g_{ch} is related to g_{chi} by eq. (4-3-28). We note that in the subthreshold regime, the channel resistance will normally be much larger than the parasitic source and drain resistances, R_s and R_d, such that $g_{ch} \approx g_{chi}$, $V_{ds} \approx V_{DS}$ and $V_{gt} \approx V_{GT}$.

The saturation current above threshold, hereafter denoted I_{sata}, was obtained in Section 4.3 using a two-piece velocity-field characteristic for uniform MESFETs. This expression can be modified to be used for a general doping profile by introducing the transconductance compression parameter, t_c, as was first proposed by Statz et. al. (1987):

$$I_{sata} = \frac{2\beta V_{gt}^2}{\left(1 + 2\beta V_{gt} R_s + \sqrt{1 + 4\beta V_{gt} R_s}\right)\left(1 + t_c V_{gt}\right)} \qquad (4\text{-}4\text{-}16)$$

The transconductance parameter, β, is taken from the basic theory of Section 4.3 and can be written as

$$\beta = \frac{2\varepsilon_s v_s W}{d\left(V_{po} + 3V_L\right)} \qquad (4\text{-}4\text{-}17)$$

where $V_L = F_s L$, F_s is the saturation electric field, and v_s is the saturation velocity.

The value of the transconductance compression factor, t_c, depends on the doping profile and may also depend on the properties of the substrate-channel interface and other factors. Based on numerous simulations, we have concluded that $t_c = 0$ usually gives a good fit for uniformly doped or ion-implanted devices with pinch-off voltages smaller than approximately 2.5 V. For MESFETs with pinch-off voltages of about 6 V, Statz et al. (1987) found that $t_c \approx 0.1$ V^{-1} gave an excellent fit. We should point out that the non-uniformity in the doping profile affects the channel conductance more dramatically than the saturation current. Indeed, the dependence of the channel conductance on the gate-source voltage is proportional to the integral of the doping profile over the undepleted channel (see eqs. (4-4-3) and (4-4-15)), while the saturation current is primarily determined by the region near the drain where a large portion of the channel is depleted. Therefore, the most important part of the dependence of the drain current on the doping profile is included into its dependence on the pinch-off voltage. Accordingly, a model where the single parameter t_c accounts for the shape of the doping profile is quite adequate. As shown in Subsection 4.4.4, this model works even for the limiting case of delta-doped MESFETs.

The long channel drain current in the subthreshold regime is given by

$$I_d \approx qWD_n \frac{n_s(0) - n_s(L)}{L}$$

$$= \frac{qWD_n n_o}{L} \exp\left(\frac{V_{GT}}{\eta V_{th}}\right)\left[1 - \exp\left(-\frac{V_{DS}}{V_{th}}\right)\right] \tag{4-4-18}$$

where we recognize the characteristic transition to the diffusion dominated saturation regime at $V_{DS} \approx V_{th}$. Hence, the saturation current in the subthreshold regime becomes (using $V_{GT} \approx V_{gt}$, and $D_n = \mu_n V_{th}$)

$$I_{satb} = \frac{q n_o \mu_n V_{th} W}{L} \exp\left(\frac{V_{gt}}{\eta V_{th}}\right) \tag{4-4-19}$$

In the spirit of the derivation of the unified gate-channel capacitance (see eq. (4-4-7)), a unified expression for the saturation current can be written as follows:

$$I_{sat} = \frac{I_{sata} I_{satb}}{I_{sata} + I_{satb}} \tag{4-4-20}$$

where, in order to obtain the correct asymptotic behavior in the subthreshold regime, we can replace V_{gt} in eq. (4-4-16) by the effective gate voltage swing (note that the dependence of I_{satb} on V_{gt} in eq. (4-4-19) has to be retained in eq. (4-4-20))

$$V_{gte} = \frac{V_{th}}{2}\left[1 + \frac{V_{gt}}{V_{th}} + \sqrt{\delta^2 + \left(\frac{V_{gt}}{V_{th}} - 1\right)^2}\right] \qquad (4\text{-}4\text{-}21)$$

We note that V_{gte} approaches asymptotically V_{th} below threshold and V_{gt} above threshold. The parameter δ determines the width of the transition region ($\sim \delta V_{th}$). Typically, $\delta = 5$ is a good choice.

Based on the result for the various regions of the I–V characteristics, we can now proceed to develop a full, unified description of the I–V characteristics.

The empirical hyperbolic tangent expression by Curtice (see eq. (4-3-27)) and a similar interpolation by Statz et al. (1987) are possible ways of describing the MESFET I–V characteristics. However, these types of expressions are only valid above threshold. In addition, they cannot always adequately reproduce the knee region of the characteristics and the dependence of the output conductance on the drain-source voltage in the saturation regime. We therefore propose an alternative description of the extrinsic I–V characteristics which is valid both above and below threshold, and which allows the knee region to be accurately modeled with a shape parameter that is easily extractable from experimental data. Furthermore, the output conductance in saturation is adequately described in terms of the drain voltage induced threshold voltage shift and, if necessary, by including an additional fitting parameter (see Section 4.4.5 for the parameter extraction procedure).

The proposed unified expression for the I–V characteristics is the same as that previously used for the MOSFET (see Section 3.9):

$$I_d = \frac{g_{ch} V_{ds}}{\left[1 + \left(V_{ds}/V_{sate}\right)^m\right]^{1/m}} \qquad (4\text{-}4\text{-}22)$$

where V_{sate} is the effective, extrinsic saturation voltage, m is a parameter that determines the shape of the characteristics in the knee region, g_{ch} is the extrinsic channel conductance at small drain-source bias, $V_{ds} = V_{DS} + R_t I_d$ is the extrinsic drain-source voltage, and $R_t = R_s + R_d$ is the sum of the source and the drain parasitic resistances. According to eq. (4-4-22), the drain current saturates at the value $I_{sat} = g_{ch}V_{sate}$. (We note that $V_{sate} = I_{sat}/g_{ch}$ is identical to the effective saturation voltage used in eqs. (4-4-10) and (4-4-11).)

An important effect which must be taken into account for an adequate description of the MESFET behavior is the dependence of the threshold voltage on the drain-source voltage (see, for example Peczalski et al. (1986a)). This dependence is analogous to that observed in MOSFETs (see Section 3.5) and the V_T versus V_{ds} relationship can be fairly accurately described by the empirical expression

$$V_T = V_{To} - \sigma V_{ds} \qquad (4\text{-}4\text{-}23)$$

where V_{To} is the threshold voltage at zero drain-source voltage and σ is a coefficient which may depend on the gate voltage swing. This translates into the following shift in V_{gt}:

$$V_{gt} = V_{gto} + \sigma V_{ds} \qquad (4\text{-}4\text{-}24)$$

where V_{gto} is the gate voltage referred to threshold at zero drain-source bias. The effects of the drain bias induced threshold voltage shift can be accounted for in the I–V characteristics (see eq. (4-4-22)) by using eq. (4-4-24) in the expressions for the saturation current and the linear channel conductance. This will result in a finite output conductance in the saturation region.

Several physical mechanisms may contribute to the threshold voltage shift. First, carrier injection into the substrate may play an important role (see Shur (1987), and Jensen et al. (1991) for more details). Another mechanism is similar to the Drain Induced Barrier Lowering (DIBL) in MOSFETs (see Section 3.6). This latter mechanism may depend on the gate voltage swing because the barrier does not exist well above threshold. Hence, for modeling purposes, σ should include a dependence on V_{gt}. After many simulations, we adopted the following empirical expression:

$$\sigma = \frac{\sigma_o}{1 + \exp\left(\dfrac{V_{gto} - V_\alpha}{V_\sigma}\right)} \qquad (4\text{-}4\text{-}25)$$

which gives $\sigma \to \sigma_o$ for $V_{gto} < V_{\sigma t}$ and $\sigma \to 0$ for $V_{gto} > V_{\sigma t}$. The voltage V_σ determines the width of the transition between the two regimes.

For a more accurate description of the saturation region, it is possible to include an additional empirical factor, $(1 + \lambda V_{ds})$, in the expression for the I–V characteristics (as is usually done in SPICE MESFET models):

$$I_d = \frac{g_{ch} V_{ds} \left(1 + \lambda V_{ds}\right)}{\left[1 + \left(V_{ds}/V_{sate}\right)^m\right]^{1/m}} \qquad (4\text{-}4\text{-}26)$$

The parameter λ is related to physical effects that are not explicitly included in our model, such as gate length modulation (corresponding to a positive value of λ) and domain formation (corresponding to a negative value of λ). As a rule of thumb (and based on the analysis given in Section 3.10), we can assume $\lambda \approx k_o/L$ where k_o is a constant.

When the experimental I–V characteristics have a finite output conductance, g_{chs}, in saturation, the saturation current is conveniently defined, for the purpose of comparison with the present model, as the intersection of the linear asymptotic behavior at small drain-source voltages $(I_d = g_{ch}V_{ds})$ and that in deep saturation $(I_d = g_{ch}V_{sate} + g_{chs}V_{ds})$. Hence, by eliminating V_{ds} from these two expressions, we can express V_{sate} in terms of the saturation current as: $V_{sate} = (1 - g_{chs}/g_{ch})I_{sat}/g_{ch}$. Normally, for well-behaved transistors, $g_{chs} \ll g_{ch}$ and we can use $V_{sate} = I_{sat}/g_{ch}$ in eq. (4-4-26). We note, however, that at sufficiently large gate voltage swing, i.e., for $V_{gt} > V_{\sigma t} + 3V_\sigma$, the effects of the drain voltage induced shift in the threshold voltage vanishes and we have $g_{chs} = \lambda g_{ch}V_{sate}$ in deep saturation, as can be shown from eq. (4-4-26). This can be used to determine λ as part of the device characterization (see Subsection 4.4.5) and to obtain the corrected relationship, $V_{sate} = (g_{ch}/I_{sat} + \lambda)^{-1}$, to be used in eq. (4-4-26) well above threshold.

In the following, the present unified MESFET model will be applied to two special cases, both of which can be described by approximate, analytical expressions. The first is a MESFET with a uniformly doped channel, and the second is the same type of device with an additional delta-doped layer. Subsequently, a prescription for the extraction of model parameters from experimental data is presented. The summary of these AIM-Spice models is given in Appendix A8.

4.4.3. Model for Uniformly Doped Channel (MESA1)

For a uniformly doped channel with doping density N_d and channel thickness d, the depletion layer thickness, d_d, is given by eq. (4-3-1), and the capacitance per unit area above threshold, c_a, at low drain-source voltage can be expressed as

$$c_a = \frac{\varepsilon_s}{d_d} = \frac{\varepsilon_s}{d\sqrt{1 - V_{GT}/V_{po}}} \tag{4-4-27}$$

where the threshold and the pinch-off voltages are given by eqs. (4-3-2) and (4-3-3), respectively. The corresponding subthreshold capacitance per unit area can be derived from eqs. (4-4-6) and (4-4-8), i.e.,

$$c_b = \frac{\varepsilon_s}{d} \exp\left(\frac{V_{GT}}{\eta V_{th}}\right) \tag{4-4-28}$$

Hence, the unified differential gate-channel capacitance per unit area at low drain-source voltage can be approximated by

$$c_{gc} = \left(\frac{1}{c_a} + \frac{1}{c_b}\right)^{-1} = \frac{\varepsilon_s}{d}\left[\sqrt{1 - \frac{V_{GT}}{V_{po}}} + \exp\left(-\frac{V_{GT}}{\eta V_{th}}\right)\right]^{-1} \tag{4-4-29}$$

We are interested in finding the linear channel conductance, g_{ch}, at very low drain-source voltage, to be used in the expression for the I–V characteristics (eq. (4-4-22)). This allows us to replace V_{GT} by V_{gt} in eq. (4-4-29). The linear channel conductance depends on the sheet density of electrons in the channel, n_s, which can be calculated using eq. (4-4-9) in combination with eq. (4-4-29). The integral in eq. (4-4-9) does not have an analytical solution in this case but, to a good approximation, the result can be expressed as

$$n_s = \frac{1}{q}\int_{-\infty}^{V_{gt}} c_{gc}\, dV'_{gt}$$

$$\approx \left[\frac{1}{N_d A\left(1 - \sqrt{1 - \dfrac{V_{gte}}{V_{po}}}\right)} + \frac{1}{n_o}\exp\left(-\frac{V_{gt}}{\eta V_{th}}\right)\right]^{-1} \tag{4-4-30}$$

where the effective gate voltage swing, V_{gte}, is defined in eq. (4-4-21). Hence, the extrinsic linear channel conductance can be written as (see eqs. (4-3-28) and (4-4-15))

$$g_{ch} = \frac{g_{chi}}{1 + g_{chi} R_t} = \frac{q n_s W \mu_n / L}{1 + R_t q n_s W \mu_n / L} \tag{4-4-31}$$

Although g_{ch} does not depend explicitly on the drain-source bias, it will have an implicit dependence on V_{ds} through the effects of the drain voltage induced threshold voltage shift when g_{ch} is used in the context of the I–V characteristics (eq. (4-4-22)). This dependence can be taken into account by using $V_{gt} = V_{gto} + \sigma V_{ds}$ in the expression for n_s.

We now compare in Figs. 4.4.2 and 4.4.3 the present model for uniformly doped MESFETs with experimental data for a device with ion-implanted channel doping.

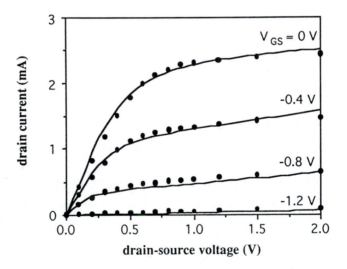

Fig. 4.4.2. Above threshold experimental (symbols) and calculated (solid lines) I–V characteristics for an ion-implanted MESFET with nominal gate length $L = 1$ μm. Other device parameters are: $W = 20$ μm, $d = 0.12$ μm, $\mu_n = 0.23$ m^2/Vs, $v_s = 1.5 \times 10^5$ m/s, $m = 2.5$, $V_{bi} = 0.75$ V, $V_{To} = -1.26$ V, $\eta = 1.73$, $R_s = R_d = 31$ ohm, $\lambda = 0.045$ V^{-1}, $t_c = 0$ V^{-1}, $\sigma_o = 0.081$, $V_\sigma = 0.1$ V, $V_{\sigma t} = 1.0$ V. (From Fjeldly and Shur (1991), and Shur et al. (1992).)

Although ion-implantation gives a non-uniform doping, the degree of non-uniformity is not very severe and we expect the model to apply if suitable effective parameters, including the transconductance compression parameter, t_c, are used.   Fig. 4.4.2 shows such a comparison for above-threshold $I-V$ characteristics, and Fig. 4.4.3 shows the same comparison in a semilog scale including the subthreshold regime.

As can be seen from the figures, our model is quite accurate in the above threshold regime and, at the same time, adequately reproduces the experimental data below threshold over several decades of drain current.   A small leakage current at large negative gate-source voltages is not accounted for in our model. This current may be caused by a small leakage of the reverse biased Schottky gate (see Conger et al. (1988) for details).   The parameters used in this calculation were obtained using the parameter extraction procedure described in Subsection 4.4.5.

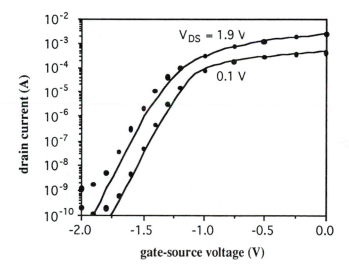

Fig. **4.4.3.** Subthreshold experimental (symbols) and calculated (solid lines) $I-V$ characteristics for ion-implanted MESFET with nominal gate length $L = 1$ μm. The device parameters are the same as in Fig. 4.4.2. (From Shur et al. (1992).)

4.4.4. Model for Channel with Delta Doping (MESA2)

We now consider a MESFET with a doping profile consisting of a uniformly doped layer of density N_{du} with thickness d_u, and a delta-doped layer with sheet density $n\delta$ at a distance $d\delta > d_u$ from the gate electrode, as illustrated in Fig. 4.4.4.

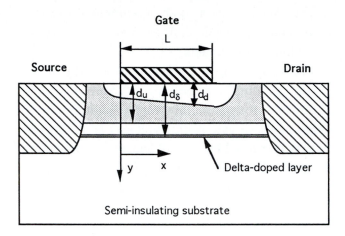

Fig. 4.4.4. Schematic representation of a MESFET with a delta-doped channel layer. (From Shur et al. (1992).)

For this structure, the pinch-off voltage can be expressed as

$$V_{po} = V_{pou} + V_{po\delta} \tag{4-4-32}$$

In eq. (4-4-32),

$$V_{pou} = \frac{qN_{du}d_u^2}{2\varepsilon_s} \tag{4-4-33}$$

is the pinch-off voltage of the uniformly doped layer alone and

$$V_{po\delta} = \frac{qn_\delta d_\delta}{\varepsilon_s} \tag{4-4-34}$$

is the pinch-off voltage of the delta-doped layer alone.

In a real device structure, the carriers are not truly confined to the delta-doped plane. First of all, the electron gas has a finite thickness on the order of

50 to 100 Å due to quantum effects.   Second, some diffusion during the
fabrication process may spread out the dopants to an effective layer thickness of
approximately 100 to 200 Å. Therefore, a box type approximation, as indicated
in Fig. 4.4.5, may be a more appropriate model for the doping profile.

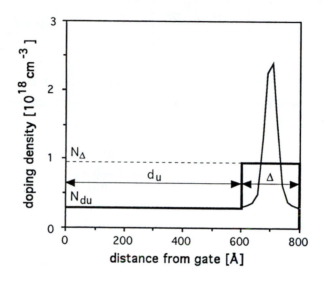

Fig. 4.4.5. Comparison of an actual doping profile of a delta-doped
MESFET (thin solid line) and a box type approximation for the profile
(thick solid line). (From Shur et al. (1992).)

Here, it is assumed that the delta doping is spread out evenly with a doping
density $N_\Delta = n_\delta/\Delta$ over a thickness Δ next to the uniformly doped layer. For
this approximation, the pinch-off voltage, $V_{po\delta}$, in eq. (4-4-34) should be
replaced by

$$V_{po\delta} = \frac{qn_\delta(d_u + \Delta/2)}{\varepsilon_s} = \frac{qN_\Delta\Delta(2d_u + \Delta)}{2\varepsilon_s} \tag{4-4-35}$$

Using the box type doping profile, the depletion layer thickness, d_d, can
now be determined from eq. (4-4-2). Hence, following the procedure used in
Subsection 4.4.3, the above-threshold differential capacitance per unit area can be
expressed as

$$c_a = \frac{\varepsilon_s}{d_d} = \frac{\varepsilon_s}{d_u} \times \begin{cases} \left[1 + \frac{N_{du}}{N_\Delta} \left(\frac{V_{po\delta} - V_{gte}}{V_{pou}} \right) \right]^{-1/2} & , \text{ for } V_{gt} \leq V_{po\delta} \\[2em] \left[\frac{V_{po} - V_{gte}}{V_{pou}} \right]^{-1/2} , & \text{ for } V_{gt} > V_{po\delta} \end{cases}$$

(4-4-36)

where V_{gte} is the effective gate voltage swing defined in eq. (4-4-21).

Correspondingly, the subthreshold differential capacitance per unit area can be written as

$$c_b = \frac{\varepsilon_s}{d_u + \Delta} \exp\left(\frac{V_{gt}}{\eta V_{th}} \right)$$

(4-4-37)

More detailed information about the distribution of charge in a delta-doped MESFET can be obtained by fitting measured C–V characteristics to the C–V characteristics calculated from eqs. (4-4-36) and (4-4-37) in combination with eq. (4-4-7). Fig. 4.4.6 shows the result of such a fit for a delta-doped MESFET, using the doping profile in Fig. 4.4.5. As can be seen from this figure, our theory reproduces fairly well the measured C–V characteristic, including the transition from the subthreshold regime into the above-threshold regime (at $V_{GS} \approx -2$ V) and the transition of the depletion boundary crossing into the low doped region (at $V_{GS} \approx 0$ V). The slope of the C–V characteristic between the two transitions is a measure of the spread of the electrons (and dopants) in the delta-doped region.

In order to calculate the current-voltage characteristics, we now have to find the sheet density, n_s, of electrons at very low drain-source voltage. This is obtained by integrating the expressions for the unified gate-channel capacitance (see eq. (4-4-9)) using eq. (4-4-7) in combination with eqs. (4-4-36) and (4-4-37). This integral does not have an analytical solution, but the following equation provides an adequate approximation:

$$n_s \approx \frac{n_{sa} n_{sb}}{n_{sa} + n_{sb}}$$

(4-4-38)

Here, n_{sa} and n_{sb} are the above threshold and the subthreshold contributions to the total electron sheet density, respectively.

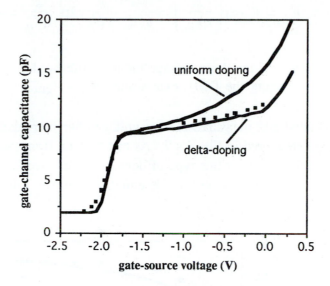

Fig. 4.4.6. Measured (symbols) and calculated (lower line) C–V characteristics for a delta doped MESFET with the doping profile shown in Fig. 4.4.5. Calculated C–V characteristic for a uniformly doped MESFET with the same geometry and pinch-off voltage is shown for comparison. (From Fjeldly and Shur (1991) and Shur et al. (1992). Experimental data provided by COMSAT Corporation).

These contributions are found by integrating each of the corresponding capacitances according to eq. (4-4-9), yielding

$$
n_{sa} = \begin{cases} N_\Delta \Delta \left\{ 1 - \dfrac{d_u}{\Delta} \left[\sqrt{1 + \dfrac{N_{du}}{N_\Delta} \left(\dfrac{V_{po\delta} - V_{gte}}{V_{pou}} \right)} - 1 \right] \right\}, & \text{for } V_{gt} \le V_{po\delta} \\[4mm] N_\Delta \Delta + N_{du} d_u \left(1 - \sqrt{\dfrac{V_{po} - V_{gte}}{V_{pou}}} \right), & \text{for } V_{gt} > V_{po\delta} \end{cases}
$$

$$(4\text{-}4\text{-}39)$$

and

$$n_{sb} = n_o \exp\left(\frac{V_{gt}}{\eta V_{th}}\right) = \frac{\varepsilon_s \eta V_{th}}{q(d_u + \Delta)} \exp\left(\frac{V_{gt}}{\eta V_{th}}\right) \qquad (4\text{-}4\text{-}40)$$

Again, the effects of the drain voltage induced threshold voltage shift are taken into account by using the DIBL expression $V_{gt} = V_{gto} + \sigma V_{ds}$ in calculating n_s.

Calculated and measured current-voltage characteristics for a delta doped MESFET are compared in Fig. 4.4.7. As can be seen from the figure, our theory works reasonably well for this type of device. Hence, we have reason to expect that our universal model will work equally well for arbitrary non-uniform profiles.

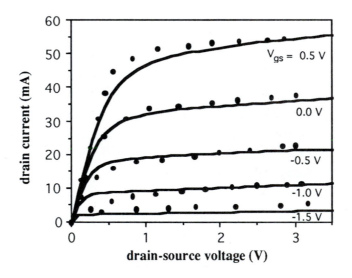

Fig. 4.4.7. Above-threshold experimental (symbols) and calculated (solid lines) current-voltage characteristics for a delta-doped MESFET with gate length $L = 1$ μm. Other device parameters are: $\varepsilon_s = 1.14 \times 10^{-10}$ F/m, $W = 150$ μm, $d_u = 0.045$ μm, $\Delta = 0.01$ μm, $N_A = 6 \times 10^{24}$ m^{-3}, $\mu_n = 0.2$ m^2/Vs, $\beta = 0.0085$ A/V^2, $m = 2.2$, $V_{bi} = 0.7$ V, $V_{To} = -2.04$ V, $\eta = 1.5$, $R_s = R_d = 2.7$ ohm, $\lambda = 0.04$ V^{-1}, $t_c = 0.001$ V^{-1}, $\sigma_o = 0.081$, $V_\sigma = 0.1$ V, $V_{\sigma t} = 1.37$ V. (From Shur et al. (1992). Experimental data from Schubert et al. (1986)).

4.4.5. Parameter Extraction

In order to apply the present model in circuit simulation, it is necessary to find a practical way of extracting the model parameters from experimental data. The accuracy of the parameter extraction depends on the quality of the model and on the accuracy and completeness of the experimental data. In real life, there are usually problems with both sides of this equation, especially for compound semiconductor devices.

Even though the analytical model developed here reflects many important effects occurring in compound semiconductor FETs, such as velocity saturation, parasitic series resistances, the dependence of the threshold voltage on the drain voltage, and finite output conductance, many other important phenomena are included indirectly, at best. For example, our model treats the negative differential mobility in GaAs and related material, resulting from intervalley transfer, only in a cursory manner through the fitting parameter λ (see eq. (4-4-26)). Overshoot and ballistic effects are included only through "effective" values of the saturation velocity which then become dependent on the gate length. Furthermore, many complicated phenomena leading to the threshold voltage dependence on the drain-source voltage are accounted for in the same empirical formula (see eqs. (4-4-23) to (4-4-25)).

On the other hand, the spread from device to device in measured MESFET characteristics is quite often fairly large. The device behavior is often dependent on illumination and may be fairly sensitive even to relatively small variations in temperature. As a consequence, such techniques as Gated Transmission Line Model (GTLM) measurements (see Baier et al. (1985)) do not always give meaningful results. This makes even an accurate determination of the effective gate length difficult. Our approach to the device characterization of GaAs MESFETs allows us to extract effective device parameters, such as the threshold voltage, V_{To}, the subthreshold ideality factor, η, the full channel resistance, R_o, the transconductance parameter, β, the source and drain series resistances, R_s and R_d, the output conductance parameter, λ, the knee shape parameter m, the threshold voltage coefficient, σ_o, and the voltages $V_{\sigma t}$ and V_σ characterizing the dependence of the parameter σ on the gate voltage swing.

In this Subsection, we describe a parameter extraction procedure which gives quite reasonable results without parameter adjustment and which can produce a nearly perfect fit with some parameter adjustment. (Such a perfect fit, however, may not be required if the spread between the characteristics of "identical" devices on the same wafer is fairly large, as is often the case.) Once these overall device parameters have been obtained, they can be used to extract the values of the effective channel doping, the effective field effect mobility and the effective saturation velocity. If the measurements are quite accurate and the device geometry is well established, the effective values of the field effect mobility at room temperature usually vary between 2000 and 4000 cm^2/Vs, and the effective values of the saturation velocity vary between 0.8×10^5 and 2×10^5 m/s, depending on the gate length, the channel thickness and the doping. However, if the effective gate length is not known very accurately (which is often the case in self-aligned, ion-implanted devices where the ion straggle in the channel may substantially affect the effective gate length) then these values may have lower or higher values due to such uncertainties. We should notice that, even in this case, we can obtain a good fit to the device characteristics and, hence, predict the circuit performance fairly accurately using the AIM-Spice simulation program. Moreover, we still retain a limited ability to predict how device characteristics will scale with changes in the doping level and profile and device geometry. In other words, our model and parameter extraction procedure attempt to give the best possible answer based on the available information.

As an example, we perform the parameter extraction for the device with the characteristics shown in Figs. 4.4.2 and 4.4.3. We start by extracting the ideality factor and the drain bias dependence of the threshold voltage from the characteristics in the subthreshold regime, as indicated in Fig. 4.4.8. From the slope of these characteristics plotted in a semilog scale, we find the value $\eta \approx 1.7$. We then estimate the threshold current in saturation, I_t, from eq. (4-4-18) using eq. (4-4-8) for electron sheet density at threshold, i.e.,

$$I_t = \frac{\varepsilon_s \mu W \eta V_{th}^2}{dL} \tag{4-4-41}$$

At this point, we do not need to know accurate values for the electron mobility, μ_n, the effective channel length, L, and the active channel thickness, d, because the threshold voltage is fairly insensitive to the exact value of I_t. (For our device, $I_t \approx 6.6 \times 10^{-6}$ A.) The intercept of the measured curves with the line $I_d = I_t$ yields the values of the threshold voltage versus drain-source voltage. A linear interpolation of this dependence yields the parameter σ_o as shown in Fig. 4.4.9. From this figure, we find $V_{To} \approx -1.26$ V and $\sigma_o \approx 0.081$.

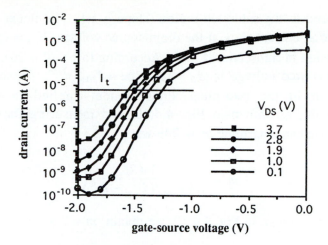

Fig. 4.4.8. Extraction of threshold voltage from subthreshold current-voltage characteristics. (From Shur et al. (1992).)

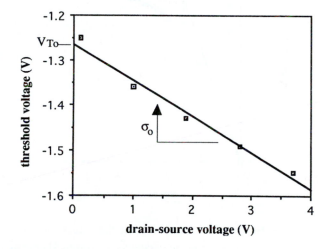

Fig. 4.4.9. Measured dependence of threshold voltage on drain-source bias. (From Shur et al. (1992).)

The next step is to find the shape parameter m for the knee region of the characteristics. For nearly ideal characteristics, with low output conductance in saturation, the extraction method is as indicated in Fig. 4.4.10. The characteristic is approximated by two linear pieces, one describing correctly the linear region at small drain-source voltages (slope g_{ch}) and one approximating the characteristic in the saturated region (slope g_{chs}). The linear pieces intersect at the saturation current, I_{sat}, at the drain-source voltage $V_{ds} = g_{ch}/I_{sat} \approx V_{sate}$ (see the discussion in Subsection 4.4.2). According to eq. (4-4-26), the real current at this drain-source voltage is $I_d = I_{sat} - \Delta I \approx I_{sat}/2^{1/m}$. Hence, by locating the intersection in the two-piece linear model applied to the experimental characteristic, as shown in Fig. 4.4.10, and measuring the currents I_{sat} and $I_{sat} - \Delta I$, we can determine m as follows:

$$m \approx \frac{\ln(2)}{\ln[\, I_{sat}/(I_{sat} - \Delta I\,)]} \qquad (4\text{-}4\text{-}42)$$

For the curve shown in Fig. 4.4.10, we obtain $m \approx 2.5$.

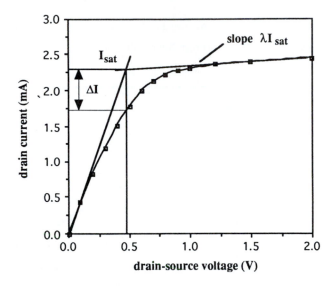

Fig. 4.4.10. Method for extracting the saturation current, I_{sat}, the shape parameter, m, and the parameter λ. (From Shur et al. (1992).)

From the output conductance in deep saturation, g_{chs}, for the largest value of V_{gs}, we also determine the value of parameter λ from the expression

$$\lambda = \frac{g_{chs}}{g_{ch}V_{sate}} \approx \frac{g_{chs}}{I_{sat}} \qquad (4\text{-}4\text{-}43)$$

(see the discussion in Subsection 4.4.2).

The next step involves the determination of the total series resistance, $R_t = R_s + R_d$, and the full channel resistance, R_o, from the linear drain conductance, g_{ch}, for different gate-source voltages. For a uniform doping profile, we have (see Hower and Bechtel (1973))

$$\frac{1}{g_{ch}} = \frac{1}{g_{chi}} + R_t = R_o G + R_t \qquad (4\text{-}4\text{-}44)$$

where

$$R_o = \frac{L}{qN_d\mu W d} \qquad (4\text{-}4\text{-}45)$$

and

$$G = \left[1 - \sqrt{(V_{bi} - V_{gs})/V_{po}}\right]^{-1} \qquad (4\text{-}4\text{-}46)$$

Hence, R_t and R_o can be determined by plotting $1/g_{ch}$ versus G if $V_{po} = V_{To} + V_{bi}$ is known. For GaAs MESFETs, a typical value of V_{bi} is 0.75 V. We should note, however, that this procedure is fairly sensitive to the exact values of V_{bi} and V_{To}. The plot of $1/g_{ch}$ versus G is shown in Fig. 4.4.11. From this figure, we obtain $R_o \approx 56.9$ ohm and $R_t \approx 62$ ohm. With an average doping density $N_d \approx 2\times10^{23}$ m^{-3}, we obtain from eq. (4-4-45) the mobility $\mu_n \approx$ 0.23 m^2/Vs.

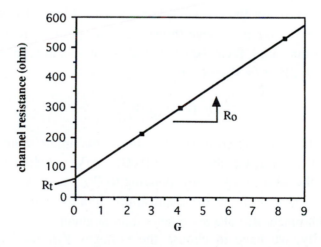

Fig. 4.4.11. Inverse channel conductance versus G (see eq. (4-4-46)). (From Shur et al (1992).)

A crude estimate of the difference between R_s and R_d can be made by measuring the difference between the normal, extrinsic device transconductance in the saturation regime, g_{ms}, and the same with the source and drain interchanged, g_{md} (at constant drain current).   Since the intrinsic trans-conductance is the same in both cases, we find

$$R_s - R_d = \frac{1}{g_{ms}} - \frac{1}{g_{md}} \qquad\qquad (4\text{-}4\text{-}47)$$

The source and drain series resistances (with somewhat limited degree of accuracy) can also be determined from gate current measurements (see **Shur** (1987) and del Alamo and Azzam (1989)).   These resistances can also be determined from small-signal measurements, see Anholt and Swirhun (1991a). For short channel devices, a very crude estimate of R_t can be obtained by using the following expression

$$\frac{I_{sat}}{g_{ch}} \approx V_L + I_{sat} R_t \qquad\qquad (4\text{-}4\text{-}48)$$

which assumes that the intrinsic saturation voltage is independent of the gate-source voltage and has the approximate value $V_L = F_s L$, which corresponds to velocity saturation in the entire channel.

Using eq. (4-4-16), it is possible to derive the transconductance parameter, β, from the value for the threshold voltage, V_{To}, (derived from the subthreshold characteristics) and the above-threshold saturation current (see Fig. 4.4.10). Well above threshold, β can be expressed as follows in terms of the intrinsic gate voltage swing, V_{GT} ($= V_{gt} - I_{sat}R_s$),

$$\beta \approx I_{sat} / V_{GT}^2 \qquad\qquad (4\text{-}4\text{-}49)$$

Using the characteristic corresponding to the maximum value of the gate-source voltage in Fig.4.4.2, we obtain $\beta \approx 0.0016$ A/V^2. (More generally, we could have fitted the whole curve corresponding to $(V_{GS})_{max}$ by adjusting β and t_c, but in our case, $t_c = 0$ is an excellent approximation which is quite typical for devices with pinch-off voltages smaller than or about 2 V.)

Finally, we have to choose the voltages $V_{\sigma t}$ and V_σ determining the dependence of the parameter σ on the gate voltage. Typically, we may choose $V_\sigma \approx 0.1$ and $V_{\sigma t}$ to be about half of the pinch-off voltage (≈ 1 V in our example). However, all parameter values can be slightly adjusted for a better fit. For the calculation presented in Figs. 4.4.2 and 4.4.3, we did not use such an adjustment. As can be seen from these figures, even this relatively crude parameter extraction procedure may be adequate for a fairly accurate reproduction of the measured data.

4.4.6. Temperature and Frequency Dependence of MESFET Parameters

The wider band gap of GaAs compared to Si makes GaAs more suitable for operating in a larger temperature range. The military range of temperatures (from −50 °C to 125 °C) is the most important for GaAs digital circuits. A detailed analysis of the temperature dependence of MESFET parameters was given by Conger (1992) (see also Conger et al. (1992)). Accurate MESFET modeling has to take into account the difference between the device temperature and the ambient temperature as well as the effects of the device temperature on the MESFET parameters. In our model, the device temperature, T, is related to the ambient temperature, T_{amb}, as follows:

$$T = T_{amb} + P_{dc}Z_{th} \tag{4-4-50}$$

where

$$Z_{th} = r_{th}W \tag{4-4-51}$$

is the thermal impedance of the device, r_{th} is the thermal impedance per unit width, and P_{dc} is the average power dissipated in the device. The value of r_{th} depends on the substrate thickness and the channel length. For a 1 μm MESFET, r_{th} varies approximately from 35 °C-mm/Watt for a 25 μm thick substrate to approximately 50 °C-mm/Watt for a 100 μm thick substrate (see, for example, Shur (1987), p. 415). The power has to be determined from a preliminary AIM-Spice run or estimated based on the values of the drain saturation current and the bias supply.

Out of all MESFET parameters, the subthreshold and gate leakage currents exhibit the most drastic temperature dependencies. According to Conger (1992), the dependence of the MESFET threshold voltage on temperature can be fairly accurately approximated by the linear dependence

$$V_{TO} = V_{TOO} - \alpha_{VT}T \tag{4-4-52}$$

A typical value of α_{VT} is on the order of 1.2 mV/°C. This dependence usually holds for the military range of temperatures. This dependence cannot be considered universal because the effects related to traps can cause a rapid variation of the threshold voltage in a narrow temperature range (see Chen et al. (1986) for details). However, in this trap-affected temperature range, the device characteristics are also sensitive to the history of applied bias and the measuring frequency so that an accurate modeling is difficult and the circuit operation is unpredictable. Under "normal" operating conditions, eq. (4-4-50) is quite adequate.

The temperature variation of other MESFET parameters can be approximated by linear functions of temperature as well, such as, for example, the transconductance parameter, β, and the saturation output conductance parameter, λ

$$\beta = \beta_o \left(1 - \frac{T}{T_\beta} \right) \tag{4-4-53}$$

$$\lambda = \lambda_o \left(1 - \frac{T}{T_\lambda} \right) \tag{4-4-54}$$

A typical value of T_β is on the order of 600°C.

It is well known that the MESFET output conductance in the saturation regime is frequency dependent. This effect is caused by traps and often results in a fairly abrupt increase in the output conductance at the transition frequency, f_λ, which is temperature dependent, as indicated in Fig. 4.4.12. As shown in Fig. 4.4.13, this dependence can be quite accurately modeled by the following expression for the saturation output conductance parameter, λ:

$$\lambda = \lambda_{lf} + \frac{\Delta\lambda}{2} \left[1 + \tanh\left(\frac{f - f_\lambda}{\Delta f} \right) \right] \tag{4-4-55}$$

where f is the frequency of MESFET operation, λ_{lf} is the low frequency value and $\lambda_{lf} + \Delta\lambda$ is the high frequency value of λ, T_f is a characteristic temperature determined by the traps, Δf is the frequency range of the transition between the low and high frequency regime given by

$$\Delta f = \Delta f_o \exp\!\left(\frac{T}{T_f}\right) \tag{4-4-56}$$

and the temperature dependence of the transition frequency is given by

$$f_\lambda = f_{\lambda o} \exp\!\left(\frac{T}{T_f}\right) \tag{4-4-57}$$

where $f_{\lambda o}$ is the low-temperature limit of f_λ.

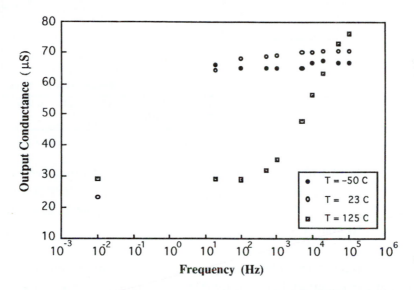

Fig. 4.4.12.   Measured output conductance versus frequency and temperature for an enhancement mode MESFET biased at $V_{ds} = 2.0$ V and $V_{gs} = 0.5$ V. (From Shur et al. (1992). Experimental data from Conger (1992).)

4.4.7. Backgating Effects in MESFET

Backgating and sidegating effects may strongly influence GaAs MESFET current-voltage characteristics (see, for example, Lundquist and Ford (1982), Goronkin et al. (1982), Biritella et al. (1982), and Shur (1987)). The term "backgating" describes the effect of the substrate bias on the MESFET characteristics. The term "sidegating" refers to the effect of a nearby device on the characteristics of a given MESFET. In a typical MESFET Direct Coupled Field effect transistor Logic (DCFL) circuit, sidegating leads to a decrease of the drain current of the load MESFET caused by the switching MESFET with the source contact biased negatively with respect to the source of the load transistor.

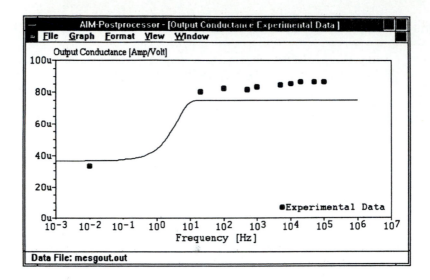

Fig. 4.4.13. Measured (symbols) and simulated (solid line) output conductance versus frequency for an enhancement mode MESFET biased at $V_{ds} = 1.0$ V and $V_{gs} = 0.5$ V. (From Shur et al. (1992). Experimental data from Conger (1992).)

Backgating and sidegating are related to a finite depletion region which exists at the boundary between the MESFET active layer and the substrate. The width of this layer depends on the density of traps and on the position of the Fermi level in the substrate and may be found using an "equivalent p-n^+ junction" model (Lundquist and Ford (1982)). This model predicts a certain dependence of the depletion width and of the threshold voltage on the substrate bias. However, in practical circuits, sidegating usually plays a more important role than backgating, and an accurate modeling of sidegating effects is quite difficult. Therefore, the AIM-Spice MESFET model utilizes an empirical equation (similar to that for a MOSFET body plot, see Section 3.7) in order to describe sidegating:

$$V_T = V_{TO} + K_s \left(V_s - V_{sg} \right) \qquad (4\text{-}4\text{-}58)$$

Here, V_{TO} is the threshold voltage unaffected by sidegating, V_s is the source potential, V_{sg} is the potential causing the sidegating or backgating (for example, the source potential of the switching FET in a DCFL inverter or the substrate potential in the backgating experiments). This approach is very similar to that used in the HSPICE program (see HSPICE User's Manual (1990)). For a typical backgating experiment, $K_s \approx 0.1$ V. Eq. (4-4-58) reproduces experimental data over a wide range of operation as demonstrated in Fig. 4.4.14. For sidegating, K_s is a function of the distance between the device and the sidegating contact. Experimental data presented in the literature predicts that K_s is inversely proportional to this distance. Usually, sidegating becomes negligible only when this distance becomes quite large (at least 30 to 40 µm).

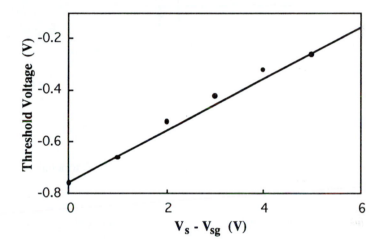

Fig. 4.4.14. Experimental (symbols) and calculated (solid line) threshold voltage dependence on sidegating voltage. (From Shur et al. (1992). Experimental data from Shoncair and Ojala (1992).)

4-5. BASIC HFET MODEL

The basic HFET charge control model is derived from the charge control concept. It was developed by Delagebeaudeauf and Linh (1982) and Lee et al. (1983), and has been reviewed, for example, by Shur (1987), Shur (1990) and Morkoç et al. (1991)). The model is considered in this Section, and the results obtained will be used in Sections 4.6 and 4.7 in order to develop more advanced HFET AIM-Spice models. We start from considering the band diagram in

Fig. 4.5.1 of a conventional HFET structure at the flat band condition.

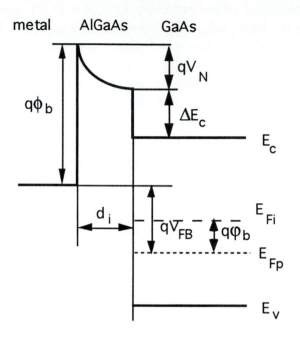

Fig. 4.5.1. Band diagram of a conventional HFET structure at flat band condition.

As can be seen from this figure, the flat-band voltage, V_{FB}, is given by

$$V_{FB} = \phi_b - V_N - (\Delta E_c + \Delta E_F)/q \qquad (4\text{-}5\text{-}1)$$

where $q\phi_b$ is the metal-semiconductor energy barrier, ΔE_c is the conduction band discontinuity, $\Delta E_F = E_c - E_{Fp}$ (in Fig. 4.5.1, we assume that the GaAs buffer layer is doped p-type with Fermi energy E_{Fp}), and

$$V_N = q\int_0^d \frac{N_d(x)}{\varepsilon_i(x)} x\,dx \qquad (4\text{-}5\text{-}2)$$

is the voltage drop across the AlGaAs layer at the flat band condition. (We note that eq. (4-5-2) has been generalized to account for the situation when both the density of ionized donors, N_d, and the composition of the AlGaAs layer can vary with distance, x.) The threshold voltage for strong inversion can be found from eq. (3-3-25):

$$V_T = V_{FB} + 2\varphi_b + \sqrt{4\varepsilon_s \, qN_a\varphi_b} \, / c_i \qquad (4\text{-}5\text{-}3)$$

where

$$c_i = \varepsilon_i \, / d_i \qquad (4\text{-}5\text{-}4)$$

is the capacitance per unit area of the AlGaAs layer and

$$\varphi_b = V_{th} \ln\!\left(\frac{N_a}{n_i} \right) \qquad (4\text{-}5\text{-}5)$$

is the potential difference between the position of the Fermi level in the GaAs layer, E_{Fp}, and the intrinsic Fermi level, E_{Fi}, N_a is the concentration of ionized acceptors in the GaAs buffer, n_i is the intrinsic carrier concentration in the GaAs layer, and V_{th} is the thermal voltage. At very low doping levels in the GaAs buffer, the last two terms on the right hand side of eq. (4-5-3) can be neglected and the threshold voltage simplifies to

$$V_T \approx \phi_b - V_N - (\Delta E_c + \Delta E_{F1}) \, / q \qquad (4\text{-}5\text{-}6)$$

where $\Delta E_{F1} = E_c - E_{Fi} - q\varphi_b$. However, the equation used for the threshold voltage in the basic HFET charge control model differs from eq. (4-5-6). In order to understand this difference, we have to consider the dependence of the position of the Fermi level in the GaAs layer on the sheet density of the two dimensional electron gas, n_s (see Fig. 4.5.2 and the discussion in Section 3.3). This dependence can be crudely approximated by a straight line for typical densities of the two dimensional gas:

$$E_F \approx E_{Fo} + a_F n_s \qquad (4\text{-}5\text{-}7)$$

where a_F is a constant that depends on the correction Δd to the gate-channel distance (see Section 3.3) and E_{Fo} is the intercept of the linear approximation with $n_s = 0$. As can be seen from Fig. 4.5.2, this approximation is not very accurate. However, we note that at large values of E_F, the numerical calculations become inaccurate because the potential well at such energies deviates from the simple triangular shape assumed here. In addition, real space transfer of carriers into the AlGaAs layer may become quite important. This effect will tend to decrease the number of carriers in the potential well compared to that of the ideal, infinitely deep potential well. Hence, the actual dependence of n_s on E_F may be closer to the crude linear approximation than indicated in Fig. 4.5.2. Using this approximation, we obtain

$$V_{To} \approx \phi_b - V_N - \left(\Delta E_c + \Delta E_{Fo}\right)/q \qquad \text{(4-5-8)}$$

where $\Delta E_{Fo} = E_c - E_{Fo}$. V_{To} can be interpreted as the threshold voltage at which the carrier density of the two-dimensional gas is extrapolated to zero.

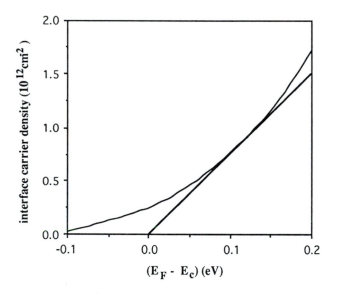

Fig. 4.5.2. Calculated dependence between Fermi level and interface electron density. E_c is the conduction band edge in the potential well at the heterointerface. (From Lee et al. (1983), © 1983 IEEE.)

As can be seen from Fig. 4.5.2, for the AlGaAs/GaAs material system at 300 K, E_{Fo} is close to E_c. This allows us to write the equation of the charge control model as follows:

$$n_s = \frac{\varepsilon_i V_{GT}}{q(d_i + \Delta d)} \qquad \text{(4-5-9)}$$

where

$$\Delta d = \frac{\varepsilon_i \, a_F}{q} \tag{4-5-10}$$

(see Lee et al. (1983), Byun et al. (1990), and Section 3.3 for more details). In the unified charge control model for HFETs, a more accurate version of eq. (4-5-3) will be used (see Section 4.6).

Performing the integration in eq. (4-5-2) for uniform doping in the AlGaAs layer, we find

$$V_{To} \approx \phi_b - \frac{qN_d \, d_d^2}{2\varepsilon_i} - \Delta E_c / q \tag{4-5-11}$$

For a delta-doped structure, we obtain

$$V_{To} \approx \phi_b - \frac{qn_\delta \, d_\delta}{\varepsilon_s} - \Delta E_c / q \tag{4-5-12}$$

where d_δ is the distance between the metal gate and the doped plane, and n_δ is the sheet concentration of donors in the doped plane. The calculated dependence of V_{To} on d_δ for uniformly doped and planar-doped devices is shown in Fig. 4.5.3. However, we should emphasize that the threshold voltage defined from subthreshold characteristics (similar to that for MOSFETs, see Sections 3.9 and 4.6) may be shifted relative to the values given by eqs. (4-5-11) and (4-5-12).

As was explained above, the dependence of the concentration of the two-dimensional gas, n_s, on the gate-source voltage can be approximated by a straight line (as predicted by eq. (4-5-3), only for the above threshold regime). On the other hand, this dependence also becomes invalid at large gate-source voltages. This can be understood by analyzing the energy band diagram of modulation-doped structures. As an example, we plot simplified band diagrams for delta-doped HFETs for different values of the gate-source voltage in Fig. 4.5.4 (from Chao et al. (1989)). As can be seen from this figure, the electron quasi-Fermi level, E_{Fn}, in the AlGaAs layer reaches the bottom of the conduction band at large gate voltage swings so that carriers are induced into the conduction band minimum. As most of these carriers (with concentration n_t per unit area) are trapped in AlGaAs, they do not contribute much to the device current.

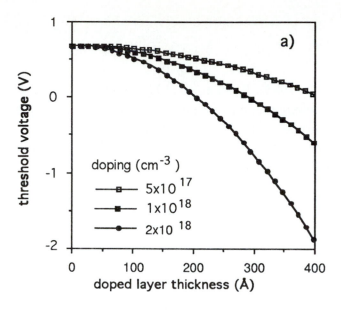

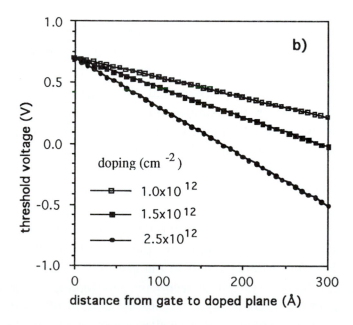

Fig. 4.5.3. Threshold voltage of conventional (a) and delta-doped (b) HFETs. (From Shur (1990)).

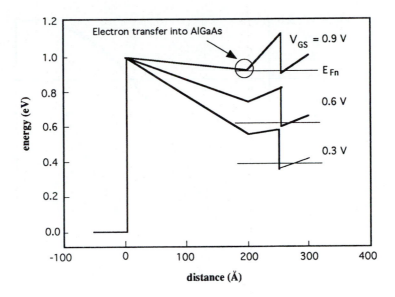

Fig. 4.5.4. Band diagrams of planar doped HEMTs at different values of gate-source voltage. At small V_{GS}, the energy separation between the electron quasi-Fermi level and the conduction band minimum in the AlGaAs charge-control layer is large and the electron concentration in the AlGaAs layer is negligible. At large V_{GS}, E_{Fn} nearly touches the conduction band minimum in AlGaAs and the transfer of electrons into this layer becomes dominant, sharply reducing the device transconductance. (From Chao et al. (1989), © 1989 IEEE.)

In Fig. 4.5.5, we show the computed dependencies of n_s and n_t on the gate-source voltage. This calculation is based on a self-consistent solution of Poisson's equation and Schrödinger's equation as described by Stern (1972). As can be seen from this figure, the concentration of electrons in the AlGaAs layer increases sharply and the slope of the n_s versus V_{GS} dependence decreases at high gate-source voltages. At such gate biases, the dependence of n_s on V_{GS} becomes sublinear and gradually saturates.

We will first consider a basic HFET model without taking the sublinear dependence of the electron sheet density on the gate-source voltage into account. In this case, the resulting model is identical to that of the MOSFET developed in Section 3.9.

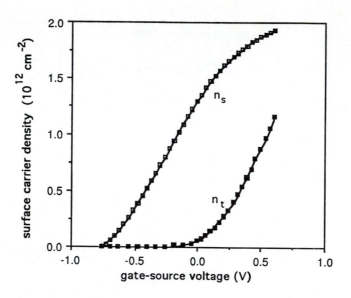

Fig. 4.5.5. Computed surface carrier densities of electrons in the two-dimensional electron gas, n_S, and in the AlGaAs layer, n_t, versus gate-source voltage. At V_{GS} close to zero, real space transfer of electrons into the AlGaAs layer becomes important, limiting the concentration of electrons induced into the quantum well. (From Chao et al. (1989), © 1989 IEEE.)

In Section 3.9, we used the following expression for the saturation current:

$$I'_{sat} = \frac{g'_{chi} \, V_{gte}}{1 + g'_{chi} R_s + \sqrt{1 + 2 g'_{chi} R_s + \left(V_{gte}/V_L\right)^2}} \qquad (4\text{-}5\text{-}13)$$

Here, $V_L = F_s L$ where F_s is the saturation field and L is the gate length, R_s is the source series resistance, $g_{chi}' = q n_s' W \mu_n / L$ is the intrinsic linear channel conductance where W is the gate width and μ_n is the low-field electron mobility, and the electron sheet density in the channel can be expressed as

$$n'_s = 2 n_o \ln\left[1 + \frac{1}{2} \exp\left(\frac{V_{gt}}{\eta V_{th}}\right)\right] \qquad (4\text{-}5\text{-}14)$$

where n_o is the electron sheet density at threshold and η is the subthreshold ideality factor (this equation is similar to an expression proposed by Ruden (1989)). We note that, in order for eq. (4-5-14) to reach the correct asymptotic value in the subthreshold regime, V_{gt} explicitly shown in eq. (4-5-13) has to be replaced by an effective gate voltage swing, V_{gte}, given by (note that the implicit dependence of g_{chi}' on V_{gt} through n_s' in eq. (4-5-14) must be retained)

$$V_{gte} = V_{th}\left[1 + \frac{V_{gt}}{2V_{th}} + \sqrt{\delta^2 + \left(\frac{V_{gt}}{2V_{th}} - 1\right)^2}\right] \qquad (4\text{-}5\text{-}15)$$

where $V_{gt} = V_{gs} - V_T$ is the extrinsic gate voltage swing, and the parameter δ determines the width of the transition region. Typically, $\delta = 3 - 5$ is a good choice.

Now, we have to take into account that the carrier concentration of the two-dimensional electron gas saturates at large gate voltage swings and cannot exceed a certain maximum value, n_{max}, which depends primarily on the material system and, to a certain extent, on the doping profile in the wide band gap semiconductor (see, for example, Lee et al. (1984)). The following approach provides good agreement with experimental data and numerical calculations. The value of the electron sheet density to be used in the expression for the intrinsic linear channel conductance can be approximated by (see Fjeldly and Shur (1991))

$$n_s = \frac{n_s'}{\left[1 + (n_s'/n_{max})^\gamma\right]^{1/\gamma}} \qquad (4\text{-}5\text{-}16)$$

where n_s' is given by eq. (4-5-14) and γ is a characteristic parameter for the transition to saturation in n_s. In the same spirit, the drain current can be expressed as

$$I_{sat} = \frac{I_{sat}'}{\left[1 + (I_{sat}'/I_{max})^\gamma\right]^{1/\gamma}} \qquad (4\text{-}5\text{-}17)$$

where I_{sat}' is given by eq. (4-5-13), and

$$I_{max} = qn_{max}v_sW \qquad (4\text{-}5\text{-}18)$$

is an upper limit for the channel current determined by n_{max} and the saturation velocity, v_s.

As was discussed in Sections 3.9 and 4.4, the I–V characteristics for FETs can be described by one unified expression which combines the linear and the saturation regions, and which includes the subthreshold regime:

$$I_{ds} \approx \frac{g_{ch} V_{ds}}{\left[1 + \left(g_{ch} V_{ds} / I_{sat} \right)^m \right]^{1/m}} \tag{4-5-19}$$

Here, m is a parameter that determines the shape of the characteristics in the knee region, $g_{ch} = g_{chi}/(1 + g_{chi}R_t)$ is the extrinsic linear channel conductance and $R_t = R_s + R_d$ is the sum of the source and the drain series resistances.

More advanced HFET models can be now developed following an approach similar to that used in Sections 3.9 and 4.4. Such HFET models are considered in the next Section.

4-6. UNIVERSAL HFET MODEL (AIM-Spice MODEL HFETA)

In this Section, we develop a new universal, analytical HFET model which describes both the below and above-threshold regimes of device operation using unified expressions, isimilar to what was done for MOSFETs in Sections 3.9 and for MESFETs in Section 4.4. We will also account for various non-ideal physical mechanisms such as drain bias induced barrier lowering (DIBL) and other factors that contribute to a finite output conductance in saturation, and the electron transfer into the semiconductor barrier layer separating the HFET channel from the gate.

In principle, the model parameters such as the effective channel mobility, the saturation velocity, the source and drain resistances, etc., are extractable from experimental data, using a procedure similar to that described for MOSFETs in Section 3.9. However, the physical mechanisms determining HFET characteristics are more complicated. In addition to the gate leakage current (considered separately in Section 4.8), HFET characteristics are strongly affected by the limitations imposed on the maximum carrier concentration in the channel by the finite energy band discontinuities (see, for example, Lee et al. (1983a) and

(1984), and Foisy et al. (1988)). This important factor is incorporated into our model. However, it increases the number of the device parameters involved and makes the parameter extraction less reliable. Nevertheless, as we demonstrate below, reasonable values of the device parameters may still be extracted from the experimental data and, with some adjustment, an excellent agreement between the model and experimental data can be obtained. Hence, our HFET model is very suitable for incorporation into circuit simulators and for the design of digital and microwave HFETs.

Throughout this Section, we refer to AlGaAs/GaAs HFETs. However, with appropriate input parameters, our model is also applicable to HFETs based on other material systems, such as AlGaAs/InGaAs/GaAs or AlInAs/InGaAs/InP.

4.6.1. Capacitance Model

As was discussed in Section 3.9, the charge control in FETs can be described in terms of a unified expression for the channel capacitance, valid for all bias voltages, with the differential gate-channel capacitance per unit area written as

$$c_{gc} = \frac{c_a c_b}{c_a + c_b} \tag{4-6-1}$$

where c_a is the above-threshold contribution and c_b is the subthreshold contribution. For HFETs, c_{gc} (through c_a) has to reflect the limitations imposed on the maximum carrier concentration in the channel, n_s, caused by the finite energy band discontinuity between the GaAs channel and the AlGaAs layer. Eqs. (4-5-16) and (4-5-17) already incorporate both the above-threshold and the subthreshold behavior of the channel charge density, including the saturation of n_s at large gate voltage swings. Hence, at small drain-source voltages, c_{gc} can be expressed as

$$c_{gc} = q \frac{dn_s}{dV_{gs}} \approx \frac{c'_{gc}}{\left[1 + (n'_s/n_{max})^\gamma\right]^{1+1/\gamma}} \tag{4-6-2}$$

where V_{gs} is the intrinsic gate-source voltage, n_{max} is the maximum electron sheet density in the channel, n_s' is the ideal unified electron sheet density (see eq. (4-5-14)) and c_{gc}' is the corresponding unified channel capacitance of an infinitely deep potential well (see Section 3.9). We note that when n_s' becomes comparable to or larger than n_{max}, c_{gs} will drop noticeably from its ideal value given by c_{gc}'.

At large gate voltage swings, the saturation in n_s is accompanied by an increase in the sheet density, n_t, of electrons in the AlGaAs layer owing to the real space transfer as was shown in Fig 4.5.5 (see Gribnikov (1973) and Hess et al. (1979)). This added charge contributes to the total differential gate capacitance per unit area, c_{gtot}, which can be represented as a parallel connection of c_{gc} and the capacitance c_{g1} associated with the AlGaAs layer. To a lowest order approximation, we can assume that the charge sheet density, qn_t, is located at a fixed distance d_1 from the gate, independent of the gate-source voltage (which is quite accurate for a delta-doped HFET). Then, this charge can be treated in full analogy with the charge density qn_s in the GaAs channel. Hence, by assuming that the onset of strong inversion in AlGaAs is characterized by a threshold voltage, V_{T1}, the capacitance c_{g1} can be written as (see Section 3.9)

$$c_{g1} = \frac{c_{a1} c_{b1}}{c_{a1} + c_{b1}} \qquad (4\text{-}6\text{-}3)$$

where

$$c_{a1} = \varepsilon_i / d_1 \qquad (4\text{-}6\text{-}4)$$

is the contribution from the strong inversion regime,

$$c_{b1} = \frac{qn_{o1}}{\eta_1 V_{th}} \exp\left(\frac{V_{gt1}}{\eta_1 V_{th}}\right) \qquad (4\text{-}6\text{-}5)$$

is the subthreshold contribution, and

$$q n_{o1} = \frac{\varepsilon_i \, \eta_1 \, V_{th}}{2 \, d_1} \qquad (4\text{-}6\text{-}6)$$

is the sheet density of inversion charge in AlGaAs at the threshold for this layer, i.e., at $V_{gt1} \equiv V_{gs} - V_{T1} = 0$. In these expressions, ε_i is the dielectric permeability of the AlGaAs, η_1 is the subthreshold ideality factor for the AlGaAs channel, and V_{th} is the thermal voltage. The threshold voltage, V_{T1}, can be estimated from the maximum sheet charge density, qn_{max}, and the threshold voltage, V_T, of the interface channel:

$$V_{T1} \approx V_T + \frac{qn_{max}}{c_i} \tag{4-6-7}$$

This approach is similar to that of Lee et al. (1983) who considered the AlGaAs layer as a parasitic MESFET with a more positive threshold voltage than the HFET threshold.

Fig. 4.6.1 shows calculated values for $C_{gc} = LW c_{gc}$ based on eq. (4-6-2) and the total gate capacitance $C_{gtot} = LW(c_{gc} + c_{g1})$ based on eqs. (4-6-2) and (4-6-3) for a typical HFET with nominal gate length $L = 1\ \mu m$ and gate width $W = 20$ mm. We notice the drop in C_{gc} and the slight increase in C_{gtot} at large gate voltage swings.

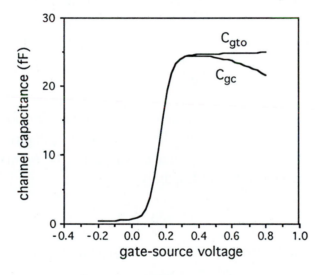

Fig. 4.6.1. Calculated gate-channel capacitance, C_{gc}, and total gate capacitance, C_{gtot}, for the same HFET as in Fig. 4.6.2. (From Fjeldly and Shur (1991).)

The capacitance model described above can be used for calculating the channel contribution to the gate-source capacitance, C_{gs}, and the gate-drain capacitance, C_{gd}, by using an approximation similar to that used in the model by Meyer (1971) (see Sections 3.9 and 4.4):

$$C_{gs} = C_f + \frac{2}{3}C_{gc}\left[1 - \left(\frac{V_{sate} - V_{dse}}{2V_{sate} - V_{dse}}\right)^2\right] \tag{4-6-8}$$

$$C_{gd} = C_f + \frac{2}{3}C_{gc}\left[1 - \left(\frac{V_{sate}}{2V_{sate} - V_{dse}}\right)^2\right] \tag{4-6-9}$$

In eqs. (4-6-8) and (4-6-9), $V_{sate} = I_{sat}/g_{ch}$ is the effective extrinsic saturation voltage, I_{sat} is the saturation drain current, g_{ch} is the extrinsic channel conductance at low drain-source bias, and V_{dse} is an effective extrinsic drain-source voltage. V_{dse} is equal to V_{ds} for $V_{ds} < V_{sat}$ and is equal to $V_{sate} = I_{sat}/g_{ch}$ for $V_{ds} > V_{sate}$. In order to obtain a smooth transition between the two regimes, we interpolate V_{dse} by the following equation:

$$V_{dse} = V_{ds}\left[1 + \left(\frac{V_{ds}}{V_{sate}}\right)^{m_c}\right]^{-1/m_c} \tag{4-6-10}$$

where m_c is a constant determining the width of the transition region between the linear and the saturation regime. The capacitance C_f in eqs. (4-6-8) to (4-6-9) is the side wall and fringing capacitance which can be estimated in terms of the capacitance of a metal line of length, W, as

$$C_f \approx \beta_c \, \varepsilon_s W \tag{4-6-11}$$

where β_c is on the order of 0.5 (see Gelmont et al. (1991)).

4.6.2. Current-Voltage Characteristics

Usually, the subthreshold current in HFETs is larger than that for Si MOSFETs. In addition to the mechanisms important in MOSFETs such as the carrier transport over the source-channel barrier, the space charge injection from the drain and source contacts into the buffer layer may play a more important role, especially in self-aligned devices (see, for example, Jensen et al. (1991) and (1991a)). This injection can be substantially reduced by using a doped p-type buffer layer (see, for example, Ruden et al. (1989)), a special p-i-p^+ buffer structure (see Kanamori et al. (1992)), or silicon nitride gate sidewalls in self-aligned structures (see Ueto et al. (1985) and Chen et al. (1987)). In addition, reverse gate leakage current may play an important role in the subthreshold region at high drain voltages (similar to that in GaAs MESFETs, see Conger et al. (1988)). However, in well-behaved HFETs, the subthreshold current, to first order, is similar to that in MOSFETs (see Sections 3.6). In the long channel limit, the subthreshold saturation current can be expressed as

$$I_{sat} = \frac{qW\mu_n V_{th} n_o}{L} \exp\left(\frac{V_{gt}}{\eta V_{th}}\right) \qquad (4\text{-}6\text{-}12)$$

Here, μ_n is the low-field electron mobility, V_{th} is the thermal voltage, n_o is the channel sheet density of electrons at threshold, η is the subthreshold ideality factor, and $V_{gt} = V_{gs} - V_T$ is the gate voltage swing where V_{gs} is the extrinsic gate-source voltage and V_T is the threshold voltage. The sheet density of channel electrons at threshold is given by (see Section 3.9)

$$n_o = \frac{\varepsilon_i \eta V_{th}}{2q(d_i + \Delta d)} \qquad (4\text{-}6\text{-}13)$$

where d_i is the thickness of the wide-gap semiconductor layer and Δd is a correction to this thickness related to the shift in the Fermi level in the inversion layer with respect to the bottom of the conduction band.

A unified expression for the saturation current, which has the correct asymptotic behavior both below and above threshold, is given in terms of eqs. (4-5-13), (4-5-15) and (4-5-17). Similarly, a unified expression for the electron sheet density in the channel, n_s, is given by eqs. (4-5-14) and (4-5-16).

As was discussed in Section 4.5, the I–V characteristics for HFETs can be described by the unified expression of eq. (4-5-19) which combines the linear and the saturation regions, and which includes the subthreshold regime. However, for a more accurate description of the saturation region, it is possible to include an additional empirical factor, $(1 + \lambda V_{ds})$, in the expression for the I–V characteristics (as usually done in SPICE FET models). The parameter λ is related to physical effects that are not explicitly included in our model, such as, for example, gate length modulation. As a rule of thumb (and based on the analysis given in Section 3.10), we can assume $\lambda \approx k_o/L$ where k_o is a constant. Hence, our model I–V characteristics can finally be expressed as:

$$I_d = \frac{g_{ch} V_{ds} \left(1 + \lambda V_{ds}\right)}{\left[1 + \left(V_{ds} / V_{sate}\right)^m\right]^{1/m}}$$

(4-6-14)

where

$$V_{sate} \approx I_{sat} / g_{ch}$$

(4-6-15)

is the effective, extrinsic saturation voltage, V_{ds} is the extrinsic drain-source voltage, m is a parameter that determines the shape of the characteristics in the knee region, $g_{ch} = g_{chi}/(1 + g_{chi}R_t)$ is the extrinsic linear channel conductance where g_{chi} is the intrinsic linear channel conductance and $R_t = R_s + R_d$ is the sum of the source and the drain series resistances. When the experimental I–V characteristics have a finite output conductance, g_{chs}, in saturation, the saturation current is conveniently defined, for the purpose of comparison with the present model, as the intersection of the linear asymptotic behavior at small drain-source voltages $(I_d = g_{ch}V_{ds})$ and in deep saturation $(I_d \approx g_{ch}V_{sate} + g_{chs}V_{ds})$. Hence, we can express V_{sate} in terms of the saturation current as: $V_{sate} = (1 - g_{chs}/g_{ch})I_{sat}/g_{ch}$. Normally, for well-behaved transistors, $g_{chs} \ll g_{ch}$ and we can use eq. (4-6-15).

We note that both the linear channel conductance (through its dependence on the electron sheet density in the channel, n_s) and the saturation current are assigned upper limits at large gate-voltage swings owing to the increase in real space transfer. These effects are described in terms of eqs. (4-5-16) to (4-5-17).

An important effect which must be taken into account for an adequate description of the HFET behavior is the dependence of the threshold voltage on the drain-source voltage, as discussed in Section 3.6. This dependence can be fairly accurately described by the equation

$$V_T = V_{To} - \sigma V_{ds}$$

(4-6-16)

where V_{To} is the threshold voltage at zero drain-source voltage and σ is a coefficient which may depend on the gate voltage swing. This translates into the following shift in V_{gt}:

$$V_{gt} = V_{gto} + \sigma V_{ds}$$

(4-6-17)

where V_{gto} is the gate voltage swing at zero drain-source bias. The effects of the drain bias induced threshold voltage shift can be accounted for in the I–V

characteristics of eq. (4-6-14)) by using eq. (4-6-17) in the expressions for the saturation current and the linear channel conductance. This will contribute to a finite output conductance in the saturation region. However, as discussed in Section 3.6, σ will depend on the gate voltage swing since the effect of drain bias induced barrier lowering (DIBL) will vanish well above threshold. Hence, for modeling purposes, we adopted the following empirical expression for σ (see Sections 3.6):

$$\sigma = \frac{\sigma_o}{1 + \exp\left(\dfrac{V_{gto} - V_{\sigma t}}{V_\sigma}\right)} \tag{4-6-18}$$

which gives $\sigma \to \sigma_o$ for $V_{gto} < V_{\sigma t}$ and $\sigma \to 0$ for $V_{gto} > V_{\sigma t}$. The voltage V_σ determines the width of the transition between the two regimes.

The quality of the present unified HFET model is illustrated in Figs. 4.6.2 and 4.6.3 for a submicrometer HFET. Fig. 4.6.2 shows the above-threshold $I–V$ characteristics and Fig. 4.6.3 shows the subthreshold characteristics in a semilog plot.

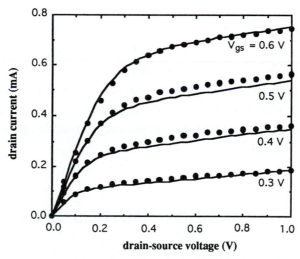

Fig. 4.6.2. Above-threshold experimental (symbols) and calculated (solid lines) HFET $I–V$ characteristics. Device parameters are: $L = 1$ μm, $W = 10$ μm, $\varepsilon_s = 1.14 \times 10^{-10}$ F/m, $d_i = 0.04$ μm, $\Delta d = 4.5 \times 10^{-9}$ m, $\mu_n = 0.4$ m^2/Vs, $v_s = 1.5 \times 10^5$ m/s, $V_{To} = 0.15$ V, $\eta = 1.28$, $R_s = R_d = 60$ ohm, $\lambda = 0.15$ V^{-1}, $n_{max} = 2 \times 10^{16}$ m^{-2}, $m = \gamma = \delta = 3.0$, $\sigma_o = 0.057$, $V_\sigma = 0.1$ V, $V_{\sigma t} = 0.3$ V. (From Fjeldly and Shur (1991).)

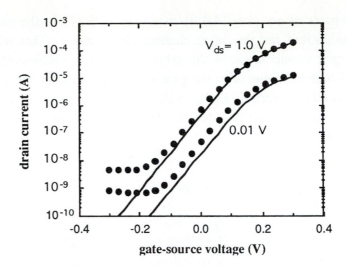

Fig. 4.6.3. Subthreshold experimental (symbols) and calculated (solid lines) *I–V* characteristics for the same device as in Fig. 4.6.2. (From Fjeldly and Shur (1991).)

As can be seen from the figures, our model reproduces quite accurately the experimental data in the entire range of bias voltages, over several decades of the current variation. A residual drain current at large negative gate-source voltages is not accounted for in our model. This current is caused by a small leakage of the reverse biased *p-n* drain-substrate junction. The parameters used in this calculation were obtained using a parameter extraction procedure similar to that described in Section 3.9.

4-7. CURRENT-VOLTAGE CHARACTERISTICS OF HFETs

4.7.1. Basic Assumptions

Analytical descriptions of Heterostructure Field Effect Transistors (HFETs) are usually based on a charge control model (see, for example, Lee et al. (1983)) which describes how the channel current is controlled by the charges induced into the channel by the gate bias. The device capacitances are found as derivatives of this charge with respect to the terminal voltages. Such a model can be modified to account for important non-ideal effects such as carrier injection into the substrate and the gate dielectric layer, channel length modulation, effect of traps, effect of the piezoelectric stress, etc. The most important parameters of the charge control model are the device threshold voltage, the gate dielectric thickness, the effective (or field-effect) electron mobility, the effective electron saturation velocity, source and drain series resistances, and parameters characterizing the gate leakage current and the subthreshold current. With a judicial choice of all these parameters, the charge control model gives a fairly good fit to measured data (see, for example, Shur (1990)).

In this Section, we present an analytical model by Moon et al. (1990) for HFETs based on the same philosophy as that used for the advanced CMOS models of Sections 3.10 and 3.11. This model allows us not only to obtain an excellent agreement with our measured data but also to develop a clear and unambiguous procedure for determining the saturation current and saturation voltage. In terms of the two-piece velocity-field characteristics by Sodini et al. (1984) (see Section 3.5), the onset of saturation in the channel is defined as the point where the electron velocity reaches the saturation velocity. In the non-saturated regime of the I–V characteristics, the drain current is then easily derived using a simplified constant capacitance model for the charge control. In the saturated regime, we divide the channel into two parts (see Fig. 4.7.1), a non-saturated part where the Gradual Channel Approximation (GCA) is still valid, and a saturated part characterized by the electrons propagating at the saturation velocity. In the saturated part of the channel, the potential distribution is derived from a simplified solution of the two-dimensional Poisson's equation for this region, as already discussed in Section 3.5. Self-consistent solutions for the device I–V characteristics in the saturation regime are then obtained by matching the solutions for the two parts of the channel.

In Subsection 4.7.2, we derive the I–V characteristics of the intrinsic HFET, neglecting the effects of the source and drain series resistances, R_s and R_d. Subsection 4.7.3 deals with the extrinsic model of the HFET, which also forms the basis of the device parameter extraction procedure discussed in Subsection 4.7.4.

The present analytical model is based on several assumptions. First of all, we neglect the gate current. This simplifies the model, but limits its validity to voltages below the turn-on voltage for the gate current. However, the effect of the gate current can be accounted for by using an appropriate equivalent circuit as discussed in Section 4.6. Also, the model is not unified since it does not

include the subthreshold regime. However, this can be remedied by using the
Unified Charge Control Model (UCCM, see Chapter 3) for the analysis instead of
the simplified constant gate-channel capacitance charge control model used here
(see Byun (1991)). A drawback with this approach is that the final model
expressions become quite complex. An alternative method is to use the simplified
procedure of adding inverse expressions for the above-threshold and subthreshold
currents, as was done for MESFETs in Section 4.4, for joining the subthreshold
and the above-threshold characteristics.

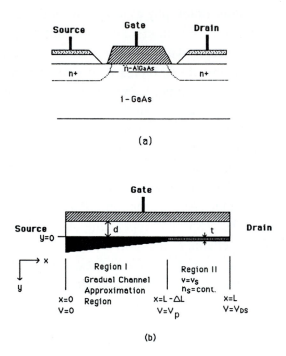

Fig. 4.7.1. Device structure (a) and schematic representation of the
conducting channel (b). (From Moon et al. (1990), © 1990 IEEE.)

4.7.2. Intrinsic Current-Voltage Characteristics

We start from the following charge control equation which describes the
sheet density of electrons, n_S, in the HFET channel:

$$qn_s = c_{gc}\left[V_{GT} - V(x)\right] \tag{4-7-1}$$

Here, $V_{GT} = V_{GS} - V_T$ is the gate voltage swing where V_{GS} is the intrinsic gate-source voltage and V_T is the threshold voltage, $V(x)$ is the position dependent channel voltage, and c_{gc} is the above-threshold gate-channel capacitance per unit area which can be expressed as

$$c_{gc} = \frac{\varepsilon_i}{d + \Delta d} \tag{4-7-2}$$

where ε_i is the dielectric permittivity and d is the thickness of the AlGaAs layer, and Δd is a correction to this thickness related to the shift in the Fermi level in the channel with respect to the bottom of the conduction band.

Moon et al. (1990) showed that the following velocity-field relationship by Sodini et al. (1984) is quite suitable for the modeling of HFETs (see also Section 3.5):

$$v(F) = \begin{cases} \dfrac{\mu_n F}{1 + \dfrac{F}{2F_s}}, & F < 2F_s \\[3mm] v_s, & F \geq 2F_s \end{cases} \tag{4-7-3}$$

where μ_n is the field effect mobility for electrons, v_s is the saturation velocity, $F_s = v_s/\mu_n$, and F is the absolute value of the longitudinal channel field.

Based on eqs. (4-7-1) and (4-7-3), the drain current in the below-saturation regime can be expressed as

$$I_d = qn_s W \mu_n \frac{F}{1 + \dfrac{F}{2F_s}} \tag{4-7-4}$$

where W is the gate width. Integrating eq. (4-7-4) over the effective gate length, L_{eff}, we obtain

$$I_d = \frac{1}{R_n} \frac{2V_{GT}V_{DS} - V_{DS}^2}{V_{DS} + 2V_L} \tag{4-7-5}$$

where V_{DS} is the intrinsic drain-source voltage, $V_L = F_s L_{eff}$,   and $R_n =$

$(Wc_{gc}v_s)^{-1}$.

The intrinsic saturation voltage, V_{SAT}, is defined as the drain-source voltage where the electron drift velocity at the drain edge of the channel becomes equal to v_s. The corresponding saturation current, I_{sat}, is obtained from eq. (4-7-5) by setting $V_{DS} = V_{SAT}$, i.e.,

$$I_{sat} = \frac{1}{R_n} \frac{2V_{GT}V_{SAT} - V_{SAT}^2}{V_{SAT} + 2V_L} \tag{4-7-6}$$

On the other hand, at the saturation point, we also have

$$I_{sat} = qn_s v_s W = \frac{1}{R_n}\left(V_{GT} - V_{SAT}\right) \tag{4-7-7}$$

From eqs. (4-7-6) and (4-7-7), we obtain the following expressions for the intrinsic saturation voltage and the saturation current:

$$V_{SAT} = \frac{2V_{GT}V_L}{V_{GT} + 2V_L} \tag{4-7-8}$$

$$I_{sat} = \frac{1}{R_n} \frac{V_{GT}^2}{V_{GT} + 2V_L} \tag{4-7-9}$$

When V_{DS} becomes larger than V_{SAT}, the onset point of saturation in the channel (marked as $x = L - \Delta L$ in Fig. 4.7.1b) moves closer to the source. This effect is called "channel length modulation". In the subthreshold regime, the channel can be divided into two regions as indicated in the figure. In Region I, $F \le 2F_s$, $v \le v_s$, and GCA is assumed to be valid. In Region II, $v = v_s$ and GCA is no longer valid owing to the rapid variation in the longitudinal electric field.

In Region I, the drain current can still be expressed by eq. (4-7-5), except that V_{DS} has to be replaced by the voltage V_p at $x = L_{eff} - \Delta L$ and V_L has to be replaced by $V_L - \Delta V_L = F_s(L_{eff} - \Delta L)$. Usually, in well-behaved devices suitable for circuit applications, the channel length modulation should be relatively small such that $\Delta L \ll L_{eff}$ and $\Delta V_L \ll V_p/2 + V_L$. Hence, we obtain

$$I_d = \frac{1}{R_n} \frac{2V_{GT}V_p - V_p^2}{V_p + 2(V_L - \Delta V_L)}$$

$$\approx \frac{1}{R_n} \frac{2V_{GT}V_p - V_p^2}{V_p + 2V_L}\left(1 + \frac{2\Delta V_L}{V_p + 2V_L}\right) \qquad (4\text{-}7\text{-}10)$$

The same current can also be derived for Region II as

$$I_d = \frac{1}{R_n}\left(V_{GT} - V_p\right) \qquad (4\text{-}7\text{-}11)$$

We note that V_p and ΔL depend on V_{DS} and V_{GT}. These quantities have to be determined self-consistently from eqs. (4-7-10) and (4-7-11), using a model for the potential distribution in Region II and requiring that the potential and the electric field be continuous at $x = L_{eff} - \Delta L$ (see Steen et al. 1990). This approach automatically ensures a continuous output conductance at the onset of saturation.

The potential distribution in Region II is the same as that derived in Section 3.5 for the MOSFET:

$$V(x) = V_p + \frac{V_\lambda}{2}\exp\left(\frac{x - L_{eff} + \Delta L}{\lambda}\right) - \frac{V_\lambda}{2}\exp\left(-\frac{x - L_{eff} + \Delta L}{\lambda}\right) \quad (4\text{-}7\text{-}12)$$

where the continuity conditions for the transition to the saturated region were used. Here, $V_\lambda = \lambda F_s$ where λ is a characteristic length of the saturated region given by

$$\lambda = \sqrt{(d + \Delta d)t} \qquad (4\text{-}7\text{-}13)$$

and t is an effective thickness of the channel region. We should note that the potential distribution given by eq. (4-7-12) is very sensitive to the magnitude of λ. In comparisons with experimental data, it is therefore convenient to treat λ as a fitting parameter rather than to use eq. (4-7-13) which itself is a result of a crude estimate.

A relationship that links ΔL and V_p with the drain-source voltage is obtained by considering eq. (4-7-12) at the drain side of the channel (i.e., at $x = L_{eff}$):

$$V_{DS} = V_p + V_\lambda \sinh\left(\frac{\Delta L}{\lambda}\right) \tag{4-7-14}$$

This expression can be solved with respect to ΔL, resulting in

$$\Delta L = \lambda \ln\left[\frac{V_{DS} - V_p}{V_\lambda} + \sqrt{\left(\frac{V_{DS} - V_p}{V_\lambda}\right)^2 + 1}\right] \tag{4-7-15}$$

In principle, I_d, V_p and ΔL can now be determined self-consistently for any bias conditions by combining eqs. (4-7-10), (4-7-11) and (4-7-15). However, this set of equations cannot, in general, be solved analytically. Hence, in order to obtain analytical solutions, Moon et al. (1990) assumed that $V_p \approx V_{SAT}$ and approximated the potential distribution in Region II by (see Section 3.5)

$$V_{DS} \approx V_p + \alpha V_\lambda\left[\exp\left(\frac{\Delta L}{\lambda}\right) - 1\right] \tag{4-7-16}$$

where the constant α is determined from the condition of continuity in the drain conductance at the onset of saturation, i.e.,

$$\alpha = \frac{8\left(V_{GT} + V_L\right)V_L}{\left(V_{GT} + 2V_L\right)^2} \tag{4-7-17}$$

This results in the following expression for the drain current in saturation:

$$I_d \approx I_{sat}\left\{1 + \frac{2V_\lambda}{V_{GT} + 2V_L} \ln\left[1 + \frac{(V_{DS} - V_{SAT})(V_{GT} + 2V_L)^2}{8V_\lambda V_L(V_{GT} + V_L)}\right]\right\} \tag{4-7-18}$$

In deep saturation, where $V_{DS} - V_{SAT} \gg V_\lambda$, the second term of the logarithm in eq. (4-7-18) becomes much larger than unity. Then the drain resistance in saturation becomes:

$$r_s = \frac{1}{\partial I_d/\partial V_{DS}}\bigg|_{\text{deep sat.}} \approx R_n \frac{(V_{DS} - V_{SAT})(V_{GT} + 2V_L)^2}{2V_\lambda V_{GT}^2} \tag{4-7-19}$$

This shows that by plotting r_S versus V_{DS}, we can determine V_{SAT} and V_λ from the intercept with $r_S = 0$ and from the slope, respectively. This procedure is the basis for the parameter extraction technique discussed in Subsection 4.7.4.

4.7.3. Extrinsic Current-Voltage Characteristics

In short channel compound semiconductor devices with gate lengths of one micrometer or less, the source and drain series resistances, R_s and R_d, are comparable to the channel resistance. Therefore, the voltage drop across these parasitic resistances can no longer be considered small compared to the external bias voltages. Moreover, for compound semiconductor devices, the determination of R_s and R_d using the Gated Transmission Line Method (see Baier et al. (1985)) is usually not as accurate as for Si MOSFETs (Chern et al. (1980)) owing to the lower degree of uniformity in the compound semiconductor device characteristics. In this Subsection, we extend our model to describe the effects of the parasitic source and drain resistances on the HFET I–V characteristics.

The relationship between the intrinsic voltages (upper case subscripts) and the extrinsic voltages (lower case subscripts) are as follows:

$$V_{DS} = V_{ds} - I_d R_t \tag{4-7-20}$$

$$V_{GS} = V_{gs} - I_d R_s \tag{4-7-21}$$

where $R_t = R_s + R_d$. Substituting eqs. (4-7-20) and (4-7-21) into eq. (4-7-5), we obtain a quadratic equation for the drain current in the non-saturated regime which has the solution

$$I_d = \frac{2V_{gt} V_{ds} - V_{ds}^2}{A + \sqrt{A^2 - B}} \tag{4-7-22}$$

where

$$A = \left(\frac{R_n}{2} - R_d\right) V_{ds} + R_t V_{gt} + R_n V_L \tag{4-7-23}$$

$$B = R_t \left(R_s - R_d + R_n\right)\left(2V_{gt} V_{ds} - V_{ds}^2\right) \tag{4-7-24}$$

The saturation current is found from the solution of another quadratic equation obtained by combining eqs. (4-7-9) and (4-7-21):

$$I_{sat} = \frac{V_{gt}^2}{R_n\left(\dfrac{V_{gt}}{2} + V_L\right) + R_s V_{gt} + \sqrt{R_n^2\left(\dfrac{V_{gt}}{2} + V_L\right)^2 + 2R_n R_s V_{gt} V_L}} \qquad (4\text{-}7\text{-}25)$$

Once I_{sat} is found as a function of V_{gt}, we can calculate V_{GT} and V_{SAT} in terms of the extrinsic voltages from eqs. (4-7-20) and (4-7-21).

The intrinsic expression for the increase in the HFET drain current caused by channel length modulation in saturation was derived in eq. (4-7-18). For reasonably well behaved HFETs, this increase should be relatively small. In other words, HFETs should be designed in such a way that λ is much smaller than L_{eff} in order to reduce the short channel effects. If this is the case, we can assume that, for $V_{DS} > V_{SAT}$, V_{GT} remain nearly the same as that for the saturation point and that $V_{ds} - V_{sat} \approx V_{DS} - V_{SAT}$ (this is the same type of approximation as used for MOSFETs in Sections 3.10 and 3.11). Hence, we can write the current in the saturation regime as follows:

$$I_d \approx I_{sat}\left[1 + \frac{2V_\lambda \ln(1 + K)}{V_{gt} + 2V_L - I_{sat} R_s}\right] \qquad (4\text{-}7\text{-}26)$$

where

$$K = \frac{(V_{ds} - V_{sat})(V_{gt} + 2V_L - I_{sat} R_s)^2}{8V_\lambda V_L (V_{gt} + V_L - I_{sat} R_s)} \qquad (4\text{-}7\text{-}27)$$

In deep saturation, the second term in the logarithmic argument in equation (4-7-26) is much larger than unity (just as for the intrinsic case). Hence, the deep saturation drain resistance in the extrinsic device can be expressed as

$$r_s = \frac{1}{\partial I_d / \partial V_{ds}}\bigg|_{\text{deep sat.}} \approx \left(\frac{V_{gt} + 2V_L}{I_{sat}} - R_s\right)\frac{(V_{ds} - V_{sat})}{2V_\lambda} \qquad (4\text{-}7\text{-}28)$$

Eq. (4-7-28) shows that by plotting r_s versus V_{ds}, we can determine V_{sat} from the intercept with $r_s = 0$. (Such a plot is shown in Fig. 4.7.3.) Moreover, we can find V_λ from the slope of this dependence once V_L and R_s are known. V_λ is an important measure of the short channel effects and the channel length modulation in the device.

In Fig. 4.7.2, we compare the present extrinsic HFET model with experimental above-threshold I–V characteristics for a conventional, self-aligned device with a nominal gate length of 1.4 μm. The comparison was done using the parameter extraction procedure described below (see Subsection 3.7.4) without additional adjustment of parameters.

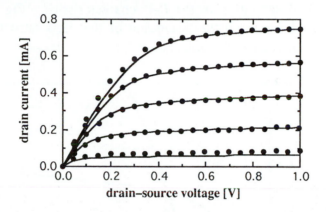

Fig. 4.7.2. Above-threshold experimental (symbols) and calculated (solid lines) I–V characteristics for an HFET with nominal gate length $L = 1.4$ μm. Other device parameters are: $W = 20$ μm, $d_i = 0.04$ μm, $\Delta d = 4.5 \times 10^{-9}$ m, $\mu_n = 1.0$ m^2/Vs, $v_s = 1.28 \times 10^5$ m/s, $V_{To} = 0.121$ V, $R_s = 162$ ohm, $R_d = 142$ ohm, $V_\lambda = 0.02$ V; $V_L = 0.155$ V $\delta L = 0.25$ μm. Top curve: $V_{gs} = 0.6$ V, step = –0.1V. (From Moon et al. (1990), © 1990 IEEE.)

As can be seen from the figure, the model is quite accurate for the 1.4 μm device. However, for a 1.0 μm HFET, the model underestimated somewhat the output conductance in saturation. This discrepancy was probably caused by additional short-channel effects such as substrate leakage current (Eastman and Shur (1979)) and Drain Induced Barrier Lowering (DIBL, see Section 3.6), not accounted for in the present model. Moon et al. (1990) proposed to increase the

p-type substrate doping substantially, from the present unintentional doping level of about 10^{16} cm^{-3}, in order reduce the excess saturation current in short-channel devices.

4.7.4. Parameter Extraction

In this Subsection, we show how to extract from experimental data the device parameters needed for a description of the HFET in terms of the present model. The device investigated is the same as that described in Fig. 4.7.2 (for further details on the device, see Moon et al. (1990)).

The basis for the parameter extraction procedure is a precise determination of the onset of saturation in the $I–V$ characteristics. Fig. 4.7.3 shows the technique used, based on the discussion of the deep saturation r_s versus V_{ds} characteristics developed in Subsection 4.7.3.

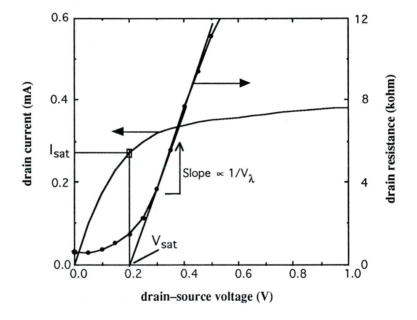

Fig. 4.7.3. Illustration of the determination of the saturation voltage and the saturation current based on eq. (4-7-28). (From Moon et al. (1990), © 1990 IEEE.)

As can be seen from the figure, there are three distinct regions in these characteristics. At $V_{ds} \leq V_{sat}$, r_s is relatively small and non-linear. However, at an intermediate voltage range, such that $V_{ds} \geq V_{sat}$, r_s shows good linearity in V_{ds}, in agreement with eq. (4-7-28). At $V_{ds} \gg V_{sat}$, r_s reaches a region of sublinear growth, indicating the presence of additional mechanisms, such as DIBL (see Section 3.6), not considered in the model. However, the region of linear behavior is wide enough to be applicable for the determination of the saturation point.

Combining eqs. (4-7-7) with eqs. (4-7-20) and (4-7-21), we obtain the following expression for the saturation current versus the gate-source voltage:

$$I_{sat} = \frac{V_{gs} - V_{sat} - V_T}{R_n - R_d} \qquad (4\text{-}7\text{-}29)$$

Once the saturation points for the various characteristics have been established, it is possible to plot I_{sat} versus $V_{gs} - V_{sat}$, as shown in Fig. 4.7.4 for HFETs with different gate lengths. According to eq. (4-7-29), these plots should give straight lines where the threshold voltage, V_T, can be determined as the intercept with $I_{sat} = 0$, and $R_n - R_d$ is the inverse slope. For the present 1.4 μm device, we find $V_T \approx 0.121$ V.

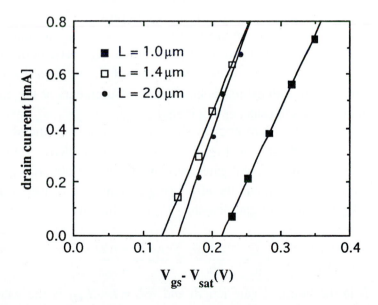

Fig. 4.7.4. Extraction of V_T and $R_n - R_d$ using eq. (4-7-29). (From Moon et al. (1990), © 1990 IEEE.)

The difference $R_d - R_s$ can be determined either from gate current measurements (see Fukui (1979) and Lee et al. (1985)), or by measuring the normal extrinsic device transconductance in the saturation regime, g_m, and its counterpart, g_{mr}, with the source and drain interchanged (at constant drain current). Since the intrinsic transconductance is the same in both cases, we find (see Section 3.9)

$$R_s - R_d = \frac{1}{g_m} - \frac{1}{g_{mr}} \tag{4-7-30}$$

Using either of these techniques, the value $R_d - R_s = 20$ ohm was obtained.

Substituting eq. (4.7.21) into eq. (4-7-9), the result can be written as follows, after some rearrangement:

$$\frac{\left(V_{gt} - I_{sat}R_s\right)^2}{I_{sat}} = R_n\left(V_{gt} - I_{sat}R_s + 2V_L\right) \tag{4-7-31}$$

This expression implies that R_n and V_L can be obtained from the slope and the intercept with $V_{gt} - I_{sat}R_s = 0$ of a plot of $(V_{gt} - I_{sat}R_s)^2/I_{sat}$ versus $V_{gt} - I_{sat}R_s$. However, this method requires that R_s is known. It is therefore necessary to obtain R_n, R_s, R_d and V_L by iteration using the known values for $R_n - R_d$ and $R_d - R_s$ in combination with eq. (4-7-31) (for details, see Moon et al. (1990)). Using such a procedure, we obtain $R_s = 162$ ohm, $R_d = 142$ ohm, and $V_L = 0.155$ V. From the extracted value of R_n, we find $v_s = 1.28 \times 10^5$ m/s, which agrees well with estimates based on measurements of the current-gain cut-off frequency (see Hikosaka et al. (1988)).

Once R_s and V_L are known, V_λ can be found from the slope of r_s versus $V_{ds} - V_{sat}$, as indicated in Fig. 4.7.3 (see eq. (4-7-28)). Taking the average value obtained from different gate-source voltages, we obtain $V_\lambda = 0.020$ V.

The mobility, μ_n, and the effective gate length, L_{eff}, can be found from the value of V_L for different gate lengths. V_L is defined as

$$V_L = \frac{v_s L_{eff}}{\mu_n} = \frac{v_s(L - \delta L)}{\mu_n} \tag{4-7-32}$$

where L is the nominal gate length and $\delta L = L - L_{eff}$ is the gate length offset Hence, if v_s and μ_n are independent of gate length, we should expect a linear relationship between V_L and L. Fig. 4.7.5 shows such a plot based on three devices with different gate lengths. Here, the slope is identical to v_s/μ_n and the intercept with $V_L = 0$ gives δL.

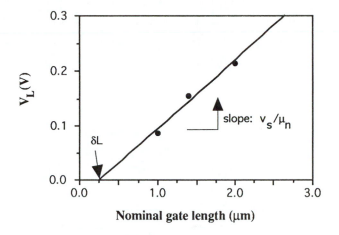

Fig. 4.7.5. Determination of mobility and gate length offset.

From this plot, we obtain $\delta L = 0.25$ μm and $\mu_n = 1.0$ m^2/Vs. The gate length offset can be introduced at many stages in the fabrication process. The main contribution in the present devices comes from the implant straggle during n^+ source and drain implantation.

4-8. GATE LEAKAGE CURRENT

As was mentioned in Section 4.1, the gate leakage current may play an important role in compound semiconductor field effect transistors where the gate and the channel are separated either by the depletion region of the Schottky contact in GaAs MESFETs or by a depleted wide band gap semiconductor in HFETs. For enhancement-mode compound semiconductor FETs, the gate current can play a dominant role and, as discussed below, may even affect the value of the "intrinsic" drain-source current, I_{ds}. Fig. 4.8.1a shows a simple conventional equivalent circuit which accounts for the gate current in GaAs MESFETs. This circuit was used in the old version of UM-SPICE (see Hyun et al. (1986)). The gate current is modeled by two equivalent Schottky diodes connected from the gate to the source and from the gate to the drain.

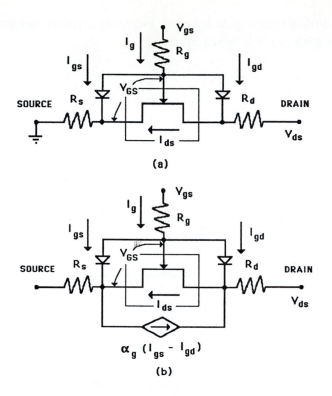

Fig. 4.8.1 HFET equivalent circuits; (a) conventional equivalent circuit, (b) equivalent circuit that takes into account the effect of the gate current on the channel current (after Ruden et al. (1989a)). (From Shur (1990).)

Using the well known diode equation, we find for the total gate current

$$I_g = J_{ss} L W \left[\exp\left(\frac{V_{GS}}{m_{gs} V_{th}} \right) + \exp\left(\frac{V_{GD}}{m_{gd} V_{th}} \right) - 2 \right] \qquad (4\text{-}8\text{-}1)$$

where J_{ss} is the reverse saturation current density which is calculated using the thermionic or thermionic-field emission theory (see Chapter 1), L and W are the gate length and gate width, V_{GS} and V_{GD} are the intrinsic gate-source and gate-drain voltages, m_{gs} and m_{gd} are the gate-source and gate-drain Schottky diode ideality factors, and V_{th} is the thermal voltage. To a first order approximation, this simple model may be adequate for a semi-quantitative description of the gate current in GaAs MESFETs.

A more accurate description was proposed by Berroth et al. (1988) who introduced effective electron temperatures at the source side and the drain side of the channel. The electron temperature at the source side of the channel is taken to be close to the lattice temperature, i.e., $T_s \approx T$, and the drain side electron temperature, T_d, is assumed to increase with the drain-source voltage to reflect the heating of the electrons in this part of the channel where the electric field is large. This effect can easily be taken into account by modifying eq. (4-8-1) to read

$$I_g = J_{gs} LW \left[\exp\left(\frac{V_{GS}}{m_{gs} V_{ths}} \right) - 1 \right] + J_{gd} LW \left[\exp\left(\frac{V_{GD}}{m_{gd} V_{thd}} \right) - 1 \right] \quad (4\text{-}8\text{-}2)$$

where J_{gs} and J_{gd} are the reverse saturation current densities for the gate-source and the gate-drain diodes, respectively, and $V_{ths} = k_B T_s / q$ and $V_{thd} = k_B T_d / q$.

In most GaAs MESFETs, the reverse gate saturation current is dependent on the reverse bias. According to Dunn (1969), this dependence can be described by the following equations:

$$J_{gs} = g_{gs} V_{gs} \exp\left(-\frac{q V_{gs} \delta_g}{k_B T_s} \right) \quad (4\text{-}8\text{-}3)$$

$$J_{gd} = g_{gd} V_{gd} \exp\left(-\frac{q V_{gd} \delta_g}{k_B T_d} \right) \quad (4\text{-}8\text{-}4)$$

where g_{gs} and g_{gd} are the reverse diode conductances and δ_g is the reverse bias conduction parameter. As was shown by Conger et al. (1988), these expressions reproduce MESFET leakage characteristics, in excellent agreement with experimental data. In forward bias, assuming the thermionic emission mechanism, we find

$$J_{gs} = \frac{A^* T_s^2}{2} \exp\left(-\frac{\Phi_b}{k_B T_s} \right) \quad (4\text{-}8\text{-}5)$$

$$J_{gd} = \frac{A^* T_d^2}{2} \exp\left(-\frac{\Phi_b}{k_B T_d} \right) \quad (4\text{-}8\text{-}6)$$

where A^* is the effective Richardson constant and Φ_b is the effective Schottky barrier height.

Berroth et al. (1988) deduced the dependence of the electron temperature at the drain, T_d, on the drain-source voltage, V_{ds}, for GaAs MESFET by using eq. (4-5-1) for the parameter extraction. They found that this temperature rises to approximately 400 K (from its equilibrium value of 300 K) at $V_{ds} \approx 1$ V and to approximately 450 K at $V_{ds} = 3$ V. Similar measurements for HFETs reported by Chen et al. (1988) showed that T_d rises to approximately 350 K (from its equilibrium value of 300 K) at $V_{ds} \approx 1$ V and practically saturates at higher values of V_{ds}. We should emphasize that these equivalent temperatures represent only an indirect characteristic of the electron distribution in the channel. Monte Carlo simulations of short channel MESFETs and HFETs show that the actual temperatures near the drain contacts of these devices are much higher (see, for example, Jensen et al. (1991) and (1991a)). However, the fact that electrons are somewhat cooler in an HFET channel can probably be attributed to transfer of hot electrons to the AlGaAs region in HFETs because of the relatively small conduction band discontinuity separating the channel from the AlGaAs layer. Hence, the gate current may "cool" the electrons in the HFET channel more effectively. This may explain why the noise figure is typically smaller for microwave HFETs than for microwave MESFETs. A clearer explanation of this difference may be obtained by using self consistent Monte Carlo simulations of the type reported by Jensen et al. (1991) and (1991a).

The possibility to change the electron temperature and, hence, the gate current in an HFET channel by changing the drain bias points out to a new hot electron regime of HFET operation. In this regime, which utilizes the real space transfer of electrons (RST) over a heterostructure barrier, the input voltage is the drain-source voltage and the output current is the gate current. This mode of operation is similar to that of the Charge Injection Transistor (CHINT), see Kastalsky and Luryi (1983). In HFETs, this regime was first observed by Shur et al. (1986). Recently, Maezawa and Mizutani (1991) showed that in this mode of operation, the HFET has a higher cut-off frequency. Martinez et al. (1992) measured extremely high HFET transconductances in this new regime, several times higher than in the conventional mode of operation.

The equivalent circuit shown in Fig. 4.8.1b was proposed by Ruden et al. (1989a) for HFETs. (We expect that the same equivalent circuit can be used for MESFETs as well.) This circuit takes into account the effect of gate current on the channel current. Indeed, the gate current is distributed along the channel, with the largest current density taking place near the source side of the channel. This leads to a redistribution of the electric field along the channel, with an increase in the field near the source side of the device and an overall decrease in the drain current. A numerical solution of coupled differential equations describing the gate and channel current distributions along the channel by Baek and Shur (1990) provided a theoretical justification for this new equivalent circuit shown in Fig. 4.8.1b. Experimental data for HFETs clearly indicate that the increase in the gate current may lead to a saturation or even a decrease in the drain current (see Shur et al. (1986)). This drop can even result in a negative differential resistance. The effect is related to the real space transfer mechanism, (see Gribnikov (1973) and Hess et al. (1979)), i.e., to the transfer of hot electrons from the channel over the barrier created by the conduction band discontinuity. A similar effect was observed in HFETs by Chen et al. (1987a).

The equivalent circuit shown in Fig. 4.8.1b can easily be implemented in AIM Spice. However, we have instead used the somewhat less accurate equivalent circuit shown in Fig. 4.8.2 because it yields considerable savings of computer time, especially for large circuits and, more importantly, it simplifies the characterization procedure.

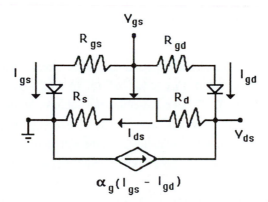

Fig. 4.8.2 Simplified FET equivalent circuit accounting for gate leakage current. (From Shur et al. (1992).)

This simplified circuit does not properly account for the voltage drop across the source series resistance caused by the gate current. However, this can be compensated for by a judicial choice of the parameter α_g in Fig. 4.8.2. Also, when this voltage drop becomes large, the electron velocity in the channel may saturate (see Lee et al. (1985) for details). Under such conditions, any equivalent circuit utilizing a constant source series resistance becomes inaccurate.

The gate leakage current in HFETs was analyzed by Ponse et al. (1985), Ruden et al. (1988), Chen et al. (1988), Ruden et al. (1989a), Baek and Shur (1990), Ruden (1989), and Schuermeyer et al. (1991) and (1992). Fig. 4.8.3 shows qualitative band diagrams of an HFET for different gate-source voltages.

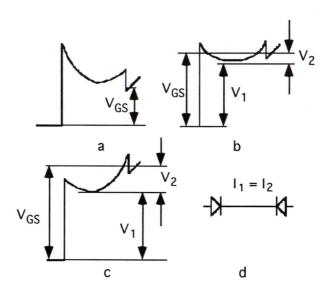

Fig. 4.8.3. Band diagrams and equivalent circuit of the metal-AlGaAs-GaAs structure; a) $V_{GS} < \phi_b - \Delta E_c/q$, b) $V_{GS} \approx \phi_b - \Delta E_c/q$, c) $V_{GS} > \phi_b - \Delta E_c/q$, d) diodes representing the Schottky barrier between the gate metal and the AlGaAs layer (diode 1) and the barrier related to the conduction band discontinuity across the heterointerface (diode 2). V_1 and V_2 are voltage drops across diodes 1 and 2, respectively.

As can be seen from the figure, at small gate-source voltages, the largest barrier separating the channel from the gate metal is at the metal-semiconductor interface. At such voltages, the gate current is primarily determined by the Schottky barrier at the metal semiconductor interface. This situation changes at $V_{GS} \approx \phi_b - \Delta E_c/q$, where $q\phi_b$ is the Schottky barrier height at the metal semiconductor interface and ΔE_c is the conduction band discontinuity. At larger V_{GS}, the gate current is limited by the conduction band discontinuity or, more precisely, by the effective barrier height equal to the difference between the bottom of the conduction band in the AlGaAs layer at the heterointerface and the electron quasi-Fermi level in the channel. This barrier height changes relatively little with the gate bias (and only as a consequence of the dependence of the quasi-Fermi level on electron concentration in the two-dimensional electron gas). In a way, this interface behaves similarly to a reverse biased Schottky diode with a voltage dependent barrier height. As was pointed out by Chen et al. (1988), the current voltage characteristic of such a diode can be described by the diode equation for a forward biased diode, but with a very large ideality factor, typically in the range from 5 to 20 for HFETs.

Based on the above argument, Chen et al. (1988) proposed an equivalent circuit for the HFET where each diode in the equivalent circuits shown in Figs. 4.8.1 and 4.8.2 is substituted by a series combination of two diodes, one representing the Schottky barrier and the other representing the conduction band discontinuity. The current in the two diodes can be written as

$$I_1 = I_{s1} \left[\exp\left(\frac{V_1}{m_1 V_{th}} \right) - 1 \right] \tag{4-8-7}$$

$$I_2 = - I_{s2} \left[\exp\left(\frac{-V_2}{m_2 V_{th}} \right) - 1 \right] \tag{4-8-8}$$

where the gate-source voltage is divided with V_1 across the Schottky barrier and V_2 across the heterointerface such that

$$V_1 + V_2 = V_{GS} \tag{4-8-9}$$

Continuity in the gate current requires

$$I_g = I_1 = I_2 \tag{4-8-10}$$

This model can be further improved by using different equivalent electron temperatures at the source and drain sides of the channel (compare with eq. (4-8-2)).

As can be seen from Fig. 4.8.4, the model based on the equivalent circuit shown in Fig. 4.8.2 agrees quite well with experimental data for *n*-channel

HFETs. The AIM-Spice HFET model utilizes this approach in conjunction with the equivalent circuit shown in Fig. 4.8.2. A more detailed physical analysis of the gate current in n-channel and p-channel HFETs based on the Unified Charge Control Model (UCCM, see Chapter 3) is given by Schuermeyer et al. (1991) and (1992).

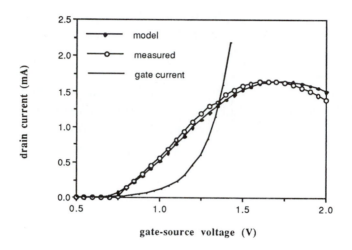

Fig. 4.8.4. Drain and gate current in a 1 μm gate length n-channel AlGaAs/GaAs MODFET. (From Ruden et al. (1989a), © 1989 IEEE.)

In self-aligned FETs, where the gate contact touches highly doped ohmic regions, the gate current may flow not only from the FET channel but from the contact regions as well. Experimental evidence that such a mechanism may be dominant in short channel self-aligned HFETs was provided by Schuermeyer et al. (1992). At the present, this additional mechanism is not directly accounted for in the AIM-Spice models. It can be modeled by introducing additional equivalent diodes representing the contributions from the source-gate and drain-gate junctions. However, such a modification of the models is only warranted if accurate measurements of the gate current are available for devices with many different gate lengths in order to separate gate-length dependent and independent contributions to the gate current.

REFERENCES

A. A. AKINWANDE, J. ZOU, M. SHUR and A. GOPINATH, "Dipole Heterostructure Field Effect Transistor", *IEEE Electron Device Letters*, EDL-11, No. 8, pp. 332-33 (1990)

R. ANHOLT and S. SWIRHUN, "Measurement and Analysis of GaAs MESFET Parasitic Capacitances", *IEEE Trans. Micr. Theory Tech.*, 39, pp. 1247-1251 (1991)

R. ANHOLT and S. SWIRHUN, "Equivalent Circuit Parameter Extraction for Cold GaAs MESFETs", *IEEE Trans. Micr. Theory Tech.*, 39, pp. 1243-1247 (1991a)

P. M. ASBECK, C. P. LEE and F. M. CHANG, *IEEE Trans. Electron Devices*, ED-31, p. 1377 (1984)

J. H. BAEK and M. SHUR, "Mechanism of Negative Transconductance in Heterostructure Field Effect Transistors", *IEEE Trans. Electron Devices*, ED-37, No. 8, pp. 1917-1921, Aug. 1990

S. M. BAIER, M. S. SHUR, K. LEE, N. C. CIRILLO and S. A. HANKA, "FET Characterization Using Gated TLM Structure", *IEEE Trans. Electron Devices*, ED-32, No. 12, pp. 2824-2829, Dec. (1985)

M. BERROTH, M. SHUR and W. HAYDL, "Experimental Studies of Hot Electron Effects in GaAs MESFETs", in *Extended Abstracts of the 20th Intern. Conf on Solid State Devices and Materials* (SSDM-88), Tokyo, pp. 255-258, Aug. (1988)

M. S. BIRITELLA, W. C. SEELBACH and H. GORONKIN, "The Effect of Backgating on Design and Performance of GaAs Integrated Circuits", *IEEE Trans. Electron Devices*, ED-29, No.7, pp. 1135-1142, July (1982)

Y. BYUN, K. LEE and M. SHUR, "Unified Charge Control Model and Subthreshold Current in Heterostructure Field Effect Transistors", *IEEE Electron Device Letters*, EDL-11, No. 1, pp. 50-53, Jan. (1990) (see erratum *IEEE Electron Device Letters*, EDL-11, No. 6, p. 273, June (1990))

Y. BYUN, *New Approach to Modeling and Characterization of Field Effect Transistors*, Ph. D. Thesis, University of Virginia (1991)

A. CAPPY, B. CARNES, R. FAUQUEMBERGUE, G. SALMER and E. CONSTANT, *IEEE Trans. Electron Devices*, ED-27, pp. 2158-2168 (1980)

P. C. CHAO, M. SHUR, , R. C. TIBERIO, K. H. G. DUH, P. M. SMITH, J. M. BALLINGALL, P. HO and A. A. JABRA, "DC and Microwave Characteristics of Sub-0.1 µm Gate-Length Planar-Doped Pseudomorphic HEMTS", *IEEE Trans. Electron Devices*, ED-36, No. 3, pp. 461-473, March (1989)

C. H. CHEN, M. SHUR and A. PECZALSKI, *IEEE Trans. Electron Devices*, ED-33, pp. 792-798 (1986)

C. H. CHEN, A. PECZALSKI, M. SHUR and H.K. CHUNG, *IEEE Trans. Electron Devices*, ED-34, No. 7, pp. 1470-1481, July (1987)

C. H. CHEN, S. BAIER, D. ARCH and M. SHUR, "A New and Simple Model for GaAs Heterojunction FET Characteristics", *IEEE Trans. Electron Devices*, ED-35, No. 5, pp. 570-577, May (1988)

Y. K. CHEN, D. C. RADULESCU, G. W. WANG, A. N. LEPORE, P. J. TASKER, L. F. EASTMAN and E. STRID, "Bias-Dependent Microwave Characteristics of an Atomic Planar-Doped AlGaAs/InGaAs/GaAs Double Heterojunction MODFET", in *Proceedings of IEEE MTT Symposium*, p. 871, Las Vegas, June (1987)

J. G. CHERN, P. CHANG, R. F. MOTTA and N. Godinho, "A New Method to Determine MOSFET Channel Length", *IEEE Electron Device Lett.*, EDL–1, No. 9, pp. 170-173, (1980)

N. C. CIRILLO, J. K. ABROKWAH and M. S. SHUR, "Self-Aligned Modulation Doped (Al,Ga)As/GaAs Field-effect Transistor", *IEEE Electron Device Lett.*, EDL-5, No. 4, pp. 129-131, April (1984)

J. CONGER, A. PECZALSKI and M. SHUR, "Subthreshold Current in GaAs MESFETs", *IEEE Electron Device Lett.*, EDL-9, No.3, pp. 128-129, March (1988)

J. CONGER, *Characterization, Modeling and Simulation of Compound Semiconductor Field-Effect Transistors and Integrated Circuits*, Ph.D. Thesis, University of Minnesota (1992)

J. CONGER, A. PECZALSKI and M. SHUR, "Temperature Modeling of GaAs Logic Circuits", unpublished (1992)

W. R. CURTICE, *IEEE Trans. Micr. Theory and Tech.*, MTT-28, p. 448 (1980)

D. DELAGEBEAUDEUF and N. T. LINH, "Metal-(n)AlGaAs-GaAs Two-Dimensional Electron Gas FET," *IEEE Trans. Electron Devices*, ED-29, No. 6, pp. 955-960 (1982)

J. A. DEL ALAMO and W. AZZAM, "A Floating Gate Transmission-Line Model Technique for Measuring Source Resistance in Heterostructure Field-Effect Transistors", *IEEE Trans. Electron Devices*, ED-36, p. 2386 (1989)

C. DUNN, *Microwave Semiconductor Devices and Their Circuit Applications*, H. A. Watson, Ed., McGraw Hill, New York (1969)

L. F. EASTMAN and M. S. SHUR, "Substrate Current in GaAs MESFET's", *IEEE Trans. Electron Devices*, ED-26, No. 9, pp. 1359-1361, Sept. (1979)

L. F. EASTMAN, S. TIWARI and M. S. SHUR, "Design Criteria for GaAs MESFET's Related to Stationary High Field Domains", *Solid State Electronics*, 23, pp. 383-389, April (1980)

R. W. H. ENGELMANN and C. A. LIEHTI, *IEEE Trans. Electron Devices*, ED-24, pp. 1288-1296 (1977)

M. FENG, C. L. LAU, V. EU and C. ITO, "Does the Two-Dimensional Electron Gas Effect Contribute to High-Frequency and High Speed Performance of Field-Effect Transistors?", *Appl. Phys. Lett.*, 57, p. 1233, (1990)

T. A. FJELDLY, "Analytical Modeling of the Stationary Domains in GaAs MESFETs", *IEEE Trans. Electron Devices*, ED-33, pp. 874-880 (1986)

T. A. FJELDLY, "Analytical Formulation of Nonstationary Electron Dynamics in Submicron Field Effect Transistors", *Superlattices and Microstructures*, 4, p. 55 (1988)

T. A. FJELDLY, A. PAULSEN and Ø. JENSEN, "A GaAs MESFET Small-Signal Equivalent Circuit Including Transmission Line Effects", *IEEE Trans. Electron Devices*, ED-36, No. 11, pp. 1557-1563, Sept. (1989)

T. A. FJELDLY, M. SHUR and K. PANDE, "Unified Model for GaAs Delta-Doped MESFETs", unpublished (1991)

T. A. FJELDLY and M. SHUR, "Unified CAD Models for HFETs and MESFETs", in *Proceedings of the 11th European Microwave Conference (Workshop Volume)*, Stuttgart, pp. 198-205, Sept. (1991)

M. C. FOISY, P. J. TASKER, B. HUGHES and L. F. EASTMAN, "The Role of Inefficient Charge Modulations in Limiting the Current Gain Cutoff Frequency of the MODFET", *IEEE Trans. Electron Devices*, ED-35, No. 7, pp. 871-878, July (1988)

H. FUKUI, "Determination of the Basic Device Parameters of a GaAs MESFET", *Bell Sys. Tech. J.*, pp. 711-797, (1979)

B. GELMONT, M. SHUR and R. J. MATTAUCH, "Capacitance-Voltage Characteristics of Microwave Schottky Diodes", *IEEE Trans. Microwave Theory and Technique*, 39, No. 5, pp. 857-863, May (1991)

H. GORONKIN, M. S. BIRITELLA, W. C. SEELBACH, AND VAITKUS, "Backgating and Light Sensitivity in Ion-Implanted GaAs Integrated Circuits", IEEE Trans. Electron Devices, ED-29, No. 5, pp. 845-850, May (1982)

Z. S. GRIBNIKOV, "Negative Differential Conductivity in a Multilayer Heterostructure," *Fiz. Tekh. Poluprovodn.*, 6, 1380 (1972) [*Sov. Phys. Semicond.*, 6, 1204 (1973)]

M. HEIBLUM, M. I. NATHAN, D. C. THOMAS and C. M. KNOEDLER, "Direct Observation of Ballistic Transport in GaAs", *Phys. Rev. Lett.*, 55, p. 2200 (1985)

K. HESS, H. MORKOÇ, H. SHICHIJO and B. G. STREETMAN, "Negative Differential Resistance through Real-Space Electron Transfer," *Appl. Phys. Lett.*, 35, p. 469 (1979)

K. HESS and C. KIZILYALLI, "Scaling and Transport Properties of High Electron Mobility Transistors", *IEDM Technical Digest*, Los Angeles, pp. 556-558, Dec. (1986)

K. HIKOSAKA, S. SASA and S. KURODA, "Current-Gain Cutoff Frequency Comparison of InGaAs HEMT's", *IEEE Electron Device Lett.*, EDL-9, No. 5, pp. 241-243, (1988)

P. HOWER and G. BECHTEL, "Current Saturation and Small Signal Characteristics of GaAs Field Effect Transistors", *IEEE Trans. Electron Devices*, ED-20, p. 213 (1973)

HSPICE User's Manual, Version H9001, Meta-software, Inc., 1300 White Oaks Rd., Campbell, CA 95008, (800)-346-5953 (1990)

C. H. HYUN, M. S. SHUR and N. C. CIRILLO, "Simulation and Design Analysis of AlGaAs/GaAs MODFET Integrated Circuits", *IEEE Trans. CAD ICAS*, CAD-5, No. 2, pp. 284-292, April (1986)

C. JACOBONI and P. LUGLI, *The Monte Carlo Method for Semiconductor Simulation*, Springer Series on Computational Microelectronics, ed. S. Selberherr, Springer-Verlag, Wien, New York (1989)

P. N. JENKINS, *Design and Simulation of Ultra High Speed GaAs Integrated Circuits*, Ph. D. Thesis, University of Minnesota (1989)

G. U. JENSEN, B. LUND, T. A. FJELDLY and M. SHUR, "Monte Carlo Simulation of Short Channel Heterostructure Field Effect Transistors", *IEEE Trans. Electron Devices*, ED-38, No. 4, pp. 840-851 (1991)

G. U. JENSEN, B. LUND, T.A. FJELDLY and M. SHUR, "Monte Carlo Simulation of Semiconductor Devices", *Computer Physics Communications*, (Proceedings of MSI Symposium on Supercomputer Simulation of Semiconductor Devices, Minneapolis, Nov. 1990), 67, No. 1, pp. 1-61 (1991a)

M. KANAMORI, G. JENSEN, M. SHUR and K. LEE, "Effect of $p\text{-}i\text{-}p^+$ Buffer on Characteristics of n-Channel Heterostructure Field Effect Transistors", *IEEE Trans. Electron Devices*, ED-39, No. 2, pp. 226-233, Feb. (1992)

A. KASTALSKY and S LURYI, "Novel Real-space Hot Electron Transfer Devices", *IEEE Electron Device Letters*, EDL-4, p. 334 (1983)

K. LEE, M. SHUR, T. DRUMMOND and H. MORKOÇ, "Current-Voltage and Capacitance-Voltage Characteristics of Modulation-Doped Field Effect Transistors," *IEEE Trans. Electron Devices*, ED-30 , No. 3, pp. 207-212, March (1983)

K. LEE, M. SHUR, T. J. DRUMMOND and H. MORKOÇ, "Electron Density of the Two-Dimensional Electron Gas in Modulation Doped Layers", *J. Appl. Phys.*, 54, No. 4, pp. 2093-2096, April (1983a)

K. LEE, M. S. SHUR, T. J. DRUMMOND and H. MORKOÇ, "Parasitic MESFET in (Al,Ga)As/GaAs Modulation Doped FET", *IEEE Trans. Electron Devices*, ED-31, No. 1, pp. 29-35, Jan. (1984)

K. LEE and M. SHUR, "π-Heterostructure Field Effect Transistors for VLSI applications", *IEEE Trans. Electron Devices*, ED-37, No. 8, pp. 1810-1820, Aug. (1990)

K. W. LEE, K. LEE, M. S. SHUR, T. VU, P. ROBERTS and M. HELIX, "Source, Drain, and Gate Resistances and Electron Saturation Velocity in Ion-implanted GaAs FETs", *IEEE Trans. Electron Devices*, ED-32, No. 5, pp. 987-992, May (1985)

A. F. J. LEVI, J. R. HAYES, P.M. PLATZMAN and W. WIEGMANN, "Injected Hot Electron Transport in GaAs", *Phys. Rev. Lett.*, 55, pp. 2071-2073 (1985)

P. F. LUNDQUIST and W. M. FORD, "Semi-Insulating GaAs Substrates", in *GaAs FET Principles and Technology*, ed. by J. V. DiLorenzo and D. D. Khadenwal, Artech House, Dedham, Mass. (1982)

K. MAEZAWA and T. MIZUTANI, "High-Frequency Characteristics of Charge-Injection Transistor-Mode Operation in AlGaAs/InGaAs/GaAs Metal-Insulator-Semiconductor Field-Effect Transistors", *Jap. J. Appl. Physics*, 30, No. 6, pp. 1190-1193 (1991)

T. J. MALONEY and J. FREY, *IEEE Trans. Electron Devices*, ED-22, pp. 357-358 (1975)

E. MARTINEZ, M. SHUR and F. L. SCHUERMEYER, Heterostructure Field Effect Transistors Operated in Hot Electron Regime, submitted to *IEEE Trans. Electron Device* (1992)

J. E. MEYER, "MOS Models and Circuit Simulation", *RCA Review*, 32, pp. 42-63, March (1971)

B. MOON, Y. BYUN, K. LEE and M. SHUR, "New Continuous Heterostructure Field Effect Transistor Model and Unified Parameter Extraction Technique", *IEEE Trans. Electron Devices*, ED-37, No. 4, pp. 908-918, April (1990)

H. MORKOÇ, H. UNLU and G. LI, *Principles and Technology of MODFETs*, vols. 1 and 2, John Wiley and Sons, New York (1991)

A. PECZALSKI, M. SHUR, C. H. HYUN, K. LEE and T. VU, "Design Analysis of GaAs Direct Coupled Field Effect Transistor Logic", *IEEE Trans. CAD ICAS*, CAD-5, No. 2, pp. 266-273, April (1986)

A. PECZALSKI, M. SHUR and C. H. CHEN, "Device Physics and Modeling for GaAs Integrated Circuits", in *Proceedings of KOSEF/NSF Joint Seminar*, Ed. C. Lee and W. Paul, Korea Science and Engineering Foundation and National Science Foundation, p. 227, Aug. (1986a)

F. PONSE, W. T. MASSELINK and H. MORKOÇ, "The Quasi-Fermi Level Bending in MODFETS and Its Effects on the FET Transfer Characteristics", *IEEE Trans. Electron Devices*, ED-32, pp. 1017-1023 (1989)

R. A. PUCEL, H. HAUS and H. Statz, in *Advances in Electronics and Electron Physics*, vol. 38, pp. 195-205, Academic Press, New York (1975)

J. G. RUCH, "Electronics Dynamics in Short Channel Field-Effect Transistors," *IEEE Trans. Electron Devices*, ED-19, pp. 652-654 (1972)

P. P. RUDEN, "Heterostructure FET Model Including Gate Leakage", *IEEE Trans. Electron Devices*, ED-37, No. 10, pp. 2267-2270, Oct. (1989)

P. P. RUDEN, C. J. HAN and M. SHUR, "Gate Current of Modulation Doped Field Effect Transistors", *J. Appl. Phys.*, 64, pp. 1541-1546, Aug. (1988)

P. P. RUDEN, M. SHUR, D. K. ARCH, R. R. DANIELS, D. E. GRIDER and T. NOHAVA, "Quantum Well *p*-Channel AlGaAs/InGaAs/GaAs Heterostructure Insulated Gate Field Effect Transistors", *IEEE Trans. Electron Devices*, ED-36, No. 11, pp. 2371-2379 (1989)

P. P. RUDEN, M. SHUR, A. I. AKINWANDE and P. JENKINS, "Distributive Nature of Gate Current and Negative Transconductance in Heterostructure Field Effect Transistors", *IEEE Trans. Electron Devices*, ED-36, No. 2, pp. 453-456, Feb. (1989a)

E. F. SCHUBERT, J. E. CUNNINGHAM and W. T. TSANG, "Self-Aligned Enhancement-Mode and Depletion-Mode GaAs Field-Effect Transistors Employing the δ-Doping Technique", *Appl. Phys. Lett.*, 49, 1729 (1986)

F. L. SCHUERMEYER, M. SHUR and D. GRIDER, "Gate Current in Self-Aligned *n*-channel Pseudomorphic Heterostructure Field-Effect Transistors", *IEEE Electron Device Lett.*, EDL-12, No. 10, pp. 571-573, Oct. (1991)

F. L. SCHUERMEYER, E. MARTINEZ, M. SHUR, D. E. GRIDER and J. NOHAVA, "Subthreshold and Above Threshold Gate Current in Heterostructure Insulated Gate Field-Effect Transistors", *Electronics Letters*, 28, No. 11, pp. 1024-1026 (1992)

F. S. SHONCAIR and P. K. OJALA, "High Temperature Electrical Characteristics of GaAs MESFET's (25 –400°C)", *IEEE Trans. on Electron Devices* , ED-39, No. 7, pp. 1551-1557, July (1992)

M. SHUR, "Influence of the Non-Uniform Field Distribution in the Channel on the Frequency Performance of GaAs FETs," *Electronics Lett.*, 12, pp. 615-616 (1976)

M. SHUR, "Analytical Model of GaAs MESFET", *IEEE Trans. Electron. Devices*, ED-25, No. 6, pp. 612-618, June (1978)

M. S. SHUR, "Low Field Mobility, Effective Saturation Velocity and Performance of Submicron GaAs MESFETs", *Electronics Lett.*, 18, p. 909, (1982)

M. SHUR, *GaAs Devices and Circuits*, Plenum, New York (1987)

M. SHUR, *Physics of Semiconductor Devices*, Prentice Hall, New Jersey (1990)

M. SHUR and L.F. EASTMAN, "Current-Voltage Characteristics, Small-Signal Parameters and Switching Times of GaAs FETs", *IEEE Trans. Electron Devices*, ED-25, No. 6, pp. 606-611, June (1978)

M. SHUR and L. F. EASTMAN, "Ballistic Transport in Semiconductors at Low Temperatures for Low Power High Speed Logic," *IEEE Trans. Electron Devices*, ED-26, pp. 1677-1683 (1979)

M. SHUR, D. K. ARCH, R. R. DANIELS and J. K. ABROKWAH, "New Negative Resistance Regime of Heterostructure Insulated Gate Transistor (HIGFET) Operation", *IEEE Electron Device Lett.*, EDL-7, No. 2, pp. 78-80, Feb. (1986)

M. SHUR, T. A. FJELDLY, T. YTTERDAL and K. LEE, "Unified GaAs MESFET Model for Circuit Simulations", *Intern. J of High Speed Electronics*, 3, No. 2, pp. 201-233, June (1992)

C. G. SODINI, P. K. KO and J. L. MOLL, "The Effect of High Fields on MOS Device and Circuit Performance," *IEEE Trans. Electron Devices*, ED-31, pp. 1386-1393 (1984)

H. STATZ, P. NEWMAN, I. W. SMITH, R. A. PUCEL and H. A. HAUS, *IEEE Trans. Electron Devices*, ED-34, 160 (1987)

T. STEEN, T. A. FJELDLY and M. SHUR, "Analytical Modeling of Heterostructure Field Effect Transistors", in *Proc. of the 14th Nordic Semiconductor Meeting*, Aarhus, Denmark, pp. 359-362, 17-20 June (1990)

F. STERN, "Self-Consistent Results for *n*-type Si Inversion Layers", *Phys. Rev.*, B-5, No. 12, pp. 4891-4899 (1972)

M. SZE, Ed., *High-Speed Semiconductor Devices,* John Wiley & Sons, New York (1990)

S. TIWARI, S. L. WRIGHT and J. BATEY, "Unpinned GaAs MOS Capacitors and Transistors", *IEEE Electron Device Letters*, EDL-9, No. 9, pp. 488-490, Sept. (1988)

P. W. TUINENGA, *SPICE. A Guide to Circuit Simulation & Analysis Using PSpice®*, Prentice Hall, New Jersey (1988)

K. UETO, T. FURUTSUKA, H. TOYOSHIMA, M. KANAMORI and A. HIGASHISAKA, *IEDM Tech. Digest*, p. 82 (1985)

T. VU, R. D. NELSON, G. M. LEE, P. C. T. ROBERTS, K. W. LEE, S. K. SWANSON, A. PECZALSKI, W. R. BETTEN, S. A. HANKA, M. J. HELIX, P. J. VOLD, G. Y. LEE, S. A. JAMISON, C. ARSENAULT, S. M. KARWASKI, B. A. NAUSED, B. K. GILBERT and M. SHUR, "Low Power 2K-Cell Gate Array and DCFL Circuits Using GaAs Self-aligned E/D MESFET's", *IEEE J. Solid-State Circuits*, 23, No. 1, pp. 224-238, Feb. (1988)

C. T. WANG, Editor, *Introduction to Semiconductor Technology. GaAs and Related Compounds*, J. Wiley and Sons, New York (1990)

R. A. WARRINER, *Solid State Electron Devices*, 1, p. 105 (1977)

R. E. WILLIAMS and D. W. SHAW, "Graded Channel FET's Improved Linearity and Noise Figure", *IEEE Trans. Electron Devices*, ED-25, p. 600 (1978)

K. YAMAGUCHI and H. KODERA, *IEEE Trans. Electron Devices*, ED-23, pp. 545-553 (1976)

T. YTTERDAL, K. LEE, M. SHUR and T. A. FJELDLY, "AIM-Spice, a New Circuit Simulator Based on a Charge Control Model", in *Proceeding of The First International Semiconductor Device Research Symposium*, Charlottesville, VA, pp. 481-484, Dec. (1991)

PROBLEMS

4-1-1. Consider a junction formed between an *n*-type GaAs MESFET channel doped at 2×10^{17} cm^{-3} and a semi-insulating substrate. Model the substrate as a GaAs layer doped with deep acceptors (with acceptor levels of 0.7 eV above the top of the valence band). Sketch the band diagram and comment on the acceptor population versus distance, the depletion region width, and the total charge in the depletion layer.

*4-2-1. Using the constant mobility model, calculate the MESFET and MOSFET transconductances in the saturation regime for devices with a gate length of 5 µm and compare their dependencies on the gate voltage swing. The threshold voltage is $V_T = -1$ V for both devices. Use default MESFET and MOSFET parameters specified in the AIM-Spice program in addition to those given here. How should we modify the MESFET design to approach the shape of the MOSFET transconductance versus gate voltage dependence?

4-3-1. Use the saturation velocity model to calculate the MESFET and MOSFET transconductances in the saturation regime for devices with a gate length of 1 µm and compare their dependencies on the gate voltage swing. The threshold voltage is $V_T = -1$ V for both devices. Use default MESFET and MOSFET parameters specified in the AIM-Spice program in addition to those given here.

4-4-1. Develop a unified model for a GaAs MESFET with an exponential doping profile:

$$N_d = N_{do} \exp(\frac{y}{y_o})$$

where y_o is the characteristic distance of the variation of the doping profile and y is the distance from the gate. Assume that $0 \leq y \leq d$ where d is the channel thickness.

*4-4-2. Develop a unified model of a GaAs MESFET with a p-type doping in the substrate. Using default MESFET parameters given in AIM-Spice, calculate and plot the device threshold voltage as a function of the substrate doping. Discuss advantages and disadvantages of a high substrate doping.

*4-4-5. How would you scale the MESFET channel doping and thickness with the gate length? Explain.

*4-4-6. What are possible advantages and disadvantages of a MESFET with a low doped region near the drain?

*4-4-7. A constant MESFET transconductance is very important for microwave applications since it allows one to reduce intermodulation distortion. Discuss how a MESFET doping profile can be tailored to obtain a region of the transfer characteristic with a nearly constant transconductance.
Hint: For simplicity, assume complete velocity saturation in the channel.

4-5-1. Using an approach similar to that developed for MOSFETs in Section 3.3, calculate the effective thickness, Δd, of the 2D electron gas of an AlGaAs/GaAs heterostructure for values of n_s between 10^{12} cm^{-2} and 5×10^{12} cm^{-2}. How does this value compare with the value of Δd for Si MOSFETs? What are the implications of this difference?

4-5-2. Calculate the dependence on the gate-source voltage of the 2D electron gas sheet density, n_s, for an AlGaAs/GaAs HFET
(a) using classical theory
(b) using the 2D gas theory.

Use default device parameters given in AIM-Spice (see Appendix A9). Compare the obtained dependencies with UCCM and estimate the threshold voltage, the ideality factor, and the effective gate-channel capacitance.

*4-5-3. Using the HFET model developed in Section 4.5, suggest a design of the HFET inverter shown in Fig. 4.5.1.

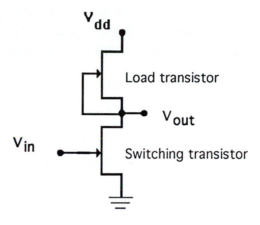

Fig.   P4.5.1

In this inverter, the threshold voltages are $V_T = 0.1$ V and $V_T = -1$ V for the switching (enhancement mode) and the load (depletion mode) transistor, respectively. Choose the thickness of the AlGaAs layer to be 300 Å. The threshold voltage for the switching transistor should be determined by the dopants in the AlGaAs layer. The more negative threshold voltage of the load transistor is determined by an additional donor implant. The density of implanted donors, N_{impl}, is given by

$$N_{impl}(y) = \frac{Q}{\sqrt{2\pi}\sigma} \exp\left[-\left(\frac{y - R_p}{\sqrt{2\pi}\sigma}\right)^2\right]$$

where y is the distance from the surface, $R_p = 700$ Å is the implant range, $\sigma = 300$ Å is the standard deviation, and Q is the implant dose. Assume that the implanted donors are not activated in AlGaAs and 100% activated in GaAs. Estimate the implant dose required.

*4-6-1. Using the AIM-Spice HFET model HFETA1, calculate the transfer curve (V_{out} versus V_{in}) of a Direct Coupled Field effect transistor Logic (DCFL) inverter (see Fig. P4.5.1) as a function of the gate length. Start from the default parameters given in the AIM-Spice program and scale the gate length down to 0.2 μm. Use the threshold voltages $V_T = 0.1$ V and $V_T = -1$ V for the switching (enhancement mode) and load (depletion mode) transistors, respectively. Scale the AlGaAs thickness and the gate width proportionally to the gate length, L. Repeat the calculation for power supply voltages V_{dd} of 3 V and 1.5 V. For each gate length, choose the ratio $(W_{dep}/L_{dep})/(W_{enh}/L_{enh})$ in such a way that the inverter logic inversion voltage is equal to $V_{dd}/2$.

4-6-2. Connect 7 inverters (described in Problem 4-6-1 using 1μm gate HFETs) into a ring oscillator as shown in Fig. P4.6.2. Calculate the output waveform for a time period of 4 ns for power supply voltages V_{dd} of 3 V, 2.5 V, 2 V, and 1.5 V. Plot the oscillation period and power dissipation as a function of V_{dd} and comment on the results.

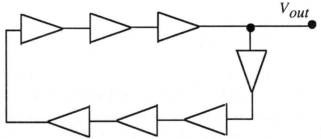

Fig. P4.6.2

Hint: You may have to set up initial values of the voltages at the inverter nodes to initiate the simulation of this circuit.

*4-6-3. Develop a unified model of an AlGaAs/GaAs HFET with a *p*-type doping in the substrate. Use default HFET parameters given in AIM-Spice to calculate and plot the device threshold voltage as a function of the substrate doping. Discuss advantages and disadvantages of a high substrate doping.

*4-6-4. Develop a model for the current-voltage characteristics of a long channel HFET in the above-threshold regime for the gate electrode shape shown in the top view in Fig. P4.6.4.

(a) Assume zero series resistance.

(b) Assume a series resistance R_{sl} per unit width.

How can the results be used for measuring the source and drain series resistances? (See . T. GLOBUS, M. SHUR, Y. BYUN and M. HACK, "New Split FET Technique for Measurements of Source Series Resistance Applied to Amorphous Silicon Thin Film Transistors", *IEEE Electron Device Letters*, EDL-13, No. 2, pp. 108-110, Feb. (1992), for comparison.)

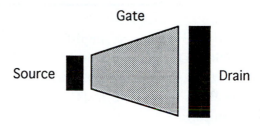

Fig. P4.6.4

4-7-1. Consider an HFET with a non-uniform doping in the wide band gap semiconductor layer. Show that the threshold voltage, V_T, for such a device can be estimated as

$$V_T = \phi_b - \frac{\Delta E_c}{q} - \frac{q}{\varepsilon_i} \int_0^d N_d(y) y \, dy$$

where $q\phi_b$ is the Schottky barrier height, ΔE_c is the conduction band discontinuity at the heterointerface, d is the thickness and ε_i is the dielectric permittivity of the wide band gap semiconductor layer. How should this equation be changed if the composition of the wide band gap material is graded?

*4-7-2. Sketch the band diagram for an Inverted HFET with the layer structure shown in Fig. P4.7.2. Choose doping levels and layer thicknesses in such a way that the device will have a threshold voltage of –0.1 V.

Hint: In case of difficulties, see K. LEE, M. SHUR, T. J. DRUMMOND and H. MORKOÇ, "Charge Control Model of Inverted GaAs-AlGaAs Modulation Doped FETs (IMODFETs)", *J. Vac. Sci. Tech.*, B2, No. 2, pp. 113-116, April-June (1984).

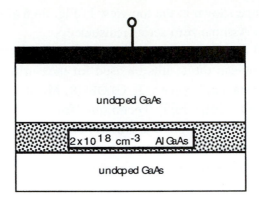

Fig. P4.7.2

*4-7-3. Fig. P4.7.3 shows a simplified layer structure for quantum well depletion (D) and enhancement (E) HFETs used in an HFET array with 6336 cells (see A. THIEDE et al., *Electronics Letters*, 28, No. 11, pp. 1005-1007, May (1992)). (Thin AlGaAs layers of 3 nm are used as etch stops.)

The gate length is 0.3 μm. The gate width is 10 μm for the E-FETs and 5 μm for D-FETs. Calculate the device current-voltage and capacitance-voltage characteristics for reasonable values of gate and drain voltages.

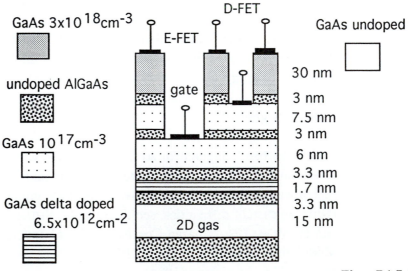

Fig. P4.7.3

4-8-1. Using the AIM-Spice MESFET model MESA1 and HFET model HFETA, calculate and compare the dependencies of the gate current on the gate-source voltage for these devices at temperatures of 100 K, 200 K, 300 K, and 400 K. Use the default parameters given in the AIM-Spice program (see Appendix A9).

*4-8-2. How will the gate leakage current in an HFET with a p^+ gate compare with that of a similar device with a Schottky gate? Explain. What would be the relationship between the threshold voltages of the two devices? How should you change the design of the device with a p^+ gate to make the threshold voltage to be equal to that of the device with a Schottky gate?

4-8-3. Sketch qualitatively the dependence of the gate current in an n-channel HFET at a fixed gate-source bias on the drain-source voltage. How will the electron heating near the drain affect this dependence?

*4-8-4. Discuss the AlGaAs/GaAs, AlGaAs/InGaAs/GaAs, and AlInAs/InGaAs/InP material systems from the point of view of possible applications in HFETs. What are their advantages and disadvantages? Which do you expect to be the fastest material system? Which one do you expect to have the smallest gate leakage current?

4-8-5. Using AIM-Spice models and default parameters, calculate inverter transfer characteristics versus V_{in} (see Fig. P4.5.1) with and without gate leakage current and discuss how the gate leakage degrades the inverter transfer characteristics. Repeat the calculation for $V_{dd} = 1.5$, 2.0, 2.5, and 3.0 V.

5

Thin Film Transistors

5-1. INTRODUCTION

Conventional electronic devices are fabricated on crystalline wafers of silicon cut from a crystal boule grown from a melt. These wafers are fragile, relatively expensive, and limited in size (currently available to approximately 8 inches in diameter). In many applications, amorphous or polycrystalline materials that may be deposited in an inexpensive, continuous process on a variety of different large-area substrates can compete with crystalline silicon or even open up completely new applications.

Spear and LeComber (1972) were the first to demonstrate that amorphous silicon films prepared by the glow discharge decomposition of silane gas (SiH_4) have a relatively low density of defect states in the energy gap. The amorphous silicon material obtained by this process is, in fact, an amorphous silicon-hydrogen alloy with a fairly large concentration of hydrogen. The hydrogen atoms tie up dangling bonds that are present in amorphous silicon in large numbers, and decrease the density of localized states in the energy gap. In amorphous silicon without hydrogen, the number of localized states is so high that the Fermi level is pinned, rendering the material unsuitable for most device applications. In amorphous Si:H, the number of localized states is reduced by many orders of magnitude. However, these states still play a dominant role in determining the transport properties of this material. The density of localized states depends not only on material preparation but also on bias stress, sample

494

illumination, and other factors. As a consequence, the behavior of a-Si:H devices is quite different from their crystalline counterparts.

Very large area (2x4 feet and larger) high quality amorphous silicon (a-Si:H) films may be inexpensively produced in a continuous process, making this material very attractive for applications in electronics and photovoltaics. Amorphous Ge:H, amorphous C:H, amorphous SiC:H, amorphous Si:F, and other amorphous materials have also been produced, opening up new opportunities for different heterostructure devices. Amorphous silicon alloy Thin Film Transistors (TFTs) have a potential to become a viable and important technology for large area, low cost integrated circuits. These circuits are currently being used to drive large area liquid crystal displays. Basic Integrated Circuits (ICs) and addressable image sensing arrays have also been implemented (see, for example, Ito et al. (1985), Thompson and Tuan (1986), Kodama et al. (1982), Hiranaka et al. (1984), and Matsumura et al. (1981)).

The main drawback of amorphous silicon and related amorphous materials is the very small mobility of electrons and holes. The effective electron mobility in a-Si thin film transistors is typically on the order of 1 cm^2/Vs. The hole mobility is even lower. In polycrystalline silicon (poly-Si) TFTs, the electron and hole mobilities are much larger (typically between 30 to 100 cm^2/Vs for electrons and 10 to 40 cm^2/Vs for holes). Integrated circuits based on poly-Si TFTs have a smaller area than a-Si ICs. However, poly-Si technology already competes with a-Si for applications in flat-panel display devices, printers, scanners, and three-dimensional Large Scale Integrated (LSI) circuits.

In this Chapter, we describe analytical models for a-Si and poly-Si TFTs implemented in AIM-Spice, and describe device characterization procedures for parameter extraction.

5-2. AMORPHOUS SILICON THIN FILM TRANSISTORS

5.2.1. Material Properties of Amorphous Silicon

Properties of a-Si:H are quite different from those of the crystalline material. a-Si:H is a direct-gap semiconductor with an energy gap close to 1.7 eV, and electron and hole band mobilities are on the order of 10 cm^2/Vs. It has a much larger absorption coefficient of light than crystalline silicon (which is especially important in photovoltaic and optoelectronic applications). As was mentioned in Section 5.1, one of the most important differences between a-Si:H and crystalline silicon is the presence of a large number of localized states in the energy gap. An approximate distribution of localized states for intrinsic (undoped) a-Si is shown in Fig. 5.2.1.

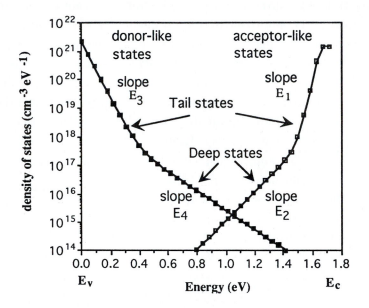

Fig. 5.2.1 Density of localized states in amorphous silicon. E1, E2, E3 and E4 are characteristic energies for the exponential variation of the density of localized states. (From Shur et al. (1989)).

The localized states in the upper half of the energy gap (closer to the bottom of the conduction band) behave as acceptor-like states, while the states in the bottom half of the energy gap behave as donor-like states. Donor-like states are positively charged when empty and neutral when filled, acceptor-like states are neutral when empty and negatively charged when filled (see Chapter1). As shown in the figure, the localized states can be divided into tail acceptor-like states, deep acceptor-like states, deep donor-like localized states, and tail donor-like states.

The distribution of the localized states is not symmetrical, with more donor-like states than acceptor-like states. Hence, the position of the Fermi level in an undoped, uniform a-Si sample in the dark, E_{Fo}, found from the neutrality condition, is shifted closer to the bottom of the conduction band, E_c. For intrinsic amorphous Si: $E_2 \approx 86\,\text{meV}$, $E_4 \approx 129\,\text{meV}$ (see Fig. 5.2.1), and $E_c - E_{Fo} \approx 620\,\text{meV}$. Dopants (typically phosphorus for n-type and boron for p-type material) can shift the position of the Fermi level and amorphous silicon p-n junctions can be made as was first shown experimentally by Spear et al. (1976). (Numerous experiments show that an additional peak in the density of the acceptor-like localized states appears in n-type doped amorphous silicon films at approximately 0.4 eV below the edge of the conduction band.) A detailed account of material properties of a-Si:H was given, for example, by Kanicki (1991).

5.2.2. Fabrication of a-Si TFTs

Hydrogenated amorphous silicon can be deposited on a variety of different substrates including glass, metal, and mylar film using a glow discharge deposition system. Intrinsic (undoped) a-Si films are deposited from silane gas (SiH_4). Adding phosphine to silane results in n-type a-Si. The films can be doped p-type by adding di-borane. A standard a-Si TFT requires only intrinsic and highly doped n-type a-Si films (for contacts). However, more advanced vertical structures (see Hack et al. (1991)) require p-type films as well.

In typical a-Si integrated circuits, TFTs (as shown in Fig. 5.2.2) are fabricated by first depositing metal gates on glass, then depositing a layer of gate dielectric (typically silicon nitride), a layer of intrinsic a-Si, and a layer of a passivating dielectric (silicon nitride). The gate insulator thickness is usually on the order of 3000 Å, even though thinner layers are possible, especially if the dielectric breakdown voltage is improved. The intrinsic amorphous silicon layer is quite thin (about 500 Å) in order to minimize the *off*-current. The passivating dielectric thickness is of the order of 1000 Å. After the deposition of these layers and spin coating with a photoresist, one shines light from the bottom of the structure (through the glass) in order to expose the photoresist. This allows one to make a passivating dielectric self-aligned with the gate metal. The next fabrication step is the deposition of a thin (≈ 100 Å) layer of highly doped n^+ a-Si required for source and drain contacts. Then the source and drain contact metal is deposited and etched using the passivating dielectric as an etch stop. This relatively basic fabrication procedure does not look so simple when we realize that for display applications, for example, we have to produce a-Si integrated circuits with sizes as large as several square feet, which puts a great demand on the fabrication equipment, even for relatively large feature sizes. Typically, the length of the deposited gate metal strip is about 10 to 15 µm. The overlaps of the source and drain contact metal over the passivating dielectric are on the order of 2 µm.

One consequence of such a design is that, at high drain voltages, the drain overlap acts as a second gate which induces electrons at the interface between the passivating dielectric and the intrinsic a-Si layer under the drain metal overlap. Hence, the effective gate length may be drain voltage dependent. However, we should note that the n^+ contact regions are separated from the inversion channel at the a-Si-gate dielectric interface by the layer of intrinsic a-Si (see Fig. 5.2.2). Therefore, we rely on charge injection through the intrinsic layer. This means that the contact resistance is non-linear and should decrease with increasing device current (see, for example, Globus et al. (1992)).

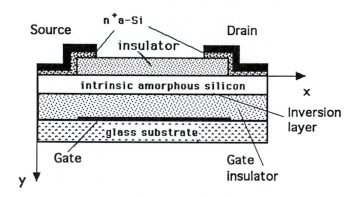

Fig. 5.2.2. Schematic diagram of an amorphous silicon Thin Film Transistor.

5.2.3. Regimes of Operation of a-Si TFTs

The position of the Fermi level with respect to the conduction band edge near the amorphous silicon-insulator interface may be changed by inducing carriers via field effect, similar to the field effect at crystalline silicon - insulator interfaces. This effect is utilized in amorphous silicon TFTs (see Fig. 5.2.2).

The low field-effect mobility in a-Si TFTs (typically 0.5 to 1 cm²/Vs) has long been an obstacle to many practical applications of this technology. However, theoretical and experimental results show that a value for the field-effect mobility close to the electron band mobility (presently estimated at 10 cm²/Vs or higher) may be achieved when the electronic charge induced in the channel is sufficiently large to fill the localized states (see Shur et al. (1985) and Leroux and Chenevas-Paule (1985)). Therefore, the operation of a-Si TFTs differs from that of crystalline MOSFETs.

In addition to the so-called "below-threshold" and "above-threshold" regimes, there are two new regimes of operation (see Shur et al. (1989)). These occur at high densities of induced charge in the a-Si TFT channel – a crystalline-like regime when the free electron concentration exceeds the localized charge concentration at the a-Si - insulator interface, and a transitional regime (between the crystalline-like and above-threshold regimes) at lower densities of induced charge when almost all localized states in the energy gap of the amorphous silicon near the interface are filled.

In the below-threshold regime, nearly all induced charge goes into the deep localized acceptor-like states in the energy gap of a-Si as well as into the surface states at the a-Si - insulator interface. With an increase in the gate-source voltage, more states are filled and the Fermi level at the a-Si-insulator interface moves closer to the conduction band. This leads to an increase in the concentration of mobile carriers in the conduction band which rises superlinearly with gate bias. There is an important difference between this regime and the subthreshold regime in crystalline MOSFETs. In a-Si TFTs, the carriers in the subthreshold regime are still induced into the channel. The very small value of the current is caused by the carriers going almost exclusively into localized states rather than into the conduction band, i.e., it takes place when the gate bias is still greater than the flat band voltage. In crystalline MOSFETs, the drain current in the subthreshold regime is small because of the large barrier between the source and drain, as was explained in detail in Section 3.6.

With a further increase in the gate-source voltage, the Fermi level enters the tail states (see Fig. 5.2.1). The characteristic energy of the exponential variation of the density of tail states, E_1, is smaller or comparable to $k_B T$ at room temperature. As a consequence, once the Fermi level is in the tail states, most of the charge is actually induced into the states above the Fermi level. Hence, the shift of the Fermi level with gate-source voltage is considerably smaller than in the below-threshold regime. This is the "above-threshold regime" (see Shur and Hack (1984)). There are two important differences between this regime in a-Si TFTs and the above-threshold regime in crystalline MOSFETs. In a-Si TFTs, most of the induced charge still goes into the tail states with only a small fraction (a few percent) going into the conduction band. At the same time, the Fermi level moves closer to the conduction band edge with increasing gate-source voltage, resulting in an increase in the field effect mobility.

When the induced charge is increased even further, we may reach a situation when the tail states at the a-Si-insulator interface are almost completely filled and the Fermi level touches the bottom of the conduction band. Any additional increase in the induced charge is divided between the charge going into the conduction band and the charge induced into the tail states farther from the a-Si-insulator interface. At first, the fraction of the mobile charge is small, but it increases with increasing gate-source voltage. We call this the "transitional regime" because it corresponds to the transition between the above-threshold regime and the crystalline-like regime. In the latter, the Fermi level at the a-Si-insulator interface has moved high enough into the conduction band so that, finally, most of the induced charge goes into the conduction band. In this case, the field-effect mobility is close to the band mobility and the operation of the a-Si TFT is truly similar to the operation of a crystalline field-effect transistor. We note that the gate-source voltage necessary to achieve the crystalline-like regime is about 50 to 100 V for a typical a-Si TFT (with a SiO_2 gate insulator thickness of approximately 1000 Å). This is too large a voltage for practical applications. However, as the material and insulator properties improve, this regime may be achieved at smaller gate voltages. (One possible solution is to use a gate insulator with a higher dielectric constant.)

5.2.4. Densities of Free and Trapped Charges. Field Effect Mobility

In amorphous silicon TFTs, the field effect mobility, μ_{FET}, is much smaller than the band mobility, μ, and it depends on the induced charge. This dependence can be evaluated by solving Poisson's equation in the direction perpendicular to the gate and channel assuming zero drain-source voltage, i.e., $V_{DS} = 0$ V. The distribution of the electric field perpendicular the TFT channel, F_y, is found from:

$$\frac{dF_y}{dy} = \frac{\rho}{\varepsilon_s} \qquad (5\text{-}2\text{-}1)$$

where y is the space coordinate (perpendicular to the gate), ε_s is the dielectric permittivity of amorphous silicon, and ρ is the space charge density:

$$\rho = -q\left(N_{loc} + n\right) \qquad (5\text{-}2\text{-}2)$$

Here, N_{loc} is the concentration of localized charge, and n is the concentration of electrons in the conduction band given by

$$n = N_c F_{1/2}(\xi) \qquad (5\text{-}2\text{-}3)$$

where $F_{1/2}$ is the Fermi-Dirac integral (see Chapter 1), N_c is the effective density of extended states in the conduction band, $\xi = (-\Delta E_{Fo} + qV)/k_B T$, and

$$\Delta E_{Fo} = E_c(bulk) - E_F \qquad (5\text{-}2\text{-}4)$$

where E_F is the equilibrium position of the Fermi level, and V is the electric potential. We choose the potential $V = 0$ V to coincide with the bottom of the conduction band far from the gate where a-Si is assumed to be in uniform equilibrium.

The density of localized electrons for n-channel devices (where the electron quasi-Fermi level in a-Si lies in the upper half of the energy gap) may be found as

$$N_{loc} = \int_{E_{Fo}}^{E_c} \frac{g_A(E)\,dE}{1 + \exp\left(\dfrac{E - E_F}{k_B T}\right)} \qquad (5\text{-}2\text{-}5)$$

where E_F is the Fermi energy and g_A is the density of localized states which may be approximated by a sum of two exponential functions representing the deep states and the tail states, respectively, (see Fig. 5.2.1):

$$g_A(E) = g_{Fo} \exp\left(\frac{E - E_{Fo}}{E_2}\right) + g_{ct} \exp\left(\frac{E_c - E}{E_1}\right) \qquad (5\text{-}2\text{-}6)$$

Here, E_{Fo} is the Fermi level in amorphous silicon at equilibrium, E_c is the bottom of the conduction band, g_{Fo} is the density of the deep localized acceptor like states at $E = E_{Fo}$, $E_2 = k_B T_2$ is the characteristic energy of the exponential variation in the density of the deep localized states, $E_1 = k_B T_1$ is the characteristic energy of the exponential variation in the density of the tail states, and g_{ct} is the density of the tail states at $E = E_c$.

Using the relationship between electric field and electric potential

$$F_y = - \, dV/dy \tag{5-2-7}$$

we find from Poisson's equation:

$$F_y^2 = - \frac{2}{\varepsilon_s} \int_0^V \rho(V')dV' \tag{5-2-8}$$

The total induced charge, $q_{ind} = qn_{ind}$, the localized charge, q_{loc}, and the free surface electron charge, qn_s, per unit area are given by

$$q_{ind} = \sqrt{2\varepsilon_s \int_0^{V_o} \rho(V')dV'} \tag{5-2-9}$$

$$q_{loc} = \int_0^{V_o} \frac{q \, N_{loc}(V')}{F_y(V')} dV' \tag{5-2-10}$$

$$n_s = \int_0^{V_o} \frac{n(V')}{F_y(V')} dV' \tag{5-2-11}$$

where V_o is equal to the surface potential of the a-Si-insulator interface (i.e., the total band bending). The potential distribution in the direction perpendicular to the gate is found from

$$y = - \int_{V_o}^V \frac{dV'}{F_y(V')} \tag{5-2-12}$$

Finally, the gate-source voltage, V_{GS}, is related to the total induced sheet charge density, q_{ind}, as follows

$$V_{GS} \approx \frac{q_{ind}}{c_i} \tag{5-2-13}$$

where

$$c_i \approx \frac{\varepsilon_i}{d} \tag{5-2-14}$$

is the insulator capacitance per unit area, ε_i is the insulator permittivity, and d is the insulator thickness.

Numerically calculated densities of free and trapped carriers, n and N_{loc}, in a-Si are shown in Fig. 5.2.3 as functions of the position of the Fermi level. This figure allows us to distinguish between the different regimes of a-Si TFT operation. When $E_{Fo} - E_v$ is less than approximately 1.4 eV (E_v corresponds to the top of the valence band), the device is in the below-threshold regime. In this regime, the free carrier density is very small and increases rapidly with increasing gate-source voltage which shifts the position of the Fermi level. When $E_{Fo} - E_v$ is less than approximately 1.7 eV and larger than 1.4 eV, the device is in the above-threshold regime where the free carrier concentration is a sizable fraction of the total induced charge. Finally, at values of $E_{Fo} - E_v$ greater than about 1.7 eV, $q n_s$ is actually greater than q_{loc}, and the device should operate in a crystalline-like regime.

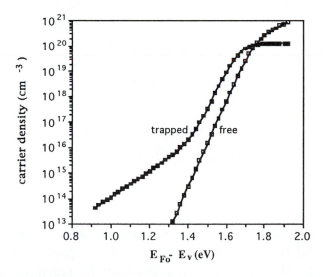

Fig. 5.2.3. Numerically calculated densities of free and trapped charges in a-Si (after Shur et al. (1989)).

The sheet density of free electrons, n_s, at the a-Si-insulator interface as a function of the gate-source voltage is shown in Fig. 5.2.4.

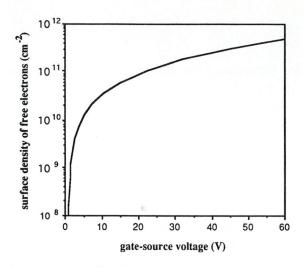

Fig. 5.2.4. Surface density of free electrons at a-Si-insulator interface as a function of gate-source voltage (after Shur et al. (1989)).

This relationship allows us to evaluate the field effect mobility as $\mu_{FET} \approx \mu n_s/n_{ind}$, i.e., as a function of the sheet density of electrons, n_{ind}, induced into the TFT channel. Shur et al. (1989) found that the field-effect mobility as a function of n_{ind} obtained from numerical calculations can be interpolated quite well by a fifth order polynomial. Here, we use a simpler model where this dependence is approximated by

$$\mu_{FET} = \mu_o \frac{m+2}{2}\left(\frac{n_{ind}}{n_o}\right)^m \qquad (5\text{-}2\text{-}15)$$

where $n_o = 10^{12}$ cm^{-2} is a convenient scaling factor, and μ_o and m are constants. This approximation is somewhat less accurate because it contains only two independent parameters (compared to six parameters in the fifth order polynomial). However, we can argue that the functional dependence of the localized states in a-Si on energy, $g_A(E)$, is not known with a great degree of accuracy and, hence, a precise agreement of the interpolation formula with the numerical calculation based on $g_A(E)$ is not overly important. At the same time, eq. (5-2-15) gives a reasonably accurate fit to the numerical calculation over a limited but important range of induced carrier concentrations (see Fig. 5.2.5).

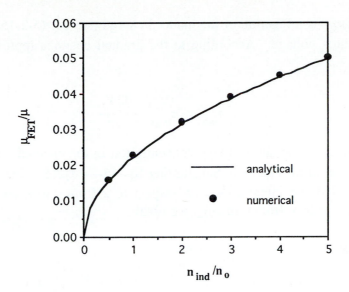

Fig. 5.2.5. Comparison of interpolated and computed dependencies of field effect mobility on the sheet density of electrons induced into the channel. Here, $m \approx 0.5$, $\mu_o \approx 0.018\,\mu$ where μ is the band mobility.

5.2.5. Current-Voltage Characteristics (AIM-Spice Model ASIA1)

The AIM-Spice model of a-Si TFTs is based on the theory of a-Si TFTs developed by Shur et al. (1989). We start from the equation relating the drain current, I_d, to the longitudinal electric field in the channel, F_x, and the concentration per unit area of free conduction band electrons in the channel, n_s,

$$I_d = -\,q\,\mu n_s\,F_x W \tag{5-2-16}$$

where μ is the band mobility, W is the gate width

$$F_x = -\frac{dV}{dx} \tag{5-2-17}$$

and V is the channel voltage.

As mentioned above, n_s can be related to the sheet density of induced electrons, n_{ind}, and the field-effect mobility, μ_{FET}, by

$$n_s = \frac{\mu_{FET}}{\mu}\,n_{ind} \tag{5-2-18}$$

Here, μ_{FET} is a function of the induced charge (see eq. (5-2-15)) and, hence, the gate-source voltage. According to the gradual channel approximation, we can write

$$n_{ind} = n_{inds} - \frac{\varepsilon_i V}{qd}$$ (5-2-19)

where n_{inds} is the sheet charge concentration in the channel per unit area at the source side of the channel. Substituting eqs. (5-2-17), (5-2-18), and (5-2-19) into eq. (5-2-16) and integrating with respect to x from 0 to L (see Fig. 5.2.2) and with respect to V from 0 to V_{DS}, we obtain

$$I_d = q\frac{W}{L} \int_0^{V_{DS}} \mu_{FET}(n_{ind})\, n_{ind}\, dV$$ (5-2-20)

which can be re-written as

$$I_d = q^2 \frac{W\,d}{L\,\varepsilon_i} \int_{n_{indd}}^{n_{inds}} \mu_{FET}(n_{ind})\, n_{ind}\, dn_{ind}$$ (5-2-21)

where the sheet density of induced electrons at the drain is given by

$$n_{indd} = n_{inds} - \frac{\varepsilon_i V_{DS}}{qd}$$ (5-2-22)

In a conventional crystalline MOSFET,

$$n_{inds} = \frac{\varepsilon_i(V_{GS} - V_T)}{qd}$$ (5-2-23)

and

$$\mu_{FET} = \mu$$ (5-2-24)

where V_T is the threshold voltage. Substitution of eqs. (5-2-23) and (5-2-24) into eq. (5-2-21) yields the standard equation for a crystalline long channel Metal-Oxide-Field Effect Transistor (MOSFET) in the linear regime. However, in the case of an a-Si TFT, n_s is not a simple linear function of the gate-source voltage. In fact, as was discussed in Subsection 5.2.3, this dependence has several different regions (corresponding to the below-threshold, above-threshold, and crystalline-like regimes of operation) which, in the relevant range of the induced charge, can be approximated by a power function (see Fig. 5.2.5 and eq. (5-2-15)).

In the gradual channel approximation, the drain saturation voltage, V_{SAT}, is reached when the value of n_{indd} is equal to zero (the pinch-off condition at the drain). Hence,

$$V_{SAT} = \frac{q n_{inds} d}{\varepsilon_i}$$

(5-2-25)

Eq. (5-2-25) reduces to the standard expression

$$V_{SAT} = V_{GS} - V_T$$

(5-2-26)

Using the pinch-off condition $(n_{indd} = 0$ at $x = L)$ in conjunction with eq. (5-2-21), we obtain the expression for the drain saturation current for an a-Si TFT

$$I_{sat} = q^2 \frac{W d}{L \varepsilon_i} \int_0^{n_{inds}} \mu_{FET} (n_{ind}) n_{ind} \, dn_{ind}$$

(5-2-27)

At small drain-source voltages, $n_{inds} - n_{indd} \ll n_{inds}$ and eq. (5-2-21) reduces to

$$I_d = q \mu_{FET} (n_{inds}) n_{inds} \frac{W}{L} V_{DS}$$

(5-2-28)

Hence, the intrinsic channel conductance, $g_{chi} = \partial I_d / \partial V_{DS}$, at small V_{DS} is given by

$$g_{chi} = q \mu_{FET} (n_{inds}) n_{inds} \frac{W}{L}$$

(5-2-29)

For a crystalline transistor, eqs. (5-2-27) and (5-2-29) reduce to the conventional formulas (see Section 3.4)

$$I_{sat} = \frac{1}{2} \frac{W}{L} \mu_n \frac{\varepsilon_i}{d} (V_{GS} - V_T)^2$$

(5-2-30)

$$g_{chi} = \frac{W}{L}\mu_n\frac{\varepsilon_i}{d}(V_{GS} - V_T)$$ (5-2-31)

Using eq. (5-2-15), we can easily evaluate the integral in eq. (5-2-21)

$$I_d = q^2\frac{Wd\mu_o}{L\varepsilon_i\,n_o^m}\left(n_{inds}^{m+2} - n_{indd}^{m+2}\right)$$ (5-2-32)

The main advantage of this approach is that we are dealing with a very limited number of parameters that can be determined directly from experimental data. Indeed, for the saturation region of the current-voltage characteristic, eq. (5-2-32) can be simplified to

$$I_{sat} = q^2\frac{Wd\mu_o}{L\varepsilon_i\,n_o^m}n_{inds}^{m+2}$$ (5-2-33)

Above threshold, n_{inds} is given by eq. (5-2-23) and, hence, eq. (5-2-33) becomes

$$I_{sat} = \frac{W\mu_o\varepsilon_i^{m+1}(V_{GS} - V_T)^{m+2}}{L(qn_o)^m\,d^{m+1}}$$ (5-2-34)

At drain-source voltages smaller than the saturation voltage, the drain current can be found from eq. (5-2-32), using (5-2-23) for n_{inds} and (5-2-22) for n_{indd}.

The model discussed above is applicable only to the above-threshold regime (i.e., for $V_{GS} > V_T$). In circuit simulation, we require models which cover the entire range of bias voltages. For crystalline MOSFETs, this requirement is met by the Unified Charge Control Model (UCCM) and by the Universal MOSFET Model (see Chapter 3). However, in a-Si TFTs, the device operation at very low current levels (typically below 0.1 µA for a 30 µm wide device) is very sensitive to temperature and the bias voltage history because the current at this level is strongly affected by the occupation of deep traps. Therefore, we will follow the approach proposed by Shur et al. (1989) who used the following empirical expressions to interpolate the surface carrier concentrations into the below-threshold regime:

$$n_{inds} = \frac{\varepsilon_i E_2}{q}f\left[\frac{q(V_G - V_S - V_T)}{E_2}\right]$$ (5-2-35)

$$n_{indd} = \frac{\varepsilon_i E_2}{q} f\left[\frac{q(V_G - V_D - V_T)}{E_2}\right] \qquad (5\text{-}2\text{-}36)$$

where V_S and V_D are the (intrinsic) voltages at the source and drain contacts, respectively, and the function $f(x)$ interpolates the induced carrier density between the two regimes, playing a role similar to eq. (3-9-17) for crystalline MOSFETs (see Section 3.9))

$$f(x) = \frac{x}{2} + \frac{1}{\pi}\tan^{-1}\left(\frac{x}{x_o}\right) + \frac{x_o}{\pi} \qquad (5\text{-}2\text{-}37)$$

Here, x_o is a characteristic dimensionless parameter determining the shape of the transition between the above-threshold and below-threshold regimes.

This model can be further improved by accounting for space-charge injection effects that are important in the saturation regime. As was shown by Shur et al. (1989), based on two-dimensional simulations and analytical calculations, an a-Si TFT in the saturation regime behaves as a series combination of an "intrinsic TFT" and an n-i-n diode formed between the drain contact and the channel at a certain distance, ΔL, from the drain contact. To first order, the output conductance in the saturation regime is given by the expression

$$g_{sat} \approx K\frac{I_{sat}}{(L - \Delta L)^2} \qquad (5\text{-}2\text{-}38)$$

where K is a constant.

Parasitic series source and drain resistances may have a considerable effect on the current-voltage characteristics of a-Si TFTs. As was discussed in Section 3.4, these resistances may be accounted for by using the following expressions relating the "intrinsic" gate-source and drain-source voltages, V_{GS} and V_{DS}, considered so far, to the "extrinsic" (measured) gate-source and drain-source voltages, V_{gs} and V_{ds}, that include voltage drops across series resistances:

$$V_{GS} = V_{gs} - I_d R_s \qquad (5\text{-}2\text{-}39)$$

$$V_{DS} = V_{ds} - I_d(R_s + R_d) \qquad (5\text{-}3\text{-}40)$$

However, we should emphasize that the source and drain contacts to the a-Si channel are usually far from ideal ohmic contacts. These contacts behave more like very leaky Schottky diodes (see Globus et al. (1992) for more details). Nevertheless, the theory reviewed above can give a fairly good fit to measured current-voltage characteristics even when constant drain and source parasitic resistances are assumed, as shown in Fig. 5.2.6 where we show typical measured and calculated current-voltage characteristics of an a-Si TFT.

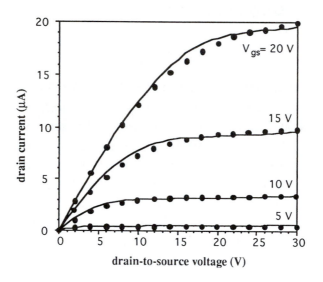

Fig. 5.2.6 Current-voltage characteristics of an a-Si TFT (from Shur et al. (1989)). Symbols: experimental data, solid lines: calculated curves.

Based on the theory reviewed above, we can now propose a semi-empirical extrinsic circuit model for a-Si TFTs similar to the Universal MOSFET Model developed in Section 3.9. The equations for this model are as follows:

$$I_{ds} = \frac{g_{ch} V_{ds} \left(1 + \lambda V_{ds}\right)}{\left[1 + \left(V_{ds} / V_{sate}\right)^m\right]^{1/m}} \qquad (5\text{-}2\text{-}41)$$

$$V_{sate} \approx I_{sat} / g_{ch} \qquad (5\text{-}2\text{-}42)$$

$$g_{ch} = \frac{g_{chi}}{1 + g_{chi}\left(R_s + R_d\right)} \qquad (5\text{-}2\text{-}43)$$

where I_{sat} and g_{chi} are defined above, V_{sate} is an effective, extrinsic saturation voltage, and the parameter λ is related to physical effects that are not explicitly included in our model (see Section 3.9). A complete summary of the equations of this model is given in Appendix A8.

5.2.6. Capacitance-Voltage Characteristics

The capacitance-voltage characteristics can be obtained using the functional form similar to Meyer's model (see Section 3.9):

$$C_{gs} = C_f + \frac{2}{3} C_{gc} \left[1 - \left(\frac{V_{sate} - V_{dse}}{2V_{sate} - V_{dse}} \right)^2 \right] \tag{5-2-44}$$

$$C_{gd} = C_f + \frac{2}{3} C_{gc} \left[1 - \left(\frac{V_{sate}}{2V_{sate} - V_{dse}} \right)^2 \right] \tag{5-2-45}$$

Here, V_{dse} is equal to V_{ds} for $V_{ds} < V_{sat}$ and is equal to V_{sate} for $V_{ds} > V_{sate}$. In order to obtain a smooth transition between the two regimes, we interpolate V_{dse} by the following equation:

$$V_{dse} = V_{ds} \left[1 + \left(\frac{V_{ds}}{V_{sate}} \right)^{m_c} \right]^{-1/m_c} \tag{5-2-46}$$

where m_c is a constant determining the width of the transition region between the linear and the saturation regime. The capacitance C_f includes both overlap and fringing capacitance components. The gate-channel capacitance, C_{gc}, is found as the capacitance of the series combination of the above-threshold capacitance, C_a, and below threshold capacitance, C_b:

$$C_{gc} = \frac{C_a C_b}{C_a + C_b} \tag{5-2-47}$$

where

$$C_a = \frac{\varepsilon_i S}{d} \tag{5-2-48}$$

and

$$C_b = \frac{C_a}{2} \exp\left(\frac{V_{gs} - V_T}{\eta E_2} \right) \qquad (5\text{-}2\text{-}49)$$

Here, η is a factor accounting for the voltage division between the semiconductor and the gate dielectric (similar to the ideality factor in crystalline FETs). Eq. (5-2-49) is obtained by assuming that the gate-channel capacitance at threshold is equal to 1/3 of the maximum gate-channel capacitance (the same condition as for MOSFETs) and assuming the shift in the position of the Fermi level in the subthreshold regime is governed by the change in the occupation of the deep acceptor-like localized states.

5-3. POLYSILICON THIN FILM TRANSISTORS

5.3.1. Fabrication of Polysilicon TFTs

Polysilicon Thin Film Transistors (poly-Si TFTs) are fabricated using crystallized amorphous silicon films. The crystallization process involves considerably higher temperatures than the deposition of amorphous silicon (a-Si is typically deposited at about 350 °C). Therefore, poly-Si TFTs earlier required expensive quartz substrates in contrast to a-Si TFTs which are fabricated on glass substrates. More recently, however, low temperature processes have been developed for poly-Si TFTs involving temperatures not higher than about 600 °C (see Thompson et al. (1990)) so that devices can be fabricated on inexpensive glass substrates.

Typically, the substrates are coated with a silicon dioxide film grown by Low Pressure Chemical Vapor Deposition (LPCVD), see Fig. 5.3.1. An amorphous silicon film, typically 1000 Å thick, is also deposited by LPCVD. This film is then crystallized, resulting in a polycrystalline film with many grains. The grain boundaries contain traps and impede the electron and hole transport. Depending on the grain size (the larger the better), the room temperature electron mobility, for example, typically varies between 30 cm^2/Vs and several hundred cm^2/Vs. The grain size increases with smaller initial nucleation rate for the grains. Nakamura et al. (1990) pointed out that this rate decreases with decreasing deposition temperature of a-Si. Hence, the deposition temperature should be low.

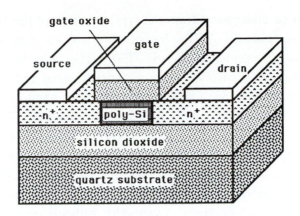

Fig. 5.3.1. Schematic diagram of poly-Si TFT.

Typically, silane (SiH_4) is used as a gas source for the a-Si deposition. However, Nakamura et al. (1990) used Si_2H_6 and were able to reduce the deposition temperature down to 445 °C. (They obtained grain sizes of about 5 μm with field effect mobilities of up to 130 cm^2/Vs for *n*-channel devices and 70 cm^2/Vs for *p*-channel devices.) The grain size strongly depended on the deposition temperature and on the annealing time. These results show that this low temperature deposition process can produce excellent devices.

An example of a relatively high temperature process, suitable for quartz substrates, is described by Lewis et al. (1990). They used a deposition (and crystallization) temperature of 600 °C, but utilized a deep silicon implant (which inhibited the grain nucleation rate) and a high temperature anneal (at 950 °C) in order to obtain a large grain size (about 2μm). In devices fabricated without the silicon ion implantation and without a high temperature anneal, the grain size was only 0.2 μm. This decrease in the grain size is commensurate with a decrease in the field effect mobility (from 95 cm^2/Vs to 50 cm^2/Vs for *n*-channel devices and from 40 cm^2/V to 15 cm^2/V for *p*-channel devices).

The next step involves the LPCVD deposition of the gate oxide with a typical thickness of 1000 Å. Highly doped polysilicon is used as the gate material. The source and drain regions are formed by ion implantation of phosphorus or boron (for *n*-channel and *p*-channel devices, respectively). Following the ion implantation and metallization for the contacts, the TFTs are exposed to a hydrogen plasma in order to passivate the grain boundaries (see Lewis et al. (1990)).

As can be understood from the typical values of the field effect mobilities given above, poly-Si TFTs have superior current drives and device transconductances compared to a-Si TFTs. The latter have typical field effect mobilities of only 1 cm^2/Vs for n-channel devices and are not used at all as p-channel devices (see Section 5.2). On the other hand, a-Si TFT technology is more mature than polysilicon TFT technology and a-Si TFT Integrated Circuits (ICs) can be fabricated with much larger areas. Also, a-Si is a very photosensitive material which leads to numerous applications in printers, copiers, imagers, and consumer electronics (see, for example, Brody (1984), Chuang et al. (1990), and Lewis et al. (1988)). Hence, both technologies will co-exist, dividing numerous applications. Eventually, polysilicon technology may gain an edge in high resolution display drivers. Both technologies have potential applications in three-dimensional integrated circuits (see, for example, Malhi et al. (1985)).

5.3.2. Model for Polysilicon TFTs

Our model for polysilicon transistors is based on the "effective medium" approach which treats the non-uniform polycrystalline sample with grain boundaries as some uniform effective medium with an effective carrier mobility and an average density of states in the energy gap (see Byun et al. (1992)). Faughan (1987) proposed such a model for poly-Si TFTs. Hack and Shaw (1990) and Hack et al. (1990) developed this model further based on two-dimensional computer modeling of both n-channel and p-channel devices. Their calculations yielded values of effective material parameters such as hole and electron mobilities, effective densities of donor-like and acceptor-like states, etc. This approach allows us to develop a simple analytical model of poly-Si TFTs which should be applicable to a whole variety of devices with mobilities ranging from 10 to 400 cm^2/Vs. We should note that such an approach has its advantages and its limitations. The advantages are the simplicity of the model and the ease of its implementation in circuit simulators and in the extraction of device parameters. The limitation of our model is its inability to relate the model parameters, such as the effective mobility and subthreshold current slope, to the material parameters such as grain size and carrier-trap-state density at grain boundaries. This problem has to be addressed separately using more sophisticated approaches which are, in general, not suitable for device design, characterization, and circuit simulation (see, for example, Guerrieri et al. (1986), Serikawa et al. (1987),

Anwar and Khondker (1987), Lin et al. (1990), Hayama and Milne (1990), and Chen and Lou (1990)). In fact, in the effective medium approach, the poly-Si TFT model is equivalent to a conventional FET model, but with the additional simplification that velocity saturation effects are usually not important because of the large gate lengths and the relatively low mobility values.

The analytical model described below is based not on the fundamental semiconductor equations but rather on empirical relations for the current-voltage characteristics which are chosen in such a way that a) the results are in good agreement with more rigorous numerical calculations, and b) they yield well established expressions for the current-voltage characteristics in limiting cases, i.e., for long-channel devices in the strong inversion regime (Shockley model) and for the subthreshold current in long-channel devices well below threshold. This approach is similar to that of our universal semi-empirical model used in Chapters 3 and 4 for Si MOSFETs, GaAs MESFETs, and HFETs.

The poly-Si TFT model is based on the Unified Charge Control Model (UCCM, see Section 3.3) and on an empirical equation proposed for long-channel HFETs (see Byun et al. (1992)).

The UCCM expression can be written as follows:

$$V_{GS} - V_T = V_{sth}\ln\left(\frac{n_{ss}}{n_o}\right) + a(n_{ss} - n_o) \qquad (5\text{-}3\text{-}1)$$

where V_{GS} is the intrinsic gate-source voltage, V_T is the threshold voltage, n_{ss} is the sheet carrier density at the source side of the channel, V_{sth} is the inverse slope of the logarithm of the subthreshold current variation with gate voltage (in conventional MOS, $V_{sth} = \eta V_{th}$, where η is the ideality factor and V_{th} is the thermal voltage, see Chapter 3), n_o is the sheet carrier density at threshold ($\approx V_{sth}/2a$), and $a \approx qd/\varepsilon_i$, where d and ε_i are the thickness and the dielectric permeability of the SiO_2 layer, respectively. Eq. (5-3-1) can be solved with respect to n_{ss} by using the approximate, analytical solution of the non-ideal diode equation discussed in Appendix A7 (see also Fjeldly et al. (1991)).

The empirical expression proposed for describing the current-voltage characteristics below saturation is

$$I_d = \frac{q\mu W}{L}\left[n_{ss}V_{DS} - \frac{V_{DS}^2}{2a\left(1 + \frac{2V_{sth}}{an_{ss}}\right)}\right] \qquad (5\text{-}3\text{-}2)$$

where V_{DS} is the intrinsic drain-source voltage, μ is the effective low-field carrier mobility, L is the gate length, and W is the gate width. This analytical model can be justified by comparing with the more precise long-channel MOSFET model discussed in Section 3.11 (see eq. (3-11-7)). Byun et al. (1992) demonstrated an excellent agreement between the two models.

Eq. (5-3-2) reduces to well known analytical equations in the limiting cases. In the strong inversion regime $(2V_{sth}/an_{ss} \ll 1)$, eq. (5-3-2) reduces to the conventional Shockley model (see Section 3.4). For the subthreshold regime $(2V_{sth}/an_{ss} \gg 1)$ and very small drain-source bias $(V_{sth} \gg V_{DS})$, eq. (5-3-2) reduces to

$$I_d = \frac{q\mu W}{L} n_{ss} V_{DS} \tag{5-3-3}$$

as expected for a long channel device.

Under ideal conditions (which are not really satisfied in a poly-Si TFT), the saturation voltage can be found from the condition

$$\frac{\partial I_d}{\partial V_{DS}} = 0 \tag{5-3-4}$$

which, in combination with eq. (5-3-2), leads to the following expression for the saturation voltage:

$$V_{SAT} = an_{ss}\left(1 + \frac{2V_{sth}}{an_{ss}}\right) \tag{5-3-5}$$

In the strong inversion regime, eq. (5-3-5) reduces to

$$V_{SAT} = an_{ss} \tag{5-3-6}$$

and the corresponding saturation current becomes

$$I_{sat} = \frac{q\mu W an_{ss}^2}{2L} \tag{5-3-7}$$

as predicted by the Shockley model.

In the subthreshold regime, we obtain

$$V_{SAT} = 2V_{sth} \tag{5-3-8}$$

$$I_{sat} = \frac{q\mu W n_{ss}}{L} V_{sth} \qquad (5\text{-}3\text{-}9)$$

(compare with the results in Section 3.6). This shows that our model describes the limiting cases corresponding to the entire range of gate and drain voltages from below to above threshold and from the linear to the saturation regime.

The condition $\partial I_{ds}/\partial V_{DS} = 0$ gives only an approximate value of V_{SAT} because the current keeps rising in the saturation region with increasing V_{DS}. The finite output conductance is partially caused by channel length modulation and by space charge injection from the channel into the drain contact (see Shur (1990)). Additionally, the saturation current increases sharply at high drain bias due to the so-called "kink" effect which may be related to impact ionization (see Hack and Shaw (1990)).

Byun et al. (1992) proposed a new continuous model for the differential output resistance, R_{DS}. As shown in Fig. 5.3.2, there are three distinct regimes in the R_{DS} versus V_{DS} dependence:

1. The linear regime $(0 \leq V_{DS} \leq V_{SAT})$, where the device characteristics are not saturated and are given by eq. (5-3-2).

2. The saturation regime $(V_{SAT} \leq V_{DS} \leq V_{Kink})$, where the current-voltage characteristics gradually saturate and the output resistance rises with increasing V_{DS}. Here, V_{Kink} is the critical voltage for the kink effect.

3. The kink regime $(V_{DS} > V_{Kink})$, where the output resistance drops dramatically due to the kink effect.

The equation for the output resistance in the linear regime is obtained by differentiating eq. (5-3-2) with respect to V_{DS}:

$$R_{DS} = \frac{\partial V_{DS}}{\partial I_d} = \frac{1}{\beta \left(n_{ss} - V_{DS}/a^* \right)} \qquad (5\text{-}3\text{-}10)$$

where

$$\beta = \frac{qW\mu}{L} \qquad (5\text{-}3\text{-}11)$$

and

$$a^* = a\left(1 + \frac{2V_{sth}}{an_{ss}} \right) \qquad (5\text{-}3\text{-}12)$$

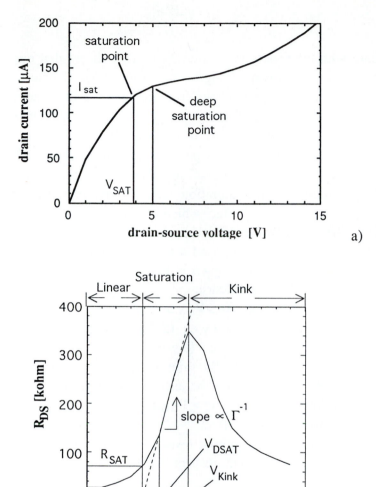

Fig. 5.3.2. Measured *I-V* characteristics (a) and drain output resistance (b) for an *n*-channel polysilicon TFT with $L = 15$ μm and $W = 50$ μm. (From Byun et al. (1992)).

To obtain the saturation voltage, we use the definition of V_{SAT} proposed in Sections 3.10 and 3.11 for submicron silicon MOSFETs, which is also applicable to poly-Si TFTs:

$$R_{SAT} = \frac{\partial V_{DS}}{\partial I_d}\bigg|_{sat} = \frac{\gamma\, V_{SAT}}{I_{sat}} \tag{5-3-13}$$

The limiting case of $\gamma \to \infty$ corresponds to ideal saturation $(R_{SAT} \to \infty)$, $\gamma = 1$ corresponds to a purely linear characteristic, and finite values, $\gamma > 1$, correspond to a gradual saturation. We chose the value of γ in such a way that V_{SAT} coincides with the voltage of the intercept with $R_{DS} = 0$, as shown in Fig. 5.3.2b (the same approach was used for MOSFETs in Sections 3.10 and 3.11). For $\gamma \to \infty$, the value of V_{SAT} is defined by eq. (5-3-5). For finite values of γ, V_{SAT} is larger and is defined by eq. (5-3-13).

This approach greatly simplifies the device characterization. Eq. (5-3-13) should be used in conjunction with eq. (5-3-2) which, at the saturation point, may be rewritten as

$$I_{sat} = \beta\left(n_{ss} - \frac{V_{SAT}}{2\,a^*}\right)V_{SAT} \tag{5-3-14}$$

We note again that the saturation voltage and current are no longer given by eqs. (5-3-6) and (5-3-7) because we now take into account the finite output conductance at the saturation point.

Solving eqs. (5-3-10), (5-3-13), and (5-3-14), we obtain the following equations for the saturation voltage, current, and resistance:

$$V_{SAT} = \frac{2(\gamma - 1)}{2\gamma - 1}\, n_{ss}\, a^* \tag{5-3-15}$$

$$I_{sat} = \frac{2\,\gamma(\gamma - 1)}{(2\gamma - 1)^2}\beta n_{ss}^2\, a^* \tag{5-3-16}$$

$$R_{SAT} = \frac{2\gamma - 1}{\beta\, n_{ss}} \tag{5-3-17}$$

As was discussed in Sections 3.10 and 3.11 for silicon n- and p-channel MOSFETs, when the drain-source voltage exceeds a certain critical value for onset of so-called deep saturation, here denoted V_{DSAT} ($> V_{SAT}$), the differential output resistance depends linearly on the drain voltage, i.e.,

$$R_{DS} = \frac{V_{DS} - V_{SAT}}{\Gamma I_{sat}} \qquad (5\text{-}3\text{-}18)$$

As can be seen from Fig. 5.3.2b, eq. (5-3-18) also applies to poly-Si TFTs although, in these devices, the range of drain-source voltages where eq. (5-3-18) is valid is relatively narrow because of the "kink" effect. The constant Γ in eq. (5-3-18) can be found by comparing with eq. (3-10-11) (see Section 3.10). In the long channel limit $(V_L \gg V_{SAT}$ in eq. (3-10-11)), $\Gamma \propto L^{-1}$.

In order to have a smooth transition from the linear to the saturation regime, i.e., with no discontinuity even in $\partial R_{DS}/\partial V_{DS}$, we choose the parameter γ in such a way that R_{DS} described by eqs. (5-3-10) and (5-3-18) has the same value and first derivative with respect to V_{DS} at the drain-source voltage which we define as the deep saturation point, V_{DSAT}. This yields

$$V_{DSAT} = \frac{a^* n_{ss} + V_{SAT}}{2} = \frac{a^* n_{ss}}{2}\left(\frac{4\gamma - 3}{2\gamma - 1}\right) \qquad (5\text{-}3\text{-}19)$$

and

$$1/\gamma = 4\left(-\Gamma + \sqrt{\Gamma^2 + \Gamma/2}\right) \qquad (5\text{-}3\text{-}20)$$

Eq. (5-3-20) allows us to find the saturation point if the value of Γ is determined from the experimental data (see Figs 5.3.2b and 5.3.3). From Fig. 5.3.3, we notice that γ is not very large even in our long-channel devices, compared to the values of this constant for crystalline FETs (see Sections 3.10 and 3.11). This necessitates an accurate modeling of the output conductance even in long-channel poly-Si TFTs.

V_{DSAT} defines the onset of the deep saturation regime (see Fig. 5.3.2). For extremely long-channel devices, $\Gamma \to 0$, $\gamma \to \infty$, and eq. (5-3-19) reduce to eq. (5-3-6), as expected.

Experimental studies of long-channel HFETs and MOSFETs show that these devices have an output resistance in the saturation regime which may be described by eq. (5-3-18) (see Chapters 3 and 4). However, we notice that eq. (5-3-18) neglects the "kink" effect and, hence, it is not appropriate for poly-Si TFTs where this effect is important (see Fig. 5.3.2). This motivated Byun et al. (1992) to develop the following empirical equation which describes the whole saturation region and accounts for the "kink" effect:

$$\frac{R_{DS}}{R_{SAT}} = \frac{\dfrac{z}{2\gamma\Gamma} + \sqrt{\left(\dfrac{z}{2\gamma\Gamma}\right)^{2} + 1}}{1 + \dfrac{u}{2\gamma\Gamma}\, z^{\kappa}} \qquad\qquad (5\text{-}3\text{-}21)$$

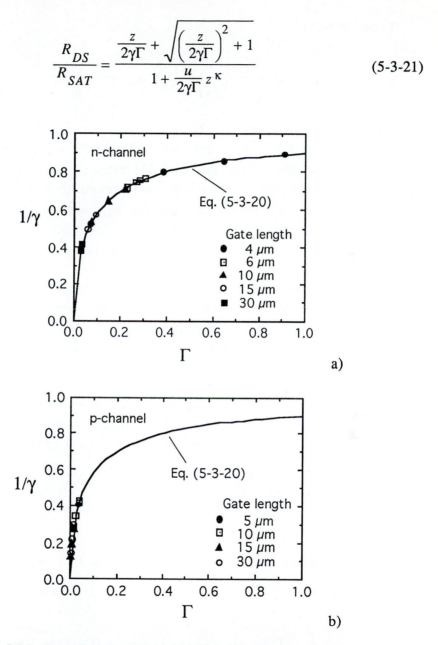

Fig. 5.3.3. Output conductance modulation factor, $1/\gamma$, versus Γ for *n*-channel (a) and *p*-channel (b) poly-Si TFTs for various gate lengths. $W = 50\ \mu m$. Calculated curves are given by eq. (5-3-20). Experimental data (symbols) are extracted from measured data. (From Byun et al. (1992).)

In eq. (5-3-21),

$$z = \frac{V_{DS} - V_{SAT}}{V_{SAT}}, \quad u = \frac{V_{GS} - V_T}{V_o} \tag{5-3-22}$$

and V_o and κ are constants. As can be seen from eq. (5-3-21), when $z \to 0$, $R_{DS} \to R_{SAT}$. When $z \gg 2\gamma\Gamma$ but $uz^n \ll 2\gamma\Gamma$, we are in the intermediate saturation regime where eq. (5-3-21) reduces to $R_{DS} \approx R_{SAT}(V_{DS} - V_{SAT})/(\gamma\Gamma V_{SAT})$, which, when combined with eq. (5-3-13), gives eq. (5-3-18). When $z \gg 2\gamma\Gamma$ and $uz^n \gg 2\gamma\Gamma$, we have $R_{DS} \propto (V_{DS} - V_{SAT})^{1-\kappa}$, describing the decrease in R_{DS} (for $\kappa \geq 2$) due to the "kink" effect (see Fig. 5.3.2). The dependence of u on the gate-source voltage and the value $n = 2$ have been chosen to fit numerous experimental data.

As shown below, the model is in an excellent agreement with experimental results. However, we should point out that this empirical approach does not allow us to explain the mechanism of the "kink" effect. This mechanism can better be studied using a two-dimensional numerical simulation based on fundamental semiconductor equations.

Using the continuous output resistance model derived above (see eqs. (5-3-10) and (5-3-21)), we can now obtain continuous current-voltage characteristics by integrating the inverse resistance over the drain-source voltage. For the linear regime (see eq. (5-3-10)), $0 \leq V_{DS} \leq V_{SAT}$, we find

$$I_d = \int_0^{V_{DS}} \frac{dV}{R_{DS}} = \beta(n_{ss} - V_{DS}/a^*)V_{DS} \tag{5-3-23}$$

For the saturation and "kink" regimes (see eq. (5-3-21)), $V_{DS} > V_{SAT}$, we obtain

$$I_d = \int_{V_{SAT}}^{V_{DS}} \frac{dV}{R_{DS}} + I_{sat} = I_{sat}(1 + Z/\gamma) \tag{5-3-24}$$

For $\kappa = 2$, (the value of κ used in our calculation), we find

$$Z = \gamma\Gamma\left(1 - \frac{\gamma\Gamma u}{2}\right)\ln\left(\frac{t}{2\gamma\Gamma}\right) + \frac{u}{16}\left(t^2 - 4\gamma^2\Gamma^2\right) -$$

$$2\gamma^3\Gamma^3\left(1 - \frac{\gamma\Gamma u}{2}\right)\left(\frac{1}{t^2} - \frac{1}{4\gamma^2\Gamma^2}\right) - 2\gamma^6\Gamma^6 u\left(\frac{1}{t^4} - \frac{1}{16\gamma^4\Gamma^4}\right) \quad (5\text{-}3\text{-}25)$$

where

$$t = z + \sqrt{z^2 + 4\gamma^2\Gamma^2} \quad\quad\quad (5\text{-}3\text{-}26)$$

As mentioned previously, this analytical model is continuous and covers the entire gate bias range including the below-threshold regime.

5.3.3. Characterization of Polysilicon TFTs

In this section, we describe characterization techniques used to extract the parameters of our model. We applied these techniques to n- and p-channel poly-Si TFTs with gate lengths 4, 5, 6, 10, 15, and 30 μm and gate width 50 μm. With such gate lengths, the contact resistance can usually be neglected, allowing us to base the parameter extraction on the intrinsic model developed above.

As shown in Fig. 5.3.2, we determine the saturation voltage (for onset of weak saturation), V_{SAT}, and the parameter Γ using a linear interpolation for the output resistance in the saturation regime. V_{SAT} is found from the intercept of the linear approximation with the V_{DS} axis, as indicated in Fig. 5.3.2b). Once V_{SAT} is determined, we obtain I_{sat} from the I–V characteristics, as shown in Fig. 5.3.2a, and Γ is given by the slope of the linear interpolation of R_{DS} (see eq. (5-3-18)). Subsequently, the value of γ is determined using eq. (5-3-20). Alternatively, γ can be determined using eq. (5-3-13) by finding I_{sat}, V_{SAT}, and R_{SAT} at the saturation point (see Fig. 5.3.2). Byun et al. (1992) showed that the two methods yield quite consistent values for γ.

The next characterization step includes the determination of the low-field mobility, μ. Usually, the low-field carrier mobility is obtained from the linear slope of the I–V characteristics at small values of the drain-source voltage (Linear Extrapolation Method). In this regime, we have

$$I_d\Big|_{V_{DS}\to 0} = \frac{q\mu W}{L} n_{ss}V_{DS} \approx \frac{q\mu W}{aL}\left(V_{GS} - V_T\right)V_{DS} \quad (5\text{-}3\text{-}27)$$

Alternatively, the mobility in a long-channel device can be determined from the dependence of the above-threshold saturation current on the gate-source voltage in the weak saturation regime, i.e., $V_{SAT} \le V_{DS} \le V_{DSAT}$, (see eq. 5-3-16):

$$I_{sat} = \frac{2q\mu W}{L} \frac{\gamma(\gamma - 1)}{(2\gamma - 1)^2} n_{ss}^2 a^*$$
(5-3-28)

Hence, using the strong inversion approximation for n_{ss} versus V_{GS}, we find a linear dependence between the square root of I_{sat} and V_{GS} in the strong inversion regime when $V_{sth} \ll a^* n_{ss}$. The mobility can be determined from the slope of this dependence.

We can also extract the mobility from the maximum value of transconductance at low drain-source bias (Transconductance Method):

$$g_{m(max)}\Big|_{V_{DS} \to 0} = \frac{q\mu W}{aL} V_{DS}$$
(5-3-29)

Still another approach is to estimate the mobility from the measured saturation point using eqs. (5-3-15) and (5-3-16) (the so-called Saturation Point Method):

$$\frac{I_{sat}}{V_{SAT}} = \frac{\beta \gamma n_s}{2(\gamma - 1/2)} \approx \frac{q\mu W}{aL(2\gamma - 1)} \gamma (V_{GS} - V_T)$$
(5-3-30)

In Fig. 5.3.4, we plot I_{sat}/V_{SAT} as a function of V_{GS} for various n- and p-channel devices. From the slopes of these plots, we calculate μ, using the values of γ obtained as shown above. The intercept with the V_{GS} axis gives the threshold voltage for each device. We found that this method gave an effective mobility which is independent of the applied gate bias, while the mobility values determined by other methods depend on the bias conditions. (Since the values of μ in polysilicon TFTs are determined by the scattering related to grain boundaries, we expect that the surface scattering will be relatively unimportant. Hence, the mobility values should be nearly independent of the gate bias, in contrast to the channel mobility in submicron MOSFETs, see Section 3.8.) The values of μ obtained by this method were between 60 and 80 cm^2/Vs for the n-channel devices and between 25 and 35 cm^2/Vs for the p-channel devices.

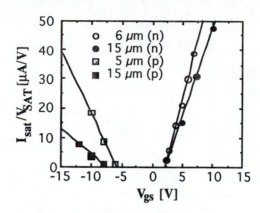

Fig. 5.3.4. I_{sat}/V_{SAT} as a function of V_{GS} for n- and p-channel devices with different gate lengths The effective low-field mobility (μ) is extracted from the slopes and the threshold voltage from the intercepts with the V_{GS}-axis (see eq. (5-3-33)). (From Byun et al. (1992).)

It is interesting to notice that the extracted parameter values for n-channel devices are more sensitive to the gate length than those for p-channel devices. This implies that the short-channel effects in p-channel devices are less noticeable and the output conductance for p-channel devices is relatively smaller than for n-channel devices. Also, we found the low-field mobility to be nearly independent of the gate length in p-channel devices (~ 30 cm^2/Vs). Similar trends were also observed for p- and n-channel MOSFETs (see Chapter 3).

In Figs. 5.3.5 through 5.3.10, we compare the measured and calculated results for poly-Si TFTs with gate lengths of 6, 10, 15, and 30 μm. As can be seen from the figures, the new analytical model for poly-Si TFTs describes both below- and above-threshold regimes in good agreement with experimental data for n- and p-channel poly-Si TFTs with 5, 6, 10, 15, and 30 μm gate lengths, with all model parameters extracted from experimental data. The deduced value of the low-field mobility of 35 cm^2/Vs for the 6 μm n-channel device is nearly half of the value for the 15 and 30 μm devices (85 cm^2/Vs). This implies that the effects of the source and drain series resistances, neglected in our model, may become noticeable in the 6 μm device. This is accounted for in our model indirectly through a reduced value of the effective mobility.

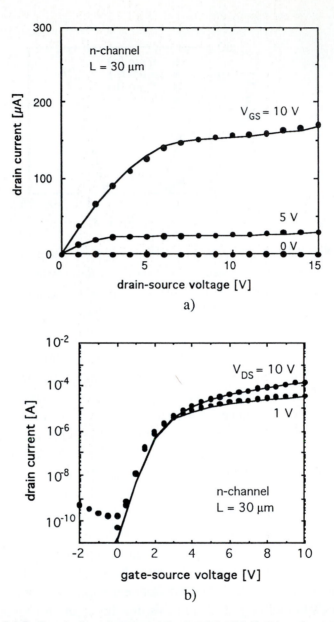

a)

b)

Fig. 5.3.5. Comparison between calculated (solid lines) and measured (symbols) results for an n-channel poly-Si TFT with $L = 30$ μm. Parameters used in the calculation: $W = 50$ μm, $d = 1000$ Å, $T = 300$ K, $V_{sth} = 0.178$ V, $V_T = 1.9$ V, μ = 85 cm²/Vs, Γ = 0.03, γ = 2.6, $V_o = 10.7$ V. (From Byun et al. (1992).)

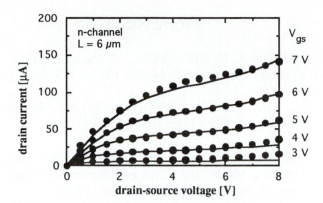

Fig. 5.3.6. Calculated (solid lines) and measured (symbols) I–V characteristics for n-channel poly-Si TFT with $L = 6\,\mu$m. Parameters used in the calculation: $W = 50\,\mu$m, $d = 1000$ Å, $T = 300$ K, $V_{sth} = 0.203$ V, $V_T = 2.15$ V, $\mu = 35$ cm^2/Vs, $\Gamma = 0.27$, $\gamma = 1.34$, $V_o = 11$ V. (From Byun et al. (1992)).

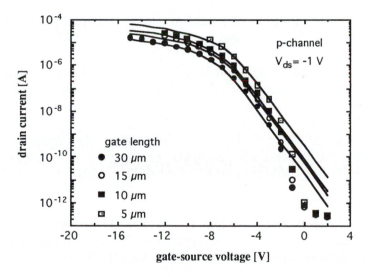

Fig. 5.3.7. Calculated (solid lines) and measured (dots) I-V characteristics for p-channel TFTs with different gate lengths. Parameters used in the calculation: $W = 50\,\mu$m, $d = 1000$ Å, $T = 300$ K. Other parameter vary with the gate length: $V_{sth} \approx 0.631$ to 0.623 V, $V_T \approx -5.48$ to -6.3 V, $\mu \approx 25$ to 35 cm^2/Vs, $\Gamma \approx 0.01$ to 0.03, $\gamma \approx 2.8$ to 4, $V_o \approx 4$ to 6.7 V. (From Byun et al. (1992).)

REFERENCES

A. F. M. ANWAR and A. N. KHONDKER, "A Model for Polysilicon MOSFET's", *IEEE Trans. Electron Devices*, ED-34, pp. 1323-1329, June (1987)

T. B. BRODY, "The Thin Film Transistor: A Late Flowering Bloom", *IEEE Trans. Electron Devices*, ED-31, pp. 1614-1628 (1984)

Y. H. BYUN, M. SHUR, M. HACK and K. LEE, "New Analytical Poly-Silicon Thin-Film Transistor Model for CAD and Parameter Characterization", *Solid State Electronics*, 35, No. 5, pp. 655-663, May (1992)

I. CHEN and F. C. LUO, "*I-V* Characteristics of Thin-Film Transistors", *J. Appl. Phys.*, 52(4), pp. 3020-3026, April (1990)

T. C. CHUANG, I-W. WU, T. Y. HUANG and A. CHIANG, "Page-Wide High Voltage Polysilicon TFT Array for Electronic Printing", *Proc. of SID Intern. Symp., Dig. of Tech. Papers*, 21, pp. 508-511, Las Vegas, Nevada (1990)

B. FAUGHAN, "Subthreshold Model of a Polycrystalline Silicon Thin-Film Field-Effect Transistor", *Appl. Phys. Lett.*, 50(5), 2 Feb. (1987)

T. A. FJELDLY, B. MOON and M. SHUR, "Approximate Analytical Solution of Generalized Diode Equation", *IEEE Trans. Electron Devices*, ED-38, No. 8, pp. 1976-1977, August (1991)

T. GLOBUS, M. SHUR, Y. BYUN and M. HACK, "New Split FET Technique for Measurements of Source Series Resistance Applied to Amorphous Silicon Thin Film Transistors", *IEEE Electron Device Lett.*, EDL-13, No. 2, pp. 108-110, Feb. (1992)

R. GUERRIERI, P. CIAMPOLINI, A. GNUDI, M. RUDAN and G. BACCARANI, "Numerical Simulation of Polycrystalline-Silicon MOSFET's", *IEEE Trans. Electron Devices*, ED-33, pp. 1201-1206, August (1986)

M. HACK and J. G. SHAW, "Numerical Simulation of Amorphous and Polysilicon TFT's", *Extended Abstracts of the 22d Intern. Conf. on Solid State Devices and Materials*, Sendai, Japan, pp. 999-1002 (1990)

M. HACK, J. G. SHAW, P. G. LeCOMBER and M. WILLUMS, "Numerical Simulation of Amorphous and Polycrystalline Silicon TFT's", *Jap. J. Appl. Phys.*, 29, No. 12, pp. L2360-2362, Dec. (1990)

M. HACK, J. G. SHAW and M. SHUR, "Vertical Thin Film Transistor and Optical Sensor Having Leakage Current Suppression Elements", *US Patent* 4,996,553, Feb. 26 (1991)

H. HAYAMA and W. I. MILNE, "A New Poly-Silicon MOS Transistor Model which Includes the Effects of Bulk Trap States in Grain Boundary Regions", *Solid-State Electronics*, 33, No. 2, pp. 279-286 (1990)

K. HIRANAKA, T. YAMAGUCHI and S. YANAGISAWA, *IEEE Electron Device Letters*, EDL-5, p. 224 (1984)

H. ITO, Y. NISHIHARA, M. NOBUE, M. FUSE, T. NAKAMURA, T. OZAWA, S. TOMIYAMA, R. WEISFIELD, H. C. TUAN and M. J. THOMPSON, *IEDM Tech. Digest, Washington DC*, p. 436, Dec. (1985)

J. KANICKI, Ed., *Physics and Applications of Amorphous and Microcrystalline Semiconductor Devices*, Artech House (1991)

T. KODAMA, N. TAKAGI, S. KAWAI, Y. NASU, S. YANAGISAWA and K. ASAMA, *IEEE Electron Device Letters*, EDL-3, p.187 (1982)

A. G. LEWIS, T. Y. HUANG, R. H. BRUCE, M. KOYANAGI, A. CHANG and I-W. WU, "Polysilicon TFTs for Analog Circuit Applications", *1988 IEDM Tech. Digest*, pp. 264-267, San Francisco, published by IEEE (1988)

A. G. LEWIS, I-W. WU, T. Y. HUANG, A. CHIANG and R. H. BRUCE, "Active Matrix Liquid Crystal Display Design Using Low and High Temperature Processed Polysilicon TFTs", *1990 IEDM Technical Digest*, San Francisco, IEEE catalog number: 78-20188, pp. 843-846 (1990)

P. S. LIN, J. Y. GUO and C. Y. WU, "A Quasi-Two-Dimensional Analytical Model for the Turn-on Characteristics of Polysilicon Thin-Film Transistors", *IEEE Trans. Electron Devices*, ED-37, pp. 666-674, March (1990)

S. D. S. MALHI, H. SHICHIJO, S. K. BANERJEE, R. SUNDARESON, M. ELAHY, G. P. POLLACK, W. RICHARDSON, A. H. SHAH, L. R. HITE, R. H. WOMACK, P. CHATTERJEE and H. WILLAN, "Characteristics and Three-Dimensional Integration of MOSFET's in Small-Grain LPCVD Polycrystalline Silicon", *IEEE Trans. Electron Devices*, ED-32, pp. 258-281 (1985)

M. MATSUMURA, H. HAYAMA, Y. NARA and K. ISHIBASHI, *Jap. J. Appl. Phys.*, 20, suppl. 20-1, p.311 (1981)

A. NAKAMURA, F. EMOTO, E. FUJII, A. YAMAMOTO, Y. UEMOTO, H. HAYASHI, Y. KATO and K. SENDA, "A High Reliability Low Operation Voltage Monolithic Active Matrix LCD by Using Advanced Solid Phase Growth Technique", *1990 IEDM Tech. Digest*, San Francisco, pp. 847-850 (1990)

T. SERIKAWA, S. SHIRAI, A. OKAMOTO and S. SUYAMA, "A Model of Current-Voltage Characteristics in Polycrystalline Silicon Thin-Film Transistors", *IEEE Trans. Electron Devices*, ED-34, pp. 321-323, Feb. (1987)

M. SHUR, *Physics of Semiconductor Devices*, Prentice Hall, New Jersey (1990)

M. SHUR and M. HACK, "Physics of Amorphous Silicon Based Alloy Field Effect Transistors", *J. Appl. Phys.*, 55, No. 10, pp. 3831-3842, May (1984)

M. SHUR, M. HACK and J. G. SHAW, "New Analytic Model for Amorphous Silicon Thin Film Transistors", *J. Appl. Phys.*, 66(7), 3371-3380, 1 Oct. (1989)

W. E. SPEAR and P. G. LeCOMBER, *J. Non-Crystal. Solids*, 8-10, p. 727 (1972)

W. E. SPEAR, P. G. LeCOMBER, S. KINMOND and M. H. BRODSKY, *Appl. Phys. Lett.*, 28, p. 105 (1976)

M. J. THOMPSON and H. C. TUAN, *IEDM Tech. Digest*, Los Angeles, p. 192, Dec. (1986)

M. J. THOMPSON, A. CHIANG, M. HACK, A. G. LEWIS, R. MARTIN and I-W. WU, "Large Area Amorphous Silicon vs. Poly Silicon Devices", *Extended Abstracts of the 22d (1990 International) Conference on Solid State Devices and Materials*, Sendai, Japan, pp. 945-950 (1990)

PROBLEMS

5-2-1. Assume the following parameters for a-Si: $E_2 \approx 86$ meV, $E_4 \approx 129$ meV, (see Fig. 5.2.1), and $\Delta E_{Fo} \approx 620$ meV. Calculate and plot the position of the electron Fermi level in a uniform a-Si sample in the dark as a function of the concentration of shallow donors for 10^{15} cm^{-3} < N_d < 10^{17} cm^{-3}. (Assume that the values of the deep acceptor-like and donor-like states extrapolated to the band edges are $g_c \approx 8 \times 10^{24}$ m^{-3}eV^{-1} and $g_v \approx 2 \times 10^{25}$ m^{-3}eV^{-1}, respectively.)

Hint: Assume that all electrons supplied by the donors fill deep acceptor-like states. Do not take into account the tail states.

*5-2-2. Estimate the source-gate capacitance related to the overlap of the source contact at the top nitride layer in a typical TFT structure (see Fig. 5.2.2). Use reasonable parameter values and explain your choices.

*5-2-3. Use the AIM-Spice a-Si TFT model ASIA1 to calculate the transfer curve (V_{out} versus V_{in}) of an Enhancement/Enhancement Logic (EEL) inverter (see Fig. P5.2.3) as a function of the gate length. Start from the default parameters given in the AIM-Spice program (see Appendix A9) and scale the gate length down to 2 μm. Use the threshold voltage $V_T = 2.5$ V for both the switching and load transistors. Scale the oxide thickness and gate width proportionally to the gate length. Repeat the calculation for the supply voltages $V_{dd} = 10$ V, 15 V, and 20 V.

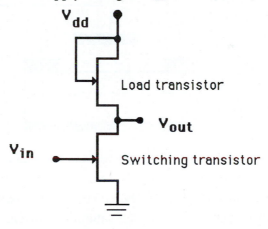

Fig. P5.2.3

*5-2-4. Consider an *n*-channel dual gate a-Si TFT. Using the AIM-Spice poly-Si TFT model, calculate and plot the device saturation current versus gate-to-source voltage, V_{gs1}, applied to the gate which is closer to the source for $0 \leq V_{gs1} < 20$ V for the values of the gate-to-source voltage, V_{gs2} applied to the gate closer to the drain of 5 V, 10 V, 15 V, and 20 V. Comment on possible applications of such a dual gate device.

*5-2-5. Roughly estimate and sketch as a function of the gate length the maximum transconductance and maximum channel current for an a-Si *p*-channel TFT. Compare using a similar plot with an a-Si *n*-channel TFT. Comment on the prospects of a-Si TFT CMOS-like technology.

*5-2-6. Consider the a-Si TFTs shown in Fig. P5.2.6. Sketch and compare qualitative plots of the saturation current versus gate-source voltage for these devices.

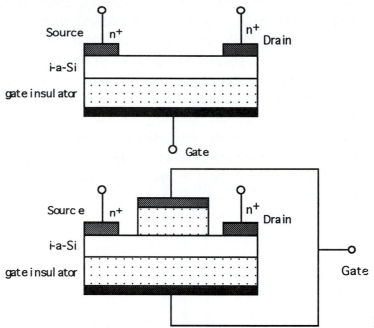

Fig. P5.2.6

*5-3-1. Use the AIM-Spice poly-Si TFT models to calculate and compare the dependencies of the drain saturation current on the gate-source voltage at $V_{ds} = 15$ V for n-channel and p-channel TFTs at temperatures of 250 K, 300 K, and 350 K. Use the default parameters given in the AIM-Spice program. How should you modify your design for TFTs operating at low temperatures, e.g. 250 K, and elevated temperatures, e.g., 350 K.

*5-3-2. Consider an n-channel dual gate poly-Si TFT. Use the AIM-Spice poly-Si TFT model to calculate and plot the device saturation current versus gate-source voltage V_{gs1} applied to the gate which is closer to the source for $0 \le V_{gs1} < 20$ V for values of the gate-source voltage V_{gs2}, applied to the gate closer to the drain of 5 V, 10 V, 15 V, and 20 V. Comment on possible applications of such a dual gate device.

*5-3-3. Use the AIM-Spice poly-Si TFT model ASIA to calculate the transfer curve (V_{out} versus V_{in}) of a Direct Coupled Field effect transistor Logic (DCFL) inverter (see Fig. P5.2.3) as a function of the gate length. Start from the default parameters given in the AIM-Spice program and scale the gate length down to 2 μm. Use the threshold voltage, $V_T = 2.5$ V for both the switching and the load transistor. Scale the oxide thickness and the gate width proportionally to the gate length. Repeat the calculation for power supply voltages $V_{dd} = 10$ V, 15 V, and 20 V. Compare with similar characteristics obtained for a-Si TFTs (see Problem 5-2-3).

*5-3-4. Use the AIM-Spice poly-Si model to show that the kink effect degrades the transfer characteristic (V_{out} versus V_{in}) of the inverter shown in Fig. P5.2.3. Explain why. How should we change the TFT design to minimize the kink effect.

Hint: Refer to Section 3.13 for comparison with crystalline MOSFETs.

*5-3-5. The equivalent circuit of a Liquid Crystal Display (LCD) pixel is shown in Fig. P5.3.5. Assume a gate width of 50 μm, a gate length of 10 μm and the capacitance $C_{LCD} = 1$ pF. How small should the TFT *off* current be in order to keep C_{LCD} charged at 30 V for longer than 1 sec? Use the AIM Spice poly-Si TFT model to simulate the response of this circuit (V_c versus time) to the gate voltage pulse shown in the figure.

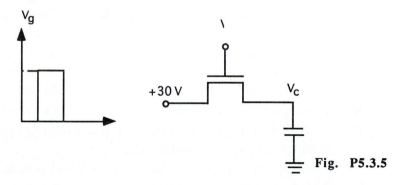

Fig. P5.3.5

6

AIM-Spice Users Manual (Windows™ 3.1Version)

6-1. INTRODUCTION

The purpose of this Chapter is to serve as a manual to our circuit simulator, Automatic Integrated Circuit Modeling Spice (AIM-Spice). This circuit simulator is based on a new version of the popular circuit simulator SPICE. The original version of Spice was developed at Berkeley in the 1970's (see Nagel (1975)) as a follow-up to the first Berkeley circuit simulator called CANCER. (Clearly, they needed a different name.) SPICE is a general purpose analog simulator which contains models for most circuit elements and can handle complex nonlinear circuits. The simulator can calculate dc operating points, perform transient analysis, locate poles and zeros for different kinds of transfer functions, find the small signal frequency response, small signal transfer functions, small signal sensitivities, and perform Fourier, noise, and distortion analyses. There are many versions and modifications of SPICE. AIM-Spice is based on the most advanced version (at the moment of writing), SPICE Version 3e.1. In Appendix A9, we provide a reference with basic information about the features of this version which are fully retained in AIM-Spice.

AIM-Spice incorporates the advanced and intermediate device models described in previous Chapters of this book. As demonstrated in those Chapters, these models can be used not only for device and circuit simulation but also for straightforward parameter extraction, which makes AIM-Spice very convenient

534

for practical applications, including such challenging tasks as yield and statistical analysis.

AIM-Spice has a simple and user friendly interface and can be used for interactive circuit simulations. AIM-Spice displays the results of the simulation in progress by plotting the output during the run. It also has extensive capabilities for post-processing data manipulation. In short, we tried to make this program as versatile and simple to use as possible.

The diskette supplied with this book contains a student version of AIM-Spice running under Microsoft Windows™ version 3.1 on IBM compatible computers. (The student version was used for most of the device and circuit simulation examples included in this book.) Information on the complete, professional version of AIM-Spice can be obtained by contacting any of the authors.

In order to make this Chapter self-contained, we give a brief introduction to Windows™-3.1 in Section 6.2. In addition, Appendix A9 provides a reference to AIM-Spice and the new AIM-Spice device models. Most of the time, this information should be sufficient for the vast majority of AIM-Spice users. However, for advanced users, we also provide references for further reading.

In Section 6.3, we give a detailed explanation of how to use AIM-Spice. Section 6.4 is a user guide for the postprocessor shipped with AIM-Spice. In Section 6.5, we give examples of AIM-Spice simulations. Finally, in Appendix A10 we explain how to implement new models in AIM-Spice.

If you have any comments and suggestions related to AIM-Spice, please write to:

Trond Ytterdal,
Department of Electrical Engineering and Computer Science
The Norwegian Institute of Technology, University of Trondheim
O. S. Bragstads plass
N-7034 Trondheim
NORWAY
E-Mail: trond.ytterdal@unit.no

6.1.1. Conventions Used in This Chapter

Before you start reading this chapter, it is important to understand the terms used below.

<u>General conventions</u>
- The word "choose" means to carry out a menu command or to click a command button in a dialog box.
- The word "select" means selecting a specific dialog box option or selecting text.
- Commands you choose are listed under the menu name preceding the command name. For example, the phrase "Choose File Open" tells you to choose the Open command from the File menu.
- The phrase "Choose OK" means that you either can click the OK button with the mouse or press the ENTER key on the keyboard to carry out the action you want.

<u>Mouse Conventions</u>

AIM-Spice requires a mouse with two buttons. Most mouse operations are performed with the left mouse button, but this can be changed with the Control Panel. The phrase "mouse button" means the left mouse button while "right" means the right mouse button.
- "Point" means to position the mouse cursor until the tip of the pointer rests on the object you want to point at.
- "Click" means to press the mouse button and immediately release it without moving the mouse.
- "Double Click" means to press and release the mouse button twice in rapid succession.
- "Drag" means to press and hold the mouse button while you move the mouse.

All these terms are fairly standard for most computers using Graphics User Interface (GUI).

6-2. INTRODUCTION TO MICROSOFT WINDOWS™-3.1

6.2.1. Basics of the Windows™-3.1 Environment

This Section contains basic information about Windows™-3.1 The best way to read this Section is in front of a computer screen, actually trying out the different operations under the Windows™ environment. For example, you can choose the word processor Write™ as a sample Windows™ application. A more detailed information about this new popular Graphics User Interface (GUI) is given, for example, by Lorenz and O'Mara (1991). Under this interface, different programs (applications) are run in different windows – rectangular areas on the screen which can be moved and resized (see Fig. 6.2.1). Some of the windows contains small pictures – called icons – representing different applications. Each application is represented by a different icon. Data or graphics files generated by a particular application are also represented by icons, each being specific for a given application. The applications can be invoked by moving the cursor on the screen to the icon representing a particular program and clicking the mouse button.

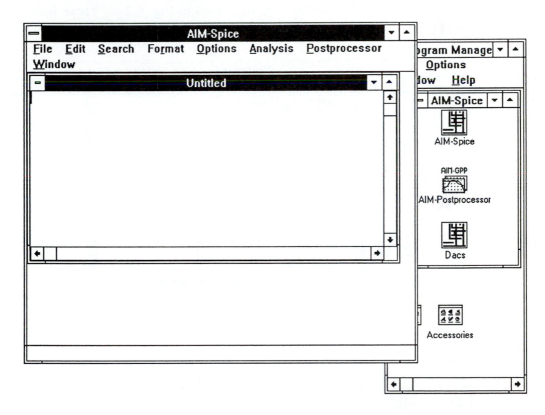

Fig. 6.2.1. Windows with icons representing different applications and data files. The position of the cursor on the screen is changed by moving the mouse. A window is selected by moving the cursor to the window and clicking the mouse button.

This environment lets you work with multiple applications simultaneously, allowing you to switch between different programs, such as a word processor, a spreadsheet, or AIM-Spice, without quitting these applications.

Other typical features of Windows™ are menus, buttons, and dialog boxes. Menus are usually located at the top of the screen. They list different options for running your program or retrieving and storing information. Menus can be selected by clicking the mouse button (i.e., the left mouse button for a standard two button mouse). Moving the cursor to a button area on the screen and clicking, you also cause the computer to perform a certain action. This action is usually explained by the text in the button or by the text in a so-called dialog box which often accompanies buttons. All this can be easily understood from the examples of the simple window screens shown in Fig. 6.2.2. These features of the GUI give you a simple and intuitive way of working with your applications.

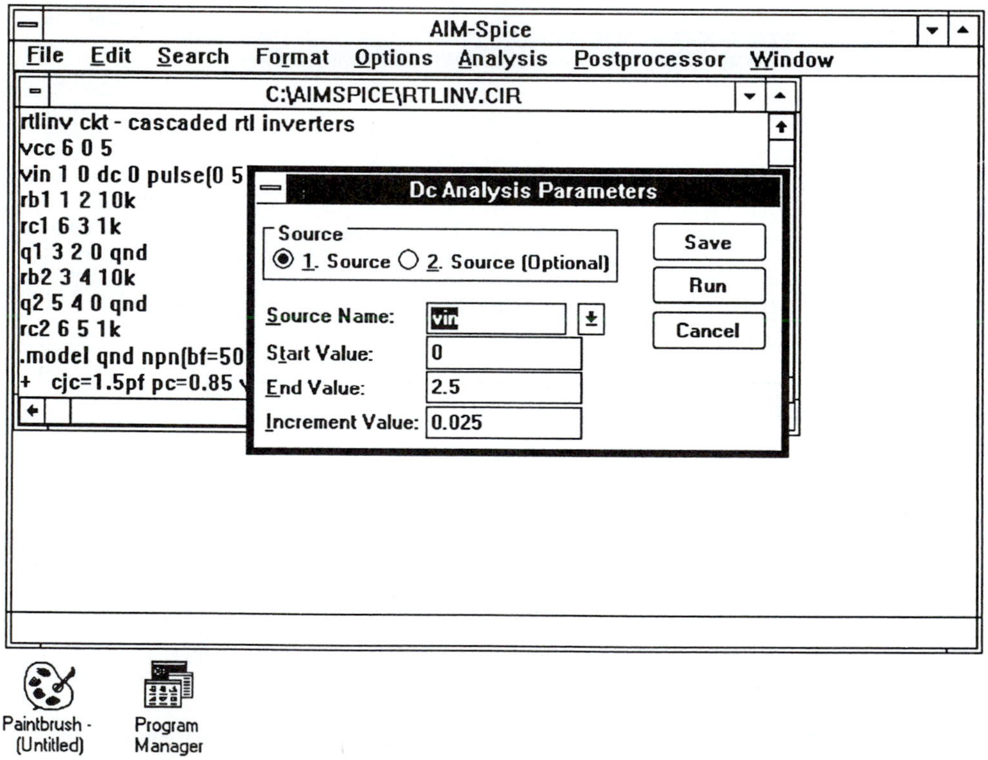

Fig. 6.2.2. Typical Window screens with menus, icons, buttons, and dialog boxes.

An important and useful feature of the Windows™ environment is the ability to easily transfer information between different applications. For example, a portion of a word processor file can be selected with a few movements of the mouse, copied into the clipboard – a reserved area in the computer memory for the purpose of sharing information between applications – and pasted into a spreadsheet. Under Windows™, the user interface is almost the same for all applications. Learning how to work with one application, you learn, at the same time, how to work with all of them. No matter which program you use – word processor, spreadsheet, communication package, data base, or AIM-Spice – you will work with menus, buttons, dialog boxes, and windows. In the next Subsection, we will take a closer look at a typical Windows™ application.

6.2.2. Working with Windows

At any given time, a user interacts with only one window, which is called the active window. Every application has a main window from which it is possible to quit the application. An application can also have additional windows, called document windows. A typical window has a title bar, a menu bar, scroll bars, and borders, as shown in Fig. 6.2.3.

We will now examine different commands and procedures for handling windows.

Selecting windows or icons

By selecting a window, you tell Windows™ that you want to work in that particular window. As mentioned above, the selected window now becomes the active window and any action you take affects this active window until another window is selected. There are two ways to select a window:

Click anywhere inside the window, except in the Maximize or Minimize boxes.
The window is brought to the front, its title bar is highlighted, and its scroll bars and other elements become visible.

Press the Alt+Esc combination to cycle through all windows until you reach the window you want, or press Ctrl+F6 to cycle through document windows.

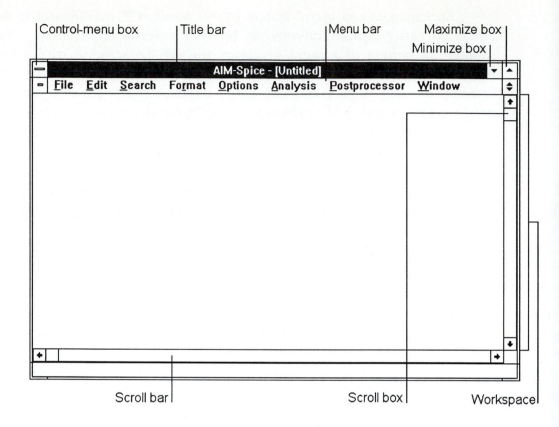

Fig. 6.2.3. Elements of a window.

Moving a window

All application windows, document windows, icons and dialog boxes with title bars can be moved into different positions on the screen as follows:

   Position the cursor in the title bar of the window. Push and hold the left hand mouse button while moving the window to the new location. Release the mouse button. To move icons you can position the cursor anywhere inside the icon.
To cancel the move, press Esc any time while performing the move.

Select the window you want to move. Then open the control menu (press Alt+Space or Alt+'-' for document windows). Select the Move command from the menu. You can now move the window with the arrow keys. When you have positioned the window, press Enter.
To cancel the move, press Esc any time while performing the move.

Resizing a window

It is often desirable to change the size of a window – for instance, if you want to position two windows beside each other.

 Select the window you want to resize. Then point to the border or a corner and drag the window until the window has the size you want.
To cancel the operation, press Esc any time before you release the mouse button.

 Select the window you want to resize. Then open the control menu (press Alt+Space or Alt+'-' for document windows). Select the Size command from the menu. You can use the arrow keys to resize the window. Press Enter when the window has the size you want.
To cancel the operation, press Esc any time while performing the resizing.

Shrinking a window to an icon

It is possible to shrink an application to an icon when you are finished using it, but want to have it available for later use.

 Select the window you want to shrink to an icon. Click the Minimize box.

 Select the window you want to shrink to an icon. Then open the control menu (press Alt+Space or Alt+'-' for document windows). Select the Minimize command from the menu.

Enlarging a window

It is possible to let an application window fill the entire screen. A document window can be enlarged to fill the entire application window except the menu bar.

       Select the window you want to enlarge. Click the Maximize box.

       Select the window you want to enlarge. Then open the control menu (press Alt+Space or Alt+'-' for document windows). Select the Maximize command from the menu.

Restoring a window or icon to its previous size

You can restore a window or icon to its previous size after maximizing or minimizing the window.

       Click the restore button of the window.

       Select the window you want to restore. Then open the control menu (press Alt+Space or Alt+'-' for document windows). Select the Restore command from the menu.

Using scroll bars

Some windows have scroll bars. Scroll bars are useful when a window contains more information that can fit inside the visible portion of the window.

To scroll:	Do this:
Up or down one line:	Click one of the scroll arrows.
Up or down one window:	Click above or below the scroll box.

Continuously: Point to one of the scroll arrows and
                       hold the mouse button until the infor-
                       mation you want comes into view.

To any position: Drag the scroll box to the position you
                       want.

To scroll:	**Do this:**
Up or down one line:	Use the arrow key for the direction you want to scroll.
Up or down one window:	PgUp or PgDn.
Left or right one window:	Ctrl+PgUp or Ctrl+PgDn.
To the top of the window:	Home.
To the bottom of the window:	End.

Closing active windows

You can close a window when you are finished working on it.

 Double-click the control-menu box of the window you
want to close.

 Press Alt+F4.

6.2.3. Working with Menus

Commands in a Windows™ application are listed in menus. All
applications have their own menus. However, one menu (called the system menu
or the control menu) is common for all applications. The symbol for the control

menu is a small box in the upper left corner of the main window. The names of the other menus are listed in the menu bar (see Fig. 6.2.3).

When you want to choose a specific command in an application, you first select a menu and then choose the command listed in the menu. Choosing a command starts an action corresponding to the command.

Selecting a menu

 Point to the name of the menu on the menu bar and click the name to open the menu.

 Press the Alt key or F10 to select the menu bar. Type the underlined letter in the menu name. It is also possible to use the arrow keys to move the highlight to the menu name you want. Press Enter to open the menu. To cancel the operation, press the Alt key or F10. Press Esc to close the menu without leaving the menu bar.

Choosing menu commands

The items listed in menus are most often commands. But they can also be characteristics you assign to graphics or text (such as bold or centered), a list of open windows or files, or the names of nested menus (menus that contain other menus).

 Click the item name.

 Press the underlined letter in the item, or use the arrow keys to move the highlight to the item you want.

Opening the control menu

Application windows, document windows, application icons, document icons and some dialog boxes have control menus.

 Click in the control-menu box or click the application icon.

Select the window or icon you want. Then press Alt+Space for application windows or icons and Alt+'-' for document windows or icons. For dialog boxes, the procedures are the same as for application windows.

Dimmed commands

A command is not accessible whenever the command name is dimmed (see Fig. 6.2.4)

Fig. 6.2.4. Dimmed command.

Accelerator keys

Some commands may have accelerator keys – combinations of key strokes which cause the same action. The accelerator keys are most often listed to the right of the command name in the menu, as shown in Fig. 6.2.5. They are often combinations of Alt or Ctrl + a function key. You can use these shortcut keys to choose menu commands without first opening the menu.

Fig. 6.2.5. Key board equivalents.

6.2.4. Dialog Boxes

Windows™ uses dialog boxes to request information from you and provide information to you. For example, when Windows™ needs additional information to carry out a command you have chosen, a dialog box requests that information. Most dialog boxes contain options, each one asking for a different type of information, see Fig. 6.2.6. After you supply all the requested information, you click on a command button to carry out the command.

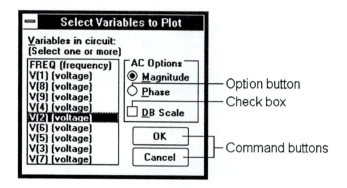

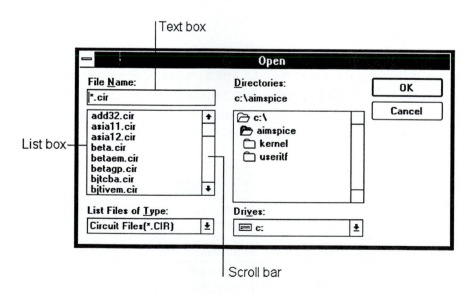

Fig. 6.2.6. Example dialog boxes.

Navigating in dialog boxes

Often you need to move around in a dialog box to complete the information needed to carry out a task. The area you select is always marked somehow – most often with a dotted box.

 Choose the option or group you want to move to.

 Press the Tab key to move the mark forward. Press Shift+Tab to move the mark backward. Or press Alt + the underlined letter on the option or the group.

Choosing options

The next few paragraphs describe each of the dialog box options.

<u>Check boxes</u>

These options give you a list of options that you can switch *on* or *off*. You can select as many or as few as you want. A check box that is currently *on* contains an X, otherwise the box is empty.

 Click the empty boxes you want to switch *on*. Click again to clear the selection.

 Press the Tab key to move the mark to the check box you want. Press the Space key to enter an X. Press the Space bar again to clear the selection.

<u>Command buttons</u>

These options, like commands in menus, start an action depending on the label of the button.

 Click the command button.

 Press the Tab key to move the mark to the button you want and press Enter.

<u>List boxes</u>

List boxes contain a list of different choices. Some list boxes let you choose more than one element, others let you choose only one.

Click the element you want to choose. To cancel a selection, click the item again.

Use the arrow keys to move the highlight to the element you want to choose and press enter to choose it. If you want to choose more than one element, press Space for every element you want to choose.

Drop-down list boxes

Drop-down list boxes are typically used when a dialog box is too small or too crowded to contain open list boxes. Initially, a drop-down list box is a rectangular box with the current selected item in the box. If you select the down arrow to the right of the box, a list of all the elements appears in the list box.

Click the down arrow to the right of the rectangular box to open the list box. Then click the item you want. If the list box contains more items than can fit inside the box, use the scroll bars to move to the item you want.

Press Alt+Down Arrow to open the list box. Use Up Arrow or Down Arrow to move to the item you want. Press Alt + Up Arrow or Down Arrow to select the item.

Option buttons

Option buttons, like check boxes, offer a list of options you can switch *on* or *off*, but the difference is that only one option button can be *on* at any time. An *on* option is marked by a black dot.

Click the option you want.

Press the Tab key to move the mark to the group you want. Then use the direction keys to select the option button you want.

<u>Text boxes</u>

Text boxes let you enter information from the keyboard as text strings.

Closing dialog boxes

When you choose one of the command buttons, the dialog box closes and the command is executed. Most of the dialog boxes contain a command button with the label Cancel. This button removes all the new information you have typed and restores any information that existed before you started editing the information in the dialog box.

The basic information about Windows™ given in this Section should be sufficient to work with AIM-Spice and many other Windows™ applications. In case of unforeseen difficulties, you may consult the book by Lorenz and O'Mara (1991) or try the computer age proven technique of "trial and error".

6-3.   AIM-Spice

As was mentioned in Section 6.1, **A**utomatic **I**ntegrated Circuit **M**odeling Spice (AIM-Spice) is based on a new version of the popular circuit simulator SPICE (Version 3.e1) developed at the University of California at Berkeley.

The AIM-Spice simulation package consists of two applications running under Microsoft Windows™-3.1 or later versions: AIM-Spice itself and a graphic postprocessor, called AIM-Postprocessor. An overview of the simulator package is shown in Fig.6.3.1.

AIM-Spice features:
- Runs under the Microsoft Windows™ Graphical Environment which gives you a simple and user friendly interface.
- Allows for Interactive Simulation Control. AIM-Spice displays simulation results in progress by plotting the output during the run with the option to cancel a simulation at any time.
- Incorporates advanced and intermediate level device models described in previous Chapters of this book. As demonstrated in those Chapters, these models can be used not only for device and circuit simulation but also for straightforward parameter extraction.

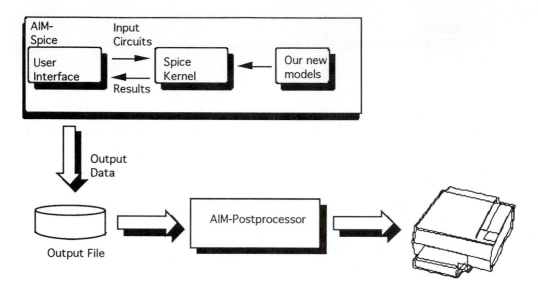

Fig. 6.3.1. Overview of the AIM-Spice simulator package.

AIM-Spice is based on Berkeley Spice Version 3.E1. This is presently the most advanced version from Berkeley. Some of the updated features of this software are listed below.

- Analyses supported: DC, AC, Transient, Transfer Function, Pole-Zero, and Noise Analysis.
- New models from Berkeley: BSIM, BSIM2, lossy transmission lines, and MOS level 6.
- Improved dc operating point analysis.

AIM-Spice is a full blown Microsoft Windows™-3.1 application and takes full advantage of all Microsoft Windows™-3.1 extensions to DOS such as: a more user friendly interface, more available memory beyond the 640 KB DOS limit, and multitasking that gives you the possibilities to switch to other applications to do meaningful work while a lengthy simulation runs in the background.

This manual gives a detailed step-by step description of how to use AIM-Spice. We refer to Appendix A9 for a full documentation on input syntax, devices and device models supported.

As mentioned in Section 6.1, the diskette supplied with this book contains a student version of AIM-Spice running under Windows™-3.1 on IBM compatible computers. (We have used the student version for most of the device and circuit simulation examples included in this book.) Information on the complete, professional version of AIM-Spice can be obtained from any of the authors.

6.3.1. Getting Started

Software requirement: AIM-Spice is a Windows application, and can only be loaded under Microsoft Windows™-3.1. This means that Windows™-3.1 must be installed and running before you can load AIM-Spice.

Hardware requirement: Minimum – IBM 286 or compatible with 2 MB RAM. Recommended – IBM 386 or compatible with math co-processor and 6 MB RAM.

AIM-Spice installation

In order to install AIM-Spice on your computer, you have to:
1) Load Windows™-3.1.
2) Insert the installation diskette.
3) Choose the File Run command in Program manager.
4) If your diskette is in drive is A, type **a:setup.exe** in the text box and then choose OK. If your diskette is in any other drive, replace 'a' by the letter designating the drive.

Running AIM-Spice

Select the AIM-Spice program group. To run AIM-Spice, double click the program icon. (If you do not have a mouse, use the arrow keys to move the highlight to the program icon and press Enter.)

The main AIM-Spice window appears together with an untitled document window, see Fig. 6.3.2. This document window is a text editor much like Notepad. This is where you enter your circuit description.

When AIM-Spice is finished loading, the text editor will be active. This is made clear to you by the blinking insertion point in the upper left corner of the window. You are now ready to type in a new circuit description.

6.3.2. Editing the Circuit Description (Netlist)

The circuit topology is described with a list of circuit elements, called the netlist. This description is made in a text editor which is integrated into AIM-Spice. In this Subsection, we take a closer look at the text editor. In the next Subsection, we discuss the rules that apply in making a netlist.

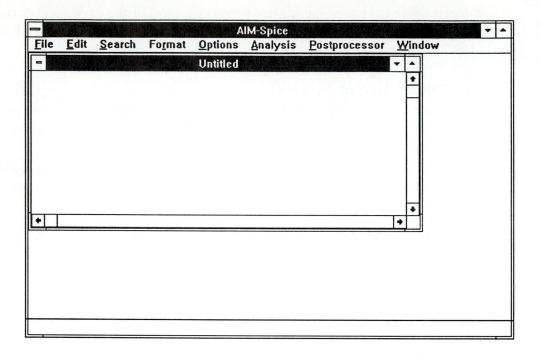

Fig. 6.3.2. Main AIM-Spice window.

If you return to the editor after doing some work in another window, the insertion point reappears at the location where you left it.

Typing and formatting text

The insertion point moves to the right when you start typing. If your typing goes beyond the right margin, the text automatically scrolls to the left so that the insertion point is always visible.

Moving the insertion point

As already mentioned, the insertion point appears in the upper left corner of the window when AIM-Spice is finished loading. You can move the insertion point anywhere you want to insert or edit text by moving the mouse cursor to the point where you want to place the insertion point and click the mouse button.

To move the insertion point with the keyboard, use the following keys or key combinations:

To move the insertion point:	Press:
to the right in a line of text	Right arrow key
to the left in a line of text	Left arrow key
up in a body of text	Up arrow key
down in a body of text	Down arrow key
to the beginning of a line of text	Home
to the end of a line of text	End
to the beginning of the netlist	Ctrl+Home
to the end of the netlist	Ctrl+End

Formatting text

The keys in the table below give you the necessary operations to type in the text exactly as you want it.

To:	Press:
insert a space	Spacebar
delete a character to the left	Backspace
delete a character to the right	Del
end a line	Enter
indent a line	Tab
insert a tab stop	Tab

To split a line, move the insertion point to the position where you want the break, and press Enter.

To join two lines, move the insertion point to the beginning of the line you want to move and press Backspace. The editor joins the line with the line above.

Scrolling

If the circuit description is longer or wider than can be shown at one time, you can scroll through the netlist. The procedures for scrolling are described in Section 6.2.

Editing text

You edit text using commands in the Edit-menu. You can delete text, move or copy text to new locations. If you change your mind after editing the text, you can use the Undo command to cancel your last edit.

Transferring text to and from other applications can be done via the clipboard. When you delete or copy text with the commands Cut or Copy, the text is placed in the Clipboard. The Paste command copies text from the Clipboard to the editor.

Selecting text

Before you use a command from the Edit-menu to edit text, you must first select the text you want the command to operate on.

1) Move the mouse cursor to the beginning of the text you want to select.
2) Drag the cursor to the end of the text you want to select. Release the mouse button.
3) To cancel the selection, press the mouse button.

1) Use the arrow keys to move the insertion point to the beginning of the text you want to select.
2) Press and hold the Shift-key while moving the insertion point to the end of the text you want to select. Release the Shift-key.
3) To cancel the selection, press one of the arrow keys.

Replacing text

When you have selected the text you want to change, you can immediately replace it by typing new text. The selected text is deleted when you type the first new character:
1) Select the text you want to replace.
2) Type the new text.

Deleting text

1) Select the text you want to delete.
2) Choose Edit Delete.
To restore deleted text, choose the Edit Undo command immediately after deleting the text.

Moving text

You can move text from one location in the editor by first copying the text to the Clipboard with Cut, and then pasting it to its new location using Paste:

1) Select the text you want to move.
2) Choose Edit Cut.
   The text is placed in the Clipboard.
3) Move the insertion point to the new location.
4) Choose Edit Paste.

Copying text

If you want to use a text more than once, you don't have to type it over each time. You can copy the text to the Clipboard with Copy, and then you can paste the text in as many places as you want by using the Paste command:

1) Select the text you want to copy.
2) Choose Edit Copy.
   The text is placed in the Clipboard.
3) Move the insertion point to the location where you want to place the text.
4) Choose Edit Paste.

Undoing an edit

The Undo command can be used to cancel the last edit you made. For example, you may have deleted text you want to keep, or you have copied text to a wrong location. If you choose the Undo command immediately after you made the mistake, the text will be restored to its state before you made the mistake:

- Choose Edit Undo.

6.3.3. Working with AIM-Spice Circuit Files

An AIM-Spice circuit consists of three main parts: the circuit description (netlist) which is placed in the text editor in a document window (called the circuit window), analysis parameters and options, and information about the last run on the circuit. This information is stored in a text file with a special format and is called a circuit file. To work with circuit files, you use commands in the File menu. These commands are used to create, open, and save circuit files in AIM-Spice format.

Opening Circuit Files

You can open new or existing circuit files in AIM-Spice. Theoretically, there is no limit to how many circuit files you can have open at one time. A new circuit window is created for every circuit file you open and the circuit description is placed in that window.

When you want to close a circuit window and you have made some changes to the circuit description or to other circuit information since you opened the file, AIM-Spice will ask you if you want to save the changes. Use the following information to determine your response:

To:	Choose:
save changes	Yes
discard changes	No
continue working in the current file	Cancel

Creating a new circuit file

   •   Choose the File New command.

A new circuit window is opened.

Opening an existing circuit file

Only files saved in the AIM-Spice format can be opened. The default extension for the names of these files are CIR. You open an existing AIM-Spice file with the following procedure:

   1) Choose the File Open command.
      The dialog box shown in Fig. 6.3.3 is displayed.
   2) Select the name of the file you want to open from the list box named Files.
   3) Choose OK.

When you use the mouse, you can open the file in one operation.

   •   Double-click the filename of the file you want to open.

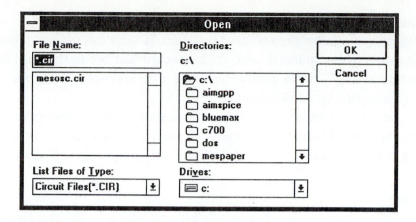

Fig. 6.3.3. File Open dialog box

Viewing a circuit file in another directory

The listbox with files only lists files with the extension CIR in the current drive and directory. It is possible to list other files as well. To list other files:

1) Select the drive, directory or the group of files you want in the listbox named Directories, or type this information in the text box at the top of the dialog box. For example, you can type *.CIR to see a list of all files with that extension.
2) Choose OK.

The listbox named Files lists the files in the drive, directory or group of files you specified.

Opening existing standard text files

It is possible to open standard text files. When you specify such files in the File Open dialog box and choose the OK command, a message box is displayed that warns you that the file you want to open is not an AIM-Spice circuit file. You can choose to cancel the operation or continue opening the file.

A standard text file is displayed in the same way as circuit files, but you have lost the benefits of the integrated environment. When you look at the menus now, you see that most of the commands are dimmed. All the menu items in the Options menu are dimmed, and only one command from the Analysis menu is available, the Run Standard Spice File command. This command runs the first analysis control line listed in the standard text file.

Importing files

Another way of loading standard text files is with the Import command in the File menu. This command converts a standard text file to the AIM-Spice circuit file format. .OPTIONS control lines and analysis control lines are converted to the format used by AIM-Spice.

After converting the file, trace through the circuit description and check if some lines need manual converting. If you try to simulate a circuit with illegal lines in the circuit description, AIM-Spice will remove these lines.

Saving a circuit file

When you create a file or you want to take a break, you can save the file and return to it later. Two commands are available for saving a file: Save As and Save.

Saving a new file:

Use Save As when you want to give a name to the new file. You can also use Save As when you want to save an old file under a new name. To save a new file:

1) Select the circuit window you want to save.
2) Choose the File Save As command.
    AIM-Spice displays the dialog box shown in Fig. 6.3.4.

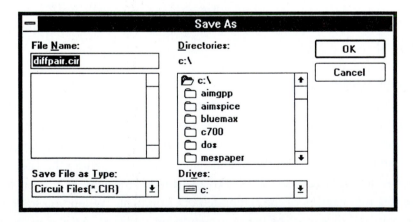

Fig. 6.3.4. Save As dialog box

3) Type the name that you want to give to the file in the text box.

If you don't give an extension, AIM-Spice will give it the extension
CIR.
4) Choose OK.
AIM-Spice saves the file on your disk.

The circuit window will remain on the screen so that you can continue working
with it. The name that you gave the circuit file now appears in the title bar of
that window.

Saving <u>Changes</u>:

You can use the Save command when you want to save changes to a circuit you
currently are working with and you want to retain the present file name. To save
changes:

- Choose the File Save command.

The file on your disk is replaced with the current version of the circuit.

6.3.4. Circuit Description in AIM-Spice

Above, we learned about how to work with the editor in AIM-Spice and
how to manage circuit files. We are now ready to learn how to define our
circuits. In this Subsection, we show how to prepare a circuit description using a
differential pair circuit as an example. The general syntax is presented after the
example.

A circuit example

Our sample circuit is shown in Fig. 6.3.5. Our first task is to name all the
nodes in the circuit diagram. A node name can be any text string, except that the
ground node must have the name "0". The most common way to name nodes is
to assign them numbers as shown in the figure.

The first line in the editor is taken as the title of the circuit. The
description of the circuit starts at line 2. The elements in the circuit are also
named, and the first letter of the name is unique for that element type. For
example, the names of resistors must start with the letter R, voltage sources with
V, bipolar transistors with Q, etc. The subsequent letters in the name can be any
alphanumeric string. The circuit nodes to which the element is connected follows
immediately after the element name. The last field of the element description is

the set of parameter values of the circuit element. If you want to continue an element description on a new line, place a '+' in column 1 of the new line.

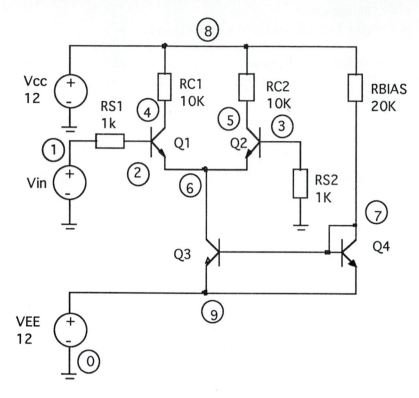

Fig. 6.3.5. Circuit example: differential pair.

The complete circuit description is shown below together with comments on some of the lines.

```
DIFFPAIR CKT - SIMPLE DIFFERENTIAL PAIR
VIN 1 0 SIN(0 0.1 5MEG 5NS) AC 1
VCC 8 0 12
VEE 9 0 -12
Q1 4 2 6 QNL
Q2 5 3 6 QNL
RS1 1 2 1K
RS2 3 0 1K
RC1 4 8 10K
```

```
RC2 5 8 10K
Q3 6 7 9 QNL
Q4 7 7 9 QNL
RBIAS 7 8 20K
.MODEL QNL NPN(BF=80 RB=100 CJS=2PF TF=0.3NS TR=6NS
+ CJE=3PF CJC=2PF VA=50)
```

In line 2, we have a voltage source with the name VIN between nodes 1 and 0. This source is specified with two values. The first, SIN(0 0.1 5MEG 5NS) is used during a Transient Analysis. The voltage source is specified in terms of the predefined function SIN which means that VIN is sinusoidal. The arguments in this function have the following meaning: offset 0 V, amplitude 0.1 V, frequency 5 MHz, and delay 5 ns with respect to time zero. AC 1 means that the source voltage will have an amplitude of 1 V during an AC Analysis (so that the output will be numerically equal to the voltage gain).

Lines 5, 6, 11 and 12 describe bipolar transistors. The last field in these lines is QNL. This is a device model specification. Bipolar transistors, like other semiconductor devices, are described in terms of specific models and associated sets of model parameters. To find the parameter values of the model QNL, we have to search for the line that starts with .MODEL, which in our case is line 14. Once the model QNL is defined, it can be used for several elements. In our circuit, this model is used for all bipolar transistors.

We note the following:

- The first line (only) is the title line, and can contain any text.
- Comment lines are marked by an '*' in the first column, and can contain any text.
- Except for the title line and subcircuit definitions, the ordering of the lines is arbitrary.
- AIM-Spice does not distinguish between upper and lower case letters.
- The number of blanks is not significant except in the title line. Commas, parentheses and tabs are equivalent to blanks.

In the rest of this Subsection, we discuss the different elements making up a complete circuit, and in the next Subsection, we focus on the set of commands available for simulating the circuit.

Names

The definition of the resistor RC1 in the circuit example is line 7 starting with `RC1`. The first field on that line is `RC1`, which is the name of the resistor. Names must start with a letter which signifies the type of circuit element being considered. The rest of the name can be any string of alphanumeric characters.

Nodes

In line 7, the two items after the name – 4 and 8 – are the nodes to which the resistor RC1 is connected. Node names are not limited to integers, but can be any alphanumeric text string. There is one exception: the ground node must have the name "0". Nodes are not treated as integers, but as text strings. Therefore, "000" and "0" are different names.

Values

The last item on line 7 is `10K`, which is the resistor value. Numerical values are written in standard floating-point notation, with optional scale suffixes. Here are some examples of legal values.

1.0   1.   1   0.5   .5   -1.0   1E6   1.6e-9

The scale suffixes follows the normal scientific notation, i.e.,

F	=	10^{-15}
P	=	10^{-12}
N	=	10^{-9}
U	=	10^{-6}
MIL	=	25.4×10^{-6}
M	=	10^{-3}
K	=	10^{+3}
G	=	10^{+9}
T	=	10^{+12}

Units are also allowed, but are ignored by AIM-Spice. All characters that are not scale suffixes can be used as units.

Circuit elements or devices

Every circuit element or device in the circuit is represented by a line not beginning with a period. All these lines have the same format:

The name of the device, followed by
two or more nodes, followed by
a model name (not all devices have this), followed by
one or more parameters

All lines that don't start with a period, except for the title line, represent circuit elements or devices. The first letter in a device name specifies the device type. Names of resistors must start with an "R", capacitors with a "C", diodes with a "D", bipolar transistors with a "Q", etc. The device type specification determines the meaning of the information on rest of the line: how many nodes, if a model name is required, and which parameters are to be specified at the end of the line.

Some of the devices allow or require a model name. A model gives you the possibility to define model parameters once, and then use that set of parameters for as many devices as you want. For example, all the transistors in the circuit example above have the same beta ($\beta = 80$). All refer to the same model, QNL, which defines β in terms of `BF=80`.

The ordering of the device lines is not significant. How they are connected is determined by the nodes. All device terminals with the same node name are connected to each other.

The rest of this paragraph presents an overview of the device types available in AIM-Spice. A complete description of the devices is found in the previous Chapters of this book, and in Appendices A8 and A9.

Passive devices

The passive devices available in AIM-Spice are resistors, inductors, capacitors, transformers, and transmission lines. They are all linear. Resistors and capacitors can have model names, but this is not required.

Semiconductor devices

The semiconductor devices available in AIM-Spice are semiconductor resistors, semiconductor capacitors, *RC* transmission lines, *p-n* diodes, Schottky diodes, heterostructure diodes, silicon bipolar transistors, Junction Field Effect Transistors (JFETs), Metal Oxide Semiconductor Field Effect Transistors (MOSFETs), compound semiconductor Metal Semiconductor Field Effect Transistors (MESFETs), Heterostructure Field Effect Transistors (HFETs), amorphous silicon Thin Film Transistors (a-Si TFTs), poly-silicon Thin Film

Transistors (poly-Si TFTs), and Heterostructure Bipolar Transistors (HBTs). All these devices require models, many of which are discussed in detail in this book. Additional information, such as device geometry, can be specified.

Voltage and current sources

These devices are the only sources generating power. There are two types of sources: controlled and independent.

Controlled sources:

All combinations of controlled sources are available in AIM-Spice: current controlled voltage source, current controlled current source, voltage controlled voltage source, and voltage controlled current source. They perform the following functions:

$$v = e \text{ x } v \qquad i = g \text{ x } v \qquad v = h \text{ x } i \qquad i = f \text{ x } i$$

where the constants e, g, h and f, represent voltage gain, transconductance, current gain and transresistance, respectively.

Independent sources:

Independent sources can have different values for different types of analysis. One value can be specified for a Transient Analysis, another for an AC Analysis and so on. A value for a DC Analysis must be prefixed by the keyword DC, the keyword for an AC Analysis is AC. For a Transient Analysis, use one of the following keywords: EXP, PULSE, PWL, SFFM, or SIN.

The voltage sources VIN, VCC, and VEE are used in the above circuit example. From the netlist, we infer that VCC and VEE have only dc values. The keyword DC can be omitted if a source only acts in a dc mode. In our case, VIN does not have a specified dc value, and it will be set to 0 V during a dc analysis. On the other hand, this source is specified for both an AC Analysis (amplitude: 1 V and phase 0 degrees) and for a Transient Analysis (sinusoidal with offset: 0 V, amplitude: 0.1 V, frequency: 5 MHz, delay: 5 ns). VCC and VEE will be assigned a value of 0 V during an ac analysis and with their specified dc values during a Transient Analysis.

The following rules apply when specifying independent sources:
- Power supplies, such as VCC and VEE, can be written without the

keyword DC.

- The inputs to the circuit, such as VIN, contain, for example, input waveforms, and clocks.
- For a voltage source without specified values for a given analysis, the values will be set to zero such that it will not influence the circuit response. However, the current through such a source can be monitored, hence, allowing the source to be used as a current meter.

Switches

Switches make it possible to change circuit connections during an analysis. They can be either voltage or current controlled. Switches require model names, SW for voltage controlled switches and CSW for current controlled switches.

Models

Many of the device types use models to specify different parameters describing the device. The .MODEL statement has the following form:

```
.MODEL NAME TYPE (PARAMETER=VALUE PARAMETER=VALUE ....)
```

The model statement in the sample circuit (see Subsection 6.3.4) is common for all the transistors in the circuit.

Appendix A9 provides a complete list of all models used by AIM-Spice. Each model has its own set of parameters. Since default values are assigned to all parameters, the specification of one or more parameter values can be omitted. For example, if only default values are used for a BJT, the model description becomes:

```
.MODEL QNL NPN()
```

Subcircuits

When a circuit contains many identical blocks or subcircuits, it is convenient to be able to write a block once and then reference it when needed. You can define a subcircuit as a "super" device and reference it many times without retyping the block. Logical elements, for example, are prime candidates for such subcircuits.

A subcircuit is defined in terms of a block of lines that start with the line .SUBCKT and ends with the line .ENDS. Between these lines, there are one or more devices, models, calls to other subcircuits, and even new subcircuit definitions. When a subcircuit is defined, it can be referenced as a device with a name that starts with the letter X.

Nodes can be defined as terminals for a subcircuit, making it possible to connect the subcircuit to the rest of the circuit. Node names used in subcircuit definitions are local names, and they will not come in conflict with global node names in the main circuit. The following is an example of a CMOS inverter subcircuit:

```
.SUBCKT INV 1 2 3
M1  3 2 1 1 MOSP W=24U L=1.4U
M2  3 2 0 0 MOSN W=12U L=1.0U
.ENDS
```

Above, we have seen how to describe circuits in terms of a netlist in AIM-Spice, and we are now ready to perform the circuit analysis.

6.3.5. Circuit Analysis with AIM-Spice

AIM-Spice supports seven different types of circuit analysis. These are:

1) Operating Point Analysis. (DC Analysis for given source values.)
2) DC Transfer Analysis. (DC Analysis for voltage or current sweeps.)
3) AC Small Signal Analysis. (Calculates frequency responses.)
4) Transient Analysis. (Calculates the time domain response.)
5) Pole-Zero Analysis. (Locates poles and zeros in the small signal transfer function.)
6) Transfer Function Analysis. (Calculates the DC small signal transfer function, input resistance and output resistance)
7) Noise Analysis. (Analysis of the device-generated noise in the circuit)

All analyses are available as commands from the Analysis menu. All analysis commands, except for the DC Operating Point Analysis, require additional control parameters to be specified. However, before choosing one of the commands in the Analysis menu, you should decide which circuit to analyze and make the corresponding circuit window the active window.

When you choose one of the commands from the Analysis window, a dialog box appears. You specify the control parameters in the dialog box before you start the simulation. All dialog boxes have one command button labeled Run. You choose this button to initiate a simulation. The dialog box for the Transient Analysis is shown in Fig. 6.3.6.

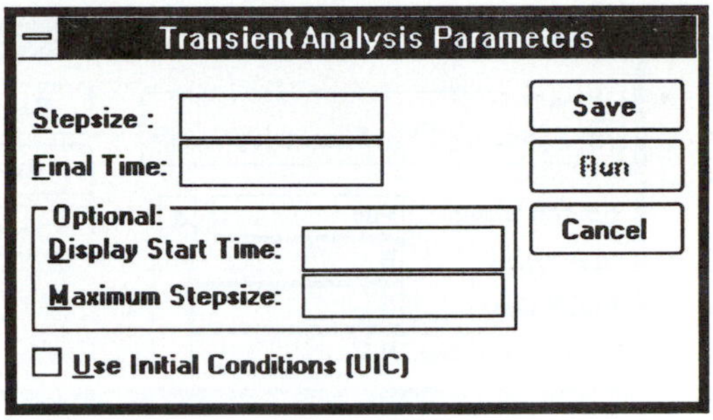

Fig. 6.3.6. Dialog box for the Transient Analysis.

The dialog boxes contain three command buttons: Cancel, Save, and Run. When you have completed the parameter fields, you can choose among these command buttons. If you choose Cancel, the parameters you entered will be discarded. If you choose Save, the parameters will be stored together with the rest of the circuit information. Run performs the same operation as Save and, in addition, initiates the analysis.

We will now discuss the control parameters for the different types of analysis available in AIM-Spice.

Operating Point

This analysis calculates the dc operating point. No control parameters are required.

DC Transfer Analysis

In this analysis, one or two source(s) (voltage or current source) are swept over a user defined interval. The dc operating point of the circuit is calculated for each value of the source(s). The DC Transfer Analysis is useful, for example, for finding the logic swing of logic gates, $I–V$ characteristics of a transistor, etc. The dialog box for the dc transfer analysis is shown in Fig. 6.3.7.

The first parameter in a dc analysis is the sweep variable(s). To specify a sweep source, open the combo box (drop-down list box) next to the source name field to see a list of all sources in the circuit, and select one of them. If you specify two sources, the first one will be in the inner loop, i.e., it varies faster than the other. The other parameters are the start, stop and increment values for the source(s).

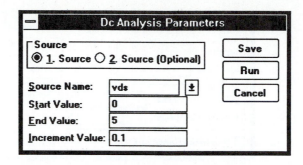

Fig. 6.3.7. Dialog box for the DC Transfer Analysis.

For example, if we want to find the *I–V* characteristics of a MOSFET, we can use the parameter values shown in the dialog box in Fig. 6.3.7. The drain-source voltage source, *vds*, is in the inner loop and sweeps from 0 to 5 V every time the gate-source voltage, *vgs*, changes value (the start, stop and increment values of *vgs* are specified by selecting the 2. Source button).

AC Small Signal Analysis

This analysis calculates the frequency response of the circuit by linearizing the circuit equations around the operating point.

The dialog box for the AC Small Signal Analysis is shown in Fig. 6.3.8.

```
┌─ Ac Analysis Parameters ──────────┐
│ ┌─Sweep:──────────────┐  ┌────────┐│
│ │ ◉ LIN  ○ OCT ○ DEC  │  │  Save  ││
│ │                     │  └────────┘│
│ │                     │  ┌────────┐│
│ │ Number of points: 200  │  Run   ││
│ │ Start Frequency:  1    └────────┘│
│ │ End Frequency:   10GHz ┌────────┐│
│ └─────────────────────┘  │ Cancel ││
│                          └────────┘│
└────────────────────────────────────┘
```

Fig. 6.3.8. Dialog box for the AC Small Signal Analysis.

The first parameter determines the number of frequencies at which the analysis is performed. The option buttons in the dialog box determine the distribution of the frequencies. If you choose LIN, the number you specify will be the total number of frequencies. If you choose OCT, the value corresponds to the number of frequencies per octave, and if you choose DEC, the value gives the number of frequencies per decade.

The selection shown in the dialog box specifies that the analysis starts at 1 Hz and ends at 10 GHz, and that the response is calculated at 200 frequencies distributed linearly in the interval.

Transient Analysis

The time domain response of the circuit is calculated from time $t = 0$ to a user defined upper time limit. The dialog box in Fig. 6.3.9 specifies a Transient Analysis which ends at 200 ns with values stored every 2 ns.

This analysis has two optional parameters. The first one specifies that the generation of output starts at a value different from zero. The second optional parameters sets an upper limit on the timesteps used by AIM-Spice.

Just as for the AC Small Signal Analysis, source values are taken from the device lines in the circuit description. In our example circuit (see Subsection 6.3.4), the voltage source VIN is specified with a sinusoidal value during a Transient Analysis.

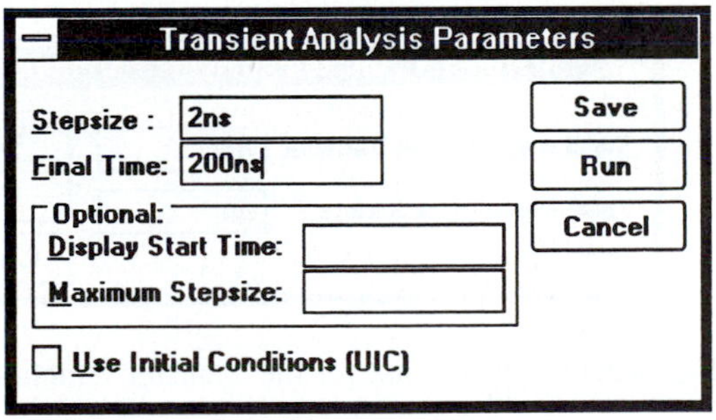

Fig. 6.3.9. Dialog box for the Transient Analysis.

Pole-Zero Analysis

AIM-Spice is able to locate poles and zeros in the small signal ac transfer function. First the dc operating point is calculated, and then the circuit is linearized around the bias point. The resulting circuit is used to locate poles and zeros.

The dialog box for the Pole-Zero Analysis is shown in Fig. 6.3.10.

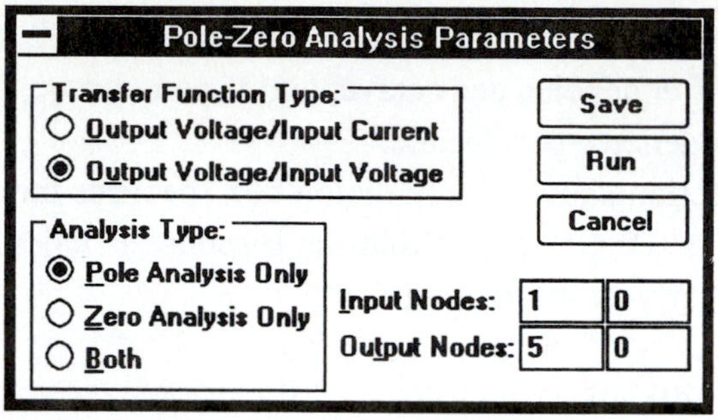

Fig. 6.3.10. Dialog box for the Pole-Zero Analysis.

In the dialog box, you specify which type of transfer function you want , if you want to locate both poles and zeros or only one kind, and the nodes defining the input and output of the circuit.

This analysis can be performed on circuits containing all types of devices except transmission lines.

Transfer Function Analysis

The dialog box for the Transfer Function Analysis is shown in Fig. 6.3.11.

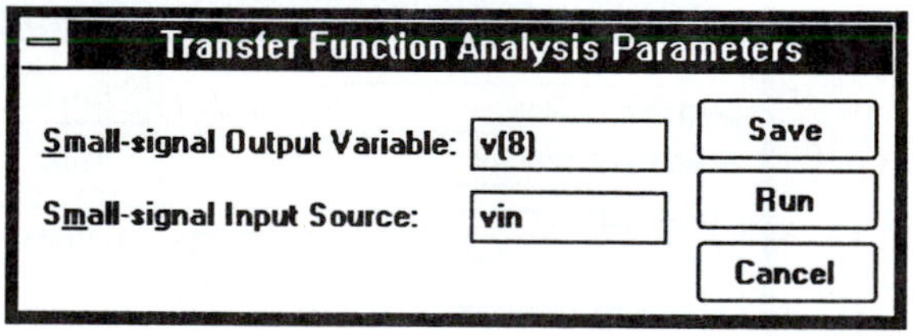

Fig. 6.3.11. Dialog box for the Transfer Function Analysis.

This analysis computes the dc small signal value of the transfer function, the input resistance, and the output resistance. In the example in the dialog box, AIM-Spice would compute the ratio v(8)/vin, the small signal input resistance for vin, and the small-signal output resistance measured across the nodes 8 and 0.

Noise Analysis

With this selection, AIM-Spice does a Noise Analysis of the circuit. The dialog box for this analysis is shown in Fig. 6.3.12.

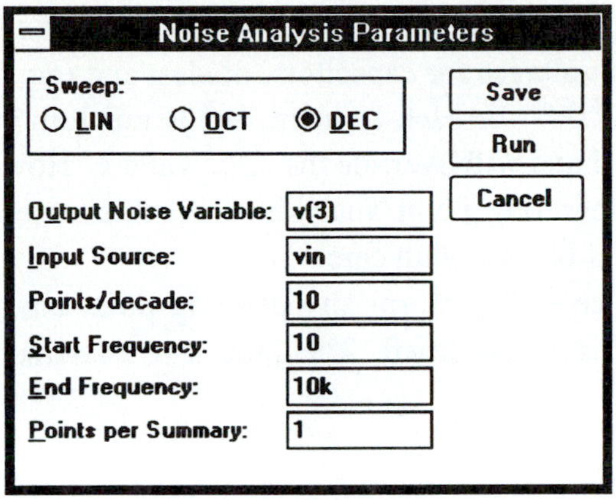

```
┌─────────────────────────────────────────────┐
│ ▬          Noise Analysis Parameters         │
├─────────────────────────────────────────────┤
│  ┌Sweep:────────────────────┐  ┌──────────┐  │
│  │ ○ LIN    ○ OCT   ◉ DEC   │  │   Save   │  │
│  └──────────────────────────┘  ├──────────┤  │
│                                │    Run   │  │
│  Output Noise Variable: │v(3) │ ├──────────┤  │
│                                │  Cancel  │  │
│  Input Source:          │vin  │ └──────────┘  │
│  Points/decade:         │10   │               │
│  Start Frequency:       │10   │               │
│  End Frequency:         │10k  │               │
│  Points per Summary:    │1    │               │
└─────────────────────────────────────────────┘
```

Fig. 6.3.12. Dialog box for the Noise Analysis.

The format of the parameter "Output Noise Variable" is V(OUTPUT<REF>), where OUTPUT is the node at which the total output noise is sought. The parameter "Input Source" is an independent source to which input noise is referred. The next three parameters are frequency information identical to the AC analysis. The last parameter is an optional integer; if specified, the noise contribution of each noise generator is produced at every "Points per Summary" frequency point.

Setting Initial Conditions

Initial conditions means currents and/or voltages specified to help locating the bias point of a circuit, or forcing the bias point to satisfy one or more conditions. One reason for giving AIM-Spice initial conditions is to select one out of two or more stable states, for example, in bistable circuits such as flip-flops.

There are three different ways of specifying initial conditions: the `.IC` statement, the `.NODESET` statement, and the `IC=` specification on individual device lines. All these statements are specified in the editor together with the netlist.

`.IC`

This control statement is used to specify initial values for a Transient

Analysis. There are two ways that this statement is interpreted, depending on whether the UIC is selected (see the dialog box for Transient Analysis in Fig. 6.3.9). If UIC is selected, the node voltages in the `.IC` statement will be used to compute initial voltages for capacitors, diodes, and transistors. This is equivalent to specifying `IC=` for each element, but is much more convenient. `IC=` can still be specified and will override the `.IC` values. However, AIM-Spice will not perform any operating point analysis when this statement is used; hence, the statement should be used with care.

AIM-Spice will perform an operating point analysis before the transient analysis if UIC is not specified. Then, the `.IC` statement has no effect.

`.NODESET`

This control statement helps AIM-Spice locate the dc operating point. Specified node voltages are used as a first guess for the dc operating point. This statement can be useful with bistable circuits. Usually, it is not needed.

`.NODESET` is active during all bias point calculations, not only with transient analyses. .IC has higher priority than `.NODESET` for a Transient Analysis.

`IC=`

Using this statement, capacitors, inductors, transmission lines, diodes, and transistors can be given initial voltage (e.g., capacitors) or current (e.g., inductors) values on the device line. The UIC option must be active in the Transient Analysis dialog box (see Fig. 6.3.9) in order to use these initial values. AIM-Spice will skip the calculation of the bias point and go directly to the Transient Analysis when the UIC option is active. All devices which have not been assigned an `IC=` value are assumed to have a zero initial value.

Options

A set of options that control different aspects of a simulation is available. These are divided among the following four logical groups corresponding to the dialog boxes shown in Figs. 6.3.13 to 6.3.16:

- General
- Analysis specific
- Device specific
- Numeric specific

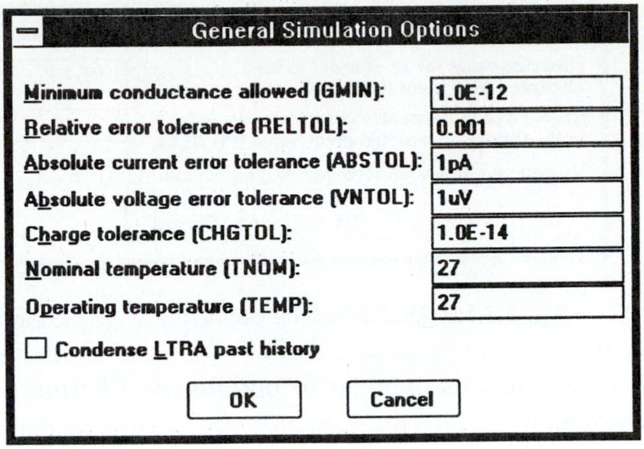

Fig. 6.3.13. Dialog box for General Simulation Options.

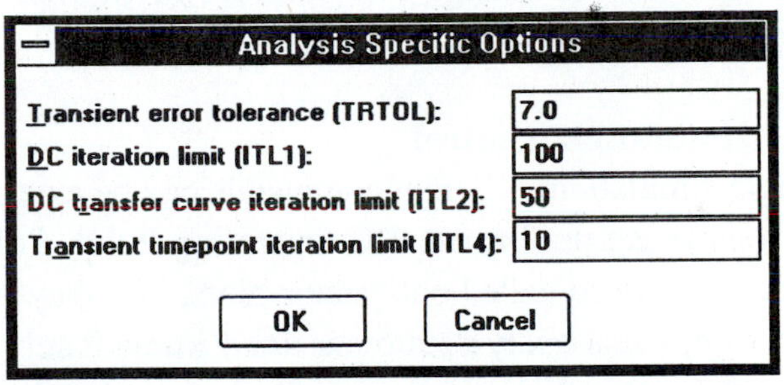

Fig. 6.3.14. Dialog box for Analysis Specific Options.

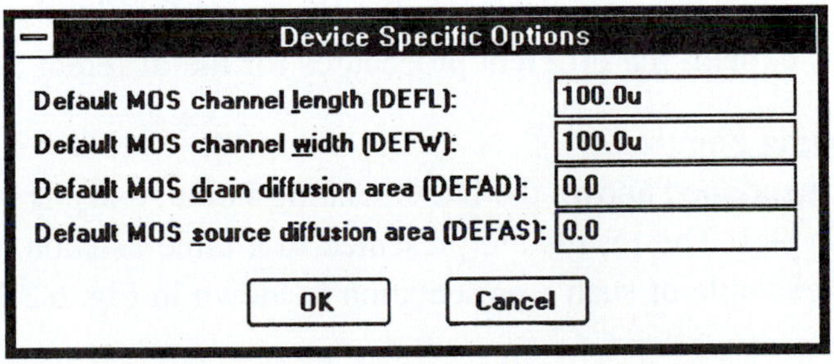

Fig. 6.3.15. Dialog box for Device Specific Options.

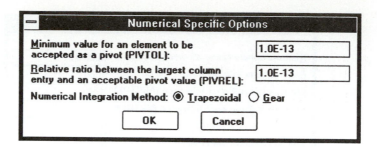

Fig. 6.3.16. Dialog box for Numerical Specific Options.

Each circuit has its own set of options, and before you reset any of the them, decide which circuit you want to work with, and make the corresponding circuit window the active window.

To reset one of the options, choose one of the commands in the Options menu. All options in a group are listed in a dialog box together with their default values. You are free to change one or more of the options. If you want to discard changes, choose the Cancel command button.

Interactive simulation control

During a simulation, different commands can be executed depending on the type of analysis being performed. DC Operating Point, Pole-Zero, and Transfer Function Analysis are so-called one-vector plots, i.e., they produce only one data point. Therefore, these analyses are executed immediately after you choose the Run command in the dialog boxes, and the results are presented after the simulation is completed. The results are presented in a table for DC Operating Point and Transfer Function Analysis, and in a graph for Pole-Zero Analysis. The other analysis types produce output during the simulation and you have to select which variables to monitor before you start the simulation. The following paragraphs explain the different procedures for the different analyses.

DC Operating Point

As mentioned above, the DC Operating Point Analysis produces a so-called one-vector plot. The results are presented in a table as soon as the simulation is over. An example of such a presentation is shown in Fig. 6.3.17.

Pole-Zero Analysis

Like the Operating Point Analysis, the Pole-Zero Analysis produces a one-vector plot. But unlike the Operating Point Analysis, this analysis presents its results in a graph. An example is shown in Fig. 6.3.18.

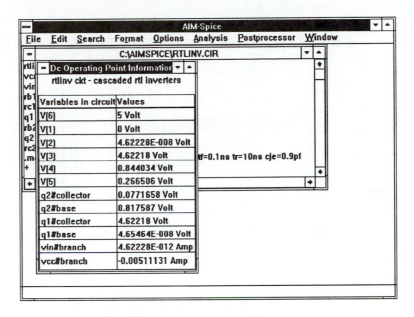

Fig. 6.3.17. Presentation of DC Operating Point results.

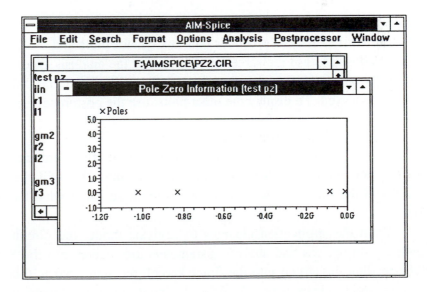

Fig. 6.3.18. Presentation of results from a Pole-Zero Analysis.

<u>Transfer Function Analysis</u>

This analysis also produces a one-vector plot, and the results are presented in a table as shown in Fig. 6.3.19.

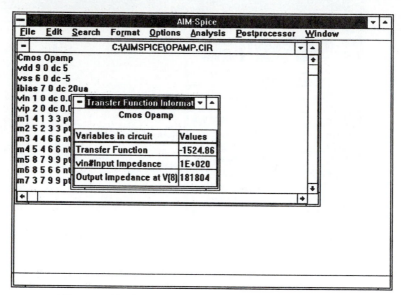

Fig. 6.3.19.   Presentation of transfer function results.

<u>Noise Analysis</u>

The Noise Analysis produces both one-vector plots and multi-vector plots. We have chosen to use the same interface as for the analysis types listed above. The results are presented after the simulation is completed and only one-vector plots are displayed. To display the other plots, use the postprocessor.

<u>AC, DC Sweep, and Transient Analysis.</u>

AIM-Spice changes mode when you choose the Run command from one of the analysis dialog boxes. The menus listed in the menu bar will change (see Fig. 6.3.20) and the status bar will give you information about what happens at any time.

When the application changes to analysis mode, the circuit description together with options and analysis parameters are loaded into the Spice kernel. While reading the circuit into the Spice kernel, the status bar will show the text "Parsing circuit, Please wait ...". After the input and error checking operation is done, the status bar shows which analysis is selected.

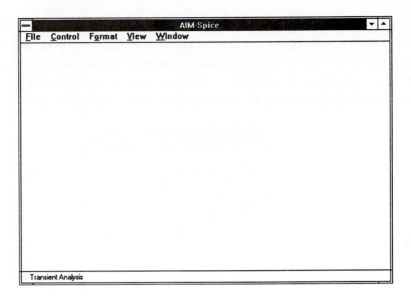

Fig. 6.3.20. AIM-Spice window in the analysis mode.

The simulation can only begin after you have specified which circuit variables to plot and the plot limits. All commands needed to do the preliminary work before starting a simulation are located in the Control menu. This menu is shown in Fig. 6.3.21.

Control	
Start Simulation	**Ctrl+S**
Select Variables to plot...	**Ctrl+V**
Analysis Limits...	**Ctrl+L**
Exit Analysis Mode	**Ctrl+E**

Fig. 6.3.21. Control menu for plot variables and limits.

Selection of variables to plot

In AIM-Spice you can open as many plot windows as you want. A plot window contains one or more circuit variables that will be plotted graphically during a simulation. To open a new plot window, choose the command Select Variables to Plot in the Control menu. This command displays the dialog box shown in Fig. 6.3.22. This dialog box is divided into two main areas. To the left is a list of all the variables in the circuit. The first element in this list is always the independent variable. The dialog box in Fig. 6.3.22 corresponds to an AC Analysis of our sample circuit (see Subsection 6.3.4). In an AC Analysis, the frequency is the independent variable. The node voltages in the circuit are listed after the independent variable. You can plot several variables in the same window by selecting more than one variable from the list.

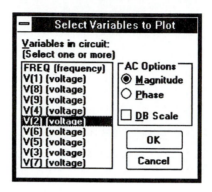

Fig. 6.3.22. Command menu dialog box for selection of plot variables.

The area to the right is active only for AC Analyses, in which case you can choose either to plot the amplitude value or the phase value. If you choose to plot the amplitude, you can select a dB-scale. This area is dimmed if other types of analyses are selected.

Choose OK to open a new plot window. The title bar of the new window indicates which variables will be plotted in the window. To open more plot windows, choose the command Select Variables to Plot again. In Fig. 6.3.23, we have opened three plot windows, two windows with only one variable, and one window containing three variables.

Note! It is not possible to display branch currents during simulation. They can only be plotted using the AIM-postprocessor (see Section 6.4).

Specifying plot limits

When you have specified which variables to plot, you have to choose limits for each of the plot windows. AIM-Spice has no way of guessing limits for the plots. You have to specify lower and upper limits for the vertical axis (y-axis) of the plots. It is wise to specify large intervals as a first guess. As the simulation progresses, the plot traces will appear on the screen, and you can reset the limits.

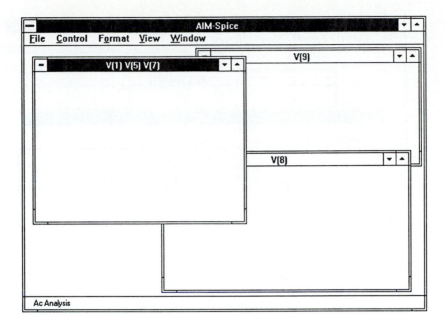

Fig. 6.3.23. Main window with three plot windows opened.

You specify *y*-axis limits by using the command Analysis Limits from the Control menu of the main window. This command displays the dialog box shown in Fig. 6.3.24. Choose OK when you have typed the limits.

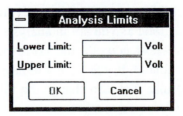

Fig. 6.3.24. Dialog box for setting plot limits.

A graph such as the one in Fig. 6.3.25 appears in the plot window that was active when you chose the Analysis Limits command. Note that you have to select the window that you want to specify limits for, before you choose the Analysis Limits command.

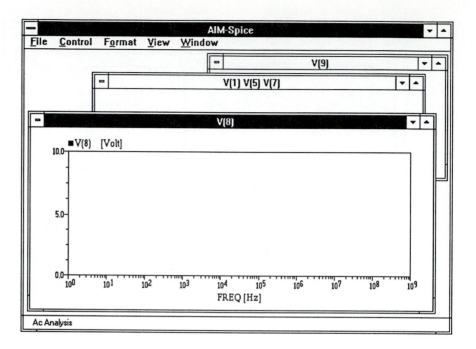

Fig. 6.3.25. Main window with graph in the active plot window after selection of *y*-axis limits.

Formatting axes and labels

When AIM-Spice creates a new graph window, it uses default axis and label formats. To change the format, use commands from the Format menu. You can use these commands as long as you are in analysis mode. To format, for example, the x-axis of a graph, first activate the corresponding plot window and then choose the X-Axis command. If you have a mouse, you may also double click the x-axis to change its format. You can change the following properties of an axis:

- Axis Type (linear or logarithmic)
- Base (when you select logarithmic axis, you have to choose which base number to use, 8 or 10).
- Minimum value
- Maximum value
- Increment value (distance between axis labels)
- Minor tics (number of tic marks between labels).

- Grid lines (you specify line styles for grid lines drawn on major and minor tic marks and if to turn grid *on* or *off*).

To format a label, use commands from the Format menu, or double click a label. You can change the following properties of labels:

- Notation (select among three numeric formats in the label: AIM-Spice scale factors, decimal or scientific).
- Number of decimal digits used in the numeric format.

Arranging plot windows

Commands for arranging the plot windows are located inside the main window in the Windows menu (see the menu bar in the main window). When you choose Cascade Windows, the plot windows will be arranged in a stack. To place a given plot window at the top of the stack, choose the title of that window from the Window menu. Choose Tile Windows to arrange all the windows side by side.

The plot windows can also be moved and resized with the mouse or with keyboard commands from the system menu.

Starting a simulation

To start a simulation, choose the command Start Simulation from the Control menu of the main window. This command is dimmed until limits are specified for all plots. Once the simulation is launched, AIM-Spice plots the selected variables in the plot windows as soon as they are available from the simulator. Fig. 6.3.26 shows a snapshot of typical simulation plots for a simulation in progress.

Stopping a simulation or resetting the plot limits

If you realize that the limits you specified are unsuitable, you can open the Control menu any time during the simulation and reset the limits. With a simulation in progress, the Control menu is slightly altered from that shown in Fig. 6.3.21. First of all, the command Start Simulation is replaced with Stop Simulation. Second, all other commands except Analysis Limits are dimmed. Hence, during a simulation, you have only two commands available: either stop the simulation or change limits.

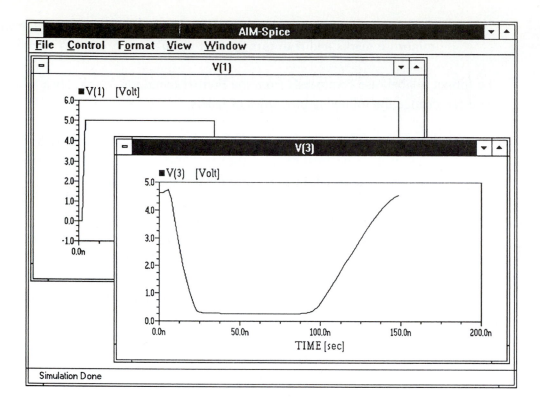

Fig. 6.3.26. Plots from a Transient Analysis in progress.

However, there is another way to reset analysis limits. In the View menu, there is the command Zoom. To make the Zoom command active, choose the command once. A check mark appears next to the command name. Now you can use the mouse to reset limits. Place the mouse cursor in the upper left corner of the new viewing rectangle. Then press and hold the right hand mouse button while you drag the mouse cursor to the lower right hand corner of the new viewing rectangle. Release the mouse button and AIM-Spice will redraw your plot with the new viewing rectangle. When you choose the Zoom command once more you deactivate the command, and the graph is redrawn with the original axis limits.

Saving results after completing a simulation

AIM-Spice indicates that a simulation is finished by displaying "Simulation Done..." in the status bar. When a simulation is over, you can tell AIM-Spice to save the results in a file by choosing the Save Plots command from the File menu of the main window. This command displays the dialog box shown in Fig. 6.3.27.

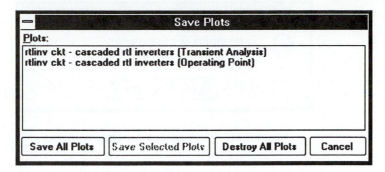

Fig. 6.3.27. Dialog box for saving of plots.

This dialog box displays a list with plots accumulated since the last time you saved plots. If you want to save all the plots in the list, choose the command Save All Plots. If you want to save only a selection of the listed plots, select the plots you want to save, and then choose the command Save Selected Plots. If you are not interested in any of the plots, choose the command Destroy All Plots. When one of the commands Save All Plots or Save Selected Plots is selected, a Save As dialog box is displayed. Complete the entries in the dialog box, and choose OK.

Hint: If you want to compare results from different simulations in the AIM-postprocessor, run all the simulations before you save the results. That way one output file contains results from all simulations. When you load this output file into the postprocessor, you are able to plot variables from different simulations in the same graph.

Exiting after completing a simulation

To leave the analysis mode, choose the command Exit Analysis Mode. The menu bar changes back to the original menu and the circuit description appears in the main window again. To exit from AIM-Spice, choose the command Exit from the File menu.

Error reporting

Error messages are written to an error file as soon as they are detected. When an error occurs, AIM-Spice interrupts the simulation and displays a message box as shown in Fig. 6.3.28.

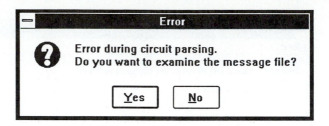

Fig. 6.3.28. Message box for error reporting.

If you choose the button labeled Yes, AIM-Spice displays the contents in the error file in a popup window as shown in Fig. 6.3.29.

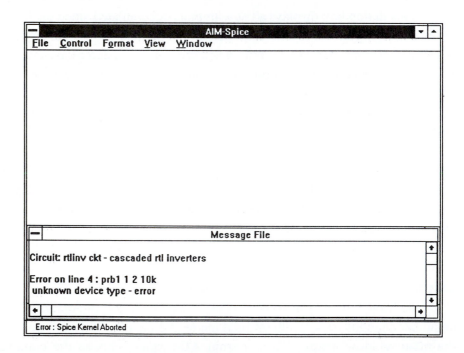

Fig. 6.3.29. Popup window with error information.

This popup window can be closed the same way you close the main application window. The error window stays at the top of all windows belonging to AIM-Spice.

To correct an error, you can quit the analysis mode and go back to the circuit window and make the necessary changes to the circuit with the error window visible. The error window automatically closes when you start another Run.

6-4. AIM-POSTPROCESSOR

The AIM-Postprocessor is an application containing routines for further processing of data obtained from the various analyses in AIM-Spice, and for graphical presentation. It works independently of the analysis part of AIM-Spice, but still within the Windows environment.

Although AIM-Spice has facilities to plot circuit variables graphically, AIM-Postprocessor has a much more powerful plotting engine including the following features:

- Plotting of sums and differences, derivatives, integrals, and mathematical functions of circuit variables.
- Cursors to select numerical values and to calculate differences between variables.
- Import of experimental data.
- Hardcopies.

This Section is divided into a tutorial on postprocessing with AIM-Postprocessor, and a complete command reference.

We assume that you have installed the Postprocessor according to the guidelines given in Section 6.3.

6.4.1. AIM-Postprocessor Tutorial

In this Subsection we give you a tutorial on how to use the AIM-Postprocessor. You will be learning to use the basics of the Postprocessor in a fast and convenient way by following the steps to produce plots for presentation. A full documentation of the Postprocessor is contained in Subsection 6.4.3, the Command Reference Subsection.

To get the most out of this tutorial, read it in front of you computer with Windows running, and follow the steps outlined below. Underlined texts are your actions and we encourage you to try out these actions.

Loading AIM-Postprocessor

Double click at the AIM-Postprocessor icon and the main window is displayed (see Fig. 6.4.1.).

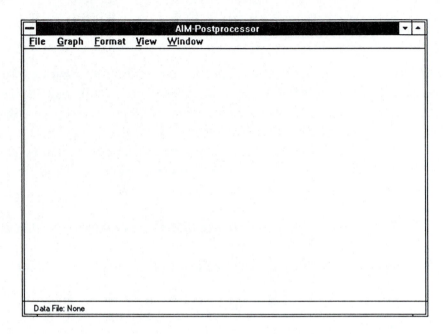

Fig. 6.4.1. The main window of AIM-Postprocessor.

The first thing to do after loading the Postprocessor is to open a data file. The data file format is a special binary format used by AIM-Spice and AIM-Postprocessor.

Opening an AIM-Spice data file

An AIM-Spice data file contains one or more plots. A plot consists of three main parts: plot information, a list of output variables, and a list of data vectors. To open an AIM-Spice data file, follow the steps below.

- Choose the File Open command.
- A standard file open dialog box is displayed. The default extension for data files is OUT. The Files list box contains a list of files in the current

directory with extension OUT.

* Select the file you want to open and choose OK. AIM-Postprocessor
  opens and reads the contents of the file.

Every plot that contains only one data vector is displayed in a table or a
graph immediately after the file is loaded. Plots from the Operating Point
Analysis, the Transfer Function Analysis, the Pole-Zero Analysis and the Noise
Analysis are such one-vector plots.

<u>Open the file "tutorial.out"</u>

Select current plot
   After the file is read, a dialog box appears containing a list of all plots
saved in the file (this dialog box is not displayed if the file contains only one
plot). This dialog box is shown in Fig. 6.4.2.

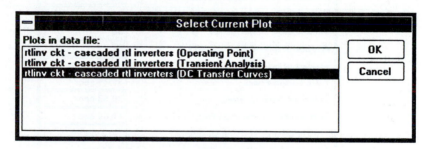

Fig. 6.4.2. The Select Current Plot dialog box.

You are asked to select one plot as the current plot. To select the current
plot, select the corresponding list box item and then choose OK. You can change
the current plot at any time by choosing the File Change Current Plot command.
   After a plot has been selected and the plot is not a one-vector plot, you are
able to create line graphs of the variables contained in that plot. If you select a
one-vector-plot as the current plot, the graph or table window with the plot is
made the active window. If the window does not exist, AIM-Postprocessor
creates a new window and displays the plot.

<u>Select the plot named "rtlinv-cascaded rtl inverters (Transient
Analysis)"</u>

The information part of a plot can be viewed at any time by choosing the View Plot Info command. This command displays information about the current plot in two dialog boxes (see Fig. 6.4.3.)

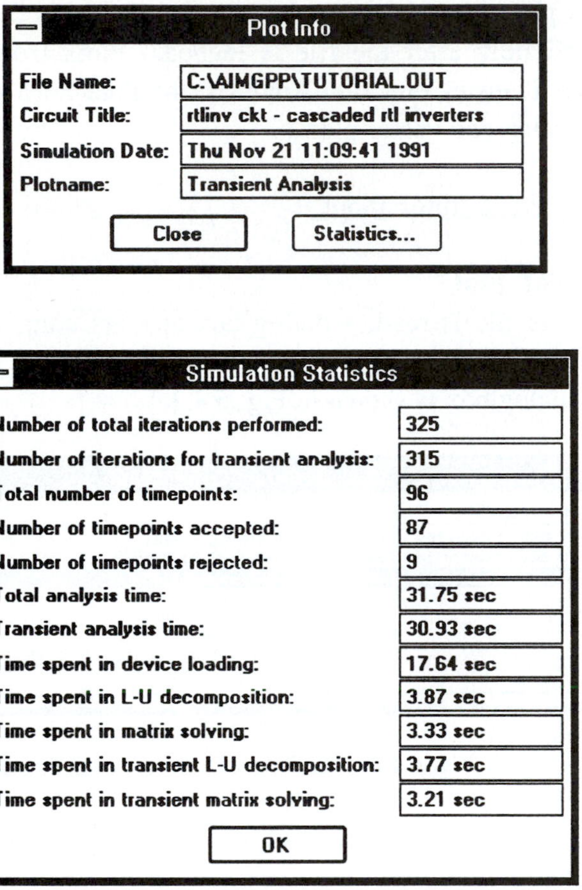

```
┌─────────────────────────────────────────────┐
│ ─              Plot Info                      │
├─────────────────────────────────────────────┤
│                                               │
│  File Name:       C:\AIMGPP\TUTORIAL.OUT      │
│  Circuit Title:   rtlinv ckt - cascaded rtl inverters │
│  Simulation Date: Thu Nov 21 11:09:41 1991    │
│  Plotname:        Transient Analysis          │
│                                               │
│       ┌──────────┐      ┌──────────────┐      │
│       │  Close   │      │ Statistics... │      │
│       └──────────┘      └──────────────┘      │
└─────────────────────────────────────────────┘
```

```
┌─────────────────────────────────────────────────┐
│ ─           Simulation Statistics                 │
├─────────────────────────────────────────────────┤
│  Number of total iterations performed:    │ 325      │
│  Number of iterations for transient analysis: │ 315   │
│  Total number of timepoints:              │ 96       │
│  Number of timepoints accepted:           │ 87       │
│  Number of timepoints rejected:           │ 9        │
│  Total analysis time:                     │ 31.75 sec │
│  Transient analysis time:                 │ 30.93 sec │
│  Time spent in device loading:            │ 17.64 sec │
│  Time spent in L-U decomposition:         │ 3.87 sec  │
│  Time spent in matrix solving:            │ 3.33 sec  │
│  Time spent in transient L-U decomposition: │ 3.77 sec │
│  Time spent in transient matrix solving:  │ 3.21 sec  │
│                 ┌──────────┐                       │
│                 │    OK    │                       │
│                 └──────────┘                       │
└─────────────────────────────────────────────────┘
```

Fig. 6.4.3. Dialog boxes that display information about the current plot.

Creating graphical plots

A graphical plot is a document window that contains a graph with one or more traces.

You create graphical plots by choosing the Graph Add Plot command. When you choose this command the Add Plot dialog box is displayed (see Fig. 6.4.4.).

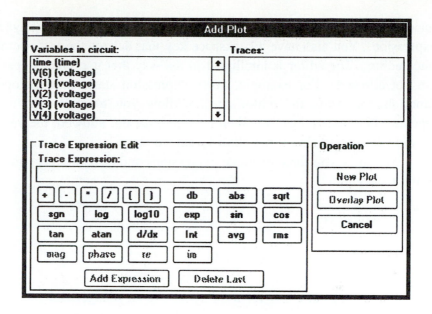

Fig. 6.4.4. The Add Plot dialog box.

The Add Plot dialog box is divided into four main fields:

- A list box with the variables in the current plot.
- A list box with the traces you want to add to the graphical plot. This list box is initially empty.
- A Trace Expression Editor (TEE) that lets you edit the traces you want.
- Three command buttons.

The following gives a description of the trace expression editing facility.

Trace Expression Editing (TEE)

Trace expression editing gives you the ability to create complex expressions including mathematical functions, derivatives and integrals of the variables in a plot. These expressions are plotted graphically when you choose the New Plot or Overlay Plot command.

TEE works much like a calculator. With a calculator you add operands, operators and functions of operands, and then you press '=' to calculate your expression. In the TEE the '=' symbol is replaced by the command Add Expression that saves the expression you have edited in the list of traces. The trace editor, as in a calculator, also has a display window that displays the current expression. You also have a backspace key that deletes the last operation in the trace editor. The editor is intelligent in the way that it excludes operations that are not allowed. For example, every expression starts with an operand or a function, therefore, the editor doesn't allow you to start with an operator. Operator buttons are inactive when operators are not allowed, the variable list is grayed when operands are not allowed, and so on.

We show you three example expressions and how to create and save them in the trace editor.

       i) V(3)
       ii) V(3)+V(5)
       iii) The average value of V(3)

Follow these steps to create the three traces shown in Fig. 6.4.5:

<u>Open the Add Plot dialog box</u>

<u>Select the variable V(3) from the list of circuit variables</u>
The variable V(3) appears in the TEE display

<u>Choose the command button Add Expression</u>
The trace is added in the trace list and the TEE display is cleared and is ready to receive a new expression

<u>Select the variable V(3) from the list of circuit variables</u>
The variable V(3) appears in the TEE display

<u>Choose the operator '+' from the operator buttons</u>
The plus-operator appears in the TEE display

<u>Select the variable V(5) from the list of circuit variables</u>
The variable V(5) appears in the TEE display

<u>Choose the command button Add Expression</u>
The trace is added in the trace list and the TEE display is cleared and is ready to receive a new expression

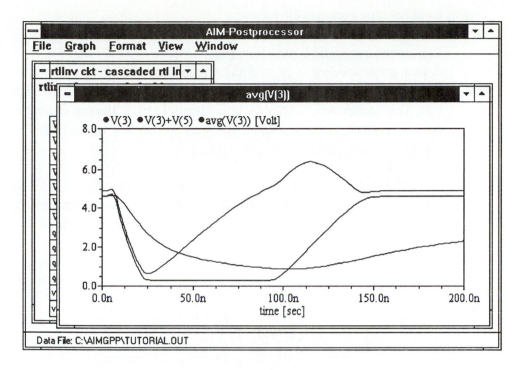

Fig. 6.4.5. The main window of AIM-Postprocessor with open graph and table windows.

Choose the function button labeled "avg"
The text "avg(" appears on the TEE display. Averages are always taken over the x-axis variables which in this case is the time (default x-axis variable in a Transient Analysis)

Select the variable V(3) from the list of circuit variables
You have now added the argument of the function and the TEE display contains "avg(V(3)"

Choose the closing parenthesis button
You have now added the closing parenthesis of the average function

Choose the command button Add Expression
The trace is added in the trace list and the TEE display is cleared and is ready to receive a new expression. This completes our examples, and the

trace list now contains three traces.

<u>Choose the command button New Plot</u>
A new graph window is created and a graph with the three traces is
displayed. Note that the Overlay Plot command was grayed in the dialog
box. This command is available only when you have already created one
or more graphs. You use it when you want to add new traces to an
already existing graph.

After the last command the main window looks like Fig. 6.4.5. On the
screen these traces can be distinguished by different colors, symbols or line
styles. It is also possible to drag the legends close to the different traces, which is
especially useful when creating hard copies.
Here, we have learned how to create a graph. Next, we will take a closer
look at the operations that prepare the graph for presentation, i.e., graph editing.

Graph editing
A graph consists of two axes, two axis labels, the trace area, text and
legends to identify the traces. Fig. 6.4.6 illustrates the different parts of a graph.
It is possible to change the appearance of all these parts with the graph editing
facilities in AIM-Postprocessor.
Note that legends, x-axis title, x-axis unit and y-axis unit are linked to the
graph and are moved whenever the trace area is moved or resized. This link is
broken when you move or resize one of these objects (the link is broken only for
the object moved or resized).
The graph edit facilities are:

- Formatting axes
- Formatting labels
- Formatting the trace area
- Adding and formatting text
- Formatting legends
- Changing x-axis expression

<u>Formatting axes</u>
To format an axis, first select a graph window as the active document
window, then choose the appropriate command from the Format menu. Note that

it is also possible to format an axis by double clicking the axis. If you choose to format the x-axis of a graph, a dialog box as the one shown in Fig. 6.4.7. appears.

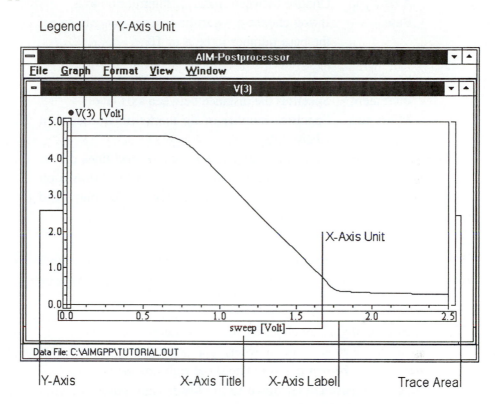

Fig. 6.4.6. Different parts of a graph.

Fig. 6.4.7. The Format Axis dialog box.

The dialog box in Fig. 6.4.7 has the following fields:

Axis Type	Choose between linear or logarithmic axis
Base	If you choose a logarithmic axis, you can select the base number to be 8 or 10
Minimum	Specifies the minimum value for the axis
Maximum	Specifies the maximum value for the axis
Increment	Specifies the distance between axis labels
Minor tics	Specifies number of tic marks drawn between labels
Grid Lines	You can specify line styles for grid lines drawn on major tic marks and on grid lines drawn on minor tic marks. The check box Display Grid turns the grid *on* or *off*

There is another and faster way of changing the axis limits, namely zooming with the mouse. To zoom with the mouse, follow the steps below:

- Turn zooming *on* by choosing the View Zoom command.
- Position the mouse cursor where you want the upper left corner of your view area to be and press and hold the right mouse button.
- Drag the mouse cursor down to the lower right corner of your view area and release the button.
- The graph is then redrawn with the new axis limits.

When you turn the zooming *off* by choosing the View Zoom command, you reset the axis limits to their values before the zoom operation.

Formatting labels

To format a label, first select a graph window as the active document window. Then choose the appropriate command from the Format menu. Note that it is also possible to format a label by double clicking the label. If you choose to format the x-axis label of a graph, a dialog box as the one shown in Fig. 6.4.8. appears.

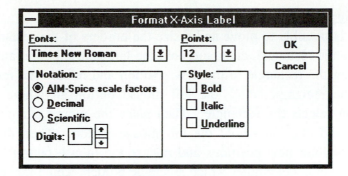

Fig. 6.4.8. The Format Axis Label dialog box.

This dialog box has the following fields:

Fonts
: Contains a list off all fonts available in the printer currently selected. The font currently used for the label is displayed in the edit box. To select another font, type the name of the font in the edit box or open the list of fonts by clicking on the down arrow in the drop down list box, and then select a font from the list.

Points
: Contains a list of point sizes. The point size currently used for label text is displayed in the edit box. To use another point size for label text, type a new point size in the edit field or open the list of point sizes by clicking on the down arrow in the drop down list box, and then select a point size from the list.

Style
: Specifies the text style used for the label. You can specify bold, italic and/or underlined text.

Notation
: Specifies the numeric format used in the label. The number 1 million is displayed as 1M if you specify AIM-Spice scale factors, 1000000 if you specify Decimal and 1E+006 if you specify Scientific. The Digits text box lets you specify how many digits to use in the numeric format.

Formatting the trace area

It is possible to move and resize the borders of the rectangular trace area. To move the rectangle, click and hold the left mouse button anywhere inside it. Then drag the rectangle to the new position and release the mouse button. To resize the rectangle, click the left mouse button anywhere inside it. This operation selects the rectangle. Then place the mouse cursor over one of the handles and click and hold the left mouse button. Then drag the rectangle borders to their new positions and release the mouse button. The handle you selected to drag determines which rectangle borders would move.

Double clicking anywhere inside the trace area restores the trace area to the default position and size. Legends, x-axis title, x-axis unit and y-axis unit are also restored to their default positions and links to the trace area are made active again.

Adding and formatting text

To add text to your graph, choose the Graph Add Text command. Now, when you place the cursor over a graph window, the cursor will change to reflect that you have selected to add text. Position the cursor on the location where you want the upper left corner of your text to be located and then click the left mouse button. The dialog box in Fig. 6.4.9. appears.

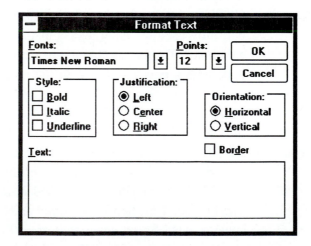

Fig. 6.4.9. The Format Text dialog box.

This dialog box has the following fields:

Fonts

Contains a list off all fonts available in the printer currently selected. The font currently used for the text is displayed in the edit box. To select another font, type the name of the font in the edit box or open the list of fonts by clicking on the down arrow in the drop down list box, and then select a font from the list.

Points

Contains a list of point sizes. The point size currently used for the text is displayed in the edit box. To use another point size for the text, type a new point size in the edit field or open the list of point sizes by clicking on the down arrow in the drop down list box, and then select a point size from the list.

Style

Specifies the style used for the text: bold, italic and/or underlined.

Justification

Specifies the horizontal justification of the text within the text rectangle.

Orientation

Specifies the orientation of the text.

Border

If selected, a rectangle is drawn around the text.

Text

Here, you type the text you want to add to the graph.

Complete the dialog box and choose OK to add the text with the format you have selected. It is possible to cancel the Add Text command by choosing the command once more before you click the mouse button.

You can format a text any time you like by double clicking the text with the left mouse button. In response, the Text Format dialog box is displayed. You now have the ability to make changes in the text you have double clicked. To make your changes visible, choose OK.

It is also possible to move and resize text. To move a text, click and hold the left mouse button anywhere in the text. Then drag the text to the new position and release the mouse button. To resize a text, click the left mouse button anywhere in the text. This operation selects the text. Then place the mouse cursor over one of the handles and click and hold the left mouse button. Then drag the rectangle borders to their new positions and release the mouse button. The handle you selected to drag determines which rectangle borders are moved.

Formatting legends

Legends are used to identify the different traces contained in a graph. The default appearance and position of legends is shown in the Fig. 6.4.10.

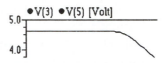

Fig. 6.4.10. Default appearance and position of graph legends.

The legends are positioned at the upper left corner of the trace area with a symbol identifying the trace by color. You have the ability to move and resize the legend, and to change its format. To move or resize the legend, follow the procedure described above for moving or resizing text.

To format a legend, point the cursor to it and double click. The dialog box in Fig. 6.4.11 appears.

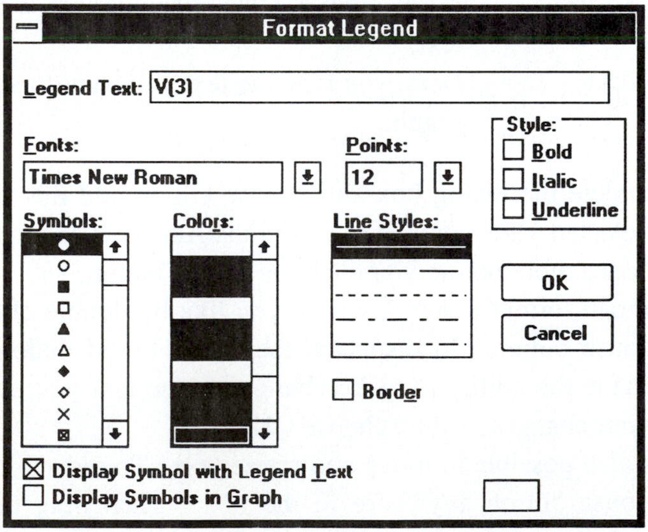

Fig. 6.4.11. The Format Legend dialog box.

The dialog box has the following fields:

Legend Text	Specifies the legend text.
Fonts	Contains a list off all fonts available on the printer currently selected. The font currently used for the legend is displayed in the edit box. To select another font, type the name of the font in the edit box or open the list of fonts by clicking the down arrow in the drop down list box, and then select a font from the list.
Points	Contains a list of point sizes. The point size currently used for the legend text is displayed in the edit box. To use another point size for the legend text, type a new point size in the edit field or open the list of point sizes by clicking the down arrow in the drop down list box, and select a point size from the list.
Style	Specifies the text style used for the legend: bold, italic and/or underlined.
Symbols	Contains a list of predefined symbols used to identify traces and data points generated by AIM-Spice.
Display Symbol with Legend Text	If selected, the currently selected symbol is drawn to the left of the legend text.
Display Symbols in Graph	If selected, the currently selected symbol is drawn on data points.
Symbol Period	Active only if Display Symbols in Graph is selected. The number you type in this field specifies the symbol frequency. If you specify 1, a symbol is drawn for every data point. If you specify 10, a symbol is drawn for every tenth data point.
Colors	Contains a list of predefined colors used to identify the traces. The color you select is used when drawing the trace and the currently selected symbol.

Line Styles   Contains a list of different line styles. AIM-
                            Postprocessor uses the line style you have
                            selected when it draws lines between data points
                            to create a line graph.

Border   If selected, a rectangle is drawn around the
                            legend.

The option Display Symbol with Legend Text has no immediately useful function. Therefore, we give you an example of when to turn this option *off*. When you use the screen as the output device, the use of different colors is the best way to identify the different traces in a graph. However, when you want to print a graph, you will probably not be able to use colors. One way of identifying the different traces without using colors is to drag the legend into the trace area and position it close to the trace. When you do so and you do not want to display a symbol next to the legend text, you turn the option Display Symbol *off* with Legend Text.

Changing the x-axis expression

The x-axis title can be changed as described previously under Creating Graphical Plots. To change the x-axis title, choose the Graph X-Axis Expression command and the dialog box in Fig. 6.4.12. appears.

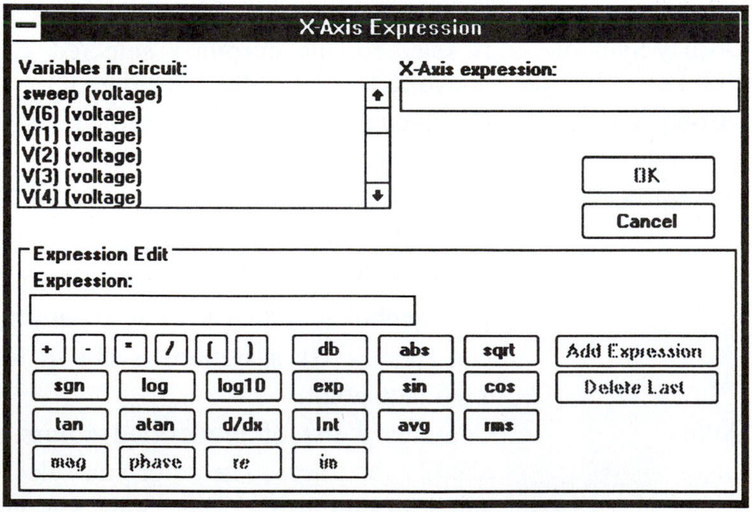

Fig. 6.4.12. The X-Axis Expression dialog box.

This dialog box is almost identical to the Add Plot dialog box. The only difference is that now you edit only one expression.

This completes the part on graph editing.

Cursors

Cursors are tools for extracting numerical values from your graphs. With two cursors, you are given the opportunity to calculate differences between traces. To turn cursors *on*, choose the View Show Cursors command. The cursor information window appears in the upper right corner of the main window as shown in Fig. 6.4.13. To remove the cursors, select the View Hide Cursors command.

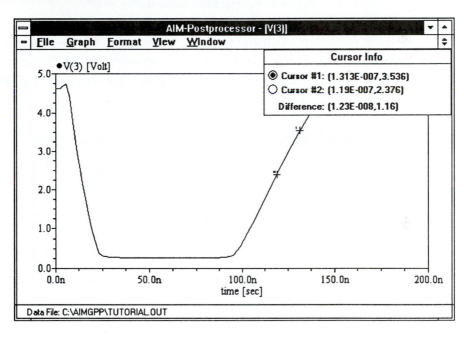

Fig. 6.4.13. The main window of AIM-Postprocessor with cursors active.

Printing

To print a graph or a table, follow the steps below.

- Format the graph or table you want to print.
- Choose File Printer Setup.
  This command lets you select a printer from a list of installed printers.

See Fig. 6.4.14.

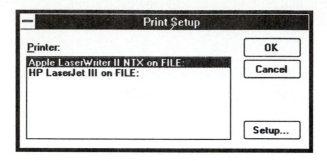

Fig. 6.4.14. The Print Setup dialog box.

- Choose the Setup command to change parameters for the printer you have selected. The dialog box displayed in response to the Setup command differs for different printers. The dialog box for an Apple Laserwriter is shown in Fig. 6.4.15.

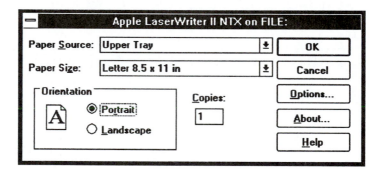

Fig. 6.4.15. The Print Setup dialog box for an Apple Laserwriter.

- Activate the document window (if not already done) that contains the graph or table you want to print.
- Choose File Print to print the graph or table.

6.4.2. Command Reference

This section gives you a description of every command and dialog box in AIM-Postprocessor. Each menu has its own set of commands. We start by reviewing the File menu (Fig. 6.4.16).

The File menu

File	
Open...	**Ctrl+F12**
Select Current Plot...	**Ctrl+S**
Print	
Printer Setup...	
Load Experimental Data...	
Export...	
Exit	**Alt+F4**
About...	

Fig. 6.4.16. The File menu.

- **Open:** Displays the dialog box in Fig. 6.4.17. This command lets you specify which AIM-Spice data file to open. This is a standard Windows File Open dialog box and Section 6.3 gives a full description of how to open files.

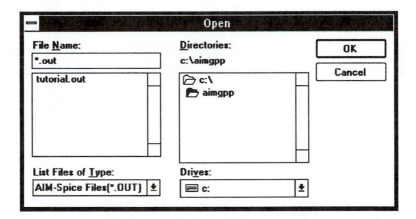

Fig. 6.4.17. The File Open dialog box.

- **Select Current Plot:** In general, an AIM-Spice data file contains more than one plot. This command lets you select the current plot. When you create graphs, the circuit variables and data vectors are taken from the current plot. This command displays the dialog box shown in Fig. 6.4.18.

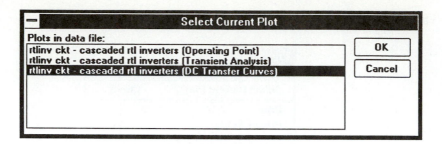

Fig. 6.4.18. The Select Current Plot dialog box.

The list box in the dialog box contains a list of all plots in the data file. You select the current plot by selecting the corresponding list box item and then choose the OK command.

- **Print:** Prints the active graph or table window.

- **Printer Setup:** Lets you specify printer and printer parameters. See Subsection 6.4.2 for a description of the printing process.

- **Load Experimental Data:** Loads a text file with experimental data. Displays a standard File Open dialog box to let you specify the file that contains the experimental data. The data are presented in the active graph window as centered symbols. The text file has the following format:

```
Number of data rows (nrows)
Number of data columns (ncol)
Legend text for column #1
Legend text for column #2
.
.
.
Legend text for column #ncol
[Experimental data formatted in nrows and ncol with one or more spaces
between columns]
```

Note that legend text must contain one text string without spaces.
An example file is shown below.

```
21
3
V(3)
V(5)
V(1)
```

0.000E+00	4.622E+00	2.665E-01	0.000E+00
1.000E-08	3.463E+00	2.674E-01	5.000E+00
2.000E-08	8.559E-01	2.879E-01	5.000E+00
3.000E-08	2.847E-01	5.395E-01	5.000E+00
4.000E-08	2.715E-01	1.277E+00	5.000E+00
5.000E-08	2.679E-01	2.085E+00	5.000E+00
6.000E-08	2.667E-01	2.866E+00	5.000E+00
7.000E-08	2.662E-01	3.576E+00	5.000E+00
8.000E-08	2.660E-01	4.185E+00	5.000E+00
9.000E-08	2.852E-01	4.671E+00	0.000E+00
1.000E-07	6.477E-01	4.922E+00	0.000E+00
1.100E-07	1.558E+00	4.729E+00	0.000E+00
1.200E-07	2.522E+00	3.773E+00	0.000E+00
1.300E-07	3.419E+00	2.318E+00	0.000E+00
1.400E-07	4.172E+00	8.201E-01	0.000E+00
1.500E-07	4.572E+00	2.887E-01	0.000E+00
1.600E-07	4.606E+00	2.728E-01	0.000E+00
1.700E-07	4.612E+00	2.686E-01	0.000E+00
1.800E-07	4.616E+00	2.673E-01	0.000E+00
1.900E-07	4.618E+00	2.668E-01	0.000E+00
2.000E-07	4.620E+00	2.666E-01	0.000E+00

After the file is read, the dialog box shown in Fig. 6.4.19 is displayed.

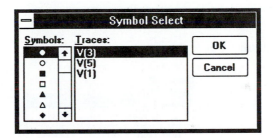

Fig. 6.4.19. The Symbol Select dialog box.

This dialog box lets you select which symbols to use with the different traces in the data file. The traces have default symbols, but you are free to use other symbols from the symbol list. To change the symbol for a given trace, follow the steps below.

    1) Select the trace for which you want to select a symbol.

    2) Select a symbol from the list of symbols.

• **Export:** Lets you export data from an AIM-Spice data file to a text file. The dialog in Fig. 6.4.20 is displayed. The list contains all variables in the current plot. Select the variables you want to export and choose the OK command. A Save As dialog box is displayed that lets you specify a filename to export to.

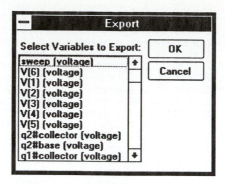

Fig. 6.4.20.   The Export dialog box.

- **Exit:** Terminates AIM-Postprocessor.

- **About:** Displays the About dialog box.

The Graph menu

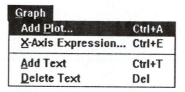

Fig. 6.4.21.   The Graph menu.

- **Add Plot:** Lets you create graphs. The command is described in detail in Subsection 6.4.1 on Creating Graphical Plots.

- **X-Axis Expression:** This command lets you change the x-axis expression. A dialog box similar to the Add Plot dialog box is displayed. The command is described in detail in Subsection 6.4.1 on Changing the X-Axis Expression.

- **Add Text:** This command lets you add text to your graphs. The command is described in detail in Subsection 6.4.1 on Adding and

Formatting Text.

- **Delete Text:** Deletes the selected text. This command can also be invoked with the accelerator key Del.

The Format menu

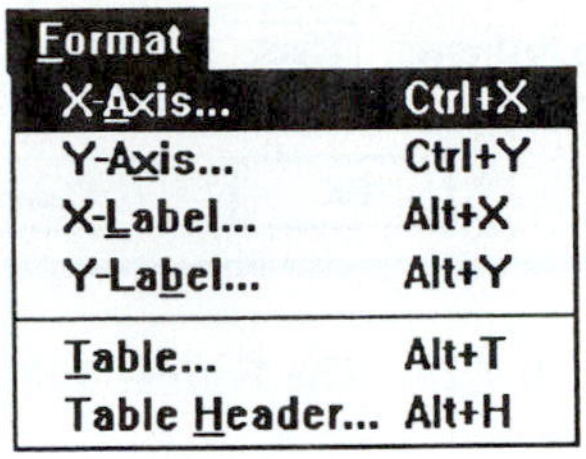

Format	
X-Axis...	Ctrl+X
Y-Axis...	Ctrl+Y
X-Label...	Alt+X
Y-Label...	Alt+Y
Table...	Alt+T
Table Header...	Alt+H

Fig. 6.4.22. The Format menu.

- **X-Axis:** Lets you change the x-axis format of the graph in the active graph window. This command is described in detail in Subsection 6.4.1 on Formatting Axes.

- **Y-Axis:** Lets you change the y-axis format of the graph in the active graph window. This command is described in detail in Subsection 6.4.1 on Formatting Axes.

- **X-Label:** Lets you change the label format for the x-axis of the graph in the active graph window. This command is described in Subsection 6.4.1 on Formatting Labels.

- **Y-Label:** Lets you change the label format for the y-axis of the graph in the active graph window. This command is described in Subsection 6.4.1 on Formatting Labels.

- **Table:** Lets you change the format of the table in the active table window. This command displays the dialog box shown in Fig. 6.4.23. The dialog box has the fields shown immediately after the Format Table dialog box.

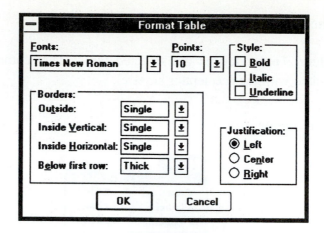

Fig. 6.4.23. The Format Table dialog box.

Fonts	Contains a list off all fonts available in the printer currently selected. The font currently used for table cells is displayed in the edit box. To select another font, type the name of the font in the edit box or open the list of fonts by clicking the down arrow in the drop down list box, and then select a font from the list.
Points	Contains a list of point sizes. The point size currently used for table cells is displayed in the edit box. To use another point size for table cells, type a new point size in the edit box or open the list of point sizes by clicking the down arrow in the drop down list box, and select a point size from the list.
Style	Specifies the text style used for table cells, bold, italic and/or underlined.
Justification	Specifies the horizontal justification of the cell text within the cell.
Borders	You can specify four types of border styles for different parts of the table: None, Single, Double and Thick.

- **Table Header:** Lets you change the table header format of the table in the active table window. This command displays the dialog box shown in Fig. 6.4.24.

```
┌─────────────────────────────────────────────────────┐
│ ▬           Format Table Header                       │
├─────────────────────────────────────────────────────┤
│                                                       │
│   ○ Left Header      ● Center Header    ○ Right Header │
│                                                       │
│   Fonts:                      Points:   ┌ Style: ──┐  │
│   ┌──────────────────────┐┌─┐ ┌──┐┌─┐   │ ⊠ Bold    │  │
│   │ Times New Roman      ││↓│ │12││↓│   │ ☐ Italic  │  │
│   └──────────────────────┘└─┘ └──┘└─┘   │ ☐ Underline│ │
│                                         └──────────┘  │
│                                                       │
│   Main Header Text: │rtlinv ckt - cascaded rtl inverters│
│   Sub Header Text:  │Operating Point                  │ │
│                                                       │
│        ┌────────┐        ┌──────────┐                 │
│        │   OK   │        │  Cancel  │                 │
│        └────────┘        └──────────┘                 │
└─────────────────────────────────────────────────────┘
```

Fig. 6.4.24. The Format Table Header dialog box.

A table has three headers, one for each type of justification. AIM-Postprocessor uses the Center Header as default when creating tables. You can also choose to use headers with left or right justification. With one header selected, the dialog box has the following fields, in addition to Fonts, Points and Style which are the same as discussed above:

Main Header   This edit field contains the text displayed as
Text   the first line of the header.

Sub Header   This edit field contains the text displayed as
Text   the second line of the header.

The View menu

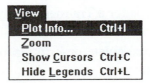

Fig. 6.4.25. The View menu.

- **Plot Info:** Displays information on the current plot. This information is presented in two dialog boxes. The second one is displayed when you select the Statistics command from the first dialog box. The dialog boxes are shown in Fig. 6.4.26.

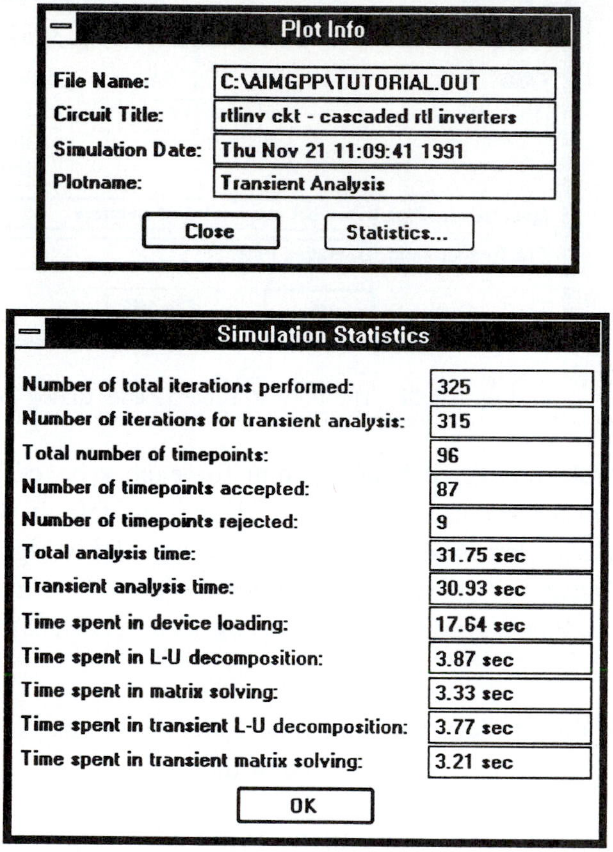

Plot Info	
File Name:	C:\AIMGPP\TUTORIAL.OUT
Circuit Title:	rtlinv ckt - cascaded rtl inverters
Simulation Date:	Thu Nov 21 11:09:41 1991
Plotname:	Transient Analysis

Close   Statistics...

Simulation Statistics	
Number of total iterations performed:	325
Number of iterations for transient analysis:	315
Total number of timepoints:	96
Number of timepoints accepted:	87
Number of timepoints rejected:	9
Total analysis time:	31.75 sec
Transient analysis time:	30.93 sec
Time spent in device loading:	17.64 sec
Time spent in L-U decomposition:	3.87 sec
Time spent in matrix solving:	3.33 sec
Time spent in transient L-U decomposition:	3.77 sec
Time spent in transient matrix solving:	3.21 sec

OK

Fig. 6.4.26. The Plot Info dialog boxes.

The Simulation Statistics dialog box differs from analysis to analysis. The one shown above is for a Transient Analysis.

- **Cursors:** For a complete description of this command, see Subsection 6.3.1 on Cursors.

- **Zoom:** This command lets you change axis limits with the mouse by drawing a rectangle which defines the limits of the axes. To perform the zooming, follow these steps:

- Turn zooming *on* by selecting the View Zoom command.
- Position the mouse cursor where you want the upper left corner of your view area to be. Press and hold the right mouse button.
- Drag the mouse cursor down to the lower right corner of your view area and release the button.

The graph is then redrawn with the new axis limits.
When you turn zooming *off* by selecting the View Zoom command once more, you reset the axis limits to their previous values.

The Window menu

Commands in the Window menu act on document windows and their icons only.

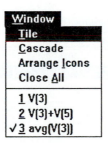

Fig. 6.4.27.   The Window menu.

- **Tile:** This command positions the windows side by side as shown in Fig. 6.4.28.

- **Cascade:** This command stacks the windows in a cascade as shown in Fig. 6.4.29.

- **Arrange Icons:** Arranges the document icons at the bottom row of the main window as shown in Fig. 6.4.30.

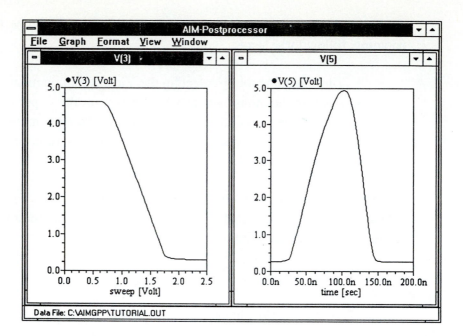

Fig. 6.4.28. The graph window positions after a Tile command.

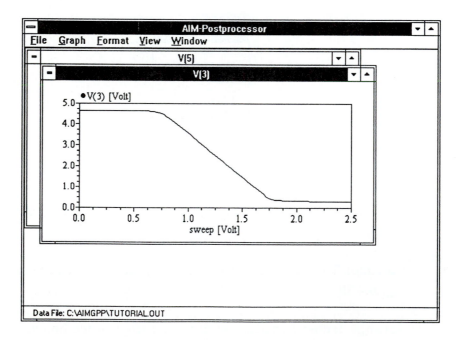

Fig. 6.4.29. The graph window positions after a Cascade command.

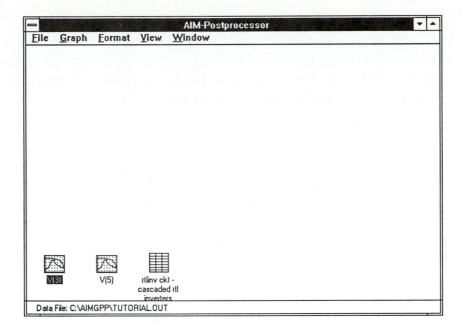

Fig. 6.4.30. Icon positions after an Arrange Icons command.

- **Close All:** Closes all graph and table windows.

  The menu items below the separator are the window titles of all the graph and table windows. When you select one of these titles, you make the corresponding graph or table window the active window. This window is placed in the front of all the other windows.

6-5.   SIMULATION EXAMPLES USING AIM-Spice

In this Section, we present simulation examples using AIM-Spice with our new models. Each example starts with a short description of the circuit together with the netlist and analysis parameters. We then present the simulation results as they appear in AIM-Spice or AIM-Postprocessor. The TFT example in Subsection 6.5.6. is discussed in more detail than the others by describing the session step by step.

In this Section we use the AIM-Spice notation for all variables.

6.5.1.   Heterostructure Diode (HDIA)

We used two diodes coupled back to back as an example of the heterostructure diode model. The voltage across these two diodes was swept from −5 V to 5 V. Below, we show the circuit (Fig. 6.5.1), the corresponding netlist (Fig. 6.5.2), and the analysis parameters (Fig. 6.5.3).

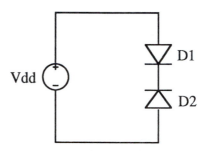

Fig. 6.5.1. Example circuit for the HDIA model.

```
HDIA Two diodes back to back (Example circuit for Ch. 6)
vdd 1 0 dc 0
d1 1 2 diode 100
d2 3 2 diode 100
* the voltage source vid is inserted to be used as an amp
* meter
vid 3 0 dc 0
.MODEL diode d level=2 bv=5 ibv=3e-7
```

Fig. 6.5.2. Circuit netlist for HDIA example.

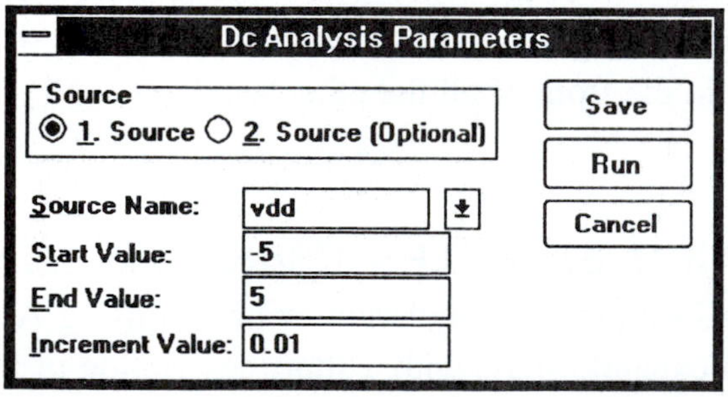

Fig. 6.5.3. DC Analysis parameters used for HDIA example.

We used default model parameters for both diodes, except for the voltage and current at breakdown where we used 5 V and 0.3 µA, respectively. The resulting current through the devices is shown in Fig. 6.5.4.

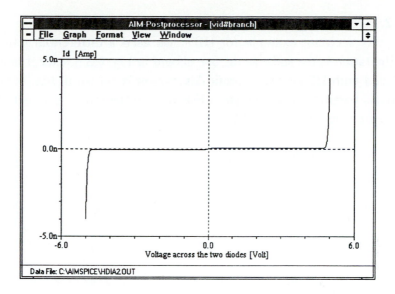

Fig. 6.5.4. A screen dump from AIM-Postprocessor showing the current versus the applied voltage through the two back-to-back diodes.

AIM-Postprocessor has a feature that lets you zoom the view area. A magnified plot around the origin is displayed in Fig. 6.5.5.

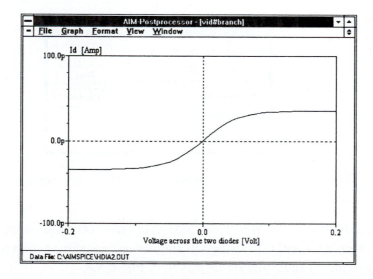

Fig. 6.5.5. Same as in Fig. 6.5.4, but here we have used the Zoom feature to magnify our curve around the origin.

6.5.2. Heterojunction Bipolar Transistor (HBTA)

As a circuit example for HBT, we used a differential pair. The circuit configuration is the same as the one shown in Fig. 6.3.4, except that the BJTs are replaced with HBTs. The circuit description is shown in Fig. 6.5.6 and the AC Analysis parameters are shown in Fig. 6.5.7. The model parameters used for the HBTs are listed in Table 6.5.1.

```
HBT differential pair
vin 1 0 dc 0 ac 1
vcc 8 0 12
vee 9 0 -12
n1 4 2 6 hbt
n2 5 3 6 hbt
rs1 1 2 100
rs2 3 0 100
rc1 4 8 10k
rc2 5 8 10k
n3 6 7 9 hbt
n4 7 7 9 hbt
rbias 7 8 20k
.model hbt hnpn(is=1e-22 bf=800 br=1 ne=2 nc=2 ise=1e-15
+ isc=0 irs1=1e-24 irs2=2e-23 irb0=1e-24 re=2 rb=10
+ rc=2 m=5 icsat=10m tf=1ps tr=10ps cje=0.1p cjc=10f)
```

Fig. 6.5.6. Circuit description for HBT differential pair.

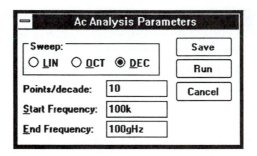

Fig. 6.5.7. AC Analysis parameters used for simulating the HBT differential pair.

The AC Analysis was run from 100 kHz to 100 GHz with the input port between nodes 1 and 0. During the simulation, we plotted the voltage at node 5 in a dB-scale. The results are displayed in Fig. 6.5.8.

Parameter Description	AIM-Spice Parameter	Value
Transport saturation current	is	1e-22 A
Ideal maximum forward beta	bf	800
Ideal maximum reverse beta	br	1
Base-emitter leakage emission coefficient	ne	2
Base-collector leakage emission coefficient	nc	2
Base-emitter leakage saturation current	ise	1e-15 A
Base-collector leakage saturation current	isc	0 A
Surface recombination current 1	irs1	1e-24 A
Surface recombination current 2	irs2	2e-23 A
Base recombination saturation current	irb0	1e-24 A
Emitter resistance	re	2 Ω
Base resistance	rb	10 Ω
Collector resistance	rc	2 Ω
Knee shape parameter	m	5
Collector saturation current	icsat	10e-3 A
Ideal forward transit time	tf	1e-12 s
Ideal reverse transit time	tr	10e-12 s
Base-emitter zero bias depletion capacitance	cje	0.1pF
Base-collector zero bias depletion capacitance	cjc	10fF

Table 6.5.1. HBT model parameters used in the simulation example.

6.5.3. MOSFET (MOSA1)

In this example, we use the universal MOSFET model described in Section 3.9 to simulate a turn-off Transient Analysis of a simple NMOS inverter circuit with an NMOS load at the output. The circuit, the circuit description, and the analysis parameters are shown in Figs. 6.5.9 to 6.5.11, respectively.

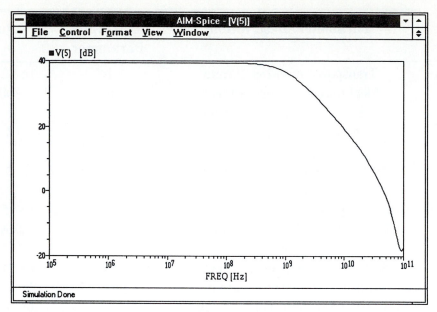

Fig. 6.5.8. Screen dump from the simulation of the HBT differential pair showing the voltage at node 5 in a dB-scale as a function of the frequency.

The subthreshold regime is emphasized by showing the turn-off recovery of the drain current of the switching transistor in a semi-log scale. The inverter and load contain altogether three identical quarter micron transistors with specifications corresponding to the transistor characteristics shown in Fig. 3.9.1.

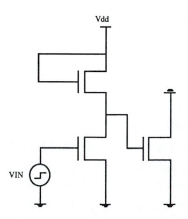

Fig. 6.5.9. Simple NMOS inverter with NMOS load at the output.

```
Simple NMOS inverter
vdd 1 0 dc 3
m1 1 1 2 0 m1 l=0.25u w=5u
rd 2 3 75
* vid is used to measure drain current
vid 3 4 dc 0
m2 4 5 6 0 m2 l=0.25u w=20u
rs 6 0 75
m3 0 2 0 0 m1 l=0.25u w=20u
vin 5 0 dc 0 pulse(0 3 5p 5p 5p 300p 600p)
.model m1 nmos(level=7 rd=75 rs=75 vmax=4e4 vto=0.44
+ tox=6.6e-9)
.model m2 nmos(level=7 rd=0 rs=0 vmax=4e4 vto=0.44
+ tox=6.6e-9)
```

Fig. 6.5.10. Circuit description for the simple inverter.

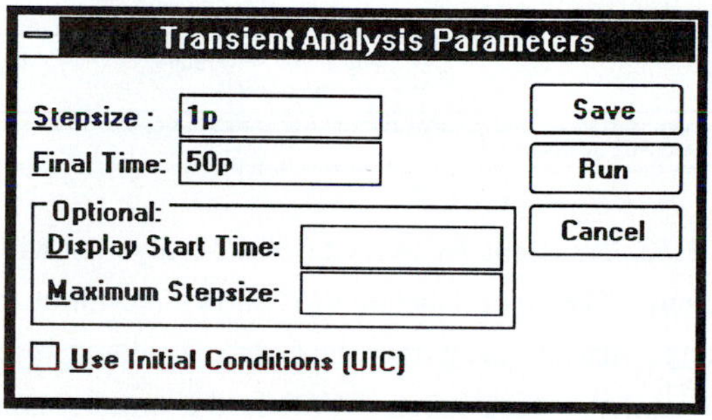

Fig. 6.5.11. Transient Analysis parameters.

Fig. 6.5.12 is a screen-dump from AIM-Postprocessor showing a composite view of relevant voltages and currents from the simulation.

6.5.4.   MESFET (MESA1)

In this example, we simulate a 3 stage ring oscillator composed of MESFET inverters with ungated loads. The model $I–V$ characteristic of the ungated load is as follows:

$$I_L = I_{Lsw} \tanh(V_L / V_{Lss})(1 + \lambda_L V_L)$$

where λ_L is the load device output conductance parameter, V_L is the drain-source voltage across the load, and V_{Lss} is an empirical parameter.

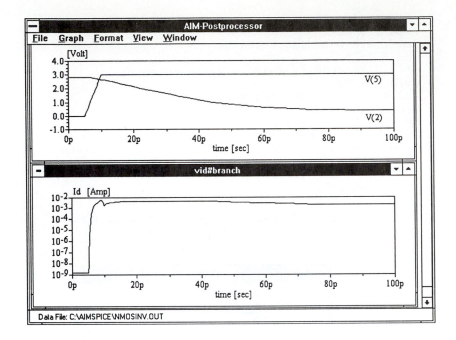

Fig. 6.5.12. Transient Analysis of NMOS inverter with NMOS load at the output. The trace labeled V(5) shows the input to the gate of the switching transistor (the gate of the inverter load transistor is fixed at V_{DD} = 3.0) and the trace labeled V(2) is the drain-source voltage of the switching transistor. The lower trace shows the drain current of the switching transistor.

The parameter V_{Lss} is given by

$$V_{Lss} = I_{Lsw} R_{offw}$$

where I_{Lsw} is the saturation current and R_{offw} is the resistance in the linear region. We used parameter values from Shur (1987) given in Table 6.5.2.

Parameter	Value
λ_L	0.027
I_{Lsw}	250 A/m
R_{offw}	2.24 Ω mm

Table 6.5.2. Ungated load parameters.

The 3 stage ring oscillator and its circuit description are shown in Figs. 6.5.13 and 6.5.14, respectively.

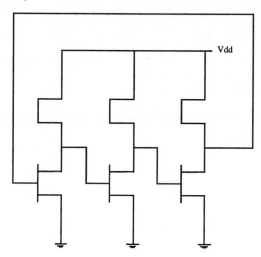

Fig. 6.5.13. 3-Stage MESFET ring oscillator with ungated loads.

```
Mesfet Ring Oscillator with ungated load/Wload=1e-6
*
* Output node is 70, 80 or 90
*
bl1 10 40 i=0.00025*tanh(v(10,40)/0.00025/2240)*
+(1+v(10,40)*0.027)
bl2 20 50 i=0.00025*tanh(v(20,50)/0.00025/2240)*
+(1+v(20,50)*0.027)
bl3 30 60 i=0.00025*tanh(v(30,60)/0.00025/2240)
+*(1+v(30,60)*0.027)

zd1 100 90 130 driver l=0.7u w=20u
zd2 110 70 140 driver l=0.7u w=20u
zd3 120 80 150 driver l=0.7u w=20u

rdl1 1 10 20
rdl2 1 20 20
rdl3 1 30 20
rsl1 40 70 20
rsl2 50 80 20
rsl3 60 90 20

rdd1 70 100 20
rdd2 80 110 20
rdd3 90 120 20
rsd1 130 0 20
rsd2 140 0 20
rsd3 150 0 20
```

```
ci1 70 0 20f
ci2 80 0 20f
ci3 90 0 20f

vdd 2000 0 dc 1.6
visrc 2000 1 dc 0

.model driver nmf level=2 jsdf=0.1 n=1.44 rd=0 rs=0
+ vs=1.9e5 mu=0.25 d=1e-7 vto=0.15 m=2 lambda=0.15
+ sigma0=0.02 vsigmat=0.5
```

Fig. 6.5.14. Circuit description of 3-Stage MESFET ring oscillator with ungated loads.

We used load widths between 1 μm and 4 μm. The ungated loads were modeled directly in the circuit description with non-linear dependent current sources (see Fig. 6.5.14).

The model parameters for the switching transistor were extracted using the procedure outlined in Section 4.4.5, and experimental data was taken from Fig. 9.3.2. in Shur (1987). Table 6.5.3 gives a complete list of parameter values.

Parameter Description	Parameter	Value
Gate length	l	0.7μm
Gate width	w	20μm
Forward saturation current density	jsdf	0.1A/m^2
Emission coefficient	n	1.44
Drain resistance	rd	20Ω
Source resistance	rs	20Ω
Saturation velocity	vs	1.9E5m/s
Low field mobility	mu	0.25m^2/Vs
Channel thickness	d	1E-7m
Threshold voltage	vto	0.15V
Knee shape parameter	m	2
Output conductance parameter	lambda	0.15V^{-1}
DIBL parameter 1	sigma0	0.02
DIBL parameter 2	vsigmat	0.5V
DIBL parameter 3	vsigma	0.1V

Table 6.5.3. MESA1 model parameters used for the 3-Stage MESFET Ring Oscillator.

The parameters for the transient analysis are shown in Fig. 6.5.15. As you can see, we have turned the UIC option *on* to force AIM-Spice to skip the bias point calculation since an oscillator does not have a quiescent operating point.

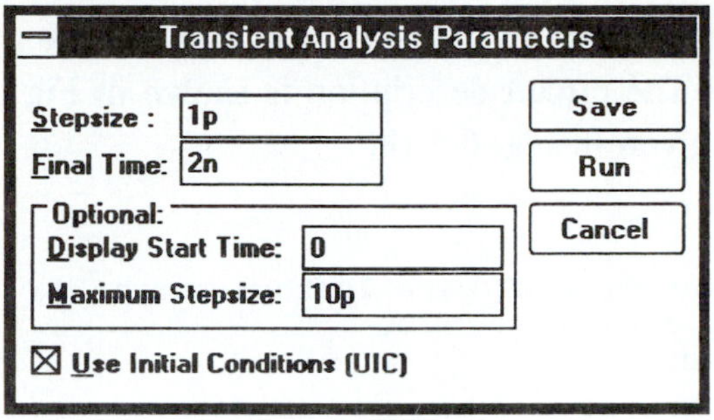

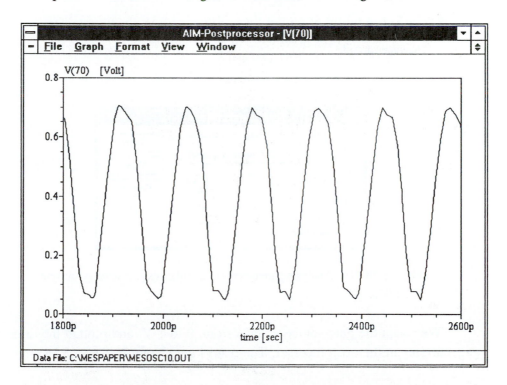

Fig. 6.5.15. Transient Analysis parameters for MESFET ring oscillator.

The output waveform of the ring oscillator is shown in Fig. 6.5.16.

Fig. 6.5.16. MESFET oscillator output waveform (see Problem 6-5-1).

6.5.5.   HFET (HFETA)

The HFET example uses the model described in Section 4.6, and the *I–V* characteristic model parameters are the same as those given in Fig. 4.6.2. In this example, we emphasize the modeling of gate leakage current. We simulated a single HFET device with the values of the gate leakage model parameters shown in Table 6.5.4. The circuit description is shown in Fig. 6.5.17 and the analysis parameters are given in Fig. 6.5.18.

```
Simulation of HFET gate leakage current
a1 4 2 5 hfet l=1u w=10u
vgs 3 0 dc 0
vds 1 0 dc 0.2
vig 3 2 dc 0
vid 1 4 dc 0
vis 5 0 dc 0
.model hfet nhfet rd=60 rs=60 rgs=8 rgd=8 js1d=1 m1d=2
+ js2d=1e11 js1s=1 m1s=2 js2s=1e11
```

Fig. 6.5.17. Circuit description for HFET device used to simulate gate leakage current.

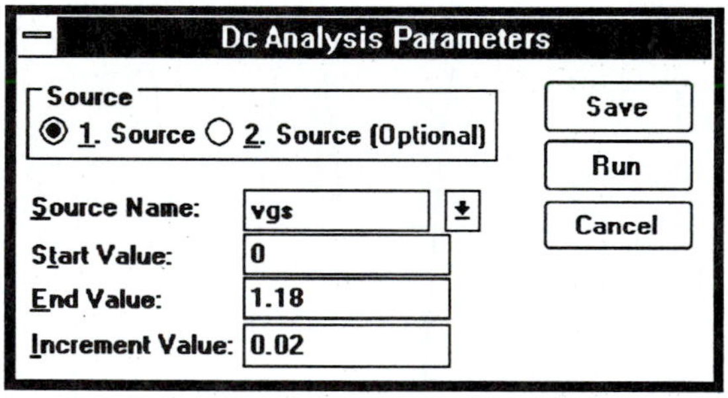

Fig. 6.5.18. DC Analysis parameters for simulating gate leakage current in HFETs.

We swept the gate-source voltage from 0 to 1 V and plotted the drain and gate currents in the postprocessor. The results are shown in Fig. 6.5.19.

Parameter Description	Parameter	Value
Gate-source ohmic resistance	rgs	8 Ohm
Gate-drain ohmic resistance	rgd	8 Ohm
Forward gate-drain diode saturation current density	js1d	1.0 A/m^2
Forward gate-drain diode ideality factor	m1d	2.0
Reverse gate-drain diode saturation current density	js2d	1e11 A/m^2
Forward gate-source diode saturation current density	js1s	1.0 A/m^2
Forward gate-source diode ideality factor	m1s	2.0
Reverse gate-source diode saturation current density	js2s	1e11 A/m^2

Table 6.5.4. Gate leakage model parameters for the HFET.

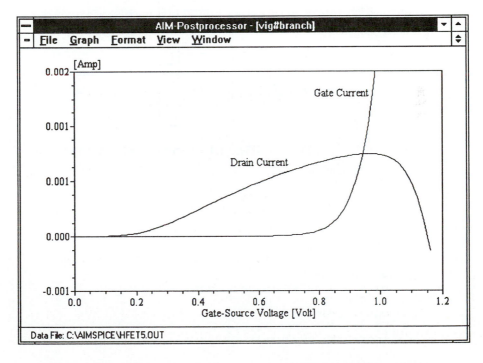

Fig. 6.5.19. HFET gate and drain currents. Drain-source bias is 0.2 V.

6.5.6. Thin Film Transistors (ASIA1)

The circuit shown in Fig. 6.5.20 (from Hack et al. (1989)) and the corresponding AIM-Spice circuit description in Fig. 6.5.21 are used as an example circuit for our Amorphous Silicon TFT model. This is an analog copy amplifier based on an a-Si TFT. A light pulse shines on the photo-diode from time $t = 5$ ms to $t = 19$ ms, and during this pulse the diode delivers a current of 4 nA to the gate, acting as a current source. Values of the circuit components are given in the caption of Fig. 6.5.20. The model parameters of the TFT different from the default values are given in Table 6.5.5.

This example is a tutorial on how to simulate a circuit and produce plots. The underlined text in this example are your actions.

Many of the menu commands have accelerator keys. If a command has an accelerator key, it is shown in parentheses after the menu command.

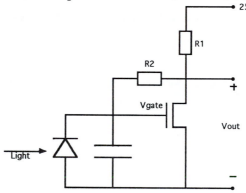

Fig. 6.5.20. a-Si TFT Amplifier. R1 = 250 MΩ, R2 = 4650 MΩ, the diode capacitance is 0.4 pF.

```
Analog copy amplifier
il 0 3 pwl(0 0 5m 0 5.01m 4n 19m 4n 19.01m 0)
cl 3 0 0.4pF
r1 1 2 250Meg
r2 2 3 4650Meg
rd 2 6 75
vid 6 4 dc 0
m1 4 3 5 0 TFT l=10u w=20u
rs 5 0 75
vdd 1 0 dc 25
.model TFT nmos level=11 rd=0 rs=0 vto=0 uo=0.5
+tox=0.3u n0=3.3e15
```

Fig. 6.5.21. AIM-Spice circuit description for the circuit in Fig.6.5.20.

Parameter Description	AIM-Spice Parameter	Value
Device length	l	10μm
Device width	w	20μm
Oxide thickness	tox	0.3μm
Scaling factor	n0	$3.3e15m^{-3}$
Mobility	uo	0.5 cm^2/Vs

Table. 6.5.5. ASIA1 model parameters used in the simulation.

The following is a step-by-step simulation session with AIM-Spice:

Load AIM-Spice

Type the circuit description shown in Fig. 6.5.21 in the untitled circuit window

Choose Transient Analysis from the Analysis menu (Ctrl+T)

Specify the analysis parameters shown in Fig. 6.5.22.

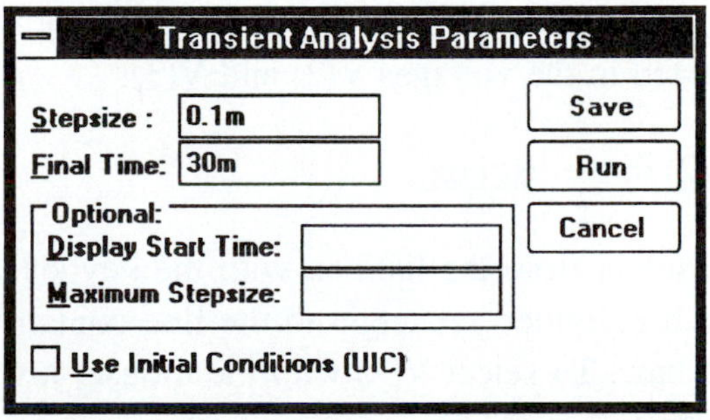

Fig. 6.5.22. Transient Analysis parameters used for the copy amplifier shown in Fig. 6.5.20.

Before we simulate our copy amplifier, it is wise to save the circuit in a circuit file. For this, we use the File Save command.

<u>Choose File Save command (Shift+F12)</u>

A Save As dialog box appears. This dialog box is displayed because the Save command has no file to save to and instead the Save As command is activated.

<u>Type the filename to which you want to save the circuit. Then choose OK</u>

We are now ready to run the simulation.

<u>Choose Transient Analysis from the Analysis menu (Ctrl+T), and choose the button labeled Run in the dialog box</u>

AIM-Spice is now entering the analysis mode. The circuit window is hidden and the status bar at the bottom of the main window displays "Parsing Circuit, Please Wait". After the circuit is loaded into the AIM-Spice kernel, the status bar displays which analysis type is about to be run. In our case it is "Transient Analysis". Before we start the simulation, we have to select circuit variables to be plotted during the simulation.

<u>Choose Select Variables to Plot from the Control menu (Ctrl+V)</u>

A dialog box with a list of all variables in the circuit is shown. The list box is a multiple selection list box. This means that you can select more than one variable to display in a single plot window. In our case, we are interested in the voltages $V(2)$ and $V(3)$.

<u>Select V(2) in the list box.</u>

To select from the list box with the keyboard, use the arrow keys to move the highlighted rectangle to the line containing $V(2)$ and then press the space bar. To select $V(2)$ with the mouse, just click on it with the left mouse button.

<u>Choose OK</u>

An empty plot window is created with a title reflecting which variables we chose to plot in that window. To make AIM-Spice draw a graph in this window, we have to specify the y-axis limits. To set the limits, we use the Analysis Limits command from the Control menu. It is wise to specify fairly large intervals as a first guess, and as the simulation progresses, the plot traces appear on the screen can be used as a guide for resetting the limits.

<u>Choose Analysis Limits from the Control menu (Ctrl+A)</u>

Specify −1.0 and 4.0 for the lower and upper limits of for V(2), respectively.

Repeat the above steps to create another plot window, now containing variable V(3). Use 0.0 and 20.0 as lower and upper limits for V(3).

Note that when you have more than one plot window and you want to use, for example, the Analysis Limits command or commands from the Format menu, you have to activate the plot window in which you want the commands to act before you choose the command. To activate a certain plot window with the mouse, just click anywhere inside it. With the keyboard it is slightly more complicated. Use Ctrl+F6 to activate one plot window after another until the one you want to be active is in front and its caption bar is highlighted.

You are now ready to start the simulation.

<u>Choose Start Simulation from the Control menu (Ctrl+S)</u>

During simulation, AIM-Spice displays "Doing Transient Analysis..." in the status bar. While running a simulation, you are able to reset the limits in one or more of the plot windows, you can cancel the simulation with the Stop Simulation command in the Control menu, or you can switch to another application to do other work while AIM-Spice runs in the background.

When the simulation is finished, AIM-Spice makes that clear to you by displaying "Simulation Done" in the status bar. You should now save the simulation results.

<u>Choose Save Plots from the File menu (Alt+F12)</u>

A dialog box with all accumulated plots since last time you saved plots or since you started AIM-Spice is displayed.

<u>Choose the button labeled Save All Plots</u>

A Save As dialog box appears. The default name for the output file is the same as the name of the circuit file, but with the extension OUT.

<u>Choose OK</u>

The next step in our tutorial is to load AIM-Postprocessor and prepare our simulation results for printing. To load the Postprocessor, we have to leave analysis mode.

<u>Choose Exit Analysis Mode from the Control menu (Ctrl+E)</u>

The plot windows are destroyed and the circuit description appears again.

<u>Choose Load Postprocessor from the Postprocessor menu</u>

The main window of AIM-Postprocessor appears in front of AIM-Spice. We are now going to load the output file which we just saved from AIM-Spice.

<u>Choose the Open command from the File menu and specify which file to open. Then choose OK</u>

AIM-Postprocessor reads the file and when the name of the file appears in the status bar, you are ready to create plots with the Add Plot command.

<u>Choose Add Plot from the Graph menu</u>

The Trace Expression Editor (TEE) appears. This editor is described in detail in Section 6.4. In the upper left part of TEE is a list of all variables.

<u>Select V(2) from the list</u>

V(2) appears in the trace expression display window of TEE. This is the expression we want, so we just add it to the list of traces with the Add Expression button.

Choose the button labeled Add Expression

V(2) appears in the previously empty list labeled "Traces:"

Choose the New Plot button

A graph window is created and our expression is plotted in the window. Repeat the above steps to create another graph window containing V(3).

We are now ready to edit the graphs. First tile the windows side by side.

Choose the Tile command from the Window menu

The next step is to change appearance of the legends.

Double click the legend V(2)

A format legend dialog box appears. We are going to change two properties of the legend, the text and the symbol next to the text. In fact, we want to hide the symbol since it is not necessary to display a legend symbol when there is only one trace in a graph.

Change the legend text of Vout. Then remove the check mark in the check box labeled Display Symbol with Legend Text, and choose OK

The graph is redrawn with its new properties. Repeat the above steps for V(3), changing its name to Vgate.

Next, we want to reduce the number of digits used on the x-axis of both graphs. Decimal digits is a property of an axis label. Therefore, we have to format the x-axis label of both graphs.

Double click the x-axis labeling, or choose the X-Label command from the Format menu (Alt+X)

A format label dialog box appears. To change the number of decimal digits, you either type in the number you want, or you use the scroll bar arrows next to the text box to set the number.

<u>Specify zero decimal digits. Then choose OK</u>

The graph is redrawn with the new label properties.

Repeat the above steps to set the number of x-axis digits of the other graph to zero as well.

This completes the tutorial. Now you may choose to print one of the graphs with the Print command from the File menu. One tip for printing: Maximize the graph window you want to print. That way you obtain better printing results.

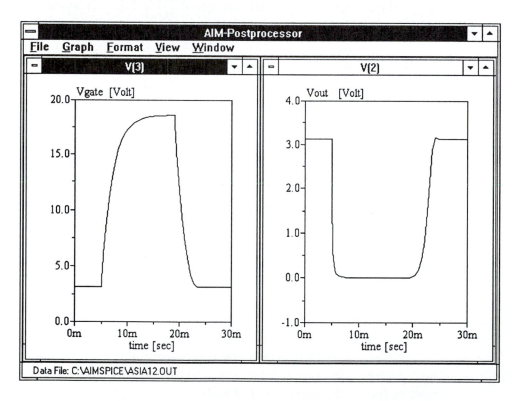

Fig. 6.5.23. Results from the simulation of the analog a-Si TFT copy amplifier. The gate voltage is shown to the left and the output voltage to the right.

REFERENCES

M. HACK, J. G. SHAW and M. SHUR, "Development of Spice Models for Amorphous Silicon Thin-Film Transistors", *Mat. Res. Soc. Symp. Proc.*, Vol. 149, p. 233 (1989)

B. JOHNSON, T. QUARLES, A. R. NEWTON, D. O. PEDERSON and A. SANGIOVANNI-VINCENTILLI, *SPICE3 Version 3e1 User's Manual*, Berkeley (1991)

L. L. LORENZ and R. M. O'MARA, *WINDOWS™-3 Companion*, Microsoft Press, Louisville, Kentucky, ISBN 0-936767-19-7 (1990)

Microsoft Windows Version 3.1 Users Guide, Microsoft Corporation (1992)

L. W. NAGEL, *SPICE 2: A Computer Program to Simulate Semiconductor Circuits*, Memorandum No. ERL-M520, Electronic Research Laboratory, College of Engineering, University of California, Berkeley (1975)

M. SHUR, *GaAs Devices and Circuits*, Plenum, New York (1987)

P. W. TUINENGA, *SPICE. A Guide to Circuit Simulation & Analysis Using PSpice®*, Prentice Hall, New Jersey (1988)

PROBLEMS

6-4-1. You cannot plot currents in AIM-Spice. This has to be done in AIM-Postprocessor. But you have to prepare the circuit description in AIM-Spice before we run the simulation. Which changes have to be made in order to be able to plot a branch current in the postprocessor? Modify our example circuit (the differential pair) from Section 6.3 in such a way that you are able to plot the collector current in transistor Q2.

6-5-1. Reproduce the waveform of the 3-stage MESFET ring oscillator from the MESFET example in Section 6.5. Use a load width of 1 μm. As mentioned in the example, you model the ungated load devices directly in the circuit description with non-linear dependent current sources (see Appendix A9 for a complete description of non-linear dependent sources). Use a power supply voltage of 1.6 V. Chances are that your first run will not reproduce oscillations. What should you do to make the circuit oscillate?

Appendices

APPENDIX A1. PHYSICAL CONSTANTS

Quantity	Symbol	Value
Avogadro number	N_{AV}	6.0221367×10^{23} 1/mol
Bohr energy	E_B	13.6060 eV
Bohr magneton	μ_B	5.78832×10^{-5} eV/T
Bohr radius	a_B	0.52917 Å
Boltzmann constant	k_B	1.38066×10^{-23} J/K
Boltzmann constant/q	k_B/q	8.61738×10^{-5} eV/K
Electronic charge	q	1.60218×10^{-19} C
Electronvolt	eV	1.60218×10^{-19} J
Fine structure constant	α	0.00729735308 ($\approx 1/137$)
Gas constant	R	1.98719 cal mol^{-1} K^{-1}
Gravitational constant	γ	6.67259×10^{-11} m^3/(kg s^2)
Impedance of free space	$1/c\varepsilon_o = \mu_o c$	376.732 Ω
Mass of electron at rest	m_e	$0.91093897 \times 10^{-30}$ kg
Mass of proton at rest	M_p	$1.6726231 \times 10^{-27}$ kg
Permeability in vacuum	μ_o	1.26231×10^{-8} H/cm ($4\pi \times 10^{-9}$)
Permittivity in vacuum	ε_o	8.85418×10^{-12} F/m ($1/\mu_o c^2$)
Planck constant	h	$6.6260755 \times 10^{-34}$ J-s
Reduced Planck constant	$\hbar = h/(2\pi)$	$1.0545727 \times 10^{-34}$ J-s
Speed of light in vacuum	c	2.99792458×10^8 m/s
Standard atmosphere		1.01325×10^5 N/m^2
Thermal voltage at 300 K	$k_B T/q$	0.025860 V
Wavelengths of visible light	λ	0.4 to 0.7 μm

APPENDIX A2.   PROPERTIES OF SILICON (Si)

Atomic number	14
Atoms/cm^3	5.02×10^{22}
Electronic shell configuration	$1s^2\ 2s^2\ 2p^6\ 3s^2\ 3p^2$
Atomic weight	28.09
Crystal structure	Diamond
Breakdown field (V/cm)	$\sim 3.0 \times 10^5$
Density (g/cm^3)	2.329 (at 298 K)
Dielectric constant	11.7
Diffusion constant (cm^2/s) (at 300 K)	37.5 (electrons) 13 (holes)
Effective density of states	
in the conduction band (cm^{-3})	2.8×10^{19} (at 300 K)
in the valence band (cm^{-3})	1.04×10^{19} (at 300 K)
Effective electron mass (in unit of m$_e$)	longitudinal : 0.92 (at 1.26 K)
	transverse : 0.19 (at 1.26 K)
	density of states : 1.28 (at 600 K)
	1.18 (at 300 K) 1.08 (at 77 K)
	1.026 (at 4.2 K)
Effective hole mass (in unit of m$_e$)	heavy hole : 0.537 (at 4.2 K)
	heavy hole : 0.49 (at 300 K)
	light hole : 0.153 (at 4.2 K)
	light hole : 0.16 (at 300 K)
	density of states : 0.591 (at 4.2 K)
	0.62 (at 77 K) 0.81 (at 300 K)
Electron Affinity (V)	4.05
Energy gap (eV)	1.12 (at 300 K) 1.17 (at 77 K)
Index of refraction	3.42
Intrinsic carrier concentration (cm^{-3})	1.02×10^{10} cm^{-3} (at 300 K)
Intrinsic Debye length (μm)	24
Intrinsic resistivity (ohm-cm)	3.16×10^5 (at 300 K)
Lattice constant (Å)	5.43107 (at 298.2 K)
Melting point (°C)	1412
Mobility (cm^2/V-s) (at 300 K)	1450 (electrons) 500 (holes)
Optical phonon energy (eV)	0.063
Specific heat (J/g-°C)	0.7
Thermal conductivity (W/cm-°C)	1.31 (at 300 K)
Thermal diffusivity (W/cm-°C)	0.9
Thermal expansion, linear (°C^{-1})	2.6×10^{-6} (at 300 K)
Young's modulus (dyn/cm^2)	1.9×10^{12} in [111] direction

APPENDIX A3.   PROPERTIES OF GALLIUM ARSENIDE (GaAs)

Crystal Structure	Zinc blende
Breakdown field (V/cm)	~4.0x10^5
Density (g/cm^3)	5.3176 (at 298 K)
Dielectric constant (κ_s)	12.9 (at 300 K)
(κ_o)	10.89 (at 300 K)
Diffusion Constant (cm^2/s) (at 300 K)	207 (electrons) 10 (holes)
Effective density of states	
in the conduction band (cm^{-3})	4.7x10^{17} (at 300 K)
in the valence band (cm^{-3})	7.0x10^{18} (at 300 K)
Effective electron mass (in units of m_e)	0.067 (0 K) 0.063 (300 K)
Effective hole mass (in units of m_e)	heavy hole : 0.51 (at T < 100 K)
	: 0.50 (at 300K)
	light hole : 0.084 (at T < 100K)
	: 0.076 (at 300K)
	density of states : 0.53
Electron Affinity (V)	4.07
Energy gap (eV)	1.424 (at 300 K)
	1.507 (at 77 K)
	1.519 (at 0 K)
Index of refraction	3.3
Intrinsic carrier concentration (cm^{-3})	2.1x10^6 (at 300 K)
Intrinsic Debye length (μm)	2250 (at 300 K)
Intrinsic resistivity (ohm-cm)	10^8 (at 300 K)
Lattice constant (Å)	5.6533 (at 300 K)
Melting point (^{o}C)	1240
Mobility (cm^2/V-s)	8500 (electrons at 300 K)
	400 (holes at 300 K)
Optical phonon energy (eV)	0.035
Specific heat (J/g-^{o}C)	0.35
Thermal conductivity (W/cm-^{o}C)	0.46
Thermal diffusivity (W/cm-^{o}C)	0.44
Thermal expansion, linear (^{o}C^{-1})	6.86x10^{-6} (at 300 K)

APPENDIX A4.   PROPERTIES OF $Al_xGa_{1-x}As$

Crystal Structure	Zinc blende
Density (g/cm^3)	5.36 - 1.6x
Dielectric constant (κ_s)	13.18 - 3.12x (at 300 K)
(κ_o)	10.89 - 2.78x (at 300 K)

Effective electron mass   0.067 + 0.083x (Γ-minimum, density of states)
(in units of m$_e$)   0.85 - 0.14x (X-minimum, density of states)

0.56 + 0.10x (L-minimum, density of states)
0.067 + 0.083x (G-minimum, conductivity)
0.32 - 0.06x (X-minimum, conductivity)
0.11 - 0.03x (L-minimum, conductivity)

Effective hole mass   heavy hole   :0.62 + 0.14x (density of states)
(in units of m$_e$)   light hole   :0.087 + 0.063x (density of states)

split-off band :0.15 + 0.09x (density of states)

Electron Affinity (V)   4.07 - 1.1x ($x < 0.45$)
3.64 - 0.14x ($0.45 < x < 1.0$)

Energy gap(eV)   1.424 + 1.247x ($x < 0.45$)
1.9 + 0.125x + 0.143x^2 ($0.45 < x < 1.0$)

Lattice constant (Å)   5.6533 + 0.0078x
Melting point (^{o}C)   1511 - 58x + 560 x^2 (solidus curve)
1511 + 1082x - 580x^2 (liquidus curve)

Mobility (cm^2/Vs), electrons $\approx$8000 – 22000x + 10000x^2 (for $x < 0.45$)
$\approx$–255 + 1160x - 720x^2 (for $x > 0.45$)

Mobility (cm^2/Vs), holes $\approx$370 – 970x + 740x^2
Thermal resistivity (cm-K/W)   2.27 + 20.83x - 30x^2
Thermal expansion, linear (10^{-6} K^{-1})   6.4 - 1.2x
Young's modulus (10^{11} dyn/cm^2)   8.53 - 0.18x
Valence band discontinuity (eV) at $\Delta E_v = 0.4\Delta E_{gg}$
  the $Al_xGa_{1-x}As/In_yGa_{1-y}As$ where ΔE_{gg} (eV) $= 1.247x + 1.5y - 0.4y^2$
  heterointerface   is the difference between Γ valleys in
$Al_xGa_{1-x}As$ and $In_yGa_{1-y}As$

Energy gap discontinuity (eV) $\Delta E_g = \Delta E_{gg}$ for $x < 0.45$
$$\Delta E_g = 0.476 + 0.125x + 0.143x^2 + 1.5y - 0.4y^2$$
for $x \geq 0.45$

Conduction band discontinuity (eV) $\Delta E_c = \Delta E_g - \Delta E_v$

APPENDIX A5. PROPERTIES OF AMORPHOUS Si (a-Si)

Fermi level in intrinsic material, $E_c - E_{Fo}$ (eV)	~0.6 - 0.7
Deep localized states density at $E = E_{FO}$ (cm^{-3}/eV)	$10^{15} - 10^{16}$
Deep states characteristic energy (meV)	86
Tail localized states density at $E = E_c$ (cm^{-3}/eV)	~2×10^{21}
Tail states characteristic energy (meV)	23
Width of tail states band (meV)	~150
Dielectric constant	~11
Diffusion constant (cm^2/s) at 300 K)	
electrons in the conduction band	0.26 - 0.52
holes in the valence band	0.13 - 0.26
Effective density of states in conduction band (cm^{-3})	~10^{19} (at 300 K)
Effective density of states in valence band (cm^{-3})	~10^{19} (at 300 K)
Energy gap (eV)	1.72 (at 300 K)
Index of refraction	3.32
Mobility (cm^2/Vs) (at 300 K)	
electrons in the conduction band	10 - 20
holes in the valence band	5 - 10

APPENDIX   A6.   PROPERTIES   OF SiO_2 AND Si_3N_4

	SiO_2	Si_3N_4
Breakdown field (V/cm)	10^7	10^7
Density (g/cm^3)	2.2	3.1
Dielectric Constant (κ_s)	3.9	7.5
Dielectric Constant (κ_o)	2.13	4.2
Energy gap (eV)	9	~5
Index of refraction	1.46	2.05
Resistivity (ohm-cm) at 300 K	10^{14} - 10^{16}	10^{14}
Resistivity (ohm-cm) at 500 K		2×10^{13}
Thermal conductivity (W/cm-K)	0.014	
Thermal expansion, linear (^{o}C^{-1})	5×10^{-7} (at 300 K)	

APPENDIX A7. APPROXIMATE SOLUTION OF THE NON-IDEAL DIODE EQUATION

The current, I, of an ideal diode in series with a resistance, R, can be expressed as a function of the applied voltage, V, by the following generalized diode equation

$$I = I_s\left[\exp\left(\frac{V - IR}{V_{th}}\right) - 1\right] \tag{A7-1}$$

where I_s is the saturation current and $V_{th} = k_B T/q$ is the thermal voltage. Expressions of this general form appear in many problems of physics and engineering where linear and exponential responses are combined. Obvious examples are photodetectors, solar cells and diodes used as circuit elements. Additional problems, not related to diodes, involving expressions of this form, include the solution of the equation for the Unified Charge Control Model (UCCM) (see Chapter 3 and Moon et al. (1991)) and the solution of equations describing the shape of Gunn domains (see Shur 1987).

In all these cases, one needs to express I explicitly as a function of V, without having to resort to iterative or other numerical routines. In the case of UCCM, such an analytical solution is needed in order to implement this model efficiently in circuit simulators, such as AIM-Spice, where explicit analytical expressions are required in both device models and in parameter extraction routines in order to avoid intolerable computational delays. Here, we derive an approximate but very precise analytical solution of eq. (A7-1).

Our approximate analytical solution and its first and second derivatives are continuous. These derivatives can be easily calculated if needed. (Our motivation for this work came from the realization that solving the generalized diode equation used in UCCM by Newton-Raphson technique slows down our circuit simulation program AIM-Spice)

For simplicity, we rewrite eq. (A7-1) in terms of the normalized variables $i = R(I + I_s)/V_{th}$ and $u = (V + RI_s)/V_{th} + \ln(RI_s/V_{th})$:

$$i = \exp(u - i) \tag{A7-2}$$

We propose to obtain an approximate solution of i as a function of u is as follows (see Fjeldly et al. (1991): First we introduce a trial function, $i_t(u)$, which reproduces roughly the correct solution for all values of u. It is required, however, that the trial function has the correct asymptotic behavior for large positive and negative values of u. A very precise solution is then obtained through a series expansion procedure by utilizing the functional properties of eq. (A7-2).

For now, we assume that a proper trial function has been established, and start by discussing the expansion procedure. We will return to the derivation of the trial function below. Fig. A7.1 illustrates the basic principle and the steps of the procedure.

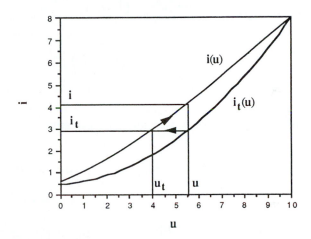

Fig. A7.1. Principle of deriving a precise approximate solution of the non-ideal diode equation. (From Fjeldly et al. (1991), © 1991 IEEE.)

For a given value of u, the trial function has the value i_t. However, by solving eq. (A7-2) with respect to u we find the exact value u_t corresponding to $i = i_t$

$$u_t = i_t + \ln(i_t)$$ (A7-3)

A solution for $i(u)$ can now be found by performing a Taylor series expansion of eq. (A7-2) about $u = u_t$

$$i(u) = i_t + \frac{di}{du}\bigg|_{u_t} (u - u_t) + \frac{1}{2}\frac{d^2 i}{du^2}\bigg|_{u_t} (u - u_t)^2 + \ldots$$ (A7-4)

Depending on the trial function, a very precise approximation to the correct solution is obtained by terminating the expansion after the first few terms. The exact derivatives to be used in the expansion are obtained from eq. (A7-2):

$$\frac{di}{du}\bigg|_{u_t} = \frac{i_t}{1+i_t} \tag{A7-5}$$

$$\frac{d^2 i}{du^2}\bigg|_{u_t} = \frac{i_t}{(1+i_t)^3} \tag{A7-6}$$

$$\frac{d^3 i}{du^3}\bigg|_{u_t} = \frac{i_t(1-2i_t)}{(1+i_t)^5} \tag{A7-7}$$

etc. To second order expansion, we can write the approximate solution as

$$i(u) \cong i_t(u)\left\{1 + \frac{u - u_t(u)}{1+i_t(u)} + \frac{1}{2}\frac{[u - u_t(u)]^2}{[1+i_t(u)]^3}\right\} \tag{A7-8}$$

We now return to the problem of finding a proper trial function, $i_t(u)$. First we consider eq. (A7-2) in the limit $i \ll 1$. In this limit, it is possible to make the expansion $\exp(-i) \approx 1 - i$ on the right hand side of eq. (A7-2), and the trial function can be written as

$$i_t(u) = \exp(u)\left[1 - \exp(u)\right] , \qquad \text{for } u \leq u_o \tag{A7-9}$$

The upper limit of validity of eq. (A7-9) can be defined by setting $\exp(u_o) = \delta$, where δ is a small number, i.e., $u_o = \ln(\delta) \approx -2.303$ for $\delta = 0.1$.

For large values of u, another asymptotic form of $i(u)$ becomes apparent. To see this, eq. (A7-2) is converted to

$$i = u - \ln(i) \tag{A7-10}$$

In the limit $u \rightarrow \infty$, we have $i \rightarrow u$. By replacing this limit for i on the right hand side of eq. (A7-10), we obtain the approximation

$$i \approx u - \ln(u) \tag{A7-11}$$

The lower limit of validity of eq. (A7-11), u_1, is found by requiring that $\ln(u_1)/u_1 = \delta$, where again δ is a small number, typically on the order of 0.1. A numerical solution with $\delta = 0.1$ gives $u_1 \approx 35.77$.

   However, a proper trial function $i_t(u)$ also has to bridge smoothly and monotonically the gap $u_o \le u \le u_1$. Furthermore, we require continuity in its functional value and its first and second derivative everywhere. We propose to achieve this by modifying eq. (A7-11) to extend its validity to the entire range $u \ge u_o$. The singularity at $u = 0$ is removed by rewriting the argument of the logarithm as follows:

$$u \rightarrow \frac{1}{2}\left[u - u_o + \sqrt{(u - u_o)^2 + c^2} \right] \tag{A7-12}$$

where c is a constant. This replacement gives the correct asymptotic behavior for large values of u and it causes the logarithm to be well-behaved everywhere. In addition, a term, $a \exp((u_o - u)/b)$, is added to the trial function in order to obtain continuity in i and its first and second derivatives at $u = u_o$. The exponential ensures that the added term vanishes in the asymptotic region. The constants a, b and c are determined numerically from the continuity conditions at $u = u_o$. Hence, the total trial function can be written as

$$i_t(u) = \begin{cases} \exp(u) \left[1 - \exp(u) \right] , & \text{for } u \le u_o \\[2em] u + a \exp\left(\dfrac{u_o - u}{b} \right) \\[1em] \quad - \ln\left[\dfrac{u - u_o}{2} + \sqrt{\left(\dfrac{u - u_o}{2} \right)^2 + \left(\dfrac{c}{2} \right)^2} \right] , & \text{for } u > u_o \end{cases} \tag{A7-13}$$

where

$$\begin{aligned} u_o &= -2.303 \\ a &= 2.221 \\ b &= 6.804 \\ c &= 1.685 \end{aligned} \tag{A7-14}$$

   The relative error in this trial function peaks at about 12 % near $u = 0$. In other regions, the error is much less and becomes negligibly small well inside the asymptotic regions $(u < u_o$ and $u > u_1)$. However, by combining eq. (A7-

13) with the series expansion discussed earlier (eqs. (A7-3) – (A7-8)), the error can be made arbitrarily small for all values of u. Fig. A7.2 shows the function $i(u)$ and Fig. A7.3 shows the u-dependence of the relative error of $i_t(u)$, and of $i_t(u)$ combined with a first and a second order series expansion.

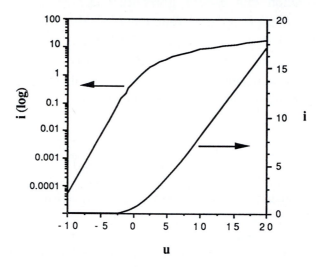

Fig. A7.2. The normalized non-ideal diode function in a logarithmic and a linear vertical scale. (From Fjeldly et al. (1991), © 1991 IEEE.)

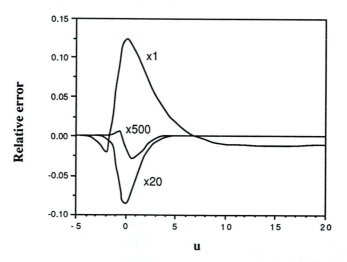

Fig. A7.3. Relative error in the trial function (x1) and in solutions based on the trial function combined with a first (x20) and a second order (x500) series expansion. (From Fjeldly et al. (1991), © 1991 IEEE.)

As can be seen from Fig. A7.3, the trial function may be accurate enough for relatively crude calculations and the analytical solutions utilizing the first and especially second order expansions provide an excellent agreement with the exact numerical solution.

A much simpler, but far less accurate trial function than eq. (A7-13) is

$$i_t(u) \approx \ln[1 + \exp(u)] \qquad (A7\text{-}15)$$

However, this trial function has the advantage of being a continuous, single expression function for all values of u.

References

T. A. FJELDLY, B. MOON and M. SHUR, "Analytical Solution of Generalized Diode Equation", *IEEE Trans. Electron Devices*, ED-38, No. 8, pp. 1976-1977, August (1991)

B. MOON, C. PARK, K. RHO, K. LEE, M. SHUR and T. A. FJELDLY, "Analytical Model for *p*-Channel MOSFETs", *IEEE Trans. Electron Devices*, ED-38 (1991)

M. SHUR, *GaAs Devices and Circuits*, Plenum, New York (1987)

APPENDIX A8.   SUMMARIES OF AIM-Spice MODELS

A8.1. Heterostructure Diode Model (HDIA)

Based on the discussion of the heterostructure diode in Section 1.10, we can write a system of model equations for this device.   For a diode with finite lengths, X_1 and X_2, of the p- and n-type regions, respectively, and with a spike in the conduction band (see Fig. 1.10.3), the characteristics of the diode can be written as follows (see Section 1.10 for details):

$$J = \left(\frac{qD_n\, n_{po}}{L_n \zeta_n} + \frac{qD_p\, p_{no}}{L_p \zeta_p} \right)\left[\exp\left(\frac{V}{V_{th}} \right) - 1 \right] \tag{A8.1-1}$$

$$\zeta_n \approx \tanh\left(\frac{X_p - x_1}{L_n} \right) + \frac{D_n}{v_n L_n}\exp\left(\frac{\Delta E_n}{q V_{th}} \right) \tag{A8.1-2}$$

$$\zeta_p \approx \tanh\left(\frac{X_n - x_2}{L_p} \right) \tag{A8.1-3}$$

$$\Delta E_n = \Delta E_c - q V_{d1} \tag{A8.1-4}$$

$$V_{d1} = \xi\left(V_{bi} - V \right) \tag{A8.1-5}$$

$$\xi = \frac{\varepsilon_1 N_d}{\varepsilon_1 N_d + \varepsilon_2 N_a} \tag{A8.1-6}$$

$$q V_{bi} = E_{g2} - \Delta E_{Fn} - \Delta E_{Fp} + \Delta E_c \tag{A8.1-7}$$

$$c_d = \sqrt{ \frac{q \varepsilon_1 \varepsilon_2\, N_a N_d}{2 V_{bi}\left(\varepsilon_1 N_a + \varepsilon_2 N_d \right)} } \tag{A8.1-8}$$

For very short diodes, the hyperbolic tangents can be expanded to first order in $(X_p - x_1)/L_n$ and $(X_n - x_2)/L_p$. Furthermore, a practical heterostructure diode construction would be of the type p^+-n or n^+-p. In the case of a p^+-n diode, we will normally have $V_{d1} \ll \Delta E_c$, $\xi \ll 1$ and $\Delta E_n \gg V_{th}$. The current density and the diffusion capacitance of such a diode can be written as

$$J \approx \left[q n_{po} \, v_n \exp\left(-\frac{\Delta E_n}{q V_{th}} \right) + \frac{q D_p \, p_{no}}{X_n - x_2} \right] \left[\exp\left(\frac{V}{V_{th}} \right) - 1 \right] \qquad (A8.1\text{-}9)$$

$$c_{dif} \approx \left[\frac{q p_{no} L_p^3}{2 V_{th} (X_n - x_1)^2} + \frac{q n_{po} v_n^2 \tau_n^2}{2 V_{th} L_n} \exp\left(-\frac{2 \Delta E_n}{q V_{th}} \right) \right] \exp\left(\frac{V_o}{V_{th}} \right) \qquad (A8.1\text{-}10)$$

A8.2. HBT Model (HBTA)

Based on the device physics of the HBT discussed in Section 2.7, we can write a simplified system of equations to model this device. This system of equations is nearly identical to the Ebers-Moll model (see Sections 2.4). However, the equation for the emitter current now also accounts for the important surface recombination term, I_{rs} (see Section 2.7 for details):

$$I_e = I_{cc} + I_{be} + I_{re} + I_{rb} + I_{rs} \qquad (A8.2\text{-}1)$$

$$I_c = I_{cc} + I_{bc} + I_{rc} \qquad (A8.2\text{-}2)$$

$$I_b = I_e - I_c \qquad (A8.2\text{-}3)$$

$$I_{be} = \frac{I_F}{\beta_N} = \frac{I_i}{\beta_N} \left[\exp\left(\frac{V_{be}}{V_{th}} \right) - 1 \right] \qquad (A8.2\text{-}4)$$

$$I_{bc} = \frac{I_R}{\beta_I} = \frac{I_i}{\beta_I} \left[\exp\left(\frac{V_{bc}}{V_{th}}\right) - 1 \right]$$ (A8.2-5)

$$I_{cc} = I_F - I_R = I_i \left[\exp\left(\frac{V_{be}}{V_{th}}\right) - \exp\left(\frac{V_{bc}}{V_{th}}\right) \right]$$ (A8.2-6)

$$I_i = -\frac{\alpha_N I_{eo}}{1 - \alpha_N \alpha_I}$$ (A8.2-7)

$$V_{be} = V_{be}^{ext} - R_e I_e - R_b I_b$$ (A8.2-8)

$$V_{bc} = V_{bc}^{ext} - R_c I_c$$ (A8.2-9)

$$I_{re} = I_{re}^o \left[\exp\left(\frac{V_{be}}{m_{re} V_{th}}\right) - 1 \right]$$ (A8.2-10)

$$I_{rb} = I_{rb}^o \left[\exp\left(\frac{V_{be}}{V_{th}}\right) - 1 \right]$$ (A8.2-11)

$$I_{rs} = A_{rs1} P_e \left[\exp\left(\frac{V_{be}}{V_{th}}\right) - 1 \right] + A_{rs2} P_e \left[\exp\left(\frac{V_{be}}{2V_{th}}\right) - 1 \right]$$ (A8.2-12)

The above equations may be further simplified in a first-order analysis by considering only the active forward mode of operation. In this case, we have

$$I_e \approx I_F \left(1 + \beta_N^{-1}\right) + I_{re} + I_{rb} + I_{rs}$$ (A8.2-13)

$$I_c \approx I_F$$ (A8.2-14)

$$\beta = \frac{I_c}{I_b} = \frac{\beta_o}{1 + \alpha_o \exp\left(-\dfrac{V_{be}}{2V_{th}}\right)} \tag{A8.2-15}$$

$$\alpha_o = \frac{I_{re}^o + A_{rs2} P_e}{I_i/\beta_N + I_{rb}^o + A_{rs1} P_e} \tag{A8.2-16}$$

$$\beta_o = \frac{I_i}{I_i/\beta_N + I_{rb}^o + A_{rs1} P_e} \tag{A8.2-17}$$

At relatively small collector currents, just as in a conventional BJT, we expect the collector current to be proportional to $\exp(V_{be}/V_{th})$ and the overall recombination current to be proportional to $\exp(V_{be}/2V_{th})$. As a consequence, β should be proportional to $I_c^{1/2}$. At very high collector current, the common emitter gain should reach a plateau if the critical current for base push out, I_{crit}, is large enough. However, in practical devices where the collector is doped relatively low in order to obtain a higher breakdown voltage, I_{crit} is not very large, and β may start dropping at large collector currents owing to the increased effective base width, $W_{eff} = W_B + \Delta W_B$. The analysis of the different terms in eq. (2-7-66) shows that $\beta \sim 1/[W_{eff} + c_2 W_{eff}^2]$ where c_2 is a constant. However, for circuit modeling, it is more appropriate to account for the base push out by introducing an empirical term into eq. (A8.2-17), keeping the effective base width constant, i.e.,

$$\beta \approx \beta_o (1 + I_{crit}/I_c) = \frac{I_i (1 + I_{crit}/I_c)}{I_i/\beta_N + I_{rb}^o + A_{rs1} P_e} \tag{A8.2-18}$$

(Usually, in the forward active mode, β_N is quite high and the first term in the denominators of eqs. (A8.2-16) to (A8.2-18) can be neglected.)

A8.3. Universal MOSFET Model (MOSA1)

Based on the calculations in Section 3.9, we can write a system of unified extrinsic equations for the MOSFET. The term unified means that the expressions for I-V and C-V characteristics cover all ranges of bias voltages, i.e., both above and below threshold, and above and below the saturation point. This is accomplished by using a single, continuous expression for the saturation current which is valid for all values of gate-source bias. The model also accounts for the drain bias induced shift in the threshold voltage and other effects related to the finite output conductance in saturation. The resulting system of equations is as follows (see Section 3.9 for details):

$$I_d = \frac{g_{ch} V_{ds} \left(1 + \lambda V_{ds}\right)}{\left[1 + \left(V_{ds} / V_{sate}\right)^m\right]^{1/m}} \tag{A8.3-1}$$

$$V_{sate} \approx I_{sat} / g_{ch} \tag{A8.3-2}$$

$$g_{ch} = \frac{g_{chi}}{1 + g_{chi}\left(R_s + R_d\right)} \tag{A8.3-3}$$

$$g_{chi} = q n_s W \mu_n / L \tag{A8.3-4}$$

$$n_s = 2 n_o \ln\left[1 + \frac{1}{2} \exp\left(\frac{V_{gt}}{\eta V_{th}}\right)\right] \tag{A8.3-5}$$

$$n_o = \frac{\varepsilon_i \eta V_{th}}{2 q\left(d_i + \Delta d\right)} \tag{A8.3-6}$$

$$I_{sat} = \frac{g_{chi} V_{gte}}{1 + g_{chi} R_s + \sqrt{1 + 2 g_{chi} R_s + \left(V_{gte} / V_L\right)^2}} \tag{A8.3-7}$$

$$V_{gte} = V_{th}\left[1 + \frac{V_{gt}}{2V_{th}} + \sqrt{\delta^2 + \left(\frac{V_{gt}}{2V_{th}} - 1\right)^2}\right] \qquad \text{(A8.3-8)}$$

$$V_{gt} = V_{gto} + \sigma V_{ds} \qquad \text{(A8.3-9)}$$

$$\sigma = \frac{\sigma_o}{1 + \exp\left(\dfrac{V_{gto} - V_\alpha}{V_\sigma}\right)} \qquad \text{(A8.3-10)}$$

$$C_{gs} = C_f + \frac{2}{3}C_{gc}\left[1 - \left(\frac{V_{sate} - V_{dse}}{2V_{sate} - V_{dse}}\right)^2\right] \qquad \text{(A8.3-11)}$$

$$C_{gd} = C_f + \frac{2}{3}C_{gc}\left[1 - \left(\frac{V_{sate}}{2V_{sate} - V_{dse}}\right)^2\right] \qquad \text{(A8.3-12)}$$

$$V_{dse} = V_{ds}\left[1 + \left(\frac{V_{ds}}{V_{sate}}\right)^{m_c}\right]^{-1/m_c} \qquad \text{(A8.3-13)}$$

$$C_{gc} = LW\left[\frac{d_i + \Delta d}{\varepsilon_s} + \frac{\eta V_{th}}{qn_o}\exp\left(-\frac{V_{gt}}{\eta V_{th}}\right)\right]^{-1} \qquad \text{(A8.3-14)}$$

A8.4. PMOS Models (PMOSA1 AND PMOSA2)

Based on the calculations in Section 3.10, we can write a system of equations for the unified long channel PMOS model (PMOSA1) and the extrinsic strong inversion PMOS model (PMOSA2). The term unified means that an expression for the *I-V* characteristics covers all ranges of bias voltages, i.e., both above and below threshold, and above and below the saturation point. For PMOSA1, this is accomplished using a single, continuous expression for the drain current. The model PMOSA2 is described in terms of separate expressions for the drain current in the linear and in the saturated regime. These expressions are joined together with a continuous value and first derivative of the current at the saturation point.

The advantages of the present models over those implemented in established circuit simulator programs such as SPICE-3 and BSIM include: analytical description of the transition between the subthreshold and the above-threshold region, much fewer adjustable parameters, an unambiguous parameter extraction technique, and a more accurate and physically justifiable description of the mobility dependence on the bias voltages.

The resulting equations are as follows (see Section 3.10 for more details):

$$\frac{1}{\mu_S} = \frac{1}{\mu_{op}} - \frac{1}{\kappa_{1p}}\left(V_{GS} + 2V_T\right) \tag{A8.4-1}$$

$$c_a \equiv \frac{q}{a} = \frac{\varepsilon_i}{d_i + \Delta d} \tag{A8.4-2}$$

$$V_{DS} = V_{ds} + I_d R_t \tag{A8.4-3}$$

$$V_{GT} = V_{gt} + I_d R_s \tag{A8.4-4}$$

For PMOSA1:

$$I_d = \frac{qWv_s}{\alpha} \frac{\eta V_{th}\left(p_S - p_D\right) + \frac{a}{2}\left(p_S^2 - p_D^2\right)}{V_L - \left(1 - \frac{V_L \mu_S}{2\,\kappa_{2p}}\right)V_{DSe}} \tag{A8.4-5}$$

$$V_{DSe} = \frac{a(p_D - p_S)}{\alpha} \tag{A8.4-6}$$

$$-V_{GT} + \alpha V_F \approx \eta V_{th} \ln\left(\frac{p_S}{p_o}\right) + a(p_s - p_o) \tag{A8.4-7}$$

$$p_o = \frac{\eta V_{th}}{2a} = \frac{\varepsilon_i \eta V_{th}}{2(d_i + \Delta d)} \tag{A8.4-8}$$

For PMOSA2:

$$I_d = \frac{V_{GT} V_{DS} - \frac{\alpha}{2} V_{DS}^2}{R_n (V_L - \zeta V_{DSe})} = \frac{2 k_1}{h_1 + \sqrt{h_1^2 - 4 g_1 k_1}} \quad \text{(linear)} \tag{A8.4-9}$$

$$\begin{aligned}
k_1 &= V_{gt} V_{ds} - \alpha V_{ds}^2/2 \\
h_1 &= R_n V_L - R_t V_{gt} + (\alpha R_t - R_s - \zeta R_n) V_{ds} \\
g_1 &= R_t (\zeta R_n + R_s - \alpha R_t /2) \\
R_n &= (c_a v_s W)^{-1}
\end{aligned} \tag{A8.4-10}$$

$$I_d = I_{sat} \left\{ 1 + \left(\frac{V_\lambda}{V_L - \zeta V_{SAT}}\right) \times \right.$$
$$\left. \times \ln\left[\sqrt{1 + \left(\frac{V_{ds} - V_{sat}}{\theta V_\lambda}\right)^2} - \frac{V_{ds} - V_{sat}}{\theta V_\lambda} \right] \right\} \quad \text{(saturated)} \tag{A8.4-11}$$

$$I_{sat} = \frac{2 k_2}{h_2 + \sqrt{h_2^2 - 4 g_2 k_2}} \tag{A8.4-12}$$

$$k_2 = \frac{2\gamma(\gamma - 1)}{\alpha(2\gamma - 1)^2} V_{gt}^2$$

$$h_2 = R_n V_L - \frac{4\gamma(\gamma - 1)}{\alpha(2\gamma - 1)^2} \left[R_s + \frac{2\gamma^2 - 2\gamma + 1}{2\gamma(\gamma - 1)} \zeta R_n \right] V_{gt} \qquad \text{(A8.4-13)}$$

$$g_2 = \frac{2[2\gamma(\alpha + 1) - \alpha]}{\alpha^2(2\gamma - 1)^2} [(\gamma - 1) R_s + \gamma\zeta R_n] \zeta R_n$$

$$V_{SAT} = \frac{2(\gamma - 1)}{\alpha(2\gamma - 1)} V_{gt} + \frac{2(\gamma - 1) R_s + 2\gamma\zeta R_n}{\alpha(2\gamma - 1)} I_{sat} \qquad \text{(A8.4-14)}$$

$$V_{sat} = V_{SAT} - I_{sat} R_t \qquad \text{(A8.4-15)}$$

$$\theta = \frac{V_{SAT} \left(V_{GT} - \frac{\alpha}{2} V_{SAT} \right)}{\alpha V_L V_{SAT} - V_L V_{GT} - \frac{\alpha}{2} \zeta V_{SAT}^2} \qquad \text{(A8.4-16)}$$

A8.5. NMOS Models (NMOSA1 AND NMOSA2)

Based on the calculations in Section 3.11, we can write a system of equations for the unified long channel NMOS model (NMOSA1) and the extrinsic strong inversion NMOS model (NMOSA2). The term unified means that an expression for the *I-V* characteristics covers all ranges of bias voltages, i.e., both above and below threshold, and above and below the saturation point. For NMOSA1, this is accomplished using a single, continuous expression for the drain current. The model NMOSA2 is described in terms of separate expressions for the drain current in the linear and in the saturated regime. These expressions are joined together with a continuous value and first derivative of the current at the saturation point.

The advantages of the present models over those implemented in established circuit simulator programs such as SPICE-3 and BSIM include: analytical description of the transition between the subthreshold and the above-threshold region, much fewer adjustable parameters, an unambiguous parameter extraction technique, and a more accurate and physically justifiable description of the mobility dependence on the bias voltages.

The resulting equations are as follows (see Section 3.11 for more details):

$$\mu_n = \mu_{on} - \kappa_{1n}\left(V_{GS} + V_T\right) + \kappa_{2n}V(x) \tag{A8.5-1}$$

$$c_a \equiv \frac{q}{a} = \frac{\varepsilon_i}{d_i + \Delta d} \tag{A8.5-2}$$

$$V_{DS} = V_{ds} - I_d R_t \tag{A8.5-3}$$

$$V_{GT} = V_{gt} - I_d R_s \tag{A8.5-4}$$

For NMOSA1:

$$I_d = \frac{qW}{\alpha L}\left(\mu_S + \frac{\kappa_2}{2}V_{DSe}\right)\left[\eta V_{th}\left(n_S - n_D\right) + \frac{a}{2}\left(n_S^2 - n_D^2\right)\right] \tag{A8.5-5}$$

$$V_{DSe} = \frac{a\left(n_S - n_D\right)}{\alpha} \tag{A8.5-6}$$

$$V_{GT} - \alpha V_F \approx \eta V_{th}\ln\left(\frac{n_S}{n_o}\right) + a\left(n_S - n_o\right) \tag{A8.5-7}$$

$$n_o = \frac{\eta V_{th}}{2a} = \frac{\varepsilon_i \eta V_{th}}{2\left(d_i + \Delta d\right)} \tag{A8.5-8}$$

For NMOSA2:

$$I_d \approx \frac{V_{GT}V_{DS} - \frac{\alpha}{2}V_{DS}^2}{R_n\sqrt{V_{DS}^2 + V_L^2}} = \frac{2k_1}{h_1 + \sqrt{h_1^2 - 4g_1k_1}} \qquad \text{(linear)} \tag{A8.5-9}$$

$$k_1 = \left(V_{gt} V_{ds} - \alpha V_{ds}^2/2 \right)\sqrt{V_{ds}^2 + V_L^2}$$

$$h_1 = \left[R_t V_{gt} - (\alpha R_t - R_s)V_{ds} + R_n\sqrt{V_{ds}^2 + V_L^2} \right]\sqrt{V_{ds}^2 + V_L^2} \quad \text{(A8.5-10)}$$

$$g_1 = R_t R_n V_{ds} + R_t (R_s - \alpha R_t/2)\sqrt{V_{ds}^2 + V_L^2}$$

$$I_d = I_{sat}\left\{ 1 + \left(\frac{V_L V_\lambda}{V_L^2 + V_{SAT}^2} \right) \times \right.$$

$$\left. \times \ln\left[\sqrt{1 + \left(\frac{V_{ds} - V_{sat}}{\theta V_\lambda} \right)^2} + \frac{V_{ds} - V_{sat}}{\theta V_\lambda} \right] \right\} \quad \text{(saturated)} \quad \text{(A8.5-11)}$$

$$I_{sat} = \frac{2 k_2}{h_2 + \sqrt{h_2^2 - 4 g_2 k_2}} \quad \text{(A8.5-12)}$$

$$k_2 = \left(\xi V_{gt} + \alpha V_L/2 \right) V_{gt}^2$$

$$h_2 = 3 R_s \xi V_{gt}^2 + \alpha R_s V_L V_{gt} + R_n\left(\xi V_{gt} + \alpha V_L \right)\sqrt{\left(\xi V_{gt} + \alpha V_L \right)^2 + V_{gt}^2}$$

$$g_2 = 3 R_s^2 \xi V_{gt} + \alpha R_s^2 V_L + 2 R_s R_n \frac{\xi\left(\xi V_{gt} + \alpha V_L \right)^2 + \left(\xi V_{gt} + \alpha V_L/2 \right)V_{gt}}{\sqrt{\left(\xi V_{gt} + \alpha V_L \right)^2 + V_{gt}^2}}$$

$$\text{(A8.5-13)}$$

$$V_{SAT} = \frac{V_L V_{GT}}{\alpha V_L + \xi V_{GT}} \approx \frac{2(\gamma - 1)}{\alpha(3\gamma - 1)} V_{GT}\left[1 - \frac{\gamma}{\gamma - 1}\left(\frac{R_n I_{sat}}{V_{GT}} \right)^2 \right] \quad \text{(A8.5-14)}$$

$$V_{sat} = V_{SAT} + I_{sat} R_t \quad \text{(A8.5-15)}$$

$$\theta = \frac{\left(\xi V_{GT} + \alpha V_L\right)\left(\xi V_{GT} + \frac{\alpha}{2} V_L\right)}{\xi \left(\xi V_{GT} + \alpha V_L\right)^2 - \frac{\alpha}{2} V_L V_{GT}} \tag{A8.5-16}$$

A8.6. MESFET Models (MESA1 and MESA2)

Based on the discussion in Section 4.4, we can write a system of unified extrinsic equations for the MESFET. Unification is accomplished by using a single, continuous expression for the saturation current which is valid for all values of gate-source bias, following a similar procedure as that used for the MOSFET in Section 3.9 (see also Appendix A8.3). The model also accounts for the drain bias induced shift in the threshold voltage and other effects related to the finite output conductance in saturation. Although the model is valid for any doping profile in the MESFET channel, specific expressions are only derived for a uniform doping (MESA1) and delta-doping (MESA2). For further details, see Section 3.9.

$$I_d = \frac{g_{ch} V_{ds} \left(1 + \lambda V_{ds}\right)}{\left[1 + \left(V_{ds} / V_{sate}\right)^m\right]^{1/m}} \tag{A8.6-1}$$

$$V_{sate} \approx I_{sat} / g_{ch} \tag{A8.6-2}$$

$$g_{ch} = \frac{g_{chi}}{1 + g_{chi} \left(R_s + R_d\right)} \tag{A8.6-3}$$

$$g_{chi} = q n_s W \, \mu_n / L \tag{A8.6-4}$$

$$I_{sat} = \frac{I_{sata} I_{satb}}{I_{sata} + I_{satb}} \tag{A8.6-5}$$

$$I_{sata} = \frac{2 \beta V_{gte}^2}{\left(1 + 2\beta V_{gte} R_s + \sqrt{1 + 4\beta V_{gte} R_s}\right)\left(1 + t_c V_{gte}\right)} \tag{A8.6-6}$$

$$I_{satb} = \frac{qn_o\mu_n V_{th}W}{L}\exp\left(\frac{V_{gt}}{\eta V_{th}}\right) \qquad \text{(A8.6-7)}$$

$$\beta = \frac{2\varepsilon_s v_s W}{d\left(V_{po} + 3V_L\right)} \qquad \text{(A8.6-8)}$$

$$V_{gte} = \frac{V_{th}}{2}\left[1 + \frac{V_{gt}}{V_{th}} + \sqrt{\delta^2 + \left(\frac{V_{gt}}{V_{th}} - 1\right)^2}\right] \qquad \text{(A8.6-9)}$$

$$n_o = \frac{\varepsilon_s \eta V_{th}}{qd} \qquad \text{(A8.6-10)}$$

$$V_{To} = V_{bi} - V_{po} \qquad \text{(A8.6-11)}$$

$$V_{gt} = V_{gto} + \sigma V_{ds} \qquad \text{(A8.6-12)}$$

$$\sigma = \frac{\sigma_o}{1 + \exp\left(\dfrac{V_{gto} - V_\alpha}{V_\sigma}\right)} \qquad \text{(A8.6-13)}$$

$$C_{gs} = C_f + \frac{2}{3}C_{gc}\left[1 - \left(\frac{V_{sate} - V_{dse}}{2V_{sate} - V_{dse}}\right)^2\right] \qquad \text{(A8.6-14)}$$

$$C_{gd} = C_f + \frac{2}{3}C_{gc}\left[1 - \left(\frac{V_{sate}}{2V_{sate} - V_{dse}}\right)^2\right] \qquad \text{(A8.6-15)}$$

$$V_{dse} = V_{ds} \left[1 + \left(\frac{V_{ds}}{V_{sate}} \right)^{m_c} \right]^{-1/m_c} \tag{A8.6-16}$$

$$C_{gc} = LW \frac{c_a c_b}{c_a + c_b} \tag{A8.6-17}$$

$$n_s \approx \frac{n_{sa} n_{sb}}{n_{sa} + n_{sb}} \tag{A8.6-18}$$

For MESA1:

$$V_{po} = \frac{qN_d d^2}{2\varepsilon_s} \tag{A8.6-19}$$

$$C_{gc} = LW \frac{\varepsilon_s}{d} \left[\sqrt{1 - \frac{V_{gt}}{V_{po}}} + \exp\left(-\frac{V_{gt}}{\eta V_{th}} \right) \right]^{-1} \tag{A8.6-20}$$

$$n_s \approx \left[\frac{1}{N_d A \left(1 - \sqrt{1 - \frac{V_{gte}}{V_{po}}} \right)} + \frac{1}{n_o} \exp\left(-\frac{V_{gt}}{\eta V_{th}} \right) \right]^{-1} \tag{A8.6-21}$$

For MESA2:

$$V_{po} = V_{pou} + V_{po\delta} \tag{A8.6-22}$$

$$V_{pou} = \frac{qN_{du} d_u^2}{2\varepsilon_s} \tag{A8.6-23}$$

$$V_{po\delta} = \frac{qN_\Delta \Delta (2d_u + \Delta)}{2\varepsilon_s} \tag{A8.6-24}$$

$$c_a = \frac{\varepsilon_s}{d_d} = \frac{\varepsilon_s}{d_u} \times \begin{cases} \left[1 + \frac{N_{du}}{N_\Delta} \left(\frac{V_{po\delta} - V_{gte}}{V_{pou}} \right) \right]^{-1/2} &, \text{ for } V_{gt} \le V_{po\delta} \\[4mm] \left[\frac{V_{po} - V_{gte}}{V_{pou}} \right]^{-1/2} &, \qquad \text{ for } V_{gt} > V_{po\delta} \end{cases} \tag{A8.6-25}$$

$$c_b = \frac{\varepsilon_s}{d_u + \Delta} \exp \left(\frac{V_{gt}}{\eta V_{th}} \right) \tag{A8.6-26}$$

$$n_s \approx \frac{n_{sa} n_{sb}}{n_{sa} + n_{sb}} \tag{A8.6-27}$$

$$n_{sa} = \begin{cases} N_\Delta \Delta \left\{ 1 - \frac{d_u}{\Delta} \left[\sqrt{1 + \frac{N_{du}}{N_\Delta} \left(\frac{V_{po\delta} - V_{gte}}{V_{pou}} \right)} - 1 \right] \right\}, \text{ for } V_{gt} \le V_{po\delta} \\[4mm] N_\Delta \Delta + N_{du} d_u \left(1 - \sqrt{\frac{V_{po} - V_{gte}}{V_{pou}}} \right), \qquad \text{ for } V_{gt} > V_{po\delta} \end{cases}$$

$$\tag{A8.6-28}$$

$$n_{sb} = \frac{\varepsilon_s \eta V_{th}}{q(d_u + \Delta)} \exp \left(\frac{V_{gt}}{\eta V_{th}} \right) \tag{A8.6-29}$$

A8.7. Universal HFET Model (HFETA)

A unified circuit model for HFETs was established in Sections 4.5 and 4.6 using the same basic procedure as for MOSFETs (Section 3.9 and Appendix A8.3) and MESFETs (Section 4.4 and Appendix A8.6). An additional feature in the present model is the incorporation of the effects of gate current. Here follows a summary of the basic model expressions for the HFET (for further details, we refer to Sections 4.5 and 4.6):

$$I_d = \frac{g_{ch} V_{ds} \left(1 + \lambda V_{ds} \right)}{\left[1 + \left(V_{ds} / V_{sate} \right)^m \right]^{1/m}} \tag{A8.7-1}$$

$$V_{sate} \approx I_{sat} / g_{ch} \tag{A8.7-2}$$

$$g_{ch} = \frac{g_{chi}}{1 + g_{chi} \left(R_s + R_d \right)} \tag{A8.7-3}$$

$$g_{chi} = q n_s W \mu_n / L \tag{A8.7-4}$$

$$n_s = \frac{n_s'}{\left[1 + \left(n_s' / n_{max} \right)^\gamma \right]^{1/\gamma}} \tag{A8.7-5}$$

$$n_s' = 2 n_o \ln\left[1 + \frac{1}{2} \exp\left(\frac{V_{gt}}{\eta V_{th}} \right) \right] \tag{A8.7-6}$$

$$n_o = \frac{\varepsilon_i \eta V_{th}}{2 q \left(d_i + \Delta d \right)} \tag{A8.7-8}$$

$$I_{sat} = \frac{I_{sat}'}{\left[1 + \left(I_{sat}' / I_{max} \right)^\gamma \right]^{1/\gamma}} \tag{A8.7-9}$$

$$I'_{sat} = \frac{g'_{chi} V_{gte}}{1 + g'_{chi} R_s + \sqrt{1 + 2 g'_{chi} R_s + \left(V_{gte}/V_L\right)^2}} \qquad \text{(A8.7-10)}$$

$$g'_{chi} = q n'_s W \mu_n / L \qquad \text{(A8.7-11)}$$

$$I_{max} = q n_{max} v_s W \qquad \text{(A8.7-12)}$$

$$V_{gte} = V_{th}\left[1 + \frac{V_{gt}}{2V_{th}} + \sqrt{\delta^2 + \left(\frac{V_{gt}}{2V_{th}} - 1\right)^2}\right] \qquad \text{(A8.7-13)}$$

$$V_{gt} = V_{gto} + \sigma V_{ds} \qquad \text{(A8.7-14)}$$

$$\sigma = \frac{\sigma_o}{1 + \exp\left(\frac{V_{gto} - V_\alpha}{V_\sigma}\right)} \qquad \text{(A8.7-15)}$$

$$C_{gs} = C_f + \frac{2}{3} C_{gc}\left[1 - \left(\frac{V_{sate} - V_{dse}}{2V_{sate} - V_{dse}}\right)^2\right] \qquad \text{(A8.7-16)}$$

$$C_{gd} = C_f + \frac{2}{3} C_{gc}\left[1 - \left(\frac{V_{sate}}{2V_{sate} - V_{dse}}\right)^2\right] \qquad \text{(A8.7-17)}$$

$$V_{dse} = V_{ds}\left[1 + \left(\frac{V_{ds}}{V_{sate}}\right)^{m_c}\right]^{-1/m_c} \qquad \text{(A8.7-18)}$$

$$C_{gc} = \frac{LW \, c'_{gc}}{\left[1 + (n'_s/n_{max})^\gamma\right]^{1+1/\gamma}}$$

(A8.7-19)

$$c'_{gc} = \left[\frac{d_i + \Delta d}{\varepsilon_i} + \frac{\eta V_{th}}{qn_o} \exp\left(-\frac{V_{gt}}{\eta V_{th}}\right)\right]^{-1}$$

(A8.7-20)

$$C_{gtot} = LW \left(c_{gc} + c_{g1}\right)$$

(A8.7-21)

$$c_{g1} = \left[\frac{d_1}{\varepsilon_i} + \frac{\eta_1 V_{th}}{qn_{o1}} \exp\left(-\frac{V_{gt1}}{\eta_1 V_{th}}\right)\right]^{-1}$$

(A8.7-22)

$$n_{o1} = \frac{\varepsilon_i \eta_1 V_{th}}{2q \, d_1}$$

(A8.7-23)

$$V_{T1} \approx V_T + \frac{qn_{max} \, d_i}{\varepsilon_i}$$

(A8.7-24)

A8.8. a-Si TFT Model (ASIA1)

Based on the theory reviewed in Section 5.2, we propose a semi-empirical circuit model of a-Si TFTs similar to that developed for MOSFETs in Section 3.9 (see also A8.3 in this Appendix). The equations for the model are as follows:

$$I_{ds} = \frac{g_{ch} V_{ds} (1 + \lambda V_{ds})}{\left[1 + (V_{ds}/V_{sate})^m\right]^{1/m}}$$

(A8.8-1)

$$V_{sate} \approx I_{sat}/g_{ch}$$

(A8.8-2)

$$g_{ch} = \frac{g_{chi}}{1 + g_{chi}(R_s + R_d)} \tag{A8.8-3}$$

$$g_{chi} = \frac{q n_{inds} W \mu_{FET}}{L} \tag{A8.8-4}$$

$$\mu_{FET} = \mu_o \frac{m+2}{2} \left(\frac{n_{inds}}{n_o}\right)^m \tag{A8.8-5}$$

$$n_{inds} = \frac{\varepsilon_i E_2}{q} f(u_{gs}) \tag{A8.8-6}$$

where

$$u_{gs} = \frac{V_{gt}}{E_2} \tag{A8.8-7}$$

$$f(x) = \frac{x\left[\pi/2 + \tan^{-1}(a_{sub} x)\right] + 1/a_{sub}}{\pi} \tag{A8.8-8}$$

$$I_{sat} = K_{sat} \frac{n_{inds}^{m+2}}{1 + q\varepsilon_i K_{sat} n_{inds}^{m+1} R_s/d_i} \tag{A8.8-9}$$

(This equation is obtain by assuming that the voltage drop across the source series resistance is small compared to V_{gt} and using a Taylor series expansion.)

$$K_{sat} = q^2 \frac{W d_i \mu_o}{L \varepsilon_i n_o^m} \tag{A8.8-10}$$

A8.9. Poly-Si TFT Model (PSIA)

Our model for polysilicon transistors is based on the "effective medium" approach which treats the non-uniform polycrystalline sample with grain boundaries as some uniform effective medium with an effective carrier mobility and an average density of states in the energy gap. In the effective medium approach, the poly-Si TFT model is equivalent to a conventional FET model, but with the additional simplification that velocity saturation effects and effects of the source and drain series resistances are usually not important because of the large gate lengths and the relatively low mobility values. This simplification allows us to develop a simpler unified model for poly-Si TFTs than that for MOSFETs (see Section 3.9 and Appendix8.3). Here follows a summary of the poly-Si TFT model expressions (for further details, see Section 5.3):

$$I_d = \begin{cases} \dfrac{q\mu W}{L}\left[n_{ss}V_{DS} - \dfrac{V_{DS}^2}{2a\left(1 + \dfrac{2V_{sth}}{an_{ss}}\right)} \right], & V_{DS} \le V_{SAT} \\[4ex] I_{sat}(1 + Z/\gamma), & V_{DS} > V_{SAT} \end{cases} \qquad \text{(A8.9-1)}$$

$$V_{sth}\ln\left(\frac{n_{ss}}{n_o}\right) + a(n_{ss} - n_o) = V_{GS} - V_T \qquad \text{(A8.9-2)}$$

$$V_{SAT} = \frac{2(\gamma - 1)}{2\gamma - 1} n_{ss} a^* \qquad \text{(A8.9-3)}$$

$$I_{sat} = \frac{2qW}{L}\mu\, \frac{\gamma(\gamma - 1)}{(2\gamma - 1)^2} n_{ss}^2 a^* \qquad \text{(A8.9-4)}$$

$$a^* = a\left(1 + \frac{2V_{sth}}{an_{ss}}\right) \qquad \text{(A8.9-5)}$$

$$a = qd_i/\varepsilon_i \qquad \text{(A8.9-6)}$$

$$1/\gamma = 4\left(-\Gamma + \sqrt{\Gamma^2 + \Gamma/2}\right) \tag{A8.9-7}$$

$$n_o \approx \frac{\varepsilon_i V_{sth}}{2q\, d_i} \tag{A8.9-8}$$

$$Z = \gamma\Gamma\left(1 - \frac{\gamma\Gamma u}{2}\right)\ln\left(\frac{t}{2\gamma\Gamma}\right) + \frac{u}{16}\left(t^2 - 4\gamma^2\Gamma^2\right) -$$
$$2\gamma^3\Gamma^3\left(1 - \frac{\gamma\Gamma u}{2}\right)\left(\frac{1}{t^2} - \frac{1}{4\gamma^2\Gamma^2}\right) - 2\gamma^6\Gamma^6 u\left(\frac{1}{t^4} - \frac{1}{16\gamma^4\Gamma^4}\right) \quad \text{(A8.9-9)}$$

$$t = z + \sqrt{z^2 + 4\gamma^2\Gamma^2} \tag{A8.9-10}$$

$$z = \frac{V_{DS} - V_{SAT}}{V_{SAT}} \tag{A8.9-11}$$

$$u = \frac{V_{GS} - V_T}{V_o} \tag{A8.9-12}$$

APPENDIX A9.   AIM-Spice REFERENCE

Background on SPICE

AIM-Spice is based on the most advanced version of SPICE (at the moment of writing) – Version 3e.1.   In this Appendix, we provide basic information about features of this version which are fully retained in AIM-Spice.   We also give specifics relevant to the additional models discussed in this book.

This Appendix contains a complete listing of all analyses, options, devices and control statements in AIM-Spice.   Analyses and Options are specified by choosing menu commands, and devices and control statements are specified in the circuit description.   This Appendix first lists all analyses and options, then devices and control statements, as well as default device parameters.   This material is a supplement to the AIM-Spice User's Manual in Chapter 6.   The reference material on original device models from Berkeley is from Johnson et al. (1991), see also Nagel (1985).

Notation Used in This Appendix

Item	Example	Description
name	M12	A name field is an alphanumeric string. It must begin with a letter and cannot contain any delimiters.
node	5000	A node field may be arbitrary character strings. The ground node must be named '0'. Node names are treated as character strings, thus '0' and '000' are different names.
scale suffix		$T=10^{12}$, $G=10^9$, $MEG=10^6$, $K=10^3$, $MIL=25.4 \cdot 10^{-6}$, $M=10^{-3}$, $U=10^{-6}$, $N=10^{-9}$, $P=10^{-12}$, $F=10^{-15}$
units suffix	V	Any letter that is not a scale factor or any letters that follows a scale suffix
value	1KHz	Floating-point number with optional scale and/or units suffixes
(text)	(option)	Comment
<item>	<OFF>	Optional item
{item}	{model}	Required item

AC Analysis

AC Analysis is used for calculating the frequency response of a circuit over a range of frequencies.

```
┌─────────────────────────────────────────────┐
│ ─      Ac Analysis Parameters                │
├─────────────────────────────────────────────┤
│ ┌─Sweep:──────────────┐   ┌──────────┐       │
│ │ ◉ LIN  ○ OCT  ○ DEC │   │   Save   │       │
│ └──────────────────────┘   └──────────┘       │
│                            ┌──────────┐       │
│ Number of points: │200  │  │   Run    │       │
│                            └──────────┘       │
│ Start Frequency:  │1     │  ┌──────────┐      │
│                             │  Cancel  │      │
│ End Frequency:    │10GHz │  └──────────┘      │
└─────────────────────────────────────────────┘
```

DEC, OCT and LIN stand for decade, octave and linear variation, respectively. The specification of number of points changes with the selection of DEC, OCT or LIN. When DEC is specified, the number of points are per decade. When OCT is specified, the number of points are per octave, and if LIN is specified, the number of points are the total number across the whole frequency range. The frequency range is specified with the Start Frequency and End Frequency parameters. Note that in order for this analysis to be meaningful, at least one independent source must be specified with an ac value.

If the circuit has only one ac input, it is convenient to set that input to unity and zero phase. Then the output variable will be the transfer function of the output variable with respect to the input.

DC Operating Point

This analysis calculates the dc operating point of a circuit. It has no parameters.

DC Transfer Curve Analysis

In a DC Transfer Curve Analysis, one or two source(s) (voltage or current sources) are swept over a user defined interval. The dc operating point of the circuit is calculated for every value of the source(s).

```
┌─────────────────────────────────────────────────┐
│ ─            Dc Analysis Parameters             │
├─────────────────────────────────────────────────┤
│ ┌─Source────────────────────────┐  ┌─────────┐  │
│ │ ◉ 1. Source ○ 2. Source (Optional) │ │  Save   │  │
│ └───────────────────────────────┘  └─────────┘  │
│                                     ┌─────────┐  │
│  Source Name:   │vds         │ ±│   │   Run   │  │
│  Start Value:   │0           │      └─────────┘  │
│  End Value:     │5           │      ┌─────────┐  │
│  Increment Value: │0.1       │      │ Cancel  │  │
│                                     └─────────┘  │
└─────────────────────────────────────────────────┘
```

Source Name is the name of an independent voltage or current source, Start Value, End Value and Increment Value are the starting, final and increment values, respectively. The example above causes the voltage source vds to be swept from 0 V to 5 V in increments of 0.1 V. A second source may optionally be specified with associated sweep parameters. In this case, the first source is swept over its range for each value of the second source. This option can be useful for obtaining semiconductor device output characteristics.

Noise Analysis

The Noise Analysis portion of AIM-Spice computes device-generated noise for a given circuit.

```
┌─────────────────────────────────────────────────┐
│ ─           Noise Analysis Parameters           │
├─────────────────────────────────────────────────┤
│ ┌─Sweep:──────────────────────┐  ┌─────────┐    │
│ │ ○ LIN     ○ OCT    ◉ DEC    │  │  Save   │    │
│ └─────────────────────────────┘  └─────────┘    │
│                                   ┌─────────┐    │
│                                   │   Run   │    │
│  Output Noise Variable: │v(3) │   └─────────┘    │
│  Input Source:          │vin  │   ┌─────────┐    │
│  Points/decade:         │10   │   │ Cancel  │    │
│  Start Frequency:       │10   │   └─────────┘    │
│  End Frequency:         │10k  │                  │
│  Points per Summary:    │1    │                  │
└─────────────────────────────────────────────────┘
```

The "Output Noise Variable" parameter has the form V(OUTPUT<,REF>) where OUTPUT is the node at which the total output noise is desired. If REF is specified, the noise voltage V(OUTPUT) – V(REF) is calculated. By default,

REF is assumed to be ground. The "Input Source" parameter is the name of an independent source to which input noise is referred. The next three parameters are the same as for AC Analysis. The last parameter is an optional integer; if specified, the noise contribution of each noise generator is produced at every "Points per Summary" frequency point.

This analysis produces two plots, one for the Noise Spectral Density curves and one for the total Integrated Noise over the specified frequency range. All noise voltages/currents are in squared units (V^2/Hz and A^2/Hz for spectral density, V^2 and A^2 for integrated noise).

Pole/Zero Analysis

The Pole-Zero Analysis computes poles and/or zeros in the small signal ac transfer function.

```
┌─────────────────────────────────────────┐
│ ─    Pole-Zero Analysis Parameters       │
│ ┌─Transfer Function Type:──────┐         │
│ │ ○ Output Voltage/Input Current│  Save   │
│ │ ◉ Output Voltage/Input Voltage│  Run    │
│ └──────────────────────────────┘         │
│ ┌─Analysis Type:──────┐         Cancel   │
│ │ ◉ Pole Analysis Only │                  │
│ │ ○ Zero Analysis Only │ Input Nodes:  1  0 │
│ │ ○ Both               │ Output Nodes: 5  0 │
│ └─────────────────────┘                  │
└─────────────────────────────────────────┘
```

You can have AIM-Spice locate only poles or only zeros. This feature is provided mainly for the case of a convergence problem in finding either the poles or the zeros, allowing at least one of the sets to be located.

Transfer Function Analysis

When you select this analysis AIM-Spice computes the DC small signal value of the transfer function, input resistance, and output resistance.

With the example shown in the dialog box below, AIM-Spice will compute the ratio v(8) to vin, the small-signal input resistance at vin, and the small-signal output resistance measured across nodes 8 and 0.

```
┌──────────────────────────────────────────┐
│ ━  │  Transfer Function Analysis Parameters│
├──────────────────────────────────────────┤
│                                 ┌────────┐ │
│ Small-signal Output Variable: │v(8)│ │ Save  │ │
│                                 └────────┘ │
│                                 ┌────────┐ │
│ Small-signal Input Source:  │vin │   │ Run   │ │
│                                 ├────────┤ │
│                                 │ Cancel │ │
│                                 └────────┘ │
└──────────────────────────────────────────┘
```

Transient Analysis

The Transient Analysis in AIM-Spice computes the time domain response of a circuit.

```
┌──────────────────────────────────────────┐
│ ━  │     Transient Analysis Parameters     │
├──────────────────────────────────────────┤
│                                 ┌────────┐ │
│ Stepsize :  │2ns        │      │  Save  │ │
│ Final Time: │200ns      │      └────────┘ │
│ ┌Optional:──────────────────┐  ┌────────┐ │
│ │Display Start Time: │      ││  │  Run   │ │
│ │Maximum Stepsize:  │      ││  └────────┘ │
│ └──────────────────────────┘  ┌────────┐ │
│                                │ Cancel │ │
│ ☐ Use Initial Conditions (UIC)  └────────┘ │
└──────────────────────────────────────────┘
```

The parameter Stepsize is the suggested computing increment, Final Time is the last time point computed. The transient analysis always starts at time zero. If you are not interested in the results until a time t_1 greater than zero, you specify t_1 for the parameter Display Start Time. The parameter Maximum Stepsize is useful when you want to limit the internal stepsize used by AIM-Spice. The Use Initial Conditions (UIC) option, when specified, indicates that the user does not want AIM-Spice to solve for the quiescent operating point before beginning the transient analysis. The values specified using IC= for the various elements are used as initial transient condition. If the .IC control line has been specified, then the node voltages on the .IC line are used to compute the initial conditions for the devices. Look at the description on the .IC control line for its interpretation when UIC is not specified.

| **Options** |

A set of options that controls different aspects of a simulation is available through the Options menu. The options are divided among four logical groups:

- General
- Analysis specific
- Device specific
- Numeric specific

Each option group has its own dialog box. The options are listed in the table below.

Option	**Description**	**Default**
GMIN	Minimum allowed conductance (1)	1.0E-12
RELTOL	Relative error tolerance (1)	0.001
ABSTOL	Absolute current error tolerance (1)	1nA
VNTOL	Absolute voltage error tolerance (1)	1μV
CHGTOL	Charge tolerance (1)	1.0E-14
TNOM	Nominal temperature. The value can be overridden by a temperature specification on any temperature dependent device model. (1)	27
TEMP	Operating temperature of the circuit. The value can be overridden by a temperature specification. (1)	27
TRYTOCOMPACT	Applicable only to the LTRA model. When specified, the simulator tries to condense LTRA transmission lines' past history of input voltages and currents (1).	Not Set
TRTOL	Transient analysis error tolerance (2)	7.0
ITL1	Maximum number of iterations in computing the dc operating point (2)	100
ITL2	Maximum number of iterations in dc transfer curve analysis (2)	50
ITL4	Transient analysis timepoint iteration limit (2)	10
DEFL	Default channel length of a MOS-transistor (3)	100μm

DEFW	Default channel width of a MOS-transistor (3)	100μm
DEFAD	Default drain diffusion area for a MOS transistor (3)	0.0
DEFAS	Default source diffusion area for a MOS transistor (3)	0.0
PIVTOL	Minimum value of an element to be accepted as a pivot element (4)	1.0E-13
PIVREL	The minimum relative ratio between the largest element in the column and an accepted pivot element (4)	1.0E-13
METHOD	Sets the numerical integration method used by AIM-Spice. Possible methods are Gear and Trapezoidal (4)	Trap

The numbers in parentheses indicate to which group they belong.

Title Line

General form:
Any text

Example:
SIMPLE DIFFERENTIAL PAIR
MOS OPERATIONAL AMPLIFIER

The title line must be the first line in the circuit description.

Comment Line

General form:
* (arbitrary text)

Example:
* MAIN CIRCUIT STARTS HERE

An asterisk in the first column indicates that this line is a comment line. Comment lines may be placed anywhere in the circuit description.

A	Heterostructure Field Effect Transistors (HFETs)

General form:
    AXXXXXXX ND NG NS MNAME <L=VALUE> <W=VALUE> <OFF>
    + <IC=VDS,VGS>

Example:
    a1 7 2 3 hfeta l=1u w=10u

ND, NG and NS are the drain, gate and source nodes, respectively. MNAME is the model name, L is the channel length, W is the channel width, and OFF indicates an optional initial value for the element in a dc analysis. The optional initial value IC=VDS,VGS is meant to be used together with UIC in a transient analysis. See the description of the .IC control statement for a better way to set transient initial conditions.

HFET Model
    .MODEL {model name} NHFET <model parameters>

The HFET model is a unified extrinsic model as described in Section 4.6. The model parameters are listed below. Note that the default model parameters correspond to the *n*-channel device discussed in Section 4.6.

Name	Parameter	Units	Default
VTO	Threshold voltage	V	0.15
RD	Drain ohmic resistance	Ω	0
RS	Source ohmic resistance	Ω	0
DI	Thickness of interface layer	m	0.04e-6
LAMBDA	Output conductance parameter	1/V	0.15
VS	Saturation velocity	m/s	1.5E5
ETA	Subthreshold ideality factor	-	1.28
M	Knee shape parameter	-	3

Name	Parameter	Units	Default
MC	Knee shape parameter	-	3
GAMMA	Knee shape parameter	-	3
SIGMA0	DIBL parameter	-	0.057
VSIGMAT	DIBL voltage parameter	V	0.3
VSIGMA	DIBL voltage parameter	V	0.1
MU	Low field mobility	m^2/Vs	0.4
DELTA	Transition width parameter	-	3
VS	Saturation velocity	m/s	1.5e5
NMAX	Maximum sheet charge density in the channel	m^{-2}	2e16
DELTAD	Thickness correction	m	4.5e-9
EPSI	Dielectric constant for interface layer	F/m	1.0841E-10
JS1D	Forward gate drain diode saturation current density	A/m^2	0
JS2D	Reverse gate drain diode saturation current density	A/m^2	0
JS1S	Forward gate source diode saturation current density	A/m^2	0
JS2S	Reverse gate source diode saturation current density	A/m^2	0
M1D	Forward gate drain diode ideality factor	-	1
M2D	Reverse gate drain diode ideality factor	-	1
M1S	Forward gate source diode ideality factor	-	1
M2S	Reverse gate source diode ideality factor	-	1
RGD	Gate-drain ohmic resistance	Ω	0
RGS	Gate-source ohmic resistance	Ω	0
ALPHAG	Drain-source correction current gain	–	0

Supported Analyses
Noise, and Pole-Zero Analysis not supported.

B	**Non-linear Dependent Sources**

General form:
    BXXXXXXX N+ N- <I=EXPR> <V=EXPR>

Example:

    b1 10 40 i=0.2*tanh(v(10,40)/0.2/2240)*(1+v(10,40)*0.027)
    b1 0 1 v=ln(sin(log(v(1,2)^2)))-v(10)^3+v(20)^v(10)
    b1 3 4 i=1
    b1 3 4 v=exp(pi^i(vcc))

N+ and N- is the positive and negative nodes, respectively. The values of the V and I parameters determine the voltages and currents across and through the device, respectively. If V is given, the device is a voltage source, and if I is given, the device is a current source.

During an AC Analysis, the source acts as a linear dependent source with a proportionality constant equal to the derivative of the source at the DC operating point.

The expressions given for V and I may be any function of voltages and currents through voltage sources in the system. The allowed functions are listed in the table below.

Function	**Description**
abs	Absolute value
acos	Inverse cosine
acosh	Inverse hyperbolic cosine
asin	Inverse sine
asinh	Inverse hyperbolic sine
atan	Inverse tangent
atanh	Inverse hyperbolic tangent
cos	Cosine
cosh	Hyperbolic cosine
exp	Exponential function
ln	Natural logarithm
log	Base 10 logarithm
sin	Sine
sinh	Hyperbolic sine
sqrt	Square root
tan	Tangent

The following operators are defined:

+   –   *   /   ^   unary –

If the argument of `log`, `ln`, or `sqrt` is less than zero, the absolute value of the argument is used. If a divisor is zero or the argument of `log` or `ln` is zero, an error will result. Other problems may occur when the argument for a function enters a region where the function is not defined.

To introduce time into an expression, you can integrate the current from a constant current source with a capacitor and use the resulting voltage. For a correct result, you have to set the initial voltage across the capacitor. Non-linear capacitors, resistors, and inductors may be simulated with the non-linear dependent source. Here is an example how to implement a non-linear capacitor:

```
* Bx: calculate f(input voltage)
Bx 1 0 v=f(v(pos,neg))
* Cx: linear capacitance
Cx 2 0 1
* Vx: Ammeter to measure current into the capacitor
Vx 2 1 DC 0 Volts
* Drive the current through Cx back into the circuit
Fx pos neg Vx 1
```

Supported Analyses
All.

C	Capacitors

General form:
    CXXXXXXX N+ N- VALUE <IC=Initial values>

Examples:
    cl 66 0 70pf
    CBYP 17 23 10U IC=3V

N+ and N- are the positive and negative element nodes respectively. VALUE is the capacitance in Farads.

The optional initial value is the initial zero time value of the capacitor voltage in volts. Note that the value is used only when the option UIC is specified in a Transient Analysis.

Semiconductor Capacitors
General form:
    CXXXXXXX N1 N2 <VALUE> <MNAME> <L=LENGTH> <W=WIDTH>
    + <IC=VALUE>

Examples:
    CMOD 3 7 CMODEL L=10U W=1U

This is a more general model for the capacitor than the one presented above. It gives you the possibility of modeling temperature effects and calculating capacitance values based on geometric and process information. VALUE if given, defines the capacitance, and information on geometry and process will be ignored. If MNAME is specified, the capacitance value is calculated based on information on process and geometry. If VALUE is not given, then MNAME and LENGTH must be specified. If WIDTH is not given, the model default width will be used.

Capacitor Model
    .MODEL {model name} C <model parameters>

The model allows calculation of the capacitance value based on information on geometry and process by the following expression:

$$C = CJ \cdot (L - NARROW) \cdot (W - NARROW) + 2 \cdot CJSW \cdot (L + W - 2 \cdot NARROW)$$

where the parameters are defined in the table below.

Name	Parameter	Unit	Default
CJ	Junction bottom capacitance	F/m^2	-
CJSW	Junction sidewall capacitance	F/m	-
DEFW	Default width	m	1e-6
NARROW	Narrowing due to side etching	m	0.0

Supported Analyses
All.

D	**Diodes**

General form:
    DXXXXXXX N+ N- N3 MNAME <AREA> <OFF> <IC=VD> <TEMP=T>

Examples:
    DBRIDGE 2 10 DIODE1
    DCLMP 3 7 DMOD 3.0 IC=0.2

N+ and N- are the positive and negative nodes, respectively. MNAME is the model name, AREA is the area factor, and OFF indicates an optional initial value during a dc analysis. If the area factor is not given, 1 is assumed. The optional initial value IC=VD is meant to be used together with an UIC in a Transient Analysis. The optional TEMP value is the temperature at which this device operates. It overrides the temperature specified as an option.

Diode Model
    .MODEL {model name} D <model parameters>

AIM-Spice has 2 diode models. Level 1 is the standard diode model supplied from Berkeley, which is the default model. Level 2 is a GaAs heterostructure diode model described in Section 1.10. To select the heterostructure diode model, specify LEVEL=2 on the model line.

Level 1 model parameters are:

Name	Parameter	Units	Default
IS	Saturation current (level 1 only)	A	1.0e-14
RS	Ohmic resistance	Ω	0
N	Emission coefficient	-	1
TT	Transit time	s	0
CJO	Zero bias junction capacitance	F	0
VJ	Junction potential	V	1
M	Grading coefficient	-	0.5
EG	Activation energy	eV	1.11
XTI	Saturation current temperature exponent	-	3.0
KF	Flicker noise coefficient	-	0
AF	Flicker noise exponent	-	1
FC	Coefficient for forward-bias depletion capacitance formula	-	0.5
BV	Reverse breakdown voltage	V	infinite
IBV	Current at breakdown voltage	A	1.0e-3
TNOM	Parameter measurement temperature	°C	27

Level 2 model parameters are (in addition to those for level 1):

Name	Parameter	Units	Default
DN	Diffusion constant for electrons	m^2/s	0.02
DP	Diffusion constant for holes	m^2/s	0.000942

Name	Parameter	Units	Default
LN	Diffusion length for electrons	m	7.21e-7
LP	Diffusion length for holes	m	8.681e-7
ND	Donor doping density	m^{-3}	7.0e24
NA	Acceptor doping density	m^{-3}	3e22
DELTAEC	Conduction band discontinuity	eV	0.6
XP	p-region width	m	1μm
XN	n-region width	m	1μm
EPSP	Dielectric constant on p-side	F/m	1.0593e-10
EPSN	Dielectric constant on n-side	F/m	1.1594e-10

Temperature Effects

Temperature appears explicitly in the exponential terms.

Temperature dependence of the saturation current in the junction diode model is determined by:

$$I_S(T_1) = I_S(T_0) \left(\frac{T_1}{T_0}\right)^{\frac{XTI}{N}} \exp\left(\frac{E_g q T_1 T_0}{Nk(T_1 - T_0)}\right)$$

where k is the Boltzmann constant, q is the electronic charge, E_g is the energy gap, XTI is the saturation current temperature exponent, and N is the emission coefficient. The last three quantities are model parameters.

For Schottky barrier diodes, the value for XTI is usually 2.

Supported Analyses
All.

E	**Linear Voltage-Controlled Voltage Sources**

General form:
    EXXXXXXX N+ N- NC+ NC- VALUE

Example:
     E1 2 3 14 1 2.0

     N+ and N− are the positive and negative nodes, respectively. NC+ and NC−
are the positive and negative controlling nodes, respectively. VALUE is the
voltage gain.

Supported Analyses
All.

| F   Linear Current-Controlled Current Sources |

General form:
     FXXXXXXX N+ N- VNAME VALUE

Example:
     F1 14 7 VIN 5

     N+ and N− are the positive and negative nodes, respectively. Current flows
from the positive node through the source to the negative node. VNAME is the
name of the voltage source where the controlling current flows. The direction of
positive control current is from the positive node through the source to the
negative node of VNAME. VALUE is the current gain.

Supported Analyses
All.

| G   Linear Voltage-Controlled Current Sources |

General form:
     GXXXXXXX N+ N- NC+ NC- VALUE

Example:
     G1 2 0 5 0 0.1MMHO

N+ and N- are the positive and negative nodes, respectively. Current flows from the positive node through the source to the negative node. NC+ and NC- are the positive and negative controlling nodes, respectively. VALUE is the transconductance in mhos.

Supported Analyses
All.

H	Linear Current-Controlled Voltage Sources

General form:
    HXXXXXXX N+ N- VNAME VALUE

Example:
    HX1 6 2 Vz 0.5K

N+ and N- are the positive and negative nodes, respectively. Current flows from the positive node through the source to the negative node. VNAME is the name of the voltage source where the controlling current flows. The direction of positive control current is from the positive node through the source to the negative node of VNAME. VALUE is the transresistance in ohms.

Supported Analyses
All.

I	Independent Current Sources

General form:
    IYYYYYYY N+ N- <<DC> DC/TRAN VALUE> <AC <ACMAG <ACPHASE>>>

Examples:
    isrc 23 21 ac 0.333 45.0 sffm(0 1 10k 5 1k)

N+ and N- are the positive and negative nodes, respectively. Positive current flows from the positive node through the source to the negative node.

DC/TRAN is the source value during a DC or a Transient Analysis. The value can be omitted if it is zero for both analyses. If the source is time invariant, its value can be prefixed with DC.

ACMAG is the amplitude and ACPHASE is the phase of the source during an AC Analysis. If ACMAG is omitted after the keyword AC, 1 is assumed. If ACPHASE is omitted, 0 is assumed.

All independent sources can be assigned time varying values during a Transient Analysis. If a source is assigned a time varying value, its value at t=0 is used during a DC Analysis. There are 5 predefined functions for time varying sources: pulse, exponent, sine, piece-wise linear, and single frequency FM. If the parameters are omitted, the default values shown below will be assumed. DT and T2 are increment time and final time in a Transient Analysis, respectively.

Pulse

General form:
    PULSE(I1 I2 TD TR TF PW PER)

Parameters	Default values	Units
I1 (initial value)	None	Amp
I2 (pulsed value)	None	Amp
TD (delay time)	0.0	seconds
TR (rise time)	DT	seconds
TF (fall time)	DT	seconds
PW (pulse width)	T2	seconds
PER (period)	T2	seconds

Example:
    IB 3 0 PULSE(1 5 1S 0.1S 0.4S 0.5S 2S)

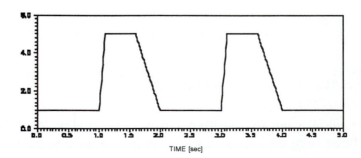

TIME [sec]

Sine
General form:
     SIN(I0 IA FREQ TD THETA)

Parameters	Default values	Units
I0 (offset)	None	Amp
IA (amplitude)	None	Amp
FREQ (frequency)	1/T2	Hz
TD (delay)	0.0	seconds
THETA (attenuation factor)	0.0	1/seconds

The shape of the waveform is:

0 < time < TD
$$I = I0$$

TD < time < T2

$$I = I0 + IA\sin(2\pi \cdot FREQ \cdot (time + TD)) \cdot \exp(-(time - TD) \cdot THETA)$$

Example:
     IB 3 0 SIN(2 2 5 1S 1)

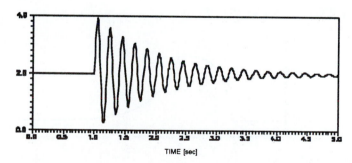

TIME [sec]

Exponent
General form:
     EXP(I1 I2 TD1 TAU1 TD2 TAU2)

Parameters	Default values	Units
I1 (initial value)	None	Amp
IA (pulsed value)	None	Amp
TD1(rise delay time)	0.0	seconds
TAU1(rise time constant)	DT	seconds
TD2 (delay fall time)	TD1+DT	seconds
TAU2 (fall time constant)	DT	seconds

The shape of the waveform is:

<u>0 < time < TD1</u>

$$I = I1$$

<u>TD1 < time < TD2</u>

$$I = I1 + (I2 - I1) \cdot \left(1 - \exp(-(time - TD1) \cdot TAU1)\right)$$

<u>TD2 < time < T2</u>

$$I = I1 + (I2 - I1) \cdot \left(1 - \exp(-(time - TD1) \cdot TAU1)\right)$$
$$+ (I1 - I2) \cdot \left(1 - \exp(-(time - TD2) \cdot TAU2)\right)$$

Example:
    IB 3 0 EXP(1 5 1S 0.2S 2S 0.5S)

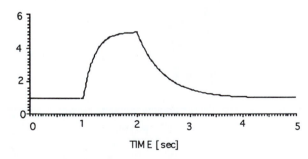

TIME [sec]

Piece-wise Linear
General form:
    PWL(T1 I1 <T2 I2 T3 I3 T4 I4 T5 I5 ....>)

Parameters and default values:

Pairs of values (T_i, I_i) specify the value of the source, I_i, at T_i. The value of the source between these values is calculated using a linear interpolation.

Example:
ICLOCK 7 5 PWL(0 0 1 0 1.2 4 1.6 2.0 2.0 5.0 3.0 1.0)

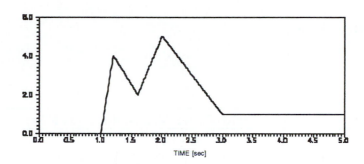

Single frequency FM

General form:
SFFM(I0 IA FC MDI FS)

Parameters	Default values	Units
I0 (offset)	None	Amp
IA (amplitude)	None	Amp
FC (carrier frequency)	1/T2	Hz
MDI (modulation index)	None	
FS (signal frequency)	1/T2	Hz

The shape of the waveform is:

$$I = I0 + IA \cdot \sin\left((2\pi \cdot FC \cdot time) + MDI \cdot \sin(2\pi \cdot FS \cdot time)\right)$$

Example:
IB 12 0 SFFM(0 1M 20K 5 1K)

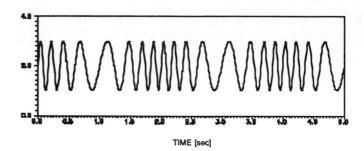

TIME [sec]

Supported Analyses
All.

J	Junction Field-Effect Transistors (JFETs)

General form:
  JXXXXXXX ND NG NS MNAME <AREA> <OFF> <IC=VDS,VGS> <TEMP=T>

Example:
  J1 7 2 3 JM1 OFF

ND, NG and NS are the drain, gate and source nodes, respectively. MNAME is the model name, AREA is the area factor, and OFF indicates an optional initial value for the element in a DC Analysis. If the area factor is omitted, 1.0 is assumed. The optional initial value IC=VDS,VGS is meant to be used together with UIC in a transient analysis. See the description of the .IC control statement for a better way to set transient initial conditions. The optional TEMP value is the temperature at which this device operates. It overrides the temperature specified in the option value.

JFET Model
  .MODEL {model name} NJF <model parameters>
  .MODEL {model name} PJF <model parameters>

Name	Parameter	Units	Default
VTO	Threshold voltage	V	-2.0
BETA	Transconductance parameter	A/V^2	1.0e-4

Name	Parameter	Units	Default
LAMBDA	Channel length modulation parameter	1/V	0
RD	Drain resistance	Ω	0
RS	Source resistance	Ω	0
CGS	Zero-bias G-S junction capacitance	F	0
CGD	Zero-bias G-D junction capacitance	F	0
PB	Gate junction potential	V	1
IS	Gate junction saturation current	A	1.0E-14
KF	Flicker noise coefficient	-	0
AF	Flicker noise exponent	-	1
FC	Coefficient for forward-bias depletion capacitance formula	-	0.5
TNOM	Parameter measurement temperature	°C	27

Temperature Effects

The temperature appears explicitly in the exponential terms.

The temperature dependence of the saturation current in the two gate junctions of the model is determined by:

$$I_S(T_1) = I_S(T_0)\exp\left[1.11\left(\frac{T_1}{T_0}-1\right)/V_{th}\right]$$

where V_{th} is the thermal voltage.

Supported Analyses
All.

K	**Coupled Inductors (Transformers)**

General form:
    KXXXXXXX LYYYYYYY LZZZZZZZ VALUE

Examples:
    k43 laa lbb 0.9999
    kxfrmr l1 l2 0.82

LYYYYYYY and LZZZZZZZ are the names of the two coupled inductors, and VALUE is the coupling coefficient, K, which must be greater than 0 and less than or equal to 1. Using the dot convention, place a dot on the first node of each inductor.

Supported Analyses
All.

| **L** | **Inductors** |

General form:
    LYYYYYYY N+ N- VALUE <IC=Initial values>

Examples:
    llink 42 69 1uh
    lshunt 23 51 10u ic=15.7ma

N+ and N- are the positive and negative element nodes respectively. VALUE is the inductance in Henries. The optional initial value is the t = 0 value of the inductor current in amps that flows from N+ through the inductor to N-. Notice that the value is used only when the option UIC is specified in a Transient Analysis.

Supported Analyses
All.

| **M** | **MOSFETs** |

General form:
    MXXXXXXX ND NG NS NB MNAME <L=VALUE> <W=VALUE> <AD=VALUE>
    + <AS=VALUE> <PD=VALUE> <PS=VALUE> <NRD=VALUE>
    + <NRS=VALUE> <OFF> <IC=VDS,VGS,VBS> <TEMP=T>

Example:
    M1 24 2 0 20 TYPE1
    m15 15 15 12 32 m w=12.7u l=207.8u
    M1 2 9 3 0 MOD1 L=10U W=5U AD=100P AS=100P PD=40U PS=40U

ND, NG, NS and NB are the drain, gate, source and bulk (substrate) nodes, respectively. MNAME is the model name, L and W are the channel length and width in meters, respectively. AD and AS are the drain and source diffusion areas in square meters. If any of L, W, AD or AS are not specified, default values are used. PD and PS are the perimeters of the drain and source diffusion areas. NRD and NRS are the relative resistivities of drain and source in number of squares, respectively. Default values of PD and PS are 0.0, while default values of NRD and NRS are 1.0. OFF indicates an optional initial value for the element in a DC Analysis. The optional initial value IC=VDS,VGS,VBS is meant to be used together with UIC in a Transient Analysis. See the description of the .IC control statement for a better way to set transient initial conditions. The optional TEMP value is the temperature at which this device operates. It overrides the temperature specified in the option value.

Note: The substrate node is ignored in level 11 and 12, and so is the VBS initial voltage.

MOSFET Model
    .MODEL {model name} NMOS <model parameters>
    .MODEL {model name} PMOS <model parameters>

AIM-Spice supports 12 MOSFET models. The parameter LEVEL selects which model to use. The default is LEVEL=1.

LEVEL=1	Shichman-Hodges
LEVEL=2	Geometric based analytical model
LEVEL=3	Semi-empirical short channel model
LEVEL=4	BSIM (Berkeley Short Channel Igfet Model)
LEVEL=5	New BSIM (BSIM2 as described in Jeng (1990))
LEVEL=6	MOS6 (as described in Sakurai and Newton (1990))
LEVEL=7	Universal extrinsic short channel MOS model (described in Section 3.9)
LEVEL=8	Unified long channel MOS model (described in Sections 3.10 and 3.11)

LEVEL=9   Short channel MOS model (described in Sections 3.10 and 3.11)

LEVEL=10   Unified intrinsic short channel model (described in Sections 3.10 and 3.11)

LEVEL=11   Unified extrinsic amorphous silicon thin film transistor model (described in Section 5.2)

LEVEL=12   Polysilicon thin film transistors model (described in Section 5.3)

Effects of charge storage are based on the model by Meyer for Levels 1, 2, 3 and 6. The BSIM models (Levels 4 and 5) have their own charge storage model. For Levels 7, 8, 9, and 10 you have the option of selecting the Meyer model or the *C-V* model described in Section 3.12. This selection is done by the model parameter CV.

Effects of a thin-oxide capacitance are treated slightly differently in Level 1. Voltage dependent capacitances are included only if TOX is specified.

A redundancy exists in specifying junction parameters. For example, the reverse current can be specified either with the IS parameter (in Amp) or with JS (in Amp/m^2). The first choice is an absolute value while the second choice is multiplied by AD and AS to give the reverse current at the drain and source junctions, respectively. The latter approach is preferred. The same is also true for the parameters CBD, CBS and CJ. Parasitic resistances can be specified in terms of RD and RS (in ohms) or RSH (in ohms/square). RSH is multiplied by number of squares NRD and NRS.

Parameters for Levels 1, 2, 3 and 6:

Name	Parameter	Units	Default
VTO	Zero-bias threshold voltage	V	0.0
KP	Transconductance parameter	A/V^2	2.0e-5
GAMMA	Bulk threshold parameter		0.0
PHI	Surface potential	V	0.6
LAMBDA	Channel length modulation (Only MOS1 and MOS2)	1/V	0.0
RD	Drain resistance	Ω	0.0
RS	Source resistance	Ω	0.0

Name	Parameter	Units	Default
CBD	Zero-bias B-S junction capacitance	F	0.0
CBS	Zero-bias B-S junction capacitance	F	0.0
IS	Bulk junction saturation current	A	1.0e-14
PB	Bulk junction potential	V	0.8
CGSO	Gate-source overlap capacitance per meter channel width	F/m	0.0
CGDO	Gate-drain overlap capacitance per meter channel width	F/m	0.0
CGBO	Gate-bulk overlap capacitance per meter channel width	F/m	0.0
RSH	Drain and source diffusion sheet resistance	Ω/Sq	0.0
CJ	Zero-bias bulk junction bottom capacitance per square-meter of junction area	F/m^2	0.0
MJ	Bulk junction bottom grading coefficient	-	0.5
CJSW	Zero-bias bulk junction sidewall capacitance per meter of junction perimeter	F/m	0.0
MJSW	Bulk junction sidewall grading coefficient	-	0.50 (level 1) 0.33 (level 2)
JS	Bulk junction saturation current per m^2 of junction area	A/m^2	0
TOX	Thin-oxide thickness	m	1.0e-7
NSUB	Substrate doping	$1/cm^3$	0.0
NSS	Surface state density	$1/cm^2$	0.0
NFS	Fast surface state density	$1/cm^2$	0.0
TPG	Type of gate material: +1 : opposite of substrate -1 : same as substrate 0 : Al gate	-	1.0

Name	Parameter	Units	Default
XJ	Metallurgical junction depth	m	0.0
LD	Lateral diffusion	m	0.0
U0	Surface mobility	cm^2/Vs	600
UCRIT	Critical field for mobility degradation (Only MOS2)	V/cm	1.0e4
UEXP	Critical field exponent in mobility degradation (Only MOS2)	-	0.0
UTRA	Transverse field coefficient (deleted for MOS2)	-	0.0
VMAX	Maximum drift velocity for carriers	m/s	0.0
NEFF	Total channel charge (fixed and mobile) coefficient (Only MOS2)	-	1.0
KF	Flicker noise coefficient	-	0.0
AF	Flicker noise exponent	-	1.0
FC	Coefficient for forward-bias depletion capacitance formula	-	0.5
DELTA	Width effect on threshold voltage (Only MOS2 and MOS3)	-	0.0
THETA	Mobility modulation (MOS3 only)	1/V	0.0
ETA	Static feedback (Only MOS3)	-	0.0
KAPPA	Saturation field factor (MOS3 only)	-	0.2
TNOM	Parameter measurement temperature	°C	27

MOSFET Levels 4 and 5 are BSIM models (Berkeley Short Channel IGFET Model). The parameters for these models should be obtained from process characterization.

Parameters marked with an '*' in the l/w column of the next table have length and width dependency. For example, for the flat band voltage, VFB, the dependence on the gate electrode geometry can be expressed in terms of the additional flat band parameters, LVFB and WVFB, measured in Volt μm:

$$VFB = VFB0 + \frac{LVFB}{L_{effective}} + \frac{WVFB}{W_{effective}}$$

where

$$L_{effective} = L_{input} - DL$$

$$W_{effective} = W_{input} - DW$$

Note that BSIM models are meant to be used together with a process characterization system. In contrast to the AIM-Spice models developed in this book, none of the parameters in the BSIM models have default values, and leaving one out is registered as an error.

BSIM (Levels 4 and 5) parameters:

Name	Parameter	Units	l/w
VFB	Flat band voltage	V	*
PHI	Surface inversion potential	V	*
K1	Body effect coefficient	$V^{1/2}$	*
K2	Drain/source depletion charge sharing coefficient	-	*
ETA	Zero-bias drain-induced barrier lowering coefficient	-	*
MUZ	Zero-bias mobility	cm^2/Vs	
DL	Shortening of channel	μm	
DW	Narrowing of channel	μm	
U0	Zero-bias transverse-field mobility degradation coefficient		*
U1	Zero-bias velocity saturation coefficient	$\mu m/V$	*
X2MZ	Sensitivity of mobility on substrate bias at $V_{ds}=0$	cm^2/V^2s	*
X2E	Sensitivity of drain-induced barrier lowering effect on substrate bias		*
X3E	Sens. of drain-induced barrier lowering effect to drain bias at $V_{ds}= V_{dd}$		*
X2U0	Sens. of transverse field mobility degradation effect to substrate bias	$1/V^2$	*
X2U1	Sens. of velocity saturation effect to substrate bias	$\mu m V^{-2}$	*

Name	Parameter	Units	l/w
MUS	Mobility at zero substrate bias and at $V_{ds}=V_{dd}$	cm^2/V^{2}s	
X2MS	Sensitivity of mobility to substrate bias at $V_{ds}=V_{dd}$	cm^2/V^{2}s	*
X3MS	Sensitivity of mobility to drain bias at $V_{ds}=V_{dd}$	μmV^{-2}	*
X3U1	Sensitivity of velocity saturation effect to drain bias at $V_{ds}=V_{dd}$	μmV^{-2}	*
TOX	Gate oxide thickness	μm	
TEMP	Temperature at which parameters were measured	C	
VDD	Measurement bias range	V	
CGDO	Gate-drain overlap capacitance per meter channel width	F/m	
CGSO	Gate-source overlap capacitance per meter channel width	F/m	
CGBO	Gate-bulk overlap capacitance per meter channel width	F/m	
XPART	Gate-oxide capacitance charge model flag	-	
N0	Zero-bias subthreshold slope coefficient	-	*
NB	Sensitivity of subthreshold slope to substrate bias	-	*
ND	Sensitivity of subthreshold slope to drain bias	-	*
RSH	Drain and source diffusion sheet resistance	Ω/Sq	
JS	Source drain junction current density	A/m^2	
PB	Built in potential of source drain junction	V	
MJ	Grading coefficient of source drain junction	-	
PBSW	Built in potential of source drain junction sidewall	V	
MJSW	Grading coefficient of source drain junction sidewall	-	
CJ	Source drain junction capacitance per unit area	F/m^2	
CJSW	Source drain junction sidewall capacitance per unit length	F/m	
WDF	Source drain junction default width	m	
DELL	Source drain junction length reduction	m	

XPART=0 selects a 40/60 drain/source partitioning of the gate charge in saturation, while XPART=1 selects a 0/100 drain/source charge partitioning (see Section 3.12).

AIM-Spice model parameters common for Levels 7, 8, 9, and 10 are listed below.

Name	Parameter	Units	Default
VTO	Zero-bias threshold voltage	V	0.0
GAMMA	Bulk threshold parameter	$V^{1/2}$	0.0
PHI	Surface potential	V	0.6
RD	Drain resistance	Ω	0.0
RS	Source resistance	Ω	0.0
CBD	Zero-bias B-D junction capacitance	F	0.0
CBS	Zero-bias B-S junction capacitance	F	0.0
IS	Bulk junction saturation current	A	1.0e-14
PB	Bulk junction potential	V	0.8
CGSO	Gate-source overlap capacitance per meter channel width	F/m	0.0
CGDO	Gate-drain overlap capacitance per meter channel width	F/m	0.0
CGBO	Gate-bulk overlap capacitance per meter channel width	F/m	0.0
RSH	Drain and source diffusion sheet resistance	Ω/Sq	0.0
CJ	Zero-bias bulk junction bottom capacitance per square-meter of junction area	F/m^2	0.0
MJ	Bulk junction bottom grading coefficient	-	0.5
CJSW	Zero-bias bulk junction sidewall capacitance per meter of junction perimeter	F/m	0.0
MJSW	Bulk junction sidewall grading coefficient	-	0.33

Name	Parameter	Units	Default
JS	Bulk junction saturation current per m^2 of junction area	A/m^2	0
TOX	Thin-oxide thickness	m	1.0e-7
NSUB	Substrate doping	1/cm^3	0.0
NSS	Surface state density	1/cm^2	0.0
NFS	Fast surface state density	1/cm^2	0.0
TPG	Type of gate material: +1 : opposite of substrate -1 : same as substrate 0 : Al gate	-	1.0
XJ	Metallurgical junction depth	m	0.0
LD	Lateral diffusion	m	0.0
U0	Surface mobility	cm^2/Vs	600
UCRIT	Critical field for mobility degradation	V/cm	1.0e4
UEXP	Critical field exponent in mobility degradation	-	0.0
VMAX	Maximum drift velocity of carriers	m/s	0.0
NEFF	Total channel charge (fixed and mobile) coefficient	-	1.0
KF	Flicker noise coefficient	-	0.0
AF	Flicker noise exponent	-	1.0
FC	Coefficient for forward-bias depletion capacitance formula	-	0.5
DELTA	Width effect on threshold voltage	-	0.0
TNOM	Parameter measurement temperature	°C	27
CV	*C-V* model to use. CV= 1 selects the model described in Section 3.12. CV= 0 selects the Meyer model.	-	0
XPART	Charge partitioning	-	0

Parameters specific for Level 7 model are:

Name	Parameter	Units	Default (NMOS)
LAMBDA	Transconductance parameter	V^{-1}	0.048
ETA	Subthreshold ideality factor	-	1.32
M	Knee shape parameter	-	4.0
SIGMA0	DIBL parameter	-	0.048
VSIGMAT	DIBL parameter	V	1.7
VSIGMA	DIBL parameter	V	0.2
DELTA	Correction to oxide thickness	m	0

Parameters specific for Level 8 model are:

Name	Parameter	Units	Default
ALPHA	Parameter accounting for the threshold dependence on the channel potential	-	1.164 (NMOS) 1.4 (PMOS)
K1	Mobility parameter	cm^2/V^2s	28 (NMOS) 1510 (PMOS)
ETA	Subthreshold ideality factor	-	1.3 (NMOS) 1.2 (PMOS)

Parameters specific for Level 9 model are:

Name	Parameter	Units	Default
ALPHA	Parameter accounting for the threshold dependence on the channel potential	-	1.2 (NMOS) 1.34 (PMOS)
XI	Saturation voltage parameter (NMOS only)	-	0.79
GAM	Saturation point parameter	-	3.0 (NMOS) 2.35 (PMOS)
K1	Mobility parameter	cm^2/V^2s	28 (NMOS) 1510 (PMOS)
LAMBDA	Characteristic length of the saturated region of the channel	m	9.94E-8 (NMOS) 1.043E-7 (OMOS)
ZETA	Velocity saturation factor (PMOS only)	-	0.34

Parameters specific for Level 10 model are:

Name	Parameter	Units	Default
ALPHA	Bulk charge factor	-	1.05 (NMOS)
			1.3 (PMOS)
ETA	Ideality factor in subthreshold	-	1.42 (NMOS)
			1.39 (PMOS)
SIGMA	Parameter accounting for DIBL effects	-	0.03 (NMOS)
			0.007 (PMOS)
K1	Mobility parameter	cm^2/V^2s	28 (NMOS)
			1510 (PMOS)
LAMBDA	Channel length modulation parameter	m	8.58E-8 (NMOS)
			7.25E-8 (PMOS)
XI	Saturation voltage factor (NMOS only)	-	0.76
ZETA	Velocity saturation factor (PMOS only)	-	0.28
GAM	Saturation point parameter	-	3.0 (NMOS)
			2.35 (PMOS)

Parameters common to Levels 11 and 12 are:

Name	Parameter	Units	Default
LEVEL	Model index	-	1
VTO	Zero-bias threshold voltage	V	0.0
GAMMA	Bulk threshold parameter	$V^{1/2}$	0.0
PHI	Surface potential	V	0.6
RD	Drain resistance	Ω	0.0
RS	Source resistance	Ω	0.0
CGSO	Gate-source overlap capacitance per meter channel width	F/m	0.0
CGDO	Gate-drain overlap capacitance per meter channel width	F/m	0.0

Name	Parameter	Units	Default
RSH	Drain and source diffusion sheet resistance	Ω/Sq	0.0
TOX	Thin-oxide thickness	m	1.0e-7
NSUB	Substrate doping	1/cm^3	0.0
NSS	Surface state density	1/cm^2	0.0
TPG	Type of gate material: +1 : opposite of substrate -1 : same as substrate 0 : Al gate	-	1.0
LD	Lateral diffusion	m	0.0
U0	Surface mobility	cm^2/Vs	600
TNOM	Parameter measurement temperature	°C	27

Parameters specific for Level 11 model are:

Name	Parameter	Units	Default
M	Knee shape parameter	-	2.5
E2	Characteristic energy	J	k_B 300.15
VSIGMA	Parameter accounting for DIBL effects	V	0.2
VSIGMAT	Parameter accounting for DIBL effects	V	1.7
SIGMA0	Parameter accounting for DIBL effects	-	0.048
LAMBDA	Transconductance parameter	1/V	0.0048
N0	Scaling factor	m^{-3}	1E18
ASUB	Fitting parameter	J/V	E2/5

Parameters specific for Level 12 model are:

Name	Parameter	Units	Default
V0	Scaling voltage	V	10.7
GAM	Saturation voltage parameter	-	0.03
ETA	Ideality factor in subthreshold regime	-	6.9

Temperature Effects

Temperature appears explicitly in the exponential terms in the equations describing current across the bulk junctions.

Temperature appears explicitly in the value of the junction potential, ϕ (in AIM-Spice PHI). The temperature dependence is given by:

$$\phi(T) = \frac{kT}{q} \log_e \left(\frac{N_a N_d}{N_i(T)^2} \right)$$

where k is Boltzmann's constant, q is the electronic charge, N_a is the acceptor impurity density, N_d is the donor impurity density, and N_i is the intrinsic carrier concentration.

Temperature appears explicitly in the value of the surface mobility μ_0 (or U0). The temperature dependence is given by:

$$\mu_0(T) = \frac{\mu_0(T_0)}{(T/T_0)^{1.5}}$$

Supported Analyses

Level 1, 2, 3:   All
Level 4:   Noise Analysis not supported
Level 5:   Noise Analysis not supported
Level 6:   AC, Noise, and Pole-Zero Analysis not supported
Level 7, 8, 9, 10, 11, 12:   Noise Analysis not supported

N	Heterojunction Bipolar Transistors (HBTs)

General form:
NXXXXXXX NC NB NE <NS> MNAME <AREA> <OFF> <IC=VBE,VCE>
+ <TEMP=T>

Example:
N23 10 24 13 NMOD IC=0.6,5.0
n2 5 4 0 nnd

NC, NB and NE are the collector, base and emitter nodes, respectively. NS is the substrate node. If this is not given, ground is assumed. MNAME is the model name, AREA is the area factor, and OFF indicates a optional initial value for the element in a dc analysis. If the area factor is omitted, 1.0 is assumed. The optional initial value `IC=VBE,VCE` is meant to be used together with UIC in a transient analysis. See the description of the .IC control statement for a better way to set transient initial conditions. The optional `TEMP` value is the temperature at which the device operates. It overrides the temperature specified in the option value.

HBT Model
.MODEL {model name} HNPN <model parameters>
.MODEL {model name} HPNP <model parameters>

The heterojunction bipolar transistor model in AIM-Spice is a modification of the Ebers-Moll bipolar transistor model.

Name	Parameter	Units	Default
IS	Transport saturation current	A	1e-16
BF	Ideal maximum forward beta	-	100
NF	Forward current emission coefficient	-	1.0
ISE	B-E leakage saturation current	A	0
NE	B-E leakage emission coefficient	-	1.2
BR	Ideal maximum reverse beta	-	1
NR	Reverse current emission coefficient	-	1
ISC	B-C leakage saturation current	A	0
NC	B-C leakage emission coefficient	-	2
RB	Base resistance	Ω	0
RE	Emitter resistance	Ω	0
RC	Collector resistance	Ω	0
CJE	B-E zero bias depletion capacitance	F	0
VJE	B-E built-in potential	V	0.75
MJE	B-E junction exponential factor	-	0.33
TF	Ideal forward transit time	s	0
XTF	Coefficient for bias dependence of TF	-	0

Name	Parameter	Units	Default
VTF	Voltage describing VBC dependence of TF	V	infinite
ITF	High current parameter for effect on TF	A	0
PTF	Excess phase at f=1.0/(TF 2π) Hz	Deg	
CJC	B-C zero bias depletion capacitance	F	0
VJC	B-C built-in potential	Volt	0.75
MJC	B-C junction exponential factor	-	0.33
XCJC	Fraction of B-C depletion capacitance connected to internal base node	-	1
TR	Ideal reverse transit-time	s	0
CJS	Zero-bias collector-substrate capacitance	F	0
VJS	Substrate junction built-in potential	V	0.75
MJS	Substrate junction exponential factor	-	0
XTB	Forward and reverse beta temperature exponent	-	0
EG	Energy gap for temperature effect on IS	eV	1.11 (Si)
XTI	Temperature exponent for effect on IS	-	3
KF	Flicker-noise coefficient	-	0
AF	Flicker-noise exponent	-	1
FC	Coefficient for forward-bias depletion capacitance formula	-	0.5
TNOM	Parameter measurement temperature	°C	27
IRB0	Base region recombination saturation current	A	0
IRS1	Surface recombination saturation current 1	A	0
IRS2	Surface recombination saturation current 2	A	0
ICSAT	Collector saturation current	A	0
M	Knee shape parameter	-	3

The modification to the Ebers-Moll model consists of two new contributions to the recombination current and a new expression for β_F.

Recombination in the base region is modeled by the expression

$$I_{rb} = IRB0\left(e^{V_{be}/V_{th}} - 1\right)$$

Surface recombination is modeled by the expression

$$I_{rs} = IRS1\left(e^{V_{be}/V_{th}} - 1\right) + IRS2\left(e^{V_{be}/2V_{th}} - 1\right)$$

where IRS1 and IRS2 are proportional to the emitter perimeter.

The expression for β_F is given by

$$\beta_F = \frac{\beta_{F0}}{\left[1 + \left(I_c / ICSAT\right)^M\right]^{1/M}}$$

Supported Analyses
Noise and Pole-Zero Analysis not supported.

O	Lossy Transmission Lines (LTRA)

General form:
OXXXXXXX N1 N2 N3 N4 MNAME

Example:
o23 1 0 2 0 lmod
ocon 10 5 20 5 interconnect

This is a two-port convolution model for single-conductor lossy transmission lines. N1 and N2 are the nodes at port 1, N3 and N4 are the nodes at port 2. It is worth mentioning that a lossy transmission line with zero loss may be more accurate than the lossless transmission.

LTRA Model
.MODEL {model name} LTRA <model parameters>

The uniform RLC/RC/LC/RG transmission line model (LTRA) models a uniform constant-parameter distributed transmission line. The RC and LC cases may also be modeled using the URC and TRA models; however, the newer LTRA model is usually faster and more accurate. The operation of the LTRA model is based on the convolution of the transmission line's impulse response with its inputs (see Roychowdhury and Pederson (1991)).

The LTRA model parameters are as follows:

Name	Parameter	Units	Default
R	Resistance/Length	Ω/m	0.0
L	Inductance/Length	H/m	0.0
C	Capacitance/Length	F/m	0.0
G	Conductance/Length	Ω^{-1}/m	0.0
LEN	Length of line	m	-
REL	Breakpoint control	-	1
ABS	Breakpoint control	-	1
NOSTEPLIMIT	Don't limit timestep to less than line delay	Flag	not set
NOCONTROL	Don't do complex timestep control	Flag	not set
LINEINTERP	Use linear interpolation	Flag	not set
MIXEDINTERP	Use linear when quadratic seems bad	Flag	not set
COMPACTREL	Special reltol for history compaction		RELTOL
COMPACTABS	Special abstol for history compaction		ABSTOL
TRUNCNR	Use Newton-Raphson method for timestep control	Flag	not set
TRUNCDONTCUT	Don't limit timestep to keep impulse-response errors low	Flag	not set

The types of lines implemented so far are: uniform transmission line with series loss only (RLC), uniform RC line, lossless transmission line (LC), and distributed series resistance and parallel conductance only (RG). Any other combination will yield erroneous results and should be avoided. The length (LEN) of the line must be specified.

Here follows a detailed description on some of the model parameters:

NOSTEPLIMIT is a flag that will remove the default restrictions of limiting time-step to less than the line delay in the RLC case.

NOCONTROL is a flag that prevents the default limiting of the time-step based on convolution error criteria in the RLC and RC cases. This speeds up the simulation but may in some cases reduce the accuracy.

LININTERP is a flag that, when set, will use linear interpolation instead of the default quadratic interpolation for calculating delayed signals.

MIXEDINTERP is a flag that, when set, uses a metric for judging whether quadratic interpolation is not applicable and if so uses linear interpolation. Otherwise it uses the default quadratic interpolation.

TRUNCDONTCUT is a flag that removes the default cutting of the time-step to limit errors in the actual calculation of impulse-response related quantities.

COMPACTREL and COMPACTABS are quantities that control the compaction of the past history of values stored for convolution. Large values of these parameters result in lower accuracy but usually increase the simulation speed. These are to be used with the TRYTOCOMPACT option.

TRUNCNR is a flag that turns *on* the use of Newton-Raphson iterations to determine an appropriate timestep in the timestep control routines. The default is a trial and error procedure cutting the previous timestep in half.

If you want to increase the speed of the simulation, follow these guidelines:

The most efficient option for increasing the speed of the simulation is REL. The default value of 1 is usually safe from the point of view of accuracy, but occasionally increases the computation time. A value greater than 2 eliminates all breakpoints and may be worth trying depending on the nature of the rest of the circuit, keeping in mind that it may not be safe from the point of view of accuracy. Breakpoints can usually be entirely eliminated if the circuit is not expected to have sharp discontinuities. Values between 0 and 1 are usually not needed, but may be used for setting a large number of breakpoints.

It is also possible to experiment with COMPACTREL when the option TRYTOCOMPACT is specified. The legal range is between 0 and 1. Larger values usually decrease the accuracy of the simulation, but in some cases improve speed. If TRYTOCOMPACT is not specified, history compaction is not attempted and the accuracy is high. The flags NOCONTROL, TRUNCDONTCUT and NOSTEPLIMIT also increase speed at the expense of accuracy in some cases.

Supported Analyses
Noise and Pole-Zero Analysis not supported.

Q	**Bipolar Junction Transistors (BJTs)**

General form:
    QXXXXXXX NC NB NE <NS> MNAME <AREA> <OFF> <IC=VBE,VCE>
    + <TEMP=T>

Example:
    Q23 10 24 13 QMOD IC=0.6,5.0
    q2 5 4 0 qnd

NC, NB and NE are the collector, base and emitter nodes, respectively. NS is the substrate node. If this is not given, ground is assumed. MNAME is the model name, AREA is the area factor, and OFF indicates a optional initial value for the element in a DC Analysis. If the area factor is omitted, 1.0 is assumed. The optional initial value IC=VBE,VCE is meant to be used together with UIC in a transient analysis. See the description of the .IC control statement for a better way to set transient initial conditions. The optional TEMP value is the temperature at which this device is to operate. It overrides the temperature specified as a option.

BJT Model
    .MODEL {model name} NPN <model parameters>
    .MODEL {model name} PNP <model parameters>

The bipolar transistor model in AIM-Spice is an adaptation of the Gummel-Poon model. In AIM-Spice, the model is extended to include high bias effects. The model automatically simplifies to Ebers-Moll if certain parameters are not given (VAF, IKF, VAR, IKR).

Name	Parameter	Units	Default
IS	Transport saturation current	A	1e-16
BF	Ideal maximum forward beta	-	100
NF	Forward current emission coefficient	-	1.0
VAF	Forward Early voltage	V	infinite
IKF	Corner for forward beta high current roll-off	A	infinite
ISE	B-E leakage saturation current	A	0
NE	B-E leakage emission coefficient	-	1.2
BR	Ideal maximum reverse beta	-	1
NR	Reverse current emission coefficient	-	1
VAR	Reverse Early voltage	V	infinite
IKR	Corner for reverse beta high current roll-off	A	infinite
ISC	B-C leakage saturation current	A	0
NC	B-C leakage emission coefficient	-	2
RB	Zero bias base resistance	Ω	0
IRB	Current where base resistance falls halfway to its minimum value	A	infinite
RBM	Minimum base resistance at high currents	Ω	RB
RE	Emitter resistance	Ω	0
RC	Collector resistance	Ω	0
CJE	B-E zero bias depletion capacitance	F	0
VJE	B-E built-in potential	V	0.75
MJE	B-E junction exponential factor	-	0.33
TF	Ideal forward transit time	s	0
XTF	Coefficient for bias dependence of TF	-	0
VTF	Voltage describing VBC dependence of TF	V	infinite
ITF	High current parameter for effect on TF	A	0

Name	Parameter	Units	Default
PTF	Excess phase at f=1.0/(TF 2π) Hz	Deg	
CJC	B-C zero bias depletion capacitance	F	0
VJC	B-C built-in potential	Volt	0.75
MJC	B-C junction exponential factor	-	0.33
XCJC	Fraction of B-C depletion capacitance connected to internal base node.	-	1
TR	Ideal reverse transit-time	s	0
CJS	zero-bias collector-substrate capacitance	F	0
VJS	Substrate junction built-in potential	V	0.75
MJS	Substrate junction exponential factor	-	0
XTB	Forward and reverse beta temperature exponent	-	0
EG	Energy gap for temperature effect on IS	eV	1.11 (Si)
XTI	Temperature exponent for effect on IS	-	3
KF	Flicker-noise coefficient	-	0
AF	Flicker-noise exponent	-	1
FC	Coefficient for forward-bias depletion capacitance formula	-	0.5
TNO M	Parameter measurement temperature	°C	27

Temperature Effects

The temperature appears explicitly in the exponential terms.

The temperature dependence of the saturation current in the model is determined by:

$$I_S(T_1) = I_S(T_0)\left(\frac{T_1}{T_0}\right)^{XTI} \exp\left(\frac{E_g q T_1 T_0}{k(T_1 - T_0)}\right)$$

where k is Boltzmann's constant, q is the electronic charge, E_g is the energy gap, and XTI is the saturation current temperature exponent. E_g and XTI are model parameters.

The temperature dependence of the forward and reverse beta is given by:

$$\beta(T_1) = \beta(T_0)\left(\frac{T_1}{T_0}\right)^{XTB}$$

where *XTB* is a user supplied model parameter. Temperature effects on beta are implemented by appropriate adjustment of the model parameters BF, ISE, BR, and ISC.

Supported Analyses
All.

R	**Resistors**

General form:
    RXXXXXXX N1 N2 VALUE

Examples:
    R1 1 2 100
    RB 1 2 10K
    RBIAS 4 8 10K

N1 and N2 are the two element nodes. VALUE is the resistance in Ohm. The value can be positive or negative, but not zero.

Semiconductor Resistors
General form:
    RXXXXXXX N1 N2 <VALUE> <MNAME> <L=LENGTH> <W=WIDTH>
    + <TEMP=T>

Example:
    rload 2 10 10K
    RMOD 3 7 RMODEL L=10U W=1U

This is a more general model for the resistor than the one presented above. It gives you the possibility to model temperature effects and calculate the resistance based on geometry and processing information. `VALUE`, if given, defines the resistance and information on geometry and processing will be ignored. If `MNAME` is specified, the resistance value is calculated based on information about the process and geometry in the model statement. If `VALUE` is not given, `MNAME` and `LENGTH` must be specified. If `WIDTH` is not given, it will be given the default value. The optional `TEMP` value is the temperature at which this device operates. It overrides the temperature specified in the option value.

Resistor Model
   .MODEL {model name} R <model parameters>

The resistor model contains process related parameters and the resistance value is a function of the temperature. The parameters are shown below.

Name	Parameter	Unit	Default
TC1	First order temperature coefficient	$\Omega/°C$	0.0
TC2	Second order temperature coefficient	$\Omega/°C^2$	0.0
RSH	Sheet resistance	Ω/Square	-
DEFW	Default width	m	1e-6
NARROW	Narrowing due to side etching	m	0.0
TNOM	Parameter measurement temperature	°C	27

The following expression is used for calculating the resistance value:

$$R = RSH \frac{L - NARROW}{W - NARROW}$$

DEFW defines a default value of W. If either RSH or L is given, a default value of 1 kOhm is used for R.

Temperature Effects
The temperature dependence of the resistance is given by:

$$R(T) = R(T_0)\left[1 + TC1(T - T_0) + TC2(T - T_0)^2\right]$$

where T is the circuit temperature, T_0 is the nominal temperature, and $TC1$ and $TC2$ is the first- and second order temperature coefficients, respectively.

Supported Analyses
All.

S	Voltage-Controlled Switch

General form:
SXXXXXXX N+ N- NC+ NC- MODEL <ON> <OFF>

Examples:
s1 1 2 3 4 switch1 ON
s2 5 6 3 0 sm2 off

N+ and N- are the positive and negative connecting nodes of the switch, respectively. NC+ and NC- are the positive and negative controlling nodes, respectively.

Switch Model
.MODEL {model name} SW <model parameters>

The switch model allows modeling of an almost ideal switch in AIM-Spice. The switch is not quite ideal since the resistance cannot change from 0 to infinity, but must have a finite positive value. The *on* and *off* resistances should therefore be chosen very small and very large, respectively, compared to the other circuit elements. The model parameters are as follows:

Name	Parameter	Units	Default
VT	Threshold voltage	V	0
VH	Hysteresis voltage	V	0
RON	*On* resistance	Ω	1
ROFF	*Off* resistance	Ω	1/GMIN

An ideal switch is highly non-linear. The use of switches can cause large discontinuities in node voltages. A rapid change such as that associated with a switching operation can cause problems with roundoff and tolerance which may lead to erroneous results or problems in selecting proper time steps. To reduce such problems, follow these steps:

Do not set the switch impedances higher and lower than necessary.
Reduce the tolerance during a Transient Analysis. This is done by specifying the value for TRTOL less than the default value of 7.0, for example, 1.0.
When switches are placed near capacitors, reduce the size of CHGTOL to, for example, 1e-16.

Supported Analyses
All.

T	**Transmission Lines (Lossless)**

General form:
```
TXXXXXXX N1 N2 N3 N4 Z0=VALUE <TD=VALUE>
+ <F=FREQ <NL=NRMLEN>> <IC=V1,I1,V2,I2>
```

Example:
```
T1 1 0 2 0 Z0=50 TD=10NS
```

N1 and N2 are the nodes for port 1, N3 and N4 are the nodes for port 2. Z0 is the characteristic impedance of the line. The length of the line can be specified in two different ways. The transmission delay, TD, can be specified directly. Alternatively, a frequency F may be given together with the normalized length of the line, NL (normalized with respect to the wavelength at the frequency F). If a frequency is specified and NL is omitted, 0.25 is assumed.

The optional initial values consists of voltages and currents at each of the two ports. Note that these values are used only if the option UIC is specified in a Transient Analysis.

Supported Analyses
Noise and Pole-Zero Analysis not supported.

U	**Uniform Distributed RC Lines (URC)**

General form:
    UXXXXXXX N1 N2 N3 MNAME L=LENGTH <N=LUMPS>

Example:
    U1 1 2 0 URCMOD L=50U
    URC2 1 12 2 UMODL L=1MIL N=6

N1 and N2 are the two nodes of the RC line itself, while N3 is the node of the capacitances. MNAME is the name of the model, LENGTH is the length of the line in meters. LUMPS, if given, is the number of segments used in modeling the RC line.

URC Model
    .MODEL {model name} URC <model parameters>

The model consists of a subcircuit expansion of the URC line into a network of lumped RC segments with internally generated nodes. The RC segments are in a geometric progression, increasing toward the middle of the URC line, with K as a proportionality constant. The number of lumped segments used, if not specified on the URC line, is determined by the following expression:

$$N = \frac{\log\left(FMAX \cdot \frac{R}{L} \cdot \frac{C}{L} \cdot 2\pi \cdot I^2 \cdot \frac{(K-1)^2}{K^2} \right)}{\log K}$$

The URC line is made up strictly of resistor and capacitor segments unless the ISPERL parameter is given a non-zero value, in which case the capacitors are replaced by reverse biased diodes with a zero-bias junction capacitance equivalent to the capacitance replaced, and with a saturation current of ISPERL amps per meter of transmission line and an optional series resistance equivalent to RSPERL ohms per meter.

Name	Parameter	Units	Default
K	Propagation constant	-	2.0
FMAX	Maximum frequency	Hz	1.0G
RPERL	Resistance per unit length	Ω/m	1000
CPERL	Capacitance per unit length	F/m	1e-15
ISPERL	Saturation current per unit length	A/m	0
RSPERL	Diode resistance per unit length	Ω/m	0

Supported Analyses
All.

V	Independent Voltage Sources

General form:
    VXXXXXXX N+ N- <<DC> DC/TRAN VALUE> <AC <ACMAG <ACPHASE>>>

Examples:
    vin 21 0 pulse(0 5 1ns 1ns 1ns 5us 10us)
    vcc 10 0 dc 6
    vmeas 12 9

N+ and N- are the positive and negative nodes, respectively. Note that the voltage source needs not to be grounded. Positive current flows from the positive node through the source to the negative node. If you insert a voltage source with a zero value, it can be used as an Ampere meter.

DC/TRAN is the source value in a DC or a Transient Analysis. The value can be omitted if it is zero for both the DC and Transient Analysis. If the source is time invariant, its value can be prefixed with DC.

ACMAG is the amplitude and ACPHASE is the phase of the source during an AC Analysis. If ACMAG is omitted after the keyword AC, 1 is assumed. If ACPHASE is omitted, 0 is assumed.

All independent sources can be assigned time varying values during a Transient Analysis. If a source is assigned a time varying value, the value at t=0 is used during a DC Analysis. There are five predefined functions for time varying sources: pulse, exponent, sine, piece-wise linear, and single frequency FM. If parameters are omitted, default values shown below will be assumed. In the tables below, DT and T2 are the incremental time and final time, respectively, in a Transient Analysis.

Pulse

General form:
    PULSE(V1 V2 TD TR TF PW PER)

Parameters	Default values	Units
V1 (initial value)	None	Volt
V2 (pulsed value)	None	Volt
TD (delay time)	0.0	seconds
TR (rise time)	DT	seconds
TF (fall time)	DT	seconds
PW (pulse width)	T2	seconds
PER (period)	T2	seconds

Example:
    VIN 3 0 PULSE(1 5 1S 0.1S 0.4S 0.5S 2S)

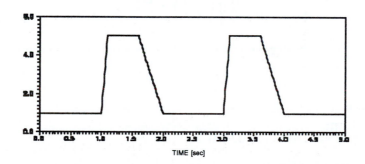

TIME [sec]

Sine

General form:
    SIN(V0 VA FREQ TD THETA)

Parameters	Default values	Units
V0 (offset)	None	Volt
VA (amplitude)	None	Volt
FREQ (frequency)	1/T2	Hz
TD (delay)	0.0	seconds
THETA(attenuation factor)	0.0	1/seconds

The shape of the waveform is:

$0 <$ time $<$ TD

$$V = V0$$

TD $<$ time $<$ T2

$$I = V0 + VA\sin(2\pi \cdot FREQ \cdot (time + TD)) \cdot \exp(-(time - TD) \cdot THETA)$$

Example:
    VIN 3 0 SIN(2 2 5 1S 1)

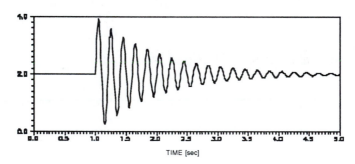

TIME [sec]

Exponent

General form:
    EXP(V1 V2 TD1 TAU1 TD2 TAU2)

Parameters	Default values	Units
V1 (initial value)	None	Volt
VA (pulsed value)	None	Volt
TD1(rise delay time)	0.0	seconds

TAU1(rise time constant)	DT	seconds
TD2 (delay fall time)	TD1+DT	seconds
TAU2 (fall time constant)	DT	seconds

The shape of the waveform is:

<u>0 < time < TD1</u>

$$V = V1$$

<u>TD1 < time < TD2</u>

$$V = V1 + (V2 - V1) \cdot \left(1 - \exp(-(time - TD1) \cdot TAU1)\right)$$

<u>TD2 < time < T2</u>

$$V = V1 + (V2 - V1) \cdot \left(1 - \exp(-(time - TD1) \cdot TAU1)\right)$$
$$+ (V1 - V2) \cdot \left(1 - \exp(-(time - TD2) \cdot TAU2)\right)$$

Example:
    VIN 3 0 EXP(1 5 1S 0.2S 2S 0.5S)

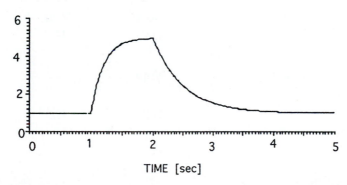

TIME [sec]

Piece-wise Linear

General form:
    PWL(T1 V1 <T2 V2 T3 V3 T4 V4 T5 V5 ....>)

Parameters and default values:
    Each pair of values (T_i, V_i) specifies the value of the source, V_i, at T_i.

The values of the source in between are calculated using a linear interpolation.

Example:
VCLOCK 7 5 PWL(0 0 1 0 1.2 4 1.6 2.0 2.0 5.0 3.0 1.0)

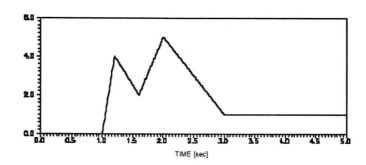

TIME [sec]

Single frequency FM

General form:
SFFM(V0 VA FC MDI FS)

Parameters	Default values	Units
V0 (offset)	None	Volt
VA (amplitude)	None	Volt
FC (carrier frequency)	1/T2	Hz
MDI (modulation index)	None	-
FS (signal frequency)	1/T2	Hz

The shape of the waveform is:

$$V = V0 + VA \cdot \sin\big((2\pi \cdot FC \cdot time) + MDI \cdot \sin(2\pi \cdot FS \cdot time)\big)$$

Example:
VIN 12 0 SFFM(0 1M 20K 5 1K)

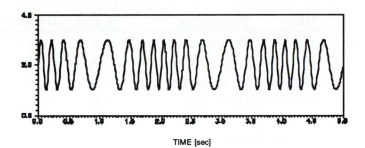

TIME [sec]

Supported Analyses
All.

W	**Current-Controlled Switch**

General form:
WYYYYYYY N+ N- VNAME MODEL <ON> <OFF>

Examples:
w1 1 2 vclock switchmod1
w2 3 0 vramp sm1 ON
wreset 5 6 vclk lossyswitch OFF

N+ and N- are positive and negative nodes of the switch, respectively. The control current is defined as the current flowing through the specified voltage source. The direction of positive control current is from the positive node through the source to the negative node.

Switch Model
.MODEL {model name} CSW <model parameters>

This switch allows the modeling of an almost ideal switch in AIM-Spice. The switch is not quite ideal since the resistance cannot change from 0 to infinity, but must have a finite positive value. The *on* and *off* resistances should therefore be chosen very small and very large, respectively, compared to the other circuit elements. The model parameters are as follows:

Name	Parameter	Units	Default
IT	Threshold current	A	0
IH	Hysteresis current	A	0
RON	On resistance	Ω	1
ROFF	Off resistance	Ω	1/GMIN

An ideal switch is highly non linear. The use of switches can cause large discontinuities in node voltages. A rapid change such as that associated with a switching operation can cause problems with roundoff and tolerance which may lead to erroneous results or problems in selecting proper time steps. To reduce such problems, follow these steps:

Do not set the switch impedances higher and lower than necessary.
Reduce the tolerance during a Transient analysis. This is done by specifying the value for TRTOL less than the default value of 7.0, for example, 1.0.
When switches are placed near capacitors, reduce the size of CHGTOL to, for example, 1e-16.

Supported Analyses
All.

Z	MESFETs

General form (Level 1):
ZXXXXXXX ND NG NS MNAME <AREA> <OFF> <IC=VDS,VGS>

General form (Level 2 and 3):
ZXXXXXXX ND NG NS MNAME <L=VALUE> <W=VALUE> <OFF>
+ <IC=VDS,VGS> <TEMP=T>

Example:
z1 7 2 3 zm1 off
z1 0 2 0 mesmod l=1u w=20u

ND, NG and NS are the drain, gate and source nodes, respectively. MNAME is the model name, AREA is the area factor, and OFF indicates an optional initial value for the element in a dc analysis. If the area factor is omitted, 1.0 is assumed. The optional initial value `IC=VDS,VGS` is meant to be used together with UIC in a Transient Analysis. See the description of the .IC control statement for a better way to set transient initial conditions. The optional `TEMP` value for Levels 2 and 3 is the temperature at which the device operates. It overrides the temperature specified in the option value.

MESFET Model
.MODEL {model name} NMF <model parameters>
.MODEL {model name} PMF <model parameters>

In AIM-Spice, three MESFET models are included. The difference is in the formulation of the *I–V* characteristics. The parameter LEVEL selects which model to use.

LEVEL=1	Statz et al. (see Section 4.3)
LEVEL=2	Unified extrinsic model for uniformly doped channel (as described in Subsection 4.4.3)
LEVEL=3	Unified extrinsic model for delta-doped channel (as described in Subsection 4.4.4)

Name	Parameter	Units	Default
VTO	Pinch-off voltage	V	-2.0 (level 1)
			-1.26 (Level 2)
			-2.04 (Level 3)
IS	Junction saturation current (Level 1 only)	A	1e-14
BETA	Transconductance parameter (Level 1 only)	A/V^2	1.0e-4
B	Doping tail extending parameter (Level 1 only)	1/V	0.3
ALPHA	Saturation voltage parameter (Level 1 only)	1/V	2
LAMBDA	Channel length modulation parameter (Level 1 only)	1/V	0
RD	Drain ohmic resistance	Ω	0
RS	Source ohmic resistance	Ω	0
CGS	Zero-bias G-S junction capacitance (Level 1 only)	F	0
CGD	Zero-bias G-D junction capacitance (Level 1 only)	F	0

Name	Parameter	Units	Default
PB	Gate junction potential (Level 1 only)	V	1
KF	Flicker noise coefficient (Level 1 only)	-	0
AF	Flicker noise exponent (Level 1 only)	-	1
FC	Coefficient for forward-bias depletion capacitance formula (Level 1 only)	-	0.5

The model parameters for MESFET Levels 2 and 3 are the same as those for Level 1, plus additional parameters listed below.

Name	Parameter	Units	Default
D	Depth of device (Level 2 only)	m	0.12μm
DU	Depth of uniformly doped layer (Level 3 only)	m	0.035μm
JSDF	Junction saturation current density at forward bias	A/m^2	1e-2
GGR	Junction conductance at reverse bias	Ω^{-1}	40
DEL	Reverse junction conductance inverse ideality factor	-	0.04
N	Junction ideality factor	-	1
LAMBDA	Output conductance parameter	1/V	0.045 (level 2) 0.04 (level 3)
LAMBDAHF	Output conductance parameter at high frequencies	1/V	0.045 (level 2) 0.04 (level 3)
VS	Saturation velocity (Level 2 only)	m/s	1.5E5
BETA	Transconductance parameter (Level 3 only)	A/V^2	0.0085
VBI	Built-in voltage (Level 3 only)	V	0.7
ETA	Subthreshold ideality factor	-	1.73 (Level 2) 1.5 (Level 3)
M	Knee shape parameter	-	2.5 (Level 2) 2.2 (Level 3)

Name	Parameter	Units	Default
MC	Knee shape parameter	-	2.5 (Level 2)
			2.2 (Level 3)
SIGMA0	DIBL parameter	-	0.081 (Level 2)
			0.02 (Level 3)
VSIGMAT	DIBL parameter	V	1.01 (Level 2)
			1.37 (Level 3)
VSIGMA	DIBL parameter	V	0.1
MU	Low field mobility	m^2/V s	0.23 (Level 2)
			0.2 (Level 3)
ND	Substrate doping (Level 2 only)	m^{-3}	3.0e23
NDU	Uniform layer doping (Level 3 only)	m^{-3}	1e22
DELTA	Transition width parameter	-	5 (Level 2)
			$5^{1/2}$ (Level 3)
TC	Transconductance compression factor	1/V	0 (Level 2)
			0.001 (Level 3)
NDELTA	Doping of delta doped layer (Level 3 only)	m^{-3}	6e24
TH	Thickness of delta doped layer (Level 3 only)	m	$0.01\mu m$
ALPHAT	Temperature coefficient for VTO	V/°C	0
TBETA	Temperature coefficient for β	°C	infinite
TLAMBDA	Temperature coefficient for LAMBDA	°C	infinite
TETA	Temperature coefficient for ETA	°C	infinite
KS	Sidegating coefficient	-	0
VSG	Sidegating voltage	V	0
TF	Characteristic temperature determined by traps	°C	TEMP
FLO	Characteristic frequency for frequency dependent output conductance	Hz	0
DELFO	Frequency range used for frequency dependent output conductance calculation	Hz	0
AG	Drain-source correction current gain	-	0

Temperature effects (Levels 2 and 3 only)

Temperature appears explicitly in the exponential terms.

The dependence of the threshold voltage on temperature is given by

$$V_T = V_{T0} - ALPHAT \cdot TEMP$$

The transconductance parameter BETA is adjusted according to

$$\beta = \beta_0 \left(1 - \frac{TEMP}{TBETA} \right)$$

where β_0 is the model parameter BETA.

The output conductance parameter LAMBDA is adjusted according to

$$\lambda = \lambda_0 \left(1 - \frac{TEMP}{TLAMBDA} \right)$$

where λ_0 is the model parameter LAMBDA.

The subthreshold ideality factor ETA is adjusted according to

$$\eta = \eta_0 \left(1 - \frac{TEMP}{TETA} \right)$$

where η_0 is the model parameter ETA.

Supported Analyses

Level 1:   All.

Level 2, 3: Noise and Pole-Zero Analysis not supported.

.SUBCKT Statement

General form:
      .SUBCKT SUBNAME N1 N2 N3 ...

Example:
      .SUBCKT OPAMP 1 2 3 4 5

A subcircuit definition starts with the `.SUBCKT` statement. `SUBNAME` is the name of the subcircuit used when referencing the subcircuit. `N1, N2, ...` are external nodes which cannot be zero. The group of elements that follows directly after the `.SUBCKT` statement defines the topology of the subcircuit. The definition must end with the `.ENDS` statement. Control statements are not allowed in a subcircuit definition. A subcircuit definition can contain other subcircuit definitions, device models, and call to other subcircuits. Note that device models and subcircuit definitions within a subcircuit definition are local to that subcircuit are not available outside. Nodes used in a subcircuit are also local, except "0" (ground) which is always global.

.ENDS Statement

General form:
.ENDS <SUBNAME>

Example:
.ENDS OPAMP

Each subcircuit definition must end with the .ENDS statement. The name of the subcircuit name can be appended to the .ENDS statement.

Call to Subcircuits

General form:
XYYYYYYY N1 <N2 N3 ...> SUBNAME

Example:
X1 2 4 17 3 1 MULTI

.NODESET Statement

General form:
.NODESET V(NODENAME)=VALUE V(NODENAME)=VALUE ...

Example:
    .NODESET V(12)=4.5 V(4)=2.23

This control statement helps AIM-Spice locating the dc operating point. Specified node voltages are used as a first guess for the dc operating point. This statement is useful when analyzing bistable circuits. Normally, .NODESET is not needed.

.IC Statement

General form:
    .IC V(NODENAME)=VALUE V(NODENAME)=VALUE ...

Example:
    .IC V(11)=5 V(1)=2.3

This control statement is used for specifying initial values of a Transient Analysis. This statement is interpreted in two ways, depending on whether UIC is specified or not.

> If UIC is specified, the node voltages in the .IC statement will be used to compute initial values for capacitors, diodes and transistors. This is equivalent to specifying IC=... for each element, only more convenient. IC=... can still be specified and will override the .IC values. AIM-Spice will not perform any operating point analysis when this statement is used, and therefore, the statement should be used with care.
> AIM-Spice will perform an operating point analysis before the transient analysis if UIC is not specified. The .IC statement has no effect.

Bugs Reported by Berkeley
    Models defined within subcircuits are not handled correctly. Models should be defined outside the subcircuit definition starting with the .SUBCKT statement and ending with .ENDS.
    Convergence problems can sometimes avoided by relaxing the maximum stepsize parameter for a transient analysis.

The base node of the bipolar transistor (BJT) is modeled incorrectly and should not be used. Use instead a semiconductor capacitor to model base effects.

Charge storage in MOS devices based on the Meyer model is incorrectly calculated.

Transient simulations of strictly resistive circuits (typical for first runs or tests) allow a time step that is too large (e.g., a sinusoidal source driving a resistor). There is no integration error to restrict the time step. Use the maximum stepsize parameter or include reactive elements.

References

M.-C. JENG, *Design and Modeling of Deep-Submicrometer MOSFET's*, ERL Memo Nos. ERL M90/90, Electronics Research Laboratory, University of California, Berkeley, October (1990)

B. JOHNSON, T. QUARLES, A. R. NEWTON, D. O. PEDERSON and A. SANGIOVANNI-VINCENTILLI, *SPICE3 Version 3e1 User's Manual*, Berkeley (1991)

L. W. NAGEL, *SPICE 2: A Computer Program to Simulate Semiconductor Circuits*, Memorandum No. ERL-M520, Electronic Research Laboratory, College of Engineering, University of California, Berkeley (1975)

J. S. ROYCHOWDHURY and D. O. PEDERSON, Efficient Transient Simulation of Lossy Interconnect, in *Proceedings of the 28th ACM/IEEE Design Automation Conference*, San Francisco, June (1991)

T. SAKURAI and A. R. NEWTON, *A simple MOSFET Model for Circuit Analysis and its applications to CMOS gate delay analysis and series-connected MOSFET Structure*, ERL Memo No. ERL M90/19, Electronics Research Laboratory, University of California, Berkeley, March (1990)

APPENDIX A10.   ADDING NEW DEVICE MODELS TO AIM-Spice

This Appendix describes how to add new device models to AIM-Spice.

Software Requirements

The professional version of the AIM-Spice is required in order to add new device models. This version is available from

Trond Ytterdal
Department of Electrical Engineering and Computer Science
The Norwegian Institute of Technology, University of Trondheim
O. S. Bragstads plass
N-7034 Trondheim, NORWAY
E-Mail: trond.ytterdal@unit.no

Once AIM-Spice is changed, it must be recompiled and relinked. These operations require the following software:

Microsoft Optimizing C/C++ Compiler version 7.0 or later.
Microsoft Windows Software Development Kit version 3.1 or later.

Adding New Device Models

Three types of changes are necessary in order to add a new device model to AIM-Spice: new routines supporting the device model, the modification of the existing interface routines to provide the information about the new device model, and the integration of the device model into the main loops of the simulation algorithms. The following gives an outline of the data structures used in AIM-Spice. Then we describe each of these required changes. The source code for the MOSFET Level 2 model included here serves as an example.

Data Structures Specific for a Device

Each device model has two internal data structures, one for the device model itself, and one for the so-called instance. For example, the first several lines of the data structure in file MOS2DEFS.H are

```
typedef struct sMOS2model {
  int MOS2modType;      /* type index of this device type */
  struct sMOS2model *MOS2nextModel;   /* pointer to next possible model
                                *in linked list */
```

```
MOS2instance * MOS2instances; /* pointer to list of instances
                             * that have this model */
IFuid MOS2modName; /* pointer to character string naming this model */
int MOS2type;        /* device type : 1 = nmos,  -1 = pmos */
int MOS2gateType;
double MOS2tnom;     /* temperature at which parms were measured */
double MOS2latDiff;
double MOS2jctSatCurDensity;    /* input - use tSatCurDens */
double MOS2jctSatCur;   /* input - use tSatCur */
double MOS2drainResistance;
```

This is the model data structure for MOSFET Level 2. The first four fields in the structure serve as a header common for all devices. They must be present exactly in the order shown above. AIM-Spice uses these four fields, for example, when traversing the list of models. The remaining fields are local to the specific device. The field `MOS2modType` contains the index number of the device type. It specifies the location of the functions used to manipulate any given model. `MOS2nextModel` is a pointer to the next model of the same type of device. Different models are stored in a linked list and this field makes it possible to traverse the list. `MOS2instances` is a pointer to the first instance of the model. `MOS2modName` is a character string naming the model.

Let us now consider the instance structure. The first lines of this structure are as follows:

```
typedef struct sMOS2instance {
  struct sMOS2model *MOS2modPtr;   /* backpointer to model */
  struct sMOS2instance *MOS2nextInstance; /* pointer to next instance of
                                          *current model*/
  IFuid MOS2name; /* pointer to character string naming this instance */
  int MOS2dNode;   /* number of the drain node of the mosfet */
  int MOS2gNode;   /* number of the gate node of the mosfet */
  int MOS2sNode;   /* number of the source node of the mosfet */
  int MOS2bNode;   /* number of the bulk node of the mosfet */
  int MOS2dNodePrime; /* number of the internal drain node */
  int MOS2sNodePrime; /* number of the internal source node */
  int MOS2mode;       /* device mode : 1 = normal, -1 = inverse */
  unsigned MOS2off :1;
  unsigned MOS2lGiven :1;
  unsigned MOS2wGiven :1;
```

This structure contains a header which consists of three fields that must be present. Additional fields, if present, must appear in a specific order. The fields are explained in the comments statements in the listing. For a MOSFET, the additional fields are `MOS2dNode`, `MOS2gNode`, `MOS2sNode` and `MOS2bNode`. This is and must be the same order in which they appear on the element line. The maximum number of external nodes is 5.

Parameter Descriptors

File MOS2.C contains two large static arrays along with other global variables. The first array contains all parameters for the device instance. The first five lines of the file listed below are parameters from the element line of MOSFET devices.

```
IFparm MOS2pTable[] = { /* parameters */
  IOP("l",      MOS2_L,    IF_REAL   , "Length"),
  IOP("w",      MOS2_W,    IF_REAL   , "Width"),
  IOP("ad",     MOS2_AD,   IF_REAL   , "Drain area"),
  IOP("as",     MOS2_AS,   IF_REAL   , "Source area"),
  IOP("pd",     MOS2_PD,   IF_REAL   , "Drain perimeter"),
```

The parameters are placed in the array by means of a macro. The array contains more than one macro for this purpose, namely IOP, OP and IP. IOP is used for the parameter which is both an input and an output parameter, OP corresponds to an output parameter only, and IP corresponds to an input parameter only. Columns 2 and 3 above contain information for functions that handle the parameters. The next array in the file is an array equivalent to the one shown above, but containing all model parameters. The rest of the variables in the file are for housekeeping purposes.

The Main Device Structure

The arrays and global variables described above are assembled into the main device structure together with pointers to device specific functions such as device loading. Each device has a header file with the name XXXXXITF.H where the string XXXXX identifies the device. This file contains the main device structure. The main device structure for MOSFET Level 2 is shown below together with a brief description of each field.

```
SPICEdev MOS2info = {
   {
   "Mos2", /* name of this type of device */
   "Level 2 MOSfet model with Meyer capacitance model",/* description */
   &MOS2nSize,    /* number of terminals */
   &MOS2nSize,    /* number of terminal names */
   MOS2names,     /* pointer to array of pointers to terminal names */
   &MOS2pTSize,   /* number of instance parameter descriptors */
   MOS2pTable,    /* array of instance parameter descriptors */
   &MOS2mPTSize,  /* number of model parameter descriptors */
   MOS2mPTable,   /* array of model parameter descriptors */
   },
   MOS2param,     /* inputs a parameter to a mosfet instance */
   MOS2mParam,    /* imnputs a parameter to a mosfet model */
```

```
    MOS2load,       /* loads the device into the matrix */
    MOS2setup,      /* preprocess device once before solution begins */
    MOS2setup,      /* setup routine specifically for pole-zero analysis */
    MOS2temp,       /* performs temperature dependent setup processing */
    MOS2trunc,      /* performs truncation error calculation */
    NULL,           /* search for device branch equations */
    MOS2acLoad,     /* loads the device into the matrix for ac analysis */
    NULL,           /* is called on acceptance of a timepoint */
    MOS2destroy,    /* destroys all models and instances */
    MOS2mDelete,    /* deletes a model and all instances of that model */
    MOS2delete,     /* deletes an instance */
    MOS2getic,      /* gets initial conditions from RHS vector */
    MOS2ask,            /* asks about device instance details */
    MOS2modAsk,     /* asks about device model details */
    MOS2pzLoad,     /* loads the device into the matrix for pz analysis */
    MOS2convTest,   /* convergence test function */
    MOS2sSetup,     /* does the setup of device sensitivity info */
    MOS2sLoad,      /* loads the device sensitivity info */
    MOS2sUpdate,    /* updates the device sensitivity info */
    MOS2sAcLoad,    /* loads the device ac sensitivity info */
    MOS2sPrint,     /* prints sensitivity info */
    NULL,           /* calculates truncation error for sens analysis */
    MOS2disto,      /* function called during a distortion analysis */
    MOS2noise,      /* function called during a noise analysis */
    &MOS2iSize,     /* size of an instance */
    &MOS2mSize      /* size of a model */
};
```

Device Specific Functions

The main device structure shown above contains pointers to the device specific functions defining the behavior of the device. Some of the function fields of the MOSFET Level 2 device are NULL because not all functions are necessary in order to define the MOSFET operation. For instance, the find branch function is NULL above. This implies that there are no branch equations for this device. MOSFET files provide a detailed example of different device specific functions.

Step One: Adding New Routines

This is the most tedious process, and it requires some insight into circuit simulation methods. The easiest strategy is to rely on the example code provided.

First, one has to determine out how many of the device specific functions are needed to define the device. The next step is to write the header files. The process is completed by writing the C source files.

Hint: Use the example code as a basis and modify it.

Step Two: Modification to Existing Interface Routines

It is not allowed to add devices starting with a new letter. Instead, new

Levels should be added to an existing device. This is done by modifying the parser routines shown next

```
INPdomodel   in INPDOMOD.C
INPfindLev   in INPFINDL.C
INP2X        in INP2X.C
```

The function `INPdoMod` has to be modified to reflect the new device to be added. In order to add a new MOSFET level, you have to locate the MOSFET part of the function which starts with

```
} else if( (strcmp(typename,"nmos")==0) || (strcmp(typename,"pmos")==0) ) {
```

Next, you add a new case statement similar to other device, but with a new level and the name given to the device in the main data structure.

This function can be completely ignored if the device does not have a ".MODEL" line associated with it.

The function `INPfindLev` currently supports up to 19 levels. It can be left unchanged if one does not add levels higher then 19.

The letter 'X' in INP2X and in INP2X.C should be substituted by the first letter for the device to be added. For example, to add a new MOSFET level, you add the line to function INP2M which can be found in the file INP2M.C.

```
/* This example adds a new MOSFET level with name "NEWMOS" */
        if( (thismodel->INPmodType != mytype ) &&
/* New line comes here */
                (thismodel->INPmodType != INPtypelook("NEWMOS") ) &&

                (thismodel->INPmodType != INPtypelook("Mos2") ) &&
                (thismodel->INPmodType != INPtypelook("Mos3") ) &&
                (thismodel->INPmodType != INPtypelook("BSIM1") ) &&
                (thismodel->INPmodType != INPtypelook("Mos5") ) &&
                (thismodel->INPmodType != INPtypelook("Mos6") ) &&
                (thismodel->INPmodType != INPtypelook("MOSA1") ) &&
                (thismodel->INPmodType != INPtypelook("NPMOSA1") ) &&
                (thismodel->INPmodType != INPtypelook("NPMOSA2") ) &&
                (thismodel->INPmodType != INPtypelook("NPMOSA3") ) &&
                (thismodel->INPmodType != INPtypelook("ASIA1") ) &&
                (thismodel->INPmodType != INPtypelook("PSIA") ) &&
                (thismodel->INPmodType != INPtypelook("BSIM2") ) ) {
        LITERR("incorrect model type")
        return;
    }
```

Step Three: Integration into the Main Loops of the Simulator

The integration of the device model into the simulator requires modification of file BCONF.C. The first part of the file is the list of define statements which makes it easy to remove devices and analysis types from the .exe file. A device or analysis type is removed by commenting out the respective define statement.

To include a new level of the MOSFET device, we first locate all positions in the file where an existing MOSFET device appears and then add the same statements for the new device model. The first occurrence of the Level 2 MOSFET is in line 95 which is an include statement. The file MOS2ITF is included (this file contains the main device structure). The next occurrence is in line 236.

```
#ifdef DEV_mos2
   &MOS2info,
#endif
```

These statements include the main device structure for MOSFET Level 2 in the array of main device structures if and only if DEV_mos2 is defined. A new device is included by adding similar statements as those described above.

Step Four: Compiling and Relinking

The C compiler and the Windows software development kit manuals explain how to set up the environment variables to point to the appropriate directories.

The source files for the new device, BCONF.C, and the changed parser code have to be compiled. The following two `make` files will assist in this process. File MKFILE1 is a basis for compiling the device specific functions, and MKFILE2 is a basis for compiling the parser and interface files. The comments in these `make` files explain how the files should be modified before the compilation. The following commands invoke the compiler

```
make mkfile1
make mkfile2
```

The response file is provided for the linker, but it has to be modified. The changed object modules must be added. Then, the linker will ignore the old object modules contained in the libraries. Also, the object modules for the new device must be added. The following command starts the linking

```
link @aimspice
```

The following error message:

```
Stack plus automatic data segment exceeds 64K
```

means that one has to comment out one or more devices in BCONF.C and recompile and relink.

The last step is to include the resources used by AIM-Spice with the command

```
rc aimspice.res
```

Index

a-Si TFTs, 497, 663
absorption coefficient, 49
ac analysis, 668
ac emitter current, 148
ac small signal analysis, 568
ac transfer function, 670
accelerator keys, 545
acceptor-like states, 496
acceptors, 21
accumulation regime, 202
acoustic deformation potential scattering, 28
active mode, BJT, 114, 126
adding new routines, 733
AIM-Postprocessor, 549, 585
AIM-Postprocessor, loading, 586
AIM-Spice HBT Model, 173
AIM-Spice installation, 551
AIM-Spice model parameters, 697
AIM-Spice models, 646
AIM-Spice reference, 667
AIM-Spice, 102, 190, 534, 549
AIM-Spice, adding device models, 730
AIM-Spice, allowed functions, 676
AIM-Spice, professional version, 730
AIM-Spice, running, 551
air bridges, 402
AlGaAs/GaAs heterostructure, 152
alloyed ohmic contacts, 93
alpha cutoff frequency, 148
$Al_xGa_{1-x}As$, 3, 95
$Al_xGa_{1-x}As$, properties, 637
ambipolar diffusion coefficient, 33
ambipolar lifetime, 64
ambipolar mobility, 34
amorphous C:H, 495
amorphous Ge:H, 495
amorphous Si, properties, 638
amorphous Si:F, 495

amorphous SiC:H, 495
amorphous silicon p-n junctions, 497
amorphous silicon TFT, 626
amorphous silicon, 494
analog copy amplifier, 626
analysis limits, 579
analysis limits, resetting, 582
Anderson model, 96
area factor, 688
atomic basis, 7
atomic configuration, 2
AuGe ohmic contacts, 402
Auger recombination, 54, 127
avalanche breakdown voltage, 138
avalanche breakdown, 76, 134
avalanche breakdown, critical voltage, 135
avalanche multiplication, 136
average depletion depth, 224
average electric field, 78
average thermal energy, 31
Avogadro number, 634
axes and labels, 580
axes, formatting, 592

backgating effects in MESFET, 439
ballistic electrons, 153
ballistic transport, 28, 153, 290, 394
band diagram, cubic semiconductors, 9
band diagram, p-n junction, 55
band diagram, planar doped HEMTs, 447
band discontinuity, 152
band gap narrowing, 152
band structure, 7
band-to-impurity recombination, 47
barrier height, 81
barrier transparency, 88
base charging time, 146

base push out, 173, 649
base region, 114
base spreading resistance, 119
base transport factor, 123, 136
base-width modulation, 127
basic HFET model, 441
basic MESFET model, 404
below-threshold regime, 499
Berkeley short channel Igfet model, 691
beta cutoff frequency, 149
bias generating circuits, 284
BICMOS, 194
binary compound, 7
bipolar junction transistor, 113, 188, 708
BJT equivalent circuit, 124
BJT model, 709
BJT technology, 120, 124
BJT, 113, 649, 708
BJT, inverse mode of operation, 127
BJT, small signal regime, 140
Bloch waves, 13
body effect constant, 279
body effect parameters, 360
body effect, 276
body plot, 283, 287, 388
Bohr energy, 22, 634
Bohr magneton, 634
Bohr radius, 22, 634
Boltzmann transport equation, 26
box type doping profile, 427
breakdown field, 77, 135
breakdown in BJTs, 134
breakdown voltage, 121, 160, 185
breakpoints, 708
Brillouin zone, 7, 222
BSIM, 288, 368, 652, 691, 694
bugs, reported by Berkeley, 729
built-in field, 159
built-in voltage, 56, 72, 80, 404
bulk junctions, 702
bulk sheet charge density, 361
buried channel PMOS, 279
buried channel, 317
buried common emitter, 121

capacitance model, HFET, 451
capacitance, M-S space charge region, 90
capacitance, p-n diode, 71
capacitive coupling, 277
capacitor model, 678
capacitor, non-linear, 677
capture cross-section, 51
carrier-trap-state density, 514
CDI technology, 120

channel conductance, MOSFET, 303
channel length modulation, 253, 323, 462, 517
characteristic diffusion time, 146
characteristic length, 254, 323, 346, 371, 463
characteristic time constant, 75, 90
characteristic tunneling length, 88
characteristic voltage, 323, 346
characterization, poly-Si TFTs, 523
charge injection transistor, 474
charge neutrality condition, 24
charge sharing model, 263
charge storage, 692
check boxes, 547
CHINT, 474
circuit description, 551, 559, 673
circuit elements, 562
circuit files, 555
circuit files, opening , 556
classical charge sheet model, 225
CMOS technology, 191
CMOS, 190
collector buried layer, 119
collector charging time, 150
collector contact, 115
collector current with open emitter, 126
collector current, 115, 124
collector region, 114
collector reverse saturation current, 126, 167
collector series resistance, 119
collector-base junction, 114, 115
collector-base recombination current, 127
collector-up configuration, 158
command buttons, 547
command reference, 602
comment line, 674
comment statements, 731
common-base configuration, 116
common-base current gain, 126, 129
common-collector configuration, 116
common-emitter configuration, 116
common-emitter current gain, 117, 151, 186
compensation , 21
compiling, 735
complementary MOSFET, 190
compositional grading, 157
compound semiconductor FETs, 393
compound semiconductors, 3
compound semicond. FET fabrication, 397
conducting channel, saturated part, 252
conduction band discontinuity, 162, 442
conduction band spike, 98, 101
conduction band, 9

constant energy surfaces, 294
contact junction depth, 273
continuity equations, 32
control menu, opening, 544
controlled sources, 564
controlling current, 683
cool CMOS, 283
counter doping, 284
coupled inductors, 689
coupling coefficient, 690
covalent bonding, 5
critical voltage, 77, 79
crystal imperfection, 25, 95
crystal symmetry, 7
crystalline-like regime, 499
current controlled switch, 721
current saturation, 247
current sources, 683
current sources, current-controlled, 682
current sources, voltage-controlled, 682
cursors, 601
cutoff frequency, 149, 160
Czochralski method, 397
C–V characteristics, MESFET, 428
C–V characteristics, MOSFET, 241

data file, 586
data structures, 730
dc analysis, 684, 688
dc operating point, 567, 574, 668, 676
dc transfer analysis, 567
DCFL inverter, 490, 533
deBroglie wavelength, 219
Debye length, 61, 201
deep acceptor-like states, 496, 530
deep donor-like localized states, 496
deep saturation, 239, 256, 323
default parameter values, 695
define statement, 735
degeneracy factor, 22, 110
degenerate semiconductor, 13
delta-doped HFET, 445
delta-doped MESFET, 428
density of localized states, 501
density of states effective mass, 16, 222
density of states, 14, 19
density of the surface states, 82
dependent sources, non-linear, 676
depletion approximation, 56, 57
depletion capacitance, 133, 145
depletion capacitance, heterojunction, 97
depletion charge, 271
depletion layer capacitance, 71, 262
depletion layer width, 83, 233

depletion region, 56, 115, 204, 234
depletion sheet charge density, 233
device instance, 732
device isolation, 121
device model, 730
device models, integration of, 735
device specific functions, 733
DHBT, 122, 156
dialog boxes, 546
dialog boxes, closing, 549
diamond structure, 6
DIBL, 260, 266, 306, 333, 421, 4 57
dielectric permittivity, 32
differential output conductance, 323
differential resistance, p-n junction, 75
differential resistance, Schottky barrier, 90
diffusion capacitance, 72, 103, 112, 647
diffusion charge, 112
diffusion coefficient, 30, 101, 115, 163
diffusion current, 65
diffusion length, 64, 101, 115, 117, 163
diffusion model, 85
dimmed commands, 545
diode equation, 66, 640
diode model, 679
diode saturation current, 66
diode series resistance, 76
diode, short, 647
diodes, 679
dipole HFETs, 401
direct gap semiconductor, 10
discrete energy levels, 21
dislocations, 7, 95
dispersion relation, 19, 108
displaced Maxwellian distribution, 36
distributed transmission line, 706
donor-like states, 496, 530
donors, 21
double degeneracy, 23
double diffusion, 121
double HBT, 122, 156
double poly silicon transistors, 122
drain bias induced charges, 270
drain induced barrier lowering, 260, 266,
306, 421, 457
drain-channel junction, 259
drift and diffusion current, 30
drift momentum, 25
drift velocity, 25, 111
drift-diffusion transport, 274
drop-down list boxes, 548
dual gate a-Si TFT, 531
dynamic base current, 132
dynamic circuits, 277

dynamic collector current, 133
dynamic emitter current, 133

Early effect, 130
Early voltage, 185, 186
Ebers-Moll equations, 166
Ebers-Moll model, 124, 647
effective additional barrier, 103
effective base width, 125
effective carrier mobility, 665
effective channel depth, 414
effective channel thickness, 230, 253, 263
effective delay time, 150
effective density of states, 15, 16
effective diffusion capacitance, 75
effective doping density, 25, 71
effective drain-source voltage, 319, 343
effective electric field, 293
effective electron temperature, 36, 46, 473
effective energy relaxation time, 37
effective extrinsic drain-source voltage, 416
effective extrinsic saturation voltage, 303, 454
effective field, 224
effective gate capacitance, 241
effective gate depletion charge, 266
effective gate length, 274, 461, 470
effective gate voltage swing, 420, 449
effective generation time, 207
effective light hole mass, 78
effective mass, 10, 25
effective medium, 665
effective momentum relaxation time, 37
effective resistance, Schottky barrier, 90
effective Richardson constant, 474
effective saturation velocity, 146
effective thickness, channel region, 463
effective transit time, 146
effective transverse field, 295
Einstein relation, 31, 111
electric flux density, 202
electron affinity, 80, 96, 197
electron capture rate, 52
electron distribution function, 36
electron emission rate, 52
electron lifetime, 53, 68
electron mobility, 341
electron sheet density, 449
electron trapping, 373
electron tunneling, 168
electron wave vector, 7
electron-electron collisions, 36
electron-hole pairs, generation, 206, 371
electron-hole scattering, 30

element line, 732
ellipsoidal equal energy surfaces, 86
ellipsometry, 216
emission from traps, 51
emitter capacitance, 145
emitter current crowding, 127, 134
emitter current, 115, 124
emitter down n-p-n transistor, 121
emitter down structure, 123
emitter injection efficiency, 136, 153
emitter periphery, 172
emitter region, 114
emitter reverse saturation current, 127, 128, 167
emitter-base junction, 115
emitter-base recombination current, 127
emitter-base voltage, 116
emitter-down structure, 158
empty trap, 50
energy band diagram, heterojunction, 97
energy band gap, 5, 9
energy band narrowing, 156
energy barrier, 81
energy diagram, M-S barrier, 80
energy gap discontinuity, 96
energy relaxation time, 26
.ENDS statement, 728
epitaxially grown MESFETs, 399
equivalent band minima, 16
equivalent biases, 362
equivalent capacitance, 76
equivalent inductance, 90
error message, 736
error reporting, 583
excess minority carrier charge, 112, 146
extrinsic carrier concentration, 25
extrinsic collector capacitance, 119
extrinsic collector-base voltage, 134
extrinsic drain conductance, 244
extrinsic emitter-base voltage, 134
extrinsic saturation voltage, 326, 349
extrinsic transconductance, 243

face-centered cubic lattice, 5
fast surface states, 204
Fermi function, 13
Fermi integral, 15
Fermi level pinning, 402
Fermi level, 13
Fermi potential pinning, 172
Fermi vector, 13
Fermi-Dirac distribution function, 13, 221
field effect mobility, 500, 504
field emission, 88

file menu, 603
filled trap, 50
first Brillouin zone, 7, 8, 109
fixed charges, 198
flat-band condition, HFET 442
flat-band voltage, 197, 279
forbidden gap, 5, 9, 21
free space permittivity, 22
frequency response, 668
fringing capacitance, 303, 417
fringing charge, 388
full channel resistance, 435

g-parameters, 142
gamma function, 41
gapless semiconductors, 10
gas constant, 634
gate charge partitioning, 367, 697
gate leakage current, 471
gate length modulation, 307
gate length offset, 471
gate transconductance, 245
gate voltage division, 212, 262
gate-channel capacitance, 214, 302
gate-drain capacitance, 302
gate-source capacitance, 302
Gauss' law, 202, 220, 262
GCA, 230
generalized UCCM, 276
generation current, 68
generation life time, 68
generation of electron-hole pairs, 49
generation rate, 32, 47
generation, 45
geometric capacitance, 76, 90
geometric magnetoresistance, 44
germanium, 2
graded base, 159
graded emitter, 168
graded semiconductor, 93
gradual channel approx., 229, 406, 506
graph editing, 592
graph menu, 606
graphical plot, 588
graphics user interface, 536, 537
Grotjohn and Hoeflinger model, 345
guard rings, 195
Gummel number, 130, 183
Gummel-Poon model, 128, 709
GaAs device technology, 397
GaAs MESFET integrated circuits, 404
GaAs MESFETs, 404
GaAs on silicon, 95
GaAs semi-insulating substrates, 397

GaAs, 2
GaAs, properties, 636

h-parameter equivalent circuit, 143
h-parameters, 142
Hall angle, 42
Hall constant, 40, 111
Hall contacts, 39
Hall effect, 38
Hall electric field, 39
Hall factor, 40, 111
Hall mobility, 41, 111
Hall voltage, 39, 111
HBT band structure, 152, 161
HBT model, 647, 703
HBT ring oscillator, 160
HBT technology, 122, 156
HBT theory, 161
HBT, 151
HBT, non-ideal effects, 170
HBTA1, 151, 173, 187
header file, 732
header, 731
heavy hole band, 9
HEMTs, 400
heterojunction barrier, 123
heterojunction bipolar technology, 122, 156
heterojunction bipolar transistor, 151, 616, 702
heterojunction interfaces, 95
heterojunction p^+-n diode, 104, 112
heterojunctions, 93
heteropolar semiconductor, 5
heterostructure diode model, 646, 680
heterostructure diode, 614, 646
heterostructure field effect transistor, 458, 674
heterostructure quantum well, 108
heterostructure, 7, 94
HFET equivalent circuit, 472
HFET inverter, 489
HFET model, 674
HFET ring oscillator, 490
HFET saturation current, 449
HFET, 458, 624, 661
HFETA, 450
HFETs, 400, 4 58
HIGFETs, 401
high bias-temperature stress test, 211
high current roll-off, 185
high electron mobility transistors, 400
high field dipole domain, 412
high injection conditions, 63
high injection effects, 127, 131
holding time, 277

hole capture rate, 52
hole emission rate, 52
hole lifetime, 53, 68
hole mobility, 317
hot electron effects, 31, 323, 370
hot electron regime, HFET operation, 474
hot electrons, 26
HSPICE, 441
hybrid-π equivalent circuit, 144
hydrogen atom-like impurity, 22
hydrogenated amorphous silicon, 190
hyperbolic tangent model, 411, 420

.IC statement, 728
ideal BJT, 114
ideal switch, 714
ideality factor, 70, 87, 262, 432, 455, 512
I²L, 119
impact ionization, 32, 76, 517
impact ionization, characteristic field, 371
imperfections, 7
implantation technology, 193
impurity band, 155
impurity solubility, 93
Imref, 46
in-plane valleys, 294
independent current sources, 683
independent sources, 564
independent voltage sources, 716
indirect gap semiconductor, 10
inductors, 690
initial conditions, 571
initial value, optional, 688
injection barrier, lowering, 271
InP substrate, 7, 123
input noise, 670
input resistance, 670
instance, 730
insulator-semiconductor interface, 197
integrated injection logic, 119
integrated noise, 670
integrated Schottky logic, 121
integrating factor, 260
interactive simulation control, 574
interband transitions, 10
intercept current, 128
interconnect metallization, 402
interface electric field, 262
interface energy barrier, 259
interface potential, 259, 268
interface recombination, 172
interface routines, modification, 733
interfacial dipole, 83
interlevel dielectric, 403

internal data structures, 730
internal photoemission technique, 96
intervalley transfer, 27
intrinsic carrier concentration, 18
intrinsic collector currents, 167
intrinsic collector-base voltage, 134
intrinsic drain conductance, 243
intrinsic emitter currents, 167
intrinsic emitter-base voltage, 134
intrinsic Fermi level, 21
intrinsic saturation voltage, 323, 346
intrinsic transconductance, 243
inverse alpha, 127
inverse common-base current gain, 128, 167
inverse common-emitter gain, 167
inverse current gain, 127
inversion charge at threshold, 214, 302
inversion layer, 204, 235
inversion regime, 199, 200, 202, 203
inversion sheet charge density, 231
inversion symmetry, 6
inversion threshold, 199, 200
inversion, 198
inverter circuit, 391
ion implanted GaAs MESFETs, 398
ionization energy, 110
ionized donors, 22
ionized impurity scattering, 28
I–V characteristic, heterojunction, 98
I–V characteristic, p-n diode, 61
I–V characteristic, Schottky diode, 85
I–V characteristics, HFET, 454, 458, 460, 465
I–V characteristics, MESFET, 420
I–V characteristics, MOSFET, 235, 239, 303, 316, 340

JFET model, 688
junction depletion regions, 125
junction diode model, 681
junction field effect transistors, 688
junction isolation, 191

Kane model, 108
kink effect, 370, 517
Kirk effect, 132, 150, 173
knee region, 420
knee shape parameter, 312, 420, 434

labels, formatting, 594
latch-up, 195
lateral BJT, 121
lattice constants, 4

lattice vibrations, 25
law of the junction, 63
LCD, 533
LDD, 386
leakage current, 236, 278
LEC, 402
LED, 395
legends, formatting, 598
light emitting diodes, 95, 395
light hole band, 9
linear extrapolation method, 523
linker, 735
liquid crystal display, 190, 495
liquid encapsulated Czochralski, 402
list boxes, 547
load line, 140
localized states, 496
LOCOS, 191
long wavelength laser technology, 123
long-base BJT, 262
longitudinal effective mass, 223
Lorentz force, 38
lossless transmission, 706
lossy transmission lines, 705
low doped drain, 386
low-field mobility, 28, 247

magnetoresistance, 38
main data structure, 734
main device structure, 732
main loops of the simulator, 735
mass filtering, 161
mass-action law, 18, 23
maximum channel sheet charge density, 451
maximum diffusion capacitance, 75
maximum electric field, 78, 84, 371
maximum oscillation frequency, 151
Maxwellian distribution, 36
MBE, 155, 394, 397
mean free path, 28, 84, 274
mean thermal velocity, 162
menu commands, choosing, 544
menu, formatting, 607
menu, selecting, 544
menus, working with, 543
MESA1, 423, 657, 659
MESA2, 425, 657, 659
MESFET models, 657, 723
MESFET parameters vs. frequency, 437
MESFET parameters vs. temperature, 437
MESFET ring oscillator, 621, 633
MESFET structures, 398
MESFET threshold voltage vs. T, 437
MESFET, 619, 657, 723

metal insulator semiconductor FET, 188
metal oxide semiconductor FET, 188
metal semiconductor FETs, 404
metal-n^+-n ohmic contact, 92
metal-semiconductor boundary, 82
metal-semiconductor contacts, 79
metal-semiconductor energy barrier, 442
metallurgical junction, 57
metallurgical width, 125
Meyer model, 302, 416, 453, 511, 692
Microsoft Windows™, 535
minority carrier diffusion constants, 125
minority carrier diffusion length, 125
MIS capacitance, 196, 205
MIS capacitance, analytical model, 217
MIS capacitor, band diagram, 197
MIS capacitor, characterization, 211
MIS capacitor, equivalent circuit, 207
MIS C–V characteristics, 208
MISFET, 188
mobile ions, 211
mobility, 25
mobility, MOSFET channel, 290
mobility, NMOS Channel, 292
mobility, PMOS Channel, 298
MOCVD, 155, 394, 397
model line, 680
model name, 674
moderate inversion, 199
molecular beam epitaxy, 155, 394, 397
momentum relaxation time, 25, 84
monolithic integrated circuits, 190
Monte Carlo method, 36
MOS capacitor, 191
MOS model, short channel, 692
MOSA1, 301
MOSFET capacitances, four terminal, 365
MOSFET capacitances, three terminal, 365
MOSFET models, 691
MOSFET saturation current, 304
MOSFET, 188, 229, 617, 690
MOSFETs, basic theory, 229
multiplication factor, 136

n-channel MOSFET, 188, 229
n-type dopants, 21
n-well CMOS, 193
narrow channel effect, 281, 388
narrow gap semiconductors, 93, 108
near equilibrium transport, 31
nearest neighbors, 2
net recombination rate, 52, 63, 68
netlist, 551
neutral impurity scattering, 28

neutral level, 81
Newton-Raphson iteration, 640, 707
NMOS Model, 345, 347, 354, 654
NMOSA, 654
NMOSA1, 342, 654, 655
NMOSA2, 347
NMOSA3, 354
nodes, 562
nodes, maximum number of external, 731
.NODESET statement, 728
noise analysis, 570, 576, 669
noise spectral density, 670
noise voltage, 669
non-degenerate semiconductor, 17
non-ideal diode equation, 210, 213, 515
non-parabolicity, 109
non-polar optical scattering, 28
non-radiative recombination, 47
non-uniform doping, MESFET, 413
nonparabolicity, 18
normal common-base current gain, 128
normal common-emitter gain, 167
np product, 18
n^+-p junction, 60, 115

object modules, 735
occupation function, 52
ohmic contact, 91
one-vector plots, 574, 587
onset of inversion, 212
onset of saturation, 250
operating point, 140
operators, 677
option buttons, 548
options menu, 672
optoelectronic devices, 95
output conductance modulation factor, 521
output conductance parameter, 438
output noise variable, 669
output resistance, 670

p-channel MOSFET, 188
p-i-p^+ buffer structure, 454
p-n junction breakdown, 76
p-n junction equivalent circuit, 75
p-n junction, 55
p-type buffer layer, 454
p-type dopants, 21
p-well CMOS, 192
p-well technology, 193
parameter descriptors, 732
parameter extraction, HFET, 468
parameter extraction, MESFET, 431

parameter extraction, NMOS, 349, 356
parameter extraction, PMOS, 327, 335
parameter extraction, MOSFET 310
parasitic bipolar elements, 194
parasitic inductance, 76
parser routines, 734
partitioning factor, 367
Pauli exclusion principle, 13
periodic crystal potential, 10
periodic table, 3
phenomenological transport equations, 36
physical magnetoresistance, 44
piezoelectric scattering, 28
pinch-off condition, 507
pinch-off, MESFET, 405
pinch-off, MOSFET, 235, 249, 371
PISCES simulation program, 256
plot limits, 578
plot windows, 577
PMOS models, 318, 322, 325, 651
PMOSA1, 318, 651
PMOSA2, 325, 652, 653
PMOSA3, 331
p^+-n diode, 66
Poisson's equation, 32, 57, 200, 223, 230, 253
polar optical scattering, 28
pole-zero analysis, 570, 574, 670
poly emitter, 122
poly-Si TFT model, 665, 692
poly-Si TFTs, 495, 512
polycrystalline film, 512
polycrystalline silicon, 495
polysilicon gate, 198, 215, 283
polysilicon thin film transistors, 512
position dependent threshold voltage, 242
power devices, 396
power gain, 151
power MESFETs, 399
primitive cell, 5
principle of detailed balance, 51
printing, 601
process characterization, 695
propagation delay, 156, 304
PSIA, 665
PSpice, 405
punch-through voltage, 138
punch-through, 134, 193, 272

QSA, 359
quantum efficiency, 95
quantum number, 19
quantum well, 18, 94, 109
quantum wire, 109

quasi neutrality condition, 33, 65
quasi-Fermi level, 35, 45, 62
quasi-static approximation, 359
quaternary compounds, 3

radiation-hard electronics, 396
radiative recombination, 47
Raytheon model, 405
real space transfer, 443, 448, 475
recessed gate MESFETs, 398
reciprocal lattice vectors, 7
reciprocal lattice, 7, 109
reciprocal space, 7
reciprocity relationship, 127
recombination centers, 95
recombination current, 69, 132, 154, 170
recombination in the base, 117, 154
recombination lifetime, 49, 54
recombination processes, 47
recombination rate, 32, 47, 111
recombination, 45
recombination, electron-hole pairs, 50
recombination, space charge region, 172
reduced effective mass, 78
refractory metal-silicon gates, 402
relaxation time model, 37
relinking, 735
resistor model, 712
resistors, 711
reverse bias conduction parameter, 473
reverse diode conductance, 473
Richardson's constant, 86, 275
run-off, 140

safe operating area, 140
saturation current, 251
saturation field, 247, 409
saturation point method, 524
saturation regime, 245
saturation velocity, 29, 247
saturation voltage parameter, 353
saturation voltage, 237, 250
saving results, 582
SBC technology, 120
Schottky barrier contact, 404
Schottky barrier diodes, 83, 681
Schottky barriers, 79
Schottky contact, 79
Schottky junction, 81
Schottky transistor logic, 121
Schrödinger equation, 223
scroll bars, 542
scrolling, 553

second breakdown, 139
self-aligned double polysilicon, 122
self-aligned HFET, 404
self-aligned MESFET, 398
semiconductor capacitance, 205, 687
semiconductor equations, 30
semiconductor physics, 1
semiconductor resistors, 712
semiconductor statistics, 13
semiconductor-insulator interface, 196
series resistance, 90
shallow acceptors, 21
shallow donors, 21
shallow hydrogen-like level, 23
SHBT, 122, 156
Shichman-Hodges, 691
Shockley equation, 66
Shockley model, 406
short channel effects, 263, 273, 281
Si-Ge heterostructure, 95
Si-SiO$_2$ MOS capacitor, 212
Si-SiO$_2$ system, 196
Si$_3$N$_4$ cap, 402
Si$_3$N$_4$, properties, 639
side wall capacitance, 303
sidegating effects, 439
silicon bipolar technology, 119
silicon nitride side walls, 398, 455
silicon p-n junction, 58
silicon, 2
silicon, properties, 635
silicon-on-insulator, 190
simple charge control model, MOSFET, 238
simplified FET equivalent circuit, 475
single HBT, 122, 156
SiO$_2$, properties, 639
small signal admittance, 104
small signal current density, 103
small signal impedance, 103
small signal p-n equivalent circuit, 75
small-signal admittance, 74
SOA, 140
Sodini's model, 291
solar cells, 95
solid state solutions, 3
solubility limit, 93
source-channel junction, 259
space charge density, 32, 230
space charge region, 56, 206
specific contact resistance, 93
SPICE analysis options, 667
SPICE device models, 667
SPICE, 288, 534, 549, 652, 667
spike-notch region, HBT 153

spin degeneracy, 14
split C–V technique, 215
split-off band, 9
square law model, 405, 410
square well potential, 18
standard diode model, 680
starting a simulation, 581
static dielectric permittivity, 22
static feedback, 277, 340
stopping a simulation, 581
strong inversion, 198, 212
subband, 20
subcircuits, 565
subcircuits, call to, 728
.SUBCKT statement, 727
substrate leakage current, 262
substrate transconductance, 245
subthreshold capacitance, 219, 423
subthreshold current, MOSFET, 258, 264, 273, 276
subthreshold regime, 257
subthreshold saturation current, 455
superlattice, 95
surface charge, 196
surface electric field, 202
surface potential, 199, 200, 202, 203, 361
surface recombination current, 154
surface recombination rate, 54
surface recombination, 47, 54, 153, 172, 647
surface scattering mobility, 293
surface state charge density, 279
surface states, 7, 82, 95, 198, 210
surface traps, 54
switch impedances, 714
switch model, 714, 722
switches, 565
symmetrical CMOS, 283

T-equivalent circuit, 145
T-gates, 399
tail acceptor-like states, 496
tail donor-like states, 496
tail states, 530
target threshold voltage, 278
ternary compounds, 3
tetrahedral bond configuration, 5
text boxes, 549
text, adding and formatting, 596
text, editing, 553
text, selecting, 554
TFT, 495
thermal breakdown, 79
thermal energy, 19
thermal generation rate, 48

thermal generation, 63
thermal impedance, 437
thermal instability, 139
thermal motion, 26
thermal velocity, 27, 84, 99
thermionic diffusion model, 162
thermionic emission current, 274
thermionic emission, 98, 473
thermionic model, 84
thermionic-emission-diffusion model, 161
thermionic-field emission, 88, 168
thin film transistors, 494, 626
thin-oxide capacitance, 692
three-dimensional integrated circuits, 514
threshold point, 219
threshold saturation current, 432
threshold voltage engineering, 277
threshold voltage shift, 272
threshold voltage shift, MESFET, 421
threshold voltage, 204, 235
threshold voltage, NMOS, 278
threshold voltage, PMOS, 278
time domain response, 671
title line, 673
total gate capacitance, HFET, 453
total gate current, 472
trace area, formatting , 596
trace expression editing, 589
transconductance compression param., 418
transconductance method, 524
transconductance parameter, 240, 249, 312, 418, 436
transconductance, BJT 144
transfer function analysis, 570, 576, 670
transient analysis, 559, 576, 671, 679, 684, 691, 717, 723
transient initial conditions, 674, 688
transient phenomena, 28
transit time, 75
transitional regime, 499
translational symmetry, 7
transmission line model, 706
transmission lines, 705
transmission lines, lossless, 715
transresistance, 683
transverse effective mass, 223, 294
transverse field, 232
transverse valleys, 294
trap level occupancy function, 51
trapped charge, 34
trapped electrons, 32
trapped holes, 32
trench isolation, 121
triangular quantum well, 222

Trofimenkov's model, 291
tunneling breakdown, 76, 79
tunneling current, 78
tunneling emitter bipolar transistor, 160
tunneling transmission coefficient, 89
turn-on voltage, 87
twin well CMOS, 194
twin well process, 193
two region model, 245
two-dimensional band structure, 108
two-dimensional electron gas, 219
two-dimensional Poisson's equation, 267
two-port linear network, 142

UCCM, 211, 226, 241, 333, 360, 508
UM-SPICE, 405, 471
unified carrier sheet charge density, 219
unified channel capacitance, 414
unified channel capacitance, MOSFET; 301
unified charge control model, 211, 241, 276, 316, 340, 508, 640
unified C–V characteristic, NMOS, 359
unified defect model, 83
unified drain current, PMOS, 333
unified I-V characteristics, MESFET, 418
unified MIS capacitance, 214
unified NMOS model, 342, 354
unified PMOS model, 331
unified saturation current, MESFET, 419
unified saturation current, MOSFET 305
unified saturation voltage, PMOS, 334
unified sheet charge density, MESFET, 416
uniform distributed RC lines, 715
unit cell, 5
universal HFET model, 450, 661
universal MESFET model, 413
universal MOSFET model, 301, 510

vacuum energy level, 96
valence band discontinuity, 96
valence band, 9

valence configuration, 2
valence electrons, 2
valence subshells, 2
velocity overshoot, 28, 123, 290, 395
velocity saturation MESFET model, 408
velocity saturation model, 247
velocity, MOSFET channel, 290
velocity-field relationship, MOSFET 246, 290
view menu, 609
voltage controlled switch, 713
voltage sources, 716
voltage sources, current-controlled, 683
voltage stress, 373

walled emitter, 122
washed emitter, 122
wave vector space, 7
weak inversion, 199
wide band gap emitter HBT, 122
wide band gap semiconductors, 93
Wigner-Seitz cell, 7
window menu, 611
window, closing active, 543
window, enlarging, 542
window, moving, 540
window, resizing, 541
window, restoring, 542
window, selecting, 539
window, shrinking, 541
windows, working with, 539
work function, 80, 197

x-axis expression, 600

y-parameters, 142

z-parameters, 142
Zener breakdown, 134
zinc blende crystal structure, 5